Green Energy and Technology

For further volumes:
http://www.springer.com/series/8059

S. M. Muyeen
Editor

Wind Energy Conversion Systems

Technology and Trends

Springer

S. M. Muyeen
Department of Electrical Engineering
The Petroleum Institute
PO Box 2533, Abu Dhabi
U.A.E.
e-mail: muyeen0809@gmail.com

ISSN 1865-3529 e-ISSN 1865-3537
ISBN 978-1-4471-2200-5 e-ISBN 978-1-4471-2201-2
DOI 10.1007/978-1-4471-2201-2
Springer London Dordrecht Heidelberg New York

British Library Cataloguing in Publication Data
A catalogue record for this book is available from the British Library

Library of Congress Control Number: 2011941710

Printed on acid-free paper

Springer is part of Springer Science+Business Media (www.springer.com)

Dedicated to My Parents

Preface

The renewable energy penetration rate to the power grid is increasing rapidly now-a-days. Wind, solar, biogas/biomass, tidal, geothermal, etc. are considered as the renewable sources of energy and among those the wind is playing the major role in world's energy market along with conventional sources of energy. The wind energy sector has already reached a matured stage due the contributions from many engineering and science disciplines in the last few decades, mainly from mechanical, electrical, electronic, computer, and aerospace. Each discipline has its own beauty and the combined efforts from scientists from different disciplines are the secret of the success of wind industry.

In this book, the present future development schemes of wind turbine generator systems are depicted based on the contribution from many renowned scientists and engineers from different disciplines. A wide verity of research results are merged together to make this book useful for students and researchers.

The chapters of the book are organized into three parts. In part I, wind energy conversion systems using different types of wind generator including necessary control schemes, are presented. Efficiency analysis of commercially available wind energy conversion systems, large scale wind generator, using superconducting material and high efficient power converter technology are the key features of this section. Part II is focused on several important issues for wind industry and transmission system operators. Grid interfacing issues, grid code, lightning strike and protection, use of energy storage options are highlighted in this section. And in the part III, the focus is given only to offshore wind power technology. Offshore wind speed observation from the space, HVDC based transmission scheme to interconnect offshore wind farm into onshore grid, hybrid offshore wind farms and marine current farms are the key issues discussed in this section. A general overview and essence of the chapters can be obtained from Chap.1 of the book.

Abu Dhabi, 31 March 2011 S. M. Muyeen

Acknowledgments

In my capacity, as the Editor of this book, first of all I would like to express my sincere appreciation to the chapter authors for their valuable contributions and enormous efforts for ensuring the quality of the materials in their chapters. Some of the results presented in to the book have already been published in international journals and appreciated in many international conferences and our thanks to those publishers for giving necessary permission to reuse the materials. Thanks go to IEEE Intellectual Property Rights Office, Risø DTU-National Laboratory of Sustainable Energy for assisting authors in various ways.

A large number of individuals including some authors of this book and organizations have assisted the authors in a variety of ways in the preparation of this work. In particular, however, we would like to thank Prof. Abdurrahim El-Keib, Dr. Ehab El-Saadany, Dr. Mohd. Hasan Ali, Dr. Stavros Papathanassiou, and Dr. S. Dutta for their tremendous support and kind suggestions throughout.

We have made use of Global Wind Energy Council (GWEC), American Wind Energy Association (AWEA), European Wind Energy Association (EWEA) publications and record our special thanks to these organizations for making documents available to us free of charge and sanctioning the permission to use some of the material therein. The Editor is grateful to Scaldis Salvage & Marine Contractors NV for providing few nice pictures used in the Introduction chapter of the book.

Finally, the Editor wishes to take this opportunity to express his gratitude to Prof. Junji Tamura for valuable suggestions to make this book successful and tremendous supports since 2002.

Contents

Part II Prime Issues for Wind Industry

Part III Offshore Trends

Contributors

Ali Abeidni was born in 1979. He finished his BS., MS. and Ph.D. in IUT, SUT and UWM respectively. He joined UW-Madison later as research associate in 2009. Hi is currently a faculty member of Khajeh Nasir Toosi University of Technology.

Asger B. Abrahamsen received the Ph.D. degree from the Technical University of Denmark (DTU) in 2003 for the work on small angle neutron scattering on the flux line lattice in the superconductor $TmNi_2B_2C$ done at Risø National Laboratory in Denmark. He conducted neutron scattering studies of superconductors and thermo electric materials in a post.doc. position at the Danish Centre for the use of Synchrotron X-ray and Neutron facilities (DANSCATT) from 2004-2006. He continued with in-situ high energy synchrotron scattering examination of the phase formation of the MgB_2 superconductor inside Fe tubes during a second Post.doc. position from 2006–2007 in the Materials Research Division at Risø National Laboratory for sustainable energy at the Technical university of Denmark (Risø DTU). Since 2007 he has been Senior Scientist in the Materials Research Division at Risø DTU with interest on characterization and applications of superconductors with special focus on superconducting wind turbine generators.

Ahmed Al-Durra obtained his B.Sc. Eng., M.Sc. Eng, and Ph.D. in 2005, 2007, 2010, respectively from The Ohio State University. For his Masters, he worked on the applications of nonlinear control theory to PEM fuel cell systems. During his Ph.D., he joined the Center of Automotive Research in OSU as a graduate research assistant, where he worked on model-based estimation and control methodologies for advance combustion engines. His research interests are application of estimation and control theory in power system stability, energy storage system (ESS), and renewable energy. Presently he is working in Electrical Engineering Department at the Petroleum Institute, Abu Dhabi, U.A.E. Dr. Ahmed is the member of IEEE.

Salvador Añó-Villalba received an M.Sc. degree in 1988 and a Ph.D. degree in 1996, both in Electrical Engineering from the Technical University of Valencia, Spain. From 1987 to 1989, he was with the R&D Department, Electronic Trafic S.A., where he was engaged in developing hardware and software for street lighting measuring and automation. In 1988, he joined the Department of Electrical Engineering of the Technical University of Valencia, where he is currently an Associate Professor. He has been a consultant to Iberdrola S.A. on high-impedance faults, load forecasting, and neutral break detection. His research interests include wind energy, electrical machines, and system analysis.

I. Azkarate was born in Durango, Spain, in 1984. She received her M.Sc. degree in telecommunications engineering from the University of the Basque Country in 2008. In 2009, she was in a work experience program at the Fraunhofer Heinrich Hertz Institute in Berlin. She joined the University of the Basque Country in 2010 as a Ph.D. student. Her research interests include digital signal processing and power quality assessment.

Soledad Bernal-Perez received its M.Sc. degree in Electrical Engineering from the Technical University of Valencia, Spain, in 1999, where she is currently working towards a Ph.D. degree. Since 2001, she has been a Radio Engineer, involved in carrying out surveys of Global Maritime Distress Safety System (GMDSS) radio installations on board of commercial ships for the main Classification Societies. Since 2003, she has been a Lecturer at the Electrical Engineering Department, Technical University of Valencia. Her research interests include electrical machine drives and the dynamics and grid integration of offshore wind farms through HVDC transmission systems.

Ramon Blasco-Gimenez obtained his B.Eng. degree from the Technical University of Valencia, Spain, in 1992, and its Ph.D. degree in Electrical and Electronic Engineering from the University of Nottingham, U.K., in 1996. From 1992 to 1995, he was a Research Assistant in the Deparment of Electrical and Electronic Engineering, University of Nottigham. In 1996, he joined the Dept. of Systems Engineering and Control of the Technical University of Valencia, where he is currently an Associate Professor. He has been a consultant to Iberdrola Renovables on integration of wind farms in weak grids. His research interests include control of motor drives, wind power generation and grid integration of renewable energy.Dr Blasco-Gimenez has been a co-recipient of the 2005 IEEE TRANSACTIONS ON INDUSTRIAL ELECTRONICS Best Paper Award. He is a registered professional engineer in Spain, Chartered Engineer (U.K.), member of the Institute of Engineering and Technology and Senior Member of the IEEE.

Zhe Chen (M'95, SM'98) received the B.Eng. and M.Sc. degrees from Northeast China Institute of Electric Power Engineering, Jilin City, China, and the Ph.D. degree from University of Durham, U.K. He was a Lecturer and then a Senior Lecturer with De Montfort University, U.K. Since 2002, Dr. Chen became a Research Professor and is now a Professor with the Institute of Energy Technology, Aalborg University, Denmark. He is the coordinator of Wind Power

System Research program at the Institute of Energy Technology, Aalborg University. His background areas are power systems, power electronics and electric machines; and his main current research areas are wind energy and modern power systems. Dr. Chen has more than 160 publications in his technical field. He is an Associate Editor (Renewable Energy) of the IEEE Transactions on Power Electronics, Guest Editor of Special Issue on Power Electronics for Wind Energy Conversion, the IEEE Transactions on Power Electronics. Dr. Chen is a Member of the Institution of Engineering and Technology (London, U.K.), and a Chartered Engineer in the U.K.

Tarek H. M. EL-Fouly was born in Cairo, Egypt in 1973. He received his B.Sc. and M.Sc. degree in Electrical Engineering from Ain shams University, Cairo, Egypt in 1996 and 2002, respectively. He received his Ph.D. degree in Electrical Engineering in 2008 from the University Of Waterloo, Waterloo, Ontario, Canada where he also worked as a Postdoctoral Fellow in electricity price forecasting, conservation and demand side management. In 2008, he joined the CanmetENERGY, Natural Resources Canada (Grid Integration of Renewable and Distributed Energy Resources Program) as a Transmission and Distribution Research Engineer where he is conducting and managing research and development activities related to active distribution networks, smart microgrids applications and remote communities. His research interests includes protection and coordination studies, integration of renewable energy resources, smart microgrid, smart remote community applications, demand side management and forecasting.

El-Saadany received his B.Sc. and M.Sc. in electrical engineering from Ain Shams University, Cairo, Egypt in 1986 and 1990, respectively, and his Ph.D. degree, also in electrical engineering in 1998 from the University of Waterloo, Waterloo, ON, Canada. Dr. El-Saadany joined the University of Waterloo as an Assistant Professor of Electrical and Computer Engineering in 2000 where currently he is a full Professor. Dr. El-Saadany's main research is in the areas of distribution system operation and control, distributed generation, smart grid applications, self-healing mechanisms, power quality and MEMS micro power generators. His research was supported by different governmental agencies and utilities such as Natural Sciences and Engineering Research Council (NSERC), Canada Foundation of Innovation (CFI), Ontario Research Fund (ORF), Ontario Center of Excellence (OCE), Natural Resources Canada, Hydro One Network and ABB. Dr. El-Saadany is a senior member in the IEEE and registered Professional Engineer in Ontario. Dr. El-Saadany's research output includes over 78 journal articles, 2 patent, over 130 conference and technical reports. As recognition of his excellent research, Dr. El-Saadany received the prestigious Early Research Award from the Government of Ontario in 2007. In 2009 Dr. El-Saadany received one of the highest research recognition in Canada and became a Canada Research Chair in Energy Systems in recognition of his contributions to the areas of Distributed Generation and Micro Grids.

Lingling Fan an assistant professor in University of South Florida (Tampa, FL) since Aug. 2009. She received the BS, MS degrees in electrical engineering from Southeast University, Nanjing, China, in 1994 and 1997, respectively. She received Ph.D. degree in electrical engineering from West Virginia University in 2001. She was a senior engineer in transmission asset management department in Midwest ISO, St. Paul, Minnesota (2001–2007) and an assistant professor in North Dakota State University (2007–2009). Her research interests include modeling and control of energy systems, large-scale power systems planning and operation.

J. J. Gutierrez was born in Barakaldo, Spain, in 1972. He received his M.Sc. degree in telecommunications engineering from the University of the Basque Country in 1996. After holding various positions in the telecommunications and electrical power sectors, he joined the University of the Basque Country (UPV/EHU) in 2003 as an assistant lecturer, obtaining his Ph.D. degree in 2009. He is currently engaged in research on digital signal processing for power quality assessment. He is also an active member on the IEC Committee SC77A/WG2.

Hany M. Hasanien received his B.Sc., M.Sc., and Ph.D degrees in Electrical Engineering from Ain Shams University, Faculty of Engineering, Cairo, Egypt, in 1999, 2004, and 2007 respectively. His Ph.D. research work focused on the performance enhancement of switched reluctance motors. Currently, he is an Assistant Professor at the Electrical Power and Machines Department, Ain Shams University. His research interests include machine design, modern control techniques, electrical drives, artificial intelligent applications on electrical machines, and renewable energy applications. Dr. Hany M. Hasanien is a senior member of the Institution of Electrical and Electronic Engineers (IEEE), and also of Power and Energy Society (PES). He published a book (co-authored with Dr. S. M. Muyeen and Prof. Dr. J. Tamura) "Switched Reluctance Machine" from Praise Worthy Prize, in February 2010. His biography has been included in "Marquis Who's Who" in the world for its 28th edition, 2011.

Bogi B. Jensen received the Ph.D. degree from Newcastle University, Newcastle upon Tyne, U.K., for his work on toroidally wound induction machines. He spent the years from 1994 to 2002 in the marine sector with roles from Engineering Cadet to Senior Field Engineer. He joined academia in 2002 as a Lecturer at the Centre of Maritime Studies and Engineering, Faroe Islands. He moved to the United Kingdom in 2004 and became a Research Associate in 2007 and a Lecturer in 2008 both at Newcastle University, Newcastle upon Tyne, U.K. He is currently Associate Professor of Electrical Machines at the Centre for Electric Technology, Department of Electrical Engineering, Technical University of Denmark (DTU), Kongens Lyngby, Denmark. His major research interests are electrical machine design, analysis, and development.

Eleni G. Karamanou received her Diploma in Electrical and Computer Engineering in 2006 and the Postgraduate Diploma in Energy Production and Management in 2009 from the National Technical University of Athens (NTUA),

Greece. She worked for the Distribution Division of the Public Power Corporation of Greece, where she was engaged in distribution equipment studies and the design of HV/MV GIS substations. She is currently working as a researcher in NTUA, in the field of renewable energy integration in autonomous island grids.

A. Lazkano was born in Azpeitia, Spain, in 1969. He received his M.Sc. degree in telecommunications engineering from the University of the Basque Country in 1993. From 1994 to 1995, he was with ETB (Basque Television). He joined the University of the Basque Country in 1995 as an assistant lecturer, obtaining his Ph.D. degree in 2001. He is currently associate professor with the Department of Electronics and Telecommunications, University of the Basque Country. His current research interests include digital signal processing applied to power systems and power quality assessment.

Wei-Jen Lee (S'85-M'85-SM'97-F'07) received the B.S. and M.S. degrees from National Taiwan University, Taipei, Taiwan, R.O.C., and the Ph.D. degree from the University of Texas, Arlington, in 1978, 1980, and 1985, respectively, all in Electrical Engineering. In 1985, he joined the University of Texas, Arlington, where he is currently a professor of the Electrical Engineering Department and the director of the Energy Systems Research Center. He has been involved in research on power flow, transient and dynamic stability, voltage stability, short circuits, relay coordination, power quality analysis, renewable energy, and deregulation for utility companies. Prof. Lee is a Fellow of IEEE and registered Professional Engineer in the State of Texas.

L. A. Leturiondo was born in Durango, Spain, in 1963. He received M.Sc. and Ph.D. degrees in electrical engineering from the University of the Basque Country, Bilbao, Spain, in 1990 and 1998, respectively. Since 1990, he has been an associate professor with the Department of Electronics and Telecommunications, University of the Basque Country. He is engaged in research on electric power systems and digital signal processing for power quality assessment and flicker measurement.

Haiwen Liu received the B.S. and and M.S. in 2003 in electrical engineering from Zhejiang University, Hangzhou, China, in 1997 and 2003 respectively, and Ph.D. degree in electrical engineering from the University of Tennessee, Knoxville, in 2009. He has been a staff engineer at Intertek, Inc. in Cortland, NY, since December 2009. His interests include design and certification of power converters and renewable energy systems.

Zhixin Miao received his BSEE from Huazhong University of Science and Technology, Wuhan, China, in 1992. He received his MSEE from the graduate school of Nanjing Automation Research Institute in 1997 and Ph.D. in Electrical Engineering from West Virginia University in 2002. He is currently with University of South Florida. Prior to joining USF in 2009, he was with the transmission asset management department in Midwest ISO, St. Paul, Minnesota from

2002 to 2009. His research interests include power system stability, microgrid and renewable energy.

Francisco Morant received his B.Eng. and M.Eng. degrees in electrical engineering and its Ph.D. degree from the Technical University of Valencia, Spain, in 1976, 1982, and 1985, respectively. He is currently a Professor at the Department of Systems Engineering and Control, Technical University of Valencia. From 1988 to 1989, he was a Guest Researcher at the Decision and Control Laboratory, Illinois University. He has been the General Director of Education and Employment of the Valencian Autonomous Government (1993–1995), and Vice-President of the Technical University of Valencia (1993–1995 and 2005–2008). His research interests include intelligent and adaptive control and fault diagnosis.

Dr. S. M. Muyeen received his B.Sc. Eng. Degree from Rajshahi University of Engineering and Technology (RUET), Bangladesh formerly known as Rajshahi Institute of Technology, in 2000 and M. Sc. Eng. and Dr. Eng. Degrees from Kitami Institute of Technology, Japan, in 2005 and 2008 respectively, all in Electrical and Electronic Engineering. His PhD research work focused on wind farm stabilization from the viewpoint of LVRT and frequency fluctuation. After completing his Ph.D. program he worked as a Postdoctoral Research Fellow under the versatile banner of Japan Society for the Promotion of Science (JSPS) from 2008–2010 at the Kitami Institute of Technology, Japan. His research interests are power system stability and control, electrical machine, FACTS, energy storage system (ESS), Renewable Energy, and HVDC system. He has published over 60 international papers. He has published two books entitled Stability Augmentation of a Grid-connected Wind Farm and Switched Reluctance Machine. He has also served as an Editor of the book entitled "Wind Power" in 2010. Dr. Muyeen is the member of IEEJ, and IEEE.

Shoji Nishikta received the B. Eng. and M. Eng. degrees from Tokyo Denki University (TDU), Tokyo, JAPAN, in 1972 and 1975, respectively, and the D. Eng. degree from Tokyo Institute of Technology (TIT), Tokyo, JAPAN, in 1984. From 1975 to 1984 he was with TIT as Research Associate. In 1984 he joined TDU, where he is currently Professor. Dr. Nishikata is a senior member of IEEE and a senior member the Institute of Electrical Engineers of Japan.

Shunsuke Oka was born in Hyogo, Japan, on 1 May 1985. He graduated from the Department of Electrical and Electronic Engineering, Kyoto University, Kyoto, Japan, in 2009. His research interests are wind power generation stability and hybrid power generation systems.

Dale Osborn received his Bachelor and Master degrees from University of Nebraska Licoln. He was the manager of planning department of NPPD. He was reactive power management manager in ABB from 1990-2000. Currently he is the principle advisor of transmission asset management department in Midwest ISO. His research interests cover power system planning, reliability, economics and reactive power device manufacturing.

Stefanos V. Papaefthymiou received the Diploma in Electrical and Computer Engineering in 2005 and the Postgraduate Diploma in Energy Production and Management in 2007 from the National Technical University of Athens (NTUA), Greece, where he is currently working towards the Ph.D. Since the beginning of 2007 he is working with the Distribution Division of the Public Power Corporation of Greece, where he is engaged in protection studies of distribution networks in island systems. His research interests lie in the field of renewable energy sources and distribution systems.

Stavros A. Papathanassiou received the Diploma in Electrical Engineering and the Ph.D. from the National Technical University of Athens (NTUA). He worked for the Distribution Division of the Public Power Corporation of Greece, in network and distributed generation studies. In 2002 he became a member of the faculty in the Electric Power Division of NTUA, where he currently serves as Assistant Professor. Since 2009 he is a Member of the Board of the Hellenic TSO. His research deals with wind turbine and PV technology and the integration of DG to the grid, including the application of storage. He is a member of the IEEE, CIGRE and the Technical Chamber of Greece.

Farhad Rachidi (IEEE Fellow, EMP Fellow) received the M.S. degree in electrical engineering and the Ph.D. degree from the Swiss Federal Institute of Technology, Lausanne, in 1986 and 1991 respectively. He worked at the Power Systems Laboratory of the same institute until 1996. In 1997, he joined the Lightning Research Laboratory of the University of Toronto in Canada and from April 1998 until September 1999, he was with Montena EMC in Switzerland. He is currently the head of the EMC Laboratory at the Swiss Federal Institute of Technology, Lausanne, Switzerland. Dr. Rachidi is the President of the International Conference on Lightning Protection (ICLP), Vice-Chair of the European COST Action on the Physics of Lightning Flash and its Effects, Associate Editor of the IEEE Transactions on Electromagnetic Compatibility and Deputy Editor-in-Chief of the Journal of Lightning Research. Farhad Rachidi is the author or coauthor of over 300 scientific papers published in reviewed journals and presented at international conferences. In 2005, he was the recipient of the IEEE Technical Achievement Award and the CIGRE Technical Committee Award. He was awarded the 2006 Blondel Medal from the French Association of Electrical Engineering, Electronics, Information Technology and Communication (SEE).

Mohammad Lutfur Rahman was born and grew up in Bangladesh. He gained bachelor, masters and Ph.D. degrees in the Philippines and Japan in the year of 2000, 2003 and 2010 respectively with the degree of Bachelor of Science in Computer Engineering (AMA Computer University), Masters in Information Technology (Technological University of the Philippines), and Doctor of Energy Science (Kyoto University). He has worked in Thailand as a lecturer in Eastern Asia University and Rajamangala University of Technology, Thanyaburi. He is currently working as the Head (Assistant Professor) in the Department of Computer Science and Engineering, State University of Bangladesh. Dr. Rahman's areas of interest are next generation power system including renewable energy sources, hybrid power system, wind power system and tidal power system.

Marcos Rubinstein received his Bachelor's degree in electronics from the Universidad Simon Bolivar, Caracas, Venezuela in 1982, and the Master's and Ph.D. degrees in electrical engineering from the University of Florida, Gainesville in 1986 and 1991. In 1992 he joined the Swiss Federal Institute of Technology in Lausanne, where he was active in the fields of electromagnetic compatibility and lightning in close cooperation with the former Swiss PTT. In 1995, he took a position at Swisscom, where he was involved in numerical electromagnetics and EMC in telecommunications and where he led a number of coordinated projects covering the fields of EMC and biological effects of electromagnetic radiation. In 2001, he moved to the University of Applied Sciences of Western Switzerland HES-SO, Yverdon-les-bains, where he is currently a professor in telecommunications and a member of the IICT institute team. His current research interests include Lightning, EMC in telecommunication systems, PLC, wireless technologies and layer-2 network security. He is the author or co-author of over 100 scientific publications in reviewed journals and international conferences. Prof. Rubinstein is the recipient of the best Master's Thesis award from the University of Florida, he received the IEEE achievement award and he is a co-recipient of NASA's recognition for innovative technological work. He is also a senior member of the IEEE, a member of the Swiss Academy of Sciences and of the International Union of Radio Science.

J. Ruiz was born in Sestao, Spain, in 1960. He received M.Sc. and Ph.D. degrees in electrical engineering from the University of the Basque Country, Bilbao, Spain, in 1983 and 1988, respectively. In 1985, he was appointed associate professor and,since 2002, was a professor with the Electronics and Telecommunications Department, University of the Basque Country. He is engaged in research on electrical power systems and digital signal processing for power-quality assessment and flicker measurement.

B. S. Rajpurohit obtained his M. Tech. and Ph.D. in Electrical Engineering from Indian Institute of Technology Roorkee and Indian Institute of Technology Kanpur, in 2005 and 2009, respective-ly. Presently, he is an Assistant Professor in the School of Computing and Electrical Engineering, Indian Institute of Technology Mandi, India. His research interests include power system harmonics, grid integra-tion of renewable energies, parameter estimation of electrical ma-chines and wind power. Dr. Rajpurohit is the member of IEEE, USA.

Johel Rodríguez-D'Derlée received the B.S. degree in Electronic Engineering from the University of Táchira, Venezuela, in 2000, and the M.Sc. degree in Mathematics and Computer Science from the University of Carabobo, Venezuela, in 2004. He is currently working towards its Ph.D. degree at the Institute of Control Systems and Industrial Computing, Technical University of Valencia, Spain, where he is involved in the research on advanced control systems for offshore wind farm and HVDC Transmission. His current research interests include advanced control techniques applied to power converters for renewable energy systems.

P. Saiz was born in Zumarraga, Spain, in 1972. She obtained her M.Sc. degree in telecommunication engineering from the University of the Basque Country (UPV/EHU), Spain, in 1996, and her Ph.D. degree in 2007, from the same University. After 6 years of professional experience in operation and engineering of GSM/GPRS/UMTS mobile networks, she joined the UPV/EHU in 2002, as an assistant lecturer in the Department of Electronics and Telecommunications at the Faculty of Engineering of Bilbao. Her current research interests include digital signal processing applied to power systems and power quality assessment.

M. R. I. Sheikh was born in Sirajgonj, Bangladesh on October 31,1967. He received his B.Sc. Engineering and M.Sc. Engineering Degree from Rajshahi University of Engineering and Technology (RUET), Bangladesh, in 1992 and 2003 respectively, all in Electrical and Electronic Engineering. He is currently an Associate Professor in the Electrical and Electronic Engineering Department, RUET. He carried out his Ph.D. Degree from Kitami Institute of Technology, Hokkaido, Kitami, Japan in September, 2010. His research interests are, Power system stability enhancement including wind generator by using SMES, FACTs devices and Load Frequency Control of multi-area power system.

Yasuyuki Shirai was born in Kyoto Prefecture Japan. He received the B.E., M.E., and D.E. degrees in electrical engineering from Kyoto University, Kyoto, Japan, in 1980, 1982, and 1988, respectively. He became an Assistant Professor in 1985, an Associate Professor in 1996, and he is now a Professor in the Graduate School of Energy Science, Kyoto University. His areas of interest are applied superconductivity to power system apparatus, next-generation power system including renewable energy sources, and energy infrastructure.

S. N. Singh was born on 5th September 1966 and obtained his M. Tech. and Ph. D. in Electrical Engineering from Indian Institute of Technology Kanpur, in 1989 and 1995, respectively. Presently, he is a Professor in the Department of Electrical Engineering, Indian Institute of Technology Kanpur, India. Dr Singh received several awards including Young Engineer Award 2000 of Indian National Academy of Engineering, Khosla Research Award of IIT Roorkee, and Young Engineer Award of CBIP New Delhi (India), 1996. Prof Singh is receipt of Humboldt Fellowship of Germany (2005, 2007) and Otto-monsted Fellowship of Denmark (2009–2010). His research interests include power system restructuring, FACTS, power system optimization and control, security analysis, wind power, etc. Prof. Singh is a Fellow of Institution of Electronics and Tele-communication Engineers (IETE) India, a Senior Member of IEEE, USA, and a Fellow of the Institution of Engineers (India). Prof Singh has published more than 285 papers in Internation-al/national journals/conferences. He has also written two books one on Electric Power Generation, Transmission and Distribution and second is Basic Electrical Engineering, published by PHI, India.

Alexander Smorgonskiy received the B.S. and M.S. degrees (with distinction) in electrical engineering from Saint-Petersburg State Polytechnic University, Saint-Petersburg, Russia, in 2007 and 2009, respectively. He is currently a Ph.D. student at Swiss Federal Institute of Technology, Lausanne, Switzerland. In 2008–2009 he was the recipient of a research scholarship from the Swiss Government.

Junji Tamura received his B.Sc. Eng. Degree from Muroran Institute of Technology, Japan, in 1979 and M.Sc. Eng. and Dr. Eng. degrees from Hokkaido University, Japan, in 1981 and 1984 respectively, all in electrical engineering. In 1984, he became a lecturer and in 1986, an associate professor at the Kitami Institute of Technology, Japan. He also holds the position of Vice President of the university. Currently he is a professor at the Kitami Institute of Technology. He also holds the position of Vice President of the university. His research areas include Rotating Electrical Machine, Power System, and Wind Energy. He has published about 100 technical papers in Transactions and international journals, presented about 150 papers in international conferences, and authored or coauthored several books and book chapters as well. He has served as conference chair, technical committee chair, member of technical committee in different conferences in domestic and international levels. He is the senior member of IEEE.

Fujio Tatsuta was born in Japan in 1959. He received the B. Eng. degree from Tokyo Denki University, Tokyo, Japan, in 1982. Since 1982, he has been with Tokyo Denki University, where he is currently an Assistant Professor. Mr. Nishikata is a member of the Institute of Electrical Engineers of Japan.

W. Timothy Liu graduated from Ohio University in 1971, with a B.S. degree in physics. He received both his M.S. and Ph.D degrees in atmospheric sciences from University of Washington in 1974 and 1978. He started as a principle investigator in satellite oceanography at the Jet Propulsion Laboratory in 1979. He has been a senior research scientist at JPL since 1993. He was the Project Scientist for three NASA scatterometer missions between 1992 and 2006. He is a fellow of the American Meteorological Society and the American Association for the Advancement of Science. His recent research interests include ocean-atmosphere interaction, water cycle, and climate.

Kuo-Hua Wang was born in Yunlin County, Taiwan, on 30 December 1982. He graduated from Department of Electrical Engineering, Feng Chia University, Taichung, in June 2005 and received his M.Sc. degree from Department of Electrical Engineering, National Chung Kung University, Tainan, Taiwan, in June 2007. His interests are simulations of wind induction generators using an HVDC transmission System to connect to a utility grid.

Li Wang (S'87-M'88-SM'05) received the Ph.D. degree from Department of Electrical Engineering, National Taiwan University, Taipei, Taiwan, in June 1988. He has been an associated professor and a professor at Department of Electrical Engineering, National Cheng Kung University, Tainan, Taiwan in 1988 and 1995, respectively. He was a visiting scholar of School of Electrical Engineering and Computer Science, Purdue University, West Lafayette, IN, USA, from February 2000 to July 2000. He was a visiting scholar of School of Electrical Engineering and Computer Science, Washington State University, Pullman, WA, USA, from August 2003 to January 2004. He was a research scholar of Energy Systems Research Center (ESRC), The University of Texas at Arlington (UTA), Arlington, TX, USA, from August 2008 to January 2009. At present, his interests include power systems dynamics, power system stability, AC machine analyses, and renewable energy. He is an IEEE Senior Member.

Lingfeng Wang is currently an assistant professor in the Department of Electrical Engineering and Computer Science at the University of Toledo (UT), Ohio, USA. He obtained his Ph.D. degree from Texas A&M University in 2008. He also holds B.S. and M.S. from Zhejiang University, China and M.S. from National University of Singapore. Before joining UT as a faculty member, he was with California ISO for regional transmission planning. Dr. Wang is the recipient of several awards for his research excellence, and he also wins several best paper awards. He is the author or coauthor of six books/research monographs and more than 100 technical publications. He served as an editorial board member for several journals and a member of technical program committee for many international conferences. He has given seminar talks in more than 25 universities/national laboratories as an invited speaker. His major research areas are power system reliability, renewable energy integration, computational intelligence, and industrial informatics.

Xiaosu Xie received a M.S. degree from the Chinese Academy of Meteorological Science in Beijing, China and a Ph.D degree in Atmospheric Sciences from University of Hawaii. She joined the Air-sea Interaction and Climate Team at JPL in 1996. Her major research interests include climate dynamics/thermodynamics, water and carbon cycle, and satellite oceanography. She is also working on developing algorithms to estimate major components of global hydrological balance and carbon cycle from satellite observations. She has involved in a number of NASA earth observing space missions, including QuikSCAT, SeaWinds, TRMM, Aqua, AMSR, and Topex/JASON.

Lie Xu received the B.Eng. degree from Zhejiang University, Hangzhou, China, in 1993, and the Ph.D. degree from the University of Sheffield, Sheffield, U.K., in 1999. He joined the School of Electronics, Electrical Engineering and Computer Science, Queen's University of Belfast, U.K. in 2004 and currently is a Senior Lecturer. He was with ALSTOM T&D in Stafford, U.K. from 2001–2003. His main interests are power electronics, wind energy generation and grid integration, and application of power electronics to power systems. Dr Xu is a senior member of the IEEE.

Liangzhong Yao received his M.Sc degree in 1989 and Ph.D degree in 1993 all in electrical power system engineering from Tsinghua University, Beijing, China. He is currently the Department Manager for Network Solutions, Smartgrids and Renewables Technologies and also a Technology Consultant and Senior Expert at

ALSTOM Grid (Former AREVA T&D) Research and Technology Centre, Stafford, United Kingdom. Prior to ALSTOM Grid, he was an Associate Professor at Tsinghua University until 1995, and was a Post Doctoral Research Associate in the Manchester Centre for Electrical Energy at University of Manchester (former UMIST), United Kingdom from 1995 to 1999, and was a Senior Power System Analyst in the Network Consulting group at ABB UK Ltd from 1999 to 2004. Dr Yao is a Fellow of the IET and also Guest Professors at both Shanghai Jiao Tong University and Sichuan University, China.

Yasuda Yoh was born in Tokyo, Japan, in 1967. He received the Graduate degree in electronics and information engineering in 1989 and the Ph.D. degree in 1994 from Yokohama National University, Kanagawa, Japan. Since 1994, he has been with the Department of Electrical Engineering and Computer Science, Kansai University, Osaka, Japan, where he is currently an Associate Professor. His current research interests are lightning protection of wind turbine and grid stability with large penetration of wind turbines. Dr. Yasuda is now an executive director of Japan Wind Energy Association. He is also a member of the European Wind Energy Association (EWEA), Institute of Electrical and Electronics Engineers (IEEE) and other domestic associations.

Hui Zhang received the B.S. and M.S. degrees in electrical engineering from Zhejiang University, Hangzhou, China, in 2000 and 2003 respectively, and Ph.D. in electrical engineering from The University of Tennessee, Knoxville, in 2007. She joined the Power Electronics and Electric Machinery Research Center in Oak Ridge National Laboratory (ORNL), Knoxville, TN, as a student member in 2005. She worked as a post-doc research Associate at The University of Tennessee and Oak Ridge National Laboratory from 2007 until 2009. Currently, she is an Assistant Professor in the Electrical Engineering Department, Tuskegee University, Alabama. Her interests include Silicon carbide power electronics, power conditioning, renewable energy systems, and motor drives, etc. Dr. Hui Zhang is a member of IEEE Power Electronics Society, Industry Application Society, and Industrial Electronics Society. She has served as a Reviewer of IEEE Transactions and the Session Chair of IEEE conferences.

Chapter 1
Introduction

S. M. Muyeen

Abstract In this chapter, first, the global wind power scenario is depicted followed by the market forecasting upto the year 2030. Then the current technology and the future trend of wind energy conversion system are discussed where the development of wind generator, blade designing, lightning protection, installation, commissioning, operation and maintenance of wind turbine generator unit are briefly stated. Some important issues such as variability and predictability of wind power, energy storage options and grid interfacing techniques are discussed as well. Prime offshore wind farm technology issues in terms of feasibility study, bulk power transmission scheme are discussed in detail. Finally, the highlights of all the chapters are given from where the flavor of the book can be obtained at a glance.

1.1 Global Wind Power Scenario

Wind energy is becoming one of the mainstream power sources in many countries around the world. According to Global Wind Energy Council (GWEC) statistics, global wind power installations increased by 35.8 GW in 2010, bringing the total installed wind energy capacity upto 194.4 GW, a 22.5% increase on the 158.7 GW installed at the end of 2009. GWEC is one of the few organizations doing an excellent job by broadcasting and forecasting regional wind power development throughout the world. In the following section, the present wind power installation scenario at the end of 2010 and some future predictions are demonstrated in light of GWEC reports [1, 2].

S. M. Muyeen (✉)
Department of Electrical Engineering, The Petroleum Institute,
P.O. Box 2533, Abu Dhabi, UAE
e-mail: smmuyeen@pi.ac.ae; muyeen0809@gmail.com

S. M. Muyeen (ed.), *Wind Energy Conversion Systems*,
Green Energy and Technology, DOI: 10.1007/978-1-4471-2201-2_1,

The regional wind power installations in 2010 are shown in Fig. 1.1 where the wind power statistics of 2009 are available as well. The installation scenarios of wind energy conversion systems of the world's top 10 countries in 2009 and 2010 are shown in Figs. 1.2 and 1.3, respectively. One important message obvious from Figs. 1.2 and1.3 is that wind power is spreading in other countries of the world along with the top ten countries which are already in good pace in installing new wind turbines. The global cumulative installed capacity from 1996 to 2010, the global annual installed capacity from 1996 to 2010, and the annual installed capacity by region from 2003 to 2010 scenarios are shown in Figs. 1.4, 1.5 and 1.6. Figure 1.4 shows an interrupted growth rate due to the worldwide economic crisis in the recent years.

1.1.1 Asia

The growth in Asian markets has been breathtaking, as more than 50% of the world's wind energy in 2010 was installed in Asia and it is the 3rd year in a row where Asia is leading the regional market on the globe. **China** was the world's largest market in 2010, adding a staggering 16.5 GW of new capacity, and slipping past the USA to become the world's leading wind power country. The Chinese market more than doubled its capacity from 12 GW in 2008 to 25.8 GW in 2009 and added 16.5 MW in 2010 to reach 42.2 GW at the end of 2010 [1]. The planning, development and construction for the "Wind Base" programme, which aims to build 138 GW of wind capacity in eight Chinese provinces, is well underway. It is expected that in its twelfth Five-Year Plan, which is expected to be adopted in March 2011, the Chinese government will increase its official target for wind power development to 200 GW by 2020.

Wind power market in **India** is now back on track after a few years of slow growth and witnessed significant growth in 2010. It comes in third behind China and the USA in terms of new installed capacity during 2010 at 2,139 MW, taking the total capacity upto 13.1 GW. The states with the highest wind power concentration are Tamil Nadu, Maharashtra, Gujarat, Rajasthan, Karnataka, Madhya Pradesh and Andhra Pradesh. In 2010 the official wind power potential estimates for India were revised upwards from 45 to 49.1 GW by the Centre for Wind Energy Technology (C-WET). However, the estimations of various industry associations and wind power producers are more optimistic, citing a potential in the range of 65–100 GW.

Other Asian countries with new capacity additions in 2010 include Japan (221 MW, for a total of 2.3 GW), South Korea (31 MW for a total of 379 MW) and Taiwan (83 MW for a total of 519 MW).

1.1.2 North America

According to American Wind Energy Association (AWEA) statistics the U.S. wind energy industry installed 5,115 MW in 2010. This is barely half of 2009s record pace, but the fourth quarter was strong, showing new momentum for 2011

GLOBAL INSTALLED WIND POWER CAPACITY (MW)– REGIONAL DISTRIBUTION

		End 2009	New 2010	End 2010
AFRICA & MIDDLE EAST	Egypt	430	120	550
	Morocco	253	33	286
	Tunisia	54	60	114
	Iran	92	0	92
	Other [1]	37	0	37
	Total	**866**	**213**	**1,079**
ASIA	China*	25,805	16,500	42,287
	India	10,926	2,139	13,065
	Japan	2,085	221	2,304
	Taiwan	436	83	519
	South Korea	348	31	379
	Philippines	33	0	33
	Other [2]	6	48	54
	Total	**39,639**	**19,022**	**58,641**
EUROPE	Germany	25,777	1,493	27,214
	Spain	19,160	1,516	20,676
	Italy	4,849	948	5,797
	France	4,574	1,086	5,660
	UK	4,245	962	5,204
	Denmark*	3,465	327	3,752
	Portugal*	3,357	345	3,702
	Netherlands*	2,223	15	2,237
	Sweden	1,560	603	2,163
	Ireland	1,310	118	1,428
	Turkey	801	528	1,329
	Greece	1,086	123	1,208
	Poland	725	382	1,107
	Austria	995	16	1,011
	Belgium	563	350	911
	Rest of Europe [3]	1,611	1,071	2,677
	Total	**76,300**	**9,883**	**86,075**
	of which EU-27 [4]	*74,919*	*9,259*	*84,074*
LATIN AMERICA & CARIBBEAN	Brazil	606	326	931
	Mexico	202	316	519
	Chile	168	4	172
	Costa Rica	123	0	123
	Caribbean	91	8	99
	Argentina	34	27	60
	Others [5]	83	23	106
	Total	**1,306**	**703**	**2,008**
NORTH AMERICA	USA	35,086	5,115	40,180
	Canada	3,319	690	4,009
	Total	**38,405**	**5,805**	**44,189**
PACIFIC REGION	Australia	1,712	167	1,880
	New Zealand	497	9	506
	Pacific Islands	12	0	12
	Total	**2,221**	**176**	**2,397**
	World total	**158,738**	**35,802**	**194,390**

1 South Africa, Cape Verde, Israel, Lebanon, Nigeria, Jordan, Kenya

2 Thailand, Bangladesh, Indonesia, Sri Lanka, Philippines, Vietnam

3 Bulgaria, Croatia, Cyprus, Czech Republic, Estonia, Faroe Islands, Finland, Hungary, Iceland, Latvia, Liechtenstein, Lithuania, Luxembourg, Malta, Norway, Romania, Russia, Slovakia, Slovenia, Switzerland, Ukraine

4 Austria, Belgium, Bulgaria, Cyprus, Czech Republic, Denmark, Estonia, Finland, France, Germany, Greece, Hungary, Ireland, Italy, Latvia, Lithuania, Luxembourg, Malta, Netherlands, Poland, Portugal, Romania, Slovakia, Slovenia, Spain, Sweden, UK

5 Colombia, Chile, Cuba

* Provisional figures

Please note:
Project decommissioning of 151 MW and rounding affect the final sums

Source: GWEC

Fig. 1.1 Regional distribution of global installed wind power capacity in MW (Source: GWEC)

(Fig. 1.7). Further wind projects are expected to start up in time to meet the new construction deadline at the end of 2011 for the Section 1603 Investment Tax Credit, which Congress recently extended by a year. Wind is increasingly appreciated for being cost-competitive with natural gas, which has helped the U.S. industry weather this latest boom-bust cycle.

Utilities are moving to lock in more wind power at long-term low rates. The nationwide capacity now totals 40,180 MW, an increase in capacity of 15% over

TOP 10 CUMULATIVE INSTALLED CAPACITY 2009

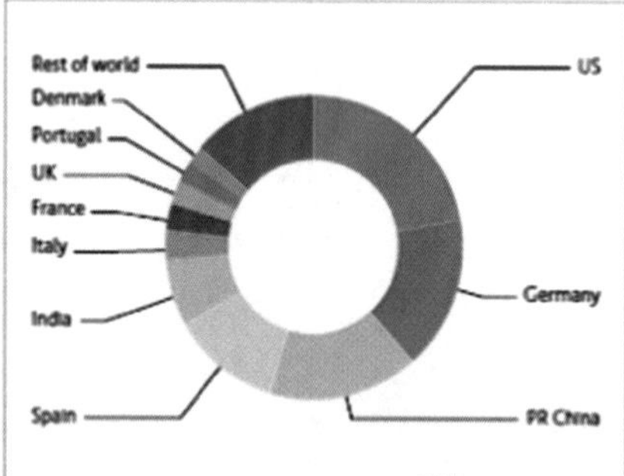

	MW	%
US	35,064	22.1
PR China	25,805	16.3
Germany	25,777	16.3
Spain	19,149	12.1
India	10,926	6.9
Italy	4,850	3.1
France	4,492	2.8
UK	4,051	2.6
Portugal	3,535	2.2
Denmark	3,465	2.2
Rest of world	21,391	13.5
Total top 10	**137,114**	**86.5**
World total	**158,505**	**100.0**

TOP 10 NEW INSTALLED CAPACITY 2009

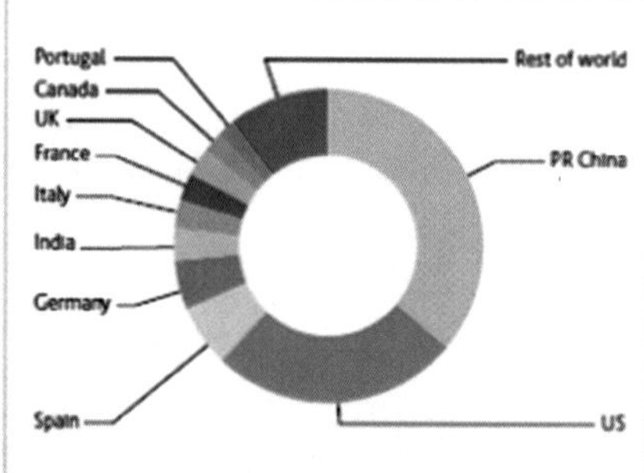

	MW	%
PR China	13,803	36.0
US	9,996	26.1
Spain	2,459	6.4
Germany	1,917	5.0
India	1,271	3.3
Italy	1,114	2.9
France	1,088	2.8
UK	1,077	2.8
Canada	950	2.5
Portugal	673	1.8
Rest of world	3,994	10.4
Total top 10	**34,349**	**89.6**
World total	**38,343**	**100.0**

Fig. 1.2 Installation scenario for the top 10 countries in 2009 (Source: GWEC)

the start of 2010. Wind power installation by each state at the end of 2010 is shown in Fig. 1.8. Texas, the leading wind power state in America for several years running, achieved a major milestone by surging past the 10,000 MW mark for total installations, with the addition of 680 MW in 2010. Texas achieved the mark thanks to aggressive pursuit of renewable energy and a renewable electricity standard passed in 1999 and strengthened in 2005. On average, wind now generates 7.8% of the electricity in the Electric Reliability Council of Texas (ERCOT), peaking as high as 25%. Other states active in pursuing targets for renewable energy last year were Illinois (498 MW added), California (455 MW), South Dakota (396 MW) and Minnesota (396 MW). Five more states doubled or more than doubled their wind power capacity in 2010. Delaware and Maryland both added their first utility-scale wind turbines in 2010. A total of 38 states now have utility-scale wind projects, and 14 of them have now installed more than 1,000 MW of wind power [3].

Canada's wind power market was also down in 2010 compared to the previous year, but it was still the second best year ever. A total of 690 MW of new wind capacity came online, compared to 950 MW in 2009, taking the total capacity to more than 4,000 MW. Ontario leads Canada's wind energy development with

TOP 10 CUMULATIVE CAPACITY DEC 2010

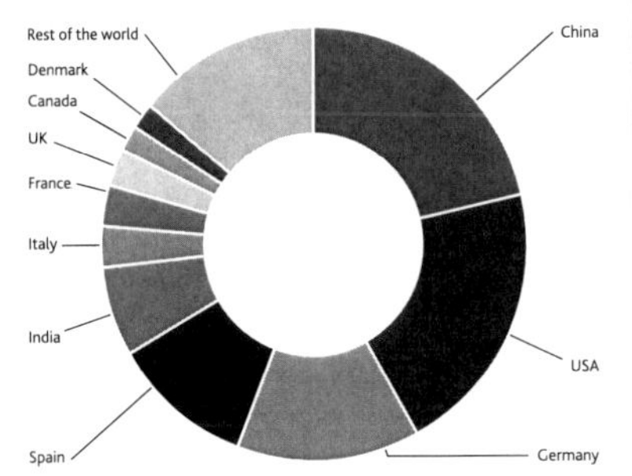

Country	MW	%
China	42,287	21.8
USA	40,180	20.7
Germany	27,214	14.0
Spain	20,676	10.6
India	13,065	6.7
Italy	5,797	3.0
France	5,660	2.9
UK	5,204	2.7
Canada	4,009	2.1
Denmark	3,752	1.9
Rest of the world	26,546	13.7
Total TOP 10	**167,844**	**86.3**
World Total	**194,390**	**100**

TOP 10 NEW INSTALLED CAPACITY JAN-DEC 2010

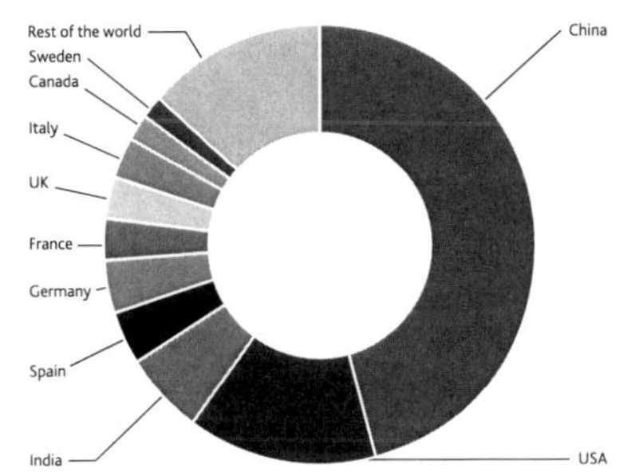

Country	MW	%
China	16,500	46.1
USA	5,115	14.3
India	2,139	6.0
Spain	1,516	4.2
Germany	1,493	4.2
France	1,086	3.0
UK	962	2.7
Italy	948	2.6
Canada	690	1.9
Sweden	603	1.7
Rest of the world	4,750	13.3
Total TOP 10	**31,052**	**86.3**
World Total	**35,802**	**100**

Fig. 1.3 Installation scenario for the top 10 countries in 2010 (Source: GWEC)

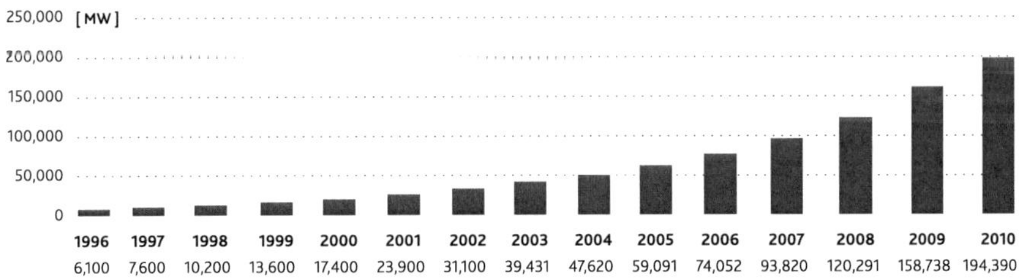

Fig. 1.4 Global cumulative installed capacity 1996–2010 (Source: GWEC)

1.5 GW of installed wind capacity. Other leading wind energy provinces include Quebec (806 MW) and Alberta (663 MW).

1.1.3 Europe

During 2010, 9,883 MW of wind power was installed across Europe, with European Union countries accounting for 9,259 MW of the total. This represents a decrease in the EU's annual wind power installations of 10% compared to 2009.

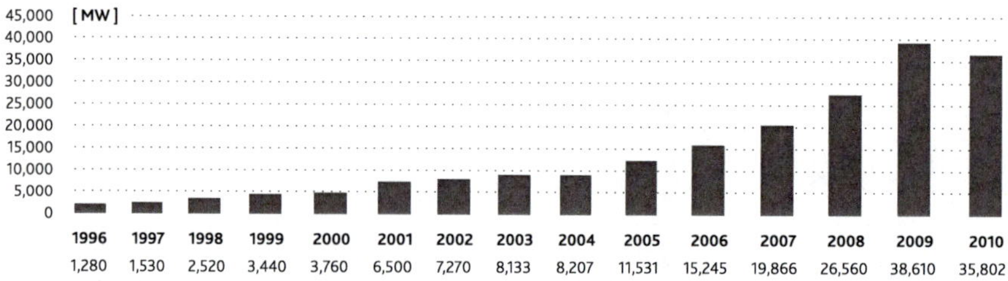

Fig. 1.5 Global annual installed capacity 1996–2010 (Source: GWEC)

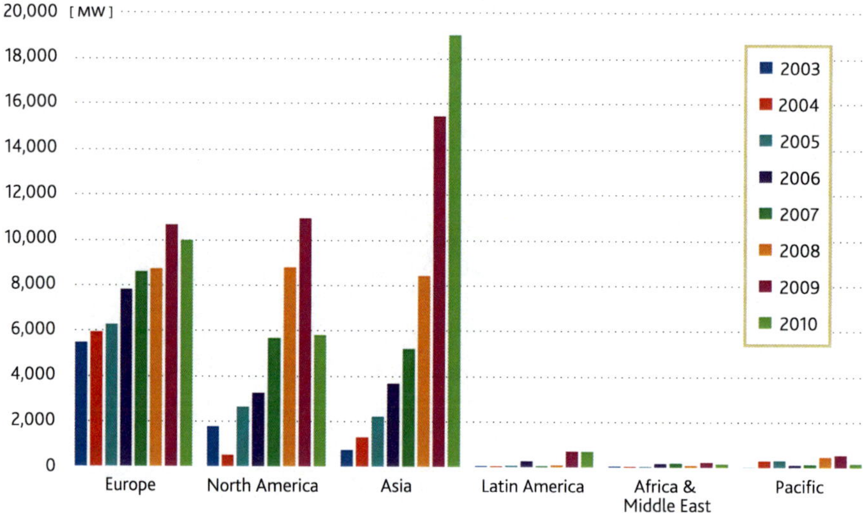

Fig. 1.6 Annual installed capacity by region 2003–2010 (Source: GWEC)

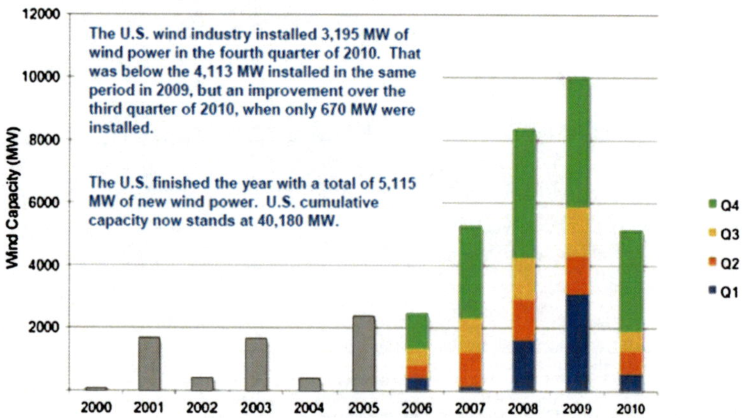

Fig. 1.7 U.S. annual and quarterly wind installations from 2000 to 2010 (Source: AWEA)

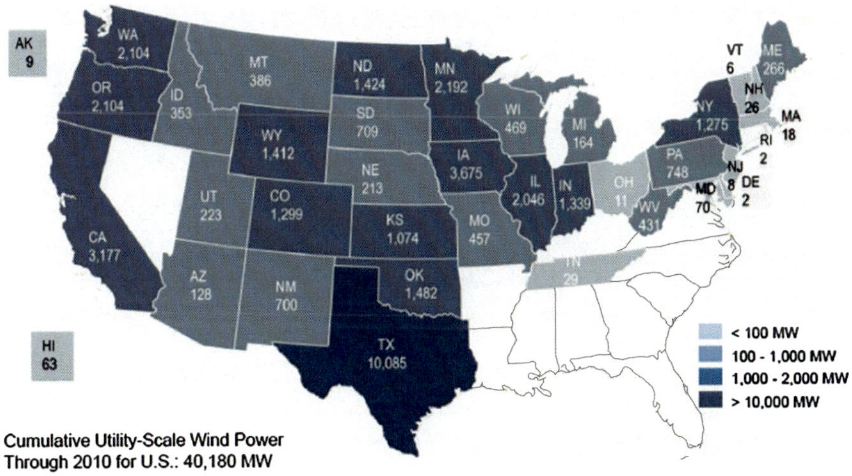

Fig. 1.8 State-wise wind installations at the end of 2010 (Source: AWEA)

Of the 9,259 MW installed in the EU, 8,377 MW were installed onshore and 883 offshore. This means that in 2010, the annual onshore market decreased by over 13% compared to the previous year, while the annual offshore market grew by 51%, and accounted for 9.5% of all capacity additions.

In terms of annual installations, Spain was the largest market in 2010, installing 1,516 MW, followed by Germany with 1,493 MW. France was the only other country to install over 1 GW (1,086 MW), followed by the UK (962 MW), Italy (948 MW), Sweden (603 MW), Romania (448 MW), Poland (382 MW), Portugal (345 MW) and Belgium (350 MW). For the first time, two new EU Member States were among the top ten largest annual markets [1].

1.1.4 Latin America

Brazil and Mexico are the leading countries in Latin America in wind power generation as can be seen from the GWEC report [1]. In 2010, Brazil added 326 MW of new capacity, slightly more than in 2009, and is now host to 931 MW of wind power. Mexico's installed wind capacity more than doubled for the second year in a row, with 316 MW of new capacity added to the existing 202 MW operating at the end of 2009. The total installed wind power capacity now amounts to 519 MW.

1.1.5 Pacific Region

At the end of 2010, 1,880 MW of wind capacity was installed in Australia, an increase of 167 MW from 2009. There are now 52 operating wind farms in the country, mostly located in South Australia (907 MW) and Victoria (428 MW).

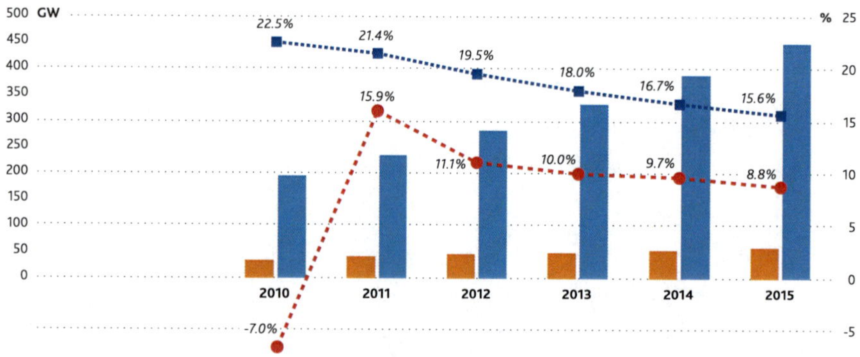

	2010	2011	2012	2013	2014	2015
Annual capacity [GW]	35.8	41.5	46.1	50.7	55.6	60.5
Cumulative capacity [GW]	194.4	235.9	282.0	332.7	388.3	448.8
Annual capacity growth rate [%]	-7.0	15.9	11.1	10.0	9.7	8.8
Cumulative capacity growth rate [%]	22.5	21.4	19.5	18.0	16.7	15.6

Fig. 1.9 Market forecast for 2010–2015 (Source: GWEC)

1.1.6 Africa and Middle East

In North Africa, the expansion of wind power continues in Morocco, Egypt and Tunisia. Egypt not only saw the largest addition of new capacity in 2010 (120 MW), bringing the total upto 550 MW but also continues to lead the region. Morocco comes in a distant second with a cumulative capacity of 286, 30 MW of which was added in 2010. Tunisia added 60 MW of new capacity in 2010, taking the total to 114 MW.

1.2 Market Forecast

GWEC predicts that at the end of 2015, global wind capacity will stand at 449 GW, up from 194 GW at the end of 2010 [1]. During 2015, 60.5 GW of new capacity will be added to the global total, compared to 35.8 GW in 2010. These are shown in Fig. 1.9. The annual growth rates during this period will average 18.2% in terms of total installed capacity, and 11.1% for annual market growth.

GWEC also estimates that Asia will remain the fastest growing market in the world, driven primarily by China, which is set to continue the rapid upscaling of its wind capacity and hold its position as the world's largest annual and cumulative market. For Asia as a whole, the annual market is expected to increase from 19 GW in 2010 to 26 GW in 2015, which would translate into a total of 116 GW of new capacity to be added over this period—far more than in any other region. In 2013, Asia is expected to overtake Europe as the region with the largest total installed capacity, and it will reach a cumulative wind power generation capacity

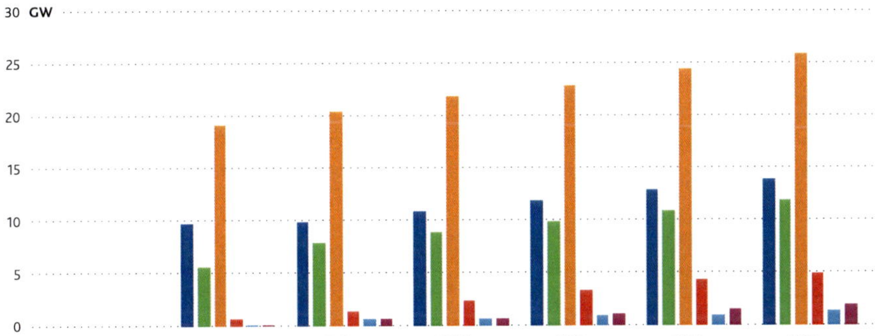

	2010	2011	2012	2013	2014	2015
Europe	9.9	10.0	11.0	12.0	13.0	14.0
North America	5.8	8.0	9.0	10.0	11.0	12.0
Asia	19.0	20.5	22.0	23.0	24.5	26.0
Latin America	0.7	1.5	2.5	3.5	4.5	5.0
Pacific	0.2	0.7	0.8	1.0	1.0	1.5
Middle East and Africa	0.2	0.8	0.8	1.2	1.6	2.0

Fig. 1.10 Regional annual market forecast for 2010–2015 (Source: GWEC)

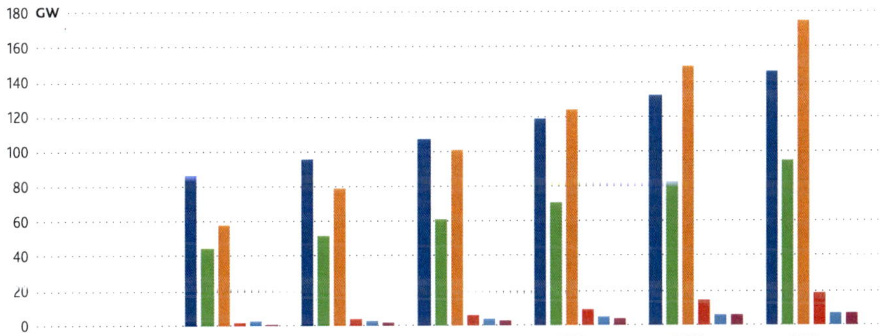

	2010	2011	2012	2013	2014	2015
Europe	86.1	96.1	107.1	119.1	132.1	146.1
North America	44.2	52.2	61.2	71.2	82.2	94.2
Asia	58.6	79.1	101.1	124.1	148.6	174.6
Latin America	2.0	3.5	6.0	9.5	14.0	19.0
Pacific	2.4	3.1	3.9	4.9	5.9	7.4
Middle East and Africa	1.1	1.9	2.7	3.9	5.5	7.5

Fig. 1.11 Regional cumulative market forecast for 2010–2015 (Source: GWEC)

of 174.6 GW by 2015. Annual and cumulative market forecasting by region from 2010 to 2015 are shown in Figs. 1.10 and 1.11, respectively.

In [2], GWEC presents three types of scenarios for wind power forecasting. The first is 'Reference scenario' based on the projections in the 2009 World Energy Outlook from the International Energy Agency (IEA). This takes into account not only existing policies and measures, but includes assumptions such as continuing electricity and gas market reform, the liberalization of cross-border energy trade and recent policies aimed at combating pollution. Second, the 'Moderate scenario'

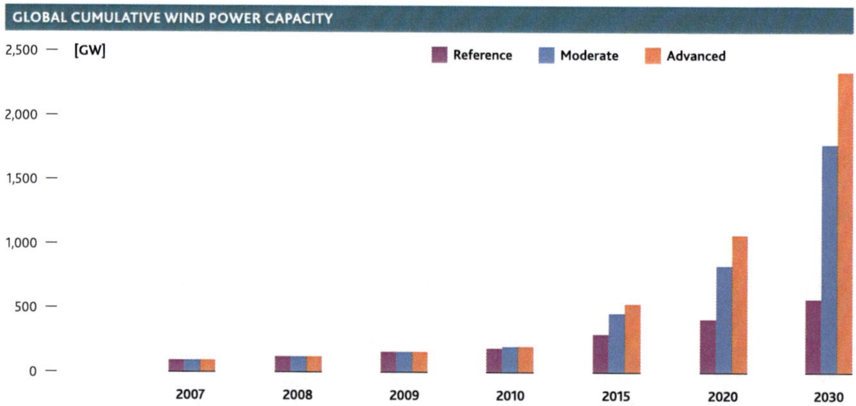

	2007	2008	2009	2010	2015	2020	2030
Reference [MW]	93,864	120,297	158,505	185,258	295,783	415,433	572,733
[TWh]	206	263	347	406	725	1,019	1,405
Moderate [MW]	93,864	120,297	158,505	198,717	460,364	832,251	1,777,550
[TWh]	206	263	347	435	1,129	2,041	4,360
Advanced [MW]	93,864	120,297	158,505	201,657	533,233	1,071,415	2,341,984
[TWh]	206	263	347	442	1,308	2,628	5,429

Fig. 1.12 Global cumulative market forecast in terms of reference, moderate and advanced scenarios (Source: GWEC)

takes into account all policy measures to support renewable energy either already enacted or in the planning stages around the world. It also assumes that the targets set by many countries for either renewables, emissions reductions and/or wind energy are successfully implemented, as well as the modest implementation of new policies aimed at pollution and carbon emission reduction, and increased energy security. It also takes into account environmental and energy policy measures that were part of many government economic stimulus packages implemented since late 2008. The third is the most ambitious, the 'Advanced scenario' which examines the extent to which this industry could grow in a best case 'wind energy vision'. The assumption here is a clear and unambiguous commitment to renewable energy as per the industry's recommendations, along with the political will necessary to carry it forward.

Wind power forecasting upto 2030 in forms of cumulative and regional breakdown are shown in Figs. 1.12 and 1.13, respectively, based on the three aforementioned scenarios.

1.3 Technological Aspects—Present and Future

Though wind energy conversion system has reached to a mature stage it is still going through a continuous development program by researcher and industry peoples from different disciplines for the improvement in both component and

REGIONAL BREAKDOWN: REFERENCE SCENARIO

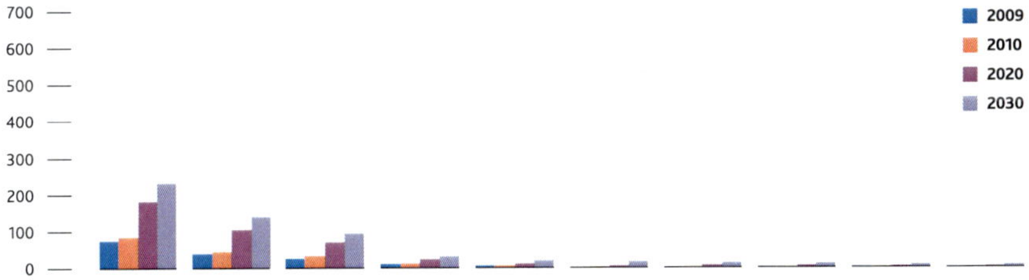

[GW]	OECD Europe	OECD North America	China	India	OECD Pacific	Non-OECD Asia	Eastern Europe/ Eurasia	Latin America	Africa	Middle East
2009	76	39	26	11	5	0,5	0,4	1	0,8	0,1
2010	86	45	33	12	5	0,6	0,9	2	1	0,1
2020	184	106	70	24	11	4	6	5	3	2
2030	234	141	95	31	19	16	13	11	8	6

REGIONAL BREAKDOWN: MODERATE SCENARIO

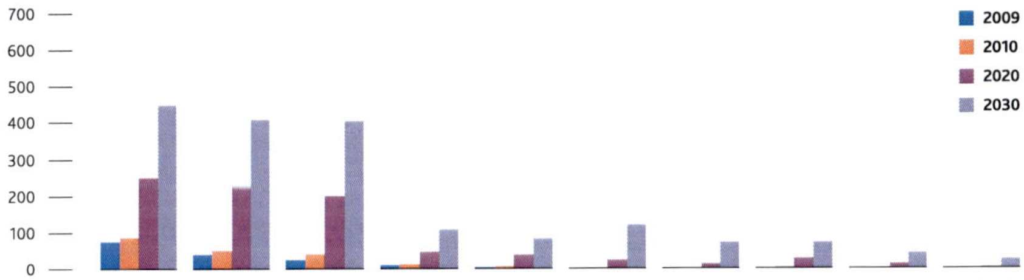

[GW]	OECD Europe	OECD North America	China	India	OECD Pacific	Non-OECD Asia	Eastern Europe/ Eurasia	Latin America	Africa	Middle East
2009	76	39	26	11	5	0,5	0,5	1	0,8	0,1
2010	87	50	40	13	6	0,7	0,9	2	1	0,1
2020	251	220	200	46	37	24	12	28	12	2
2030	447	407	404	108	81	120	71	72	42	25

REGIONAL BREAKDOWN: ADVANCED SCENARIO

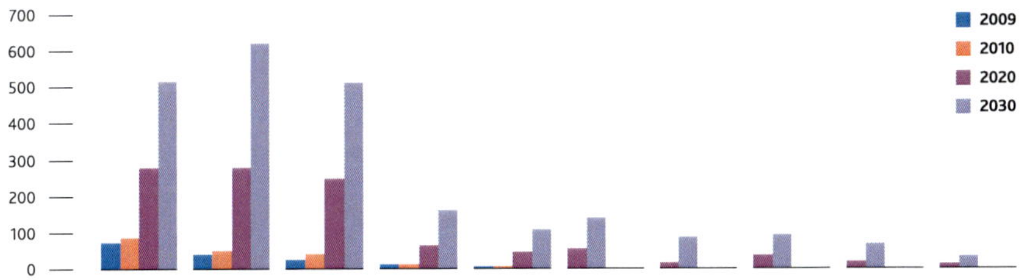

[GW]	OECD Europe	OECD North America	China	India	OECD Pacific	Non-OECD Asia	Eastern Europe/ Eurasia	Latin America	Africa	Middle East
2009	76	39	26	11	5	0,5	0,5	1	0,8	0,1
2010	87	50	41	13	6	0,7	0,9	2	1	0,1
2020	279	279	250	65	48	55	16	37	19	11
2030	515	622	513	161	109	140	87	93	67	34

Fig. 1.13 Global cumulative market forecast in terms of reference, moderate and advanced scenarios (Source: GWEC)

system level. Few of the technical issues in terms of present and future development of this technology are discussed in the following sections.

1.3.1 Wind Turbine Generator Unit

Wind turbine manufacturers are trying to incorporate the recent technology in drive train layout, turbine blade design and structural improvement to reduce the total mass and net cost and to increase energy extraction efficiency and lifetime as well. Gearbox is one of the components that causes the downtime of wind turbine generator systems the most, and therefore, gearless or one or two gearing stage with multi-pole generator based scheme is going to be a popular trend in the wind industry. Elimination of carbon fibre reinforcement from turbine blade might be a good attempt which is under consideration by blade manufacturers. Lightning protection scheme including eddy current loss minimization should be focused more in future blade designing. This is because the lightning strike in one wind turbine may not only hamper wind power extracting from that unit, but the nearby units as well.

At the generator end, although the doubly fed induction generator is dominating the wind industry, permanent magnet synchronous generators may play a vital role in the near future due to the flexibility of gearless operation as mentioned earlier. The size of the wind generator can be reduced significantly by using superconducting material, where the research focus can be targeted, especially when we are thinking about a wind turbine generator system of more than 5 MW.

1.3.2 Power Electronic Converter Technology

The issue of energy conversion from wind power nowadays involves the presence of power electronic devices. In a power electronic inverter and converter the commonly used devices are the diode, thyristor, gate turn-on thyristor (GTO) or insulated gate turn-on thyristor (IGBT) and in some cases the integrated gate-commutated thyristor (IGCT). The conduction and switching losses in these device modules have been reduced greatly and therefore the losses in high power converters/inverters have also reduced significantly. As a result, the full rating inverter/converter-based wind energy conversion system using permanent magnet synchronous generator is becoming popular. The same reason is behind the popularity of HVDC-based offshore wind farms. However, the wind power industry is looking forward to further loss reduction of high power converters and the discovery of a silicon carbide (SiC) power switch has added extra pace in this development.

Another important consideration is the reliability of the megawatt class wind energy conversion system using full or partial rated frequency converter. Loss of the frequency converter results in loss of total generation and operation of frequency converter at low power due to low wind speed may result in higher loss. Considering these issues the parallel operation of the frequency converter might be a feasible option which GAMESA has already implemented successfully, though more research is required for overall control strategy including fault ride-through characteristics of multiple units.

1.3.3 Offshore Wind Farm

In the recent years offshore wind farms show the most prominent market opportunities, since they are likely to offer comparatively higher productivity than projects based on onshore wind turbines, due to the higher speed of offshore wind. It is expected that while the average wind speed on the shore is 7 m/s, offshore speed ranges between 9 and 10 m/s. The average size of the offshore wind farm is increasing e.g., the average offshore wind farm size in Europe in 2010 exceeds 150 MW [4], therefore, the total accumulation will be on a big scale. Many large-scale offshore wind farms are under construction and many more are in the planning stage.

Numerous technological challenges exist for successful installation, commissioning and operation of offshore wind farms. The average distance from the shore is increasing compared to previous years. As per the EWEA report [4], the average distance from the shore increased in 2010 by 12.7–27.1 km, substantially less, however, than the average of 35.7 km for projects currently under construction. As a consequence, the average water depth is increasing, e.g., Average water depth in 2010 was 17.4 m, a 5.2 m increase from 2009, with projects under construction in water depth averaging 25.5 m. These raise the issues of foundations, installation methods, bulk power transmission system, etc.

The preferred foundation type for the offshore project is likely to be a multi-member design such as tripod, tripile or jacket because of the suitability of deep water. In general it is observed that at the present 2–3 MW class wind turbine generator units are dominating in offshore wind farms, however, the 5 MW class is going to be more popular in the near future. It is not an easy task to install large wind turbines in the sea where many technical challenges are needed to be overcome, especially in the deep sea. Figures 1.14a and b show the installation of a 5 MW class wind turbine generator unit at the Beatrice Offshore Wind farm by Scaldis Salvage & Marine Contractors NV.

The floating type offshore wind farm concept might be another innovative idea in the wind industry. A lot of research work is ongoing to make it a success. The variability of sea wind can be optimally utilized using the floating wind farm concept.

1.3.4 Operation and Maintenance

A downtime of the single or multiple megawatt class wind turbine generator results in loss of revenue. Therefore operation and maintenance (O & M) is a big

Fig. 1.14 a, b Installation of a 5 MW offshore wind turbine generator at the Beatrice Offshore Wind farm (Courtesy: Scaldis Salvage & Marine Contractors NV)

issue for wind farm owners of transmission system operators, both for onshore and offshore wind farms. The general picture in a wind farm is that 2 or 3 operators are engaged in O & M services for few tens of wind turbine generator units for routine checkup and for bringing back the units in operation which are out of services. In the case of offshore wind farms O & M facilities can be housed at a nearby port or floating substation depending on the distance of the wind farm from the shore. For distant offshore wind farms sometimes helicopter accessibility from the maintenance substation gives better flexibility to O & M personnel for quick repairing.

However, it is noted that remote condition monitoring based on data analysis from sensors may give a much better option and may become the standard in the future for operation and maintenance of large-scale wind farms both on offshore and onshore platforms. It will save the effort and cost of accessing the turbine unit by manual inspection. A project work carried out in Risø National Laboratory for Sustainable Energy in 2008 gave an important outline on the sensor-based remote monitoring scheme; more such works are required to be performed with support from the real industry. The Risø project focused on a structural health monitoring system for wind turbine blades [5]. The project was targeted at creating knowledge that will allow sensor signals to be used for remotely identifying the position of damage, the damage type and severity, and generate a structural condition assessment of the wind turbine blades that can integrate with existing SCADA tools to improve management of large offshore wind farms, and optimize the manual inspection/maintenance effort. The concept can be extended for remote monitoring of generator, gearbox and other sensitive equipments which widen the research opportunity in this area.

1.3.5 Moderate and Bulk Power Transmission

Economical feasibility is one of the key issues behind the selection of transmission technology (HVDC or HVAC) for grid interconnection of offshore wind farms. The distance of the offshore wind farm from the shore is an important factor to determine the transmission system as transmission line loss has a direct relation to it. High power converter technology has progressed tremendously, losses in converter have reduced significantly, and therefore, big companies are offering complete packages for bulk offshore power transmission using HVDC technology. There is a chance that future offshore wind farms will move to the HVDC-based transmission system, especially with the spread of the supergrid concept these days. Some of the other issues like multi-terminal scheme for offshore wind farm, loss minimization in HVDC stations, control scheme including fault ride-through capability augmentation should be focused more to make this technology viable. Moderate level DC-based onshore wind farms may attract the grid operator where full-bridge DC–DC converters can help in boosting the generator side voltage to higher levels using high frequency transformers.

1.3.6 Variability and Predictability

There is no argument about the variability of wind power but about intermittency! Wind power is often described as an "intermittent" energy source, and therefore unreliable. In fact, at the power system level, wind energy does not start and stop at irregular intervals, hence the term "intermittent" is misleading. The output of aggregated wind capacity is variable, just as the power system itself is inherently variable. Since wind power production is dependent on the wind, the output of a turbine and a wind farm varies over time, under the influence of meteorological fluctuations. These variations occur on all time scales: by seconds, minutes, hours, days, months, seasons and years. Understanding and predicting these variations is essential for successfully integrating wind power into the power system and to use it most efficiently [6].

Predictability is key in managing wind power variability, and significant advances have been made in improving forecasting methods. Today, wind power prediction is quite accurate for aggregated wind farms and large areas. Using increasingly sophisticated weather forecasts, wind power generation models and statistical analysis, it is possible to predict generation from five min to hourly intervals over timescales upto 72 h in advance, and for seasonal and annual periods. Using the current tools, the forecast error for a single wind farm is between 10 and 20% of the power output for a forecast horizon of 36 h. For regionally aggregated wind farms the forecast error is in the order of 10% for a day ahead and less than 5% for 1–4 h in advance [6]. However, there is further scope to work on the prediction scheme of wind speed for making it as precise as possible and there is room to integrate the prediction strategy in the control of wind energy conversion system.

1.3.7 Energy Storage Option

To cope with the variability of wind power and to meet the requirement of transmission system operators (TSO) or grid companies,the wind industry is extensively dependent on energy storage options. There is increasing interest in both large-scale storage implemented at transmission level, and in smaller scale dedicated storage embedded in distribution networks. The range of storage technologies is potentially wide. For large-scale storage, pumped hydro accumulation storage (PACPACPAC) is the most common and best known technology, which can also be done underground. Another technology option available for large scale is compressed air energy storage (CAES). On a decentralized scale storage options include flywheels, batteries, possibly in combination with electric vehicles, fuel cells, electrolysis and super-capacitors. Furthermore, an attractive solution consists of the installation of heat boilers at selected combined heat and power locations (CHP) in order to increase the operational flexibility of these units [6].

Hybrid power stations composed of wind energy conversion system, hydrogen generation, and fuel cell and energy storage system have a good potential in the future energy market. A lot of work has to be done in the area of a coordinated control scheme for hybrid power stations where the lifetimes of electrolyzer, fuel cell, energy storage devices should be considered as well. For the floating type offshore wind farm ESS may play a vital role to make this technology successful.

1.3.8 Grid Code

In response to the increasing demands from network operators, for example to stay connected to the system during a fault event, the most recent wind turbine designs have been substantially improved. The majority of MW-size turbines being installed today are capable of meeting the most severe grid code requirements, with advanced features including fault ride-through capability. This enables them to assist in keeping the power system stable when disruptions occur. Modern wind farms are moving towards becoming wind energy power plants that can be actively controlled [6]. Grid codes developed in many countries are more or less similar. They focus on policies such as transient Fault Ride Through (FRT), active power control, reactive power issues, power quality and voltage regulations. Huge penetration of wind power to the grid in the near future may lead to a change in the present grid codes available in different countries and wider regions.

1.4 Wind Power Explained in this Book

As mentioned in the preface the chapters in this book are organized into three sections. The key issues mentioned in each chapter are highlighted briefly in the following sections. In some cases, the technical challenges, future trends and further research scopes are pointed out.

It is expected that a large number of wind generators are going to be connected to the grid in the near future as can be seen from the statistics of the past few years. As different types of wind generators are commercially available in the market, it is very important to know what type of wind generator we are going to install in a particular region. This is because the overall efficiency of the wind turbine generator system depends on the losses of its various components which are generally in the form of copper loss, iron loss, stray load loss, windage loss, bearing loss and in some cases the losses in various power electronic devices. The author in Chap. 2 has made a significant contribution in the loss calculation method of different wind turbine generator systems based on wind speed. The details of the loss calculation methods for induction generator, doubly fed induction generator and permanent magnet synchronous generator are explained. A comparative study in light of loss and efficiency of different types of generators are also incorporated. This study is even helpful to find out the capacity factor of wind farms in an efficient and easy way.

In the sea the wind condition is better than on shore and therefore much energy extraction from wind is expected from a single wind turbine generator unit. However, the larger the generating capacity the larger the structural cost is, and at the same time the generator should be operated at lower speed due to the large diameter of the turbine blade. This issue initiates the idea of using superconducting material in wind generators and the authors in Chap. 3 report on this emerging technology. In the direct drive wind turbine operated at variable speed the superconducting generator can be applied successfully, and the wind industry may accept this trend in the near future.

The wind industry is extensively dependent on power electronic devices and without the rapid development of power converters it would not be possible to build large-scale wind turbine generating units. The maximization of power capture from the wind also would not be possible. The power electronic switches we use in the power converter/inverter are the conventional thyristor, GTOs, or IGBT and in some cases IGCT. The device which can offer low loss, fast-switching, higher blocking voltage, etc. are most attractive for the converter/inverter technology of wind energy conversion system. Size and compactness are other issues under consideration. Silicon carbide (SiC) power switches have the potential to meet all the aforementioned issues and may lead the converter/inverter technology that suits wind energy conversion system. Authors of Chap. 4 have focused on this promising technology.

Chapter 5 is focused on a cost-effective grid interfacing scheme of a wind farm composed of variable speed wind generators. The thyristor rectifiers ensure the maximum power capture from individual wind generators and are connected to a common DC transmission system and finally one thyristor inverter is used for grid interfacing. For reactive power compensation of the grid side, synchronous compensator using duplex reactor is considered. Detailed modelling and control strategy of the proposed system are discussed and extensive simulation analyses are also performed.

The application of a switched reluctance generator (SRG) as variable speed wind generator is demonstrated in Chap. 6. The switched reluctance generator has some inherent characteristics such as simple construction, robustness and lower cost and it might be a good choice to use it as a small wind generator. The construction and modelling of SRG, grid interfacing using asymmetric half-bridge converter, dc-link and inverter are discussed in detail. Speed control including maximum power point tracking scheme is demonstrated as well. It seems from the simulation results that SRG can be operated at variable speeds to maximize the power capture from wind and can be interconnected to the grid with the help of a power electronic converter using the available technology in the power industry.

Numerous control relevant issues are involved with the individual components of a wind turbine generator system for standalone and grid connected systems. To apply the control in a precise way, it is essential to know the dynamics of individual components. The author in Chap. 7 has made efforts towards developing the dynamics of the different parts of a wind energy conversion system. The dynamics of a power converter which are used frequently with commercially available wind turbines are also discussed in the chapter and finally, the control algorithms are presented in light of system dynamics.

Power quality is one of the key points for successful grid-interfacing of a wind turbine generator system. In Chap. 8, the authors focused on voltage flicker which is an important area to be considered for large-scale wind farm grid interconnection. The primary reasons for voltage flicker arise from tower shadow, blade pitching, yaw errors and in some cases from turbulence, wind shear or variability of wind speed. It might not be an easy task to measure the voltage flicker and the authors in this chapter attempt to find a successful way to measure voltage flicker caused by wind turbines.

An important and timely study on wind power forecasting in terms of wind speed prediction is presented by the authors in Chap. 9. Wind speed prediction is an important area to be focused on which can be integrated with modern control systems of wind turbine generator units or controlling the entire wind farm as per the requirement of transmission system operators (TSOs). The authors focused on wind speed forecasting based on the Grey predictor rolling model which can be successfully used for hourly wind power prediction. As wind power penetration level to the grid is increasing rapidly, this will allow the utility companies to resolve the issues of power quality, load management, system stability etc. In general, it can be said that the more precisely we are able to predict the wind, the more the increase in the possibility of wind power penetration.

The word "Lightning" is a major concern for wind turbine and blade manufacturers. Length and material of the blade, structure height, local terrain elevation are important factors for lightning protection and the authors in Chap. 10 have pointed and discussed on these important issues. The effect of turbine rotating blades on lightning is also taken into consideration. The presence of carbon reinforced plastics (CRP) in the blades may raise a new problem which is pointed out. The energy dissipation issue caused by CRP is discussed as well.

In Chap. 11, effects of "back-flow-surge" on wind turbines due to winter lightning are reported by the authors, which is very important for wind farm designers. The nearby wind turbines that were not struck by lightning can also be affected easily by "back-flow-surge" and therefore, a special lightning protection scheme is very important at the design level. The surge protection device of the lightning-struck turbine and other turbines nearby and even far from the striking point can be damaged. Possible solutions of the aforementioned problems are also pointed out by the authors.

Many countries have adopted their own grid code considering the huge penetration of wind power into the grid. When the wind power was not on a large scale, shutdown of a small unit was not a big issue for power grid companies. However, shutdown of a megawatt class wind turbine or a group of megawatt class wind turbines in a wind farm may cause system instability similar to a conventional power plant when it shuts down. Therefore, a wind farm has to comply with the grid code and the authors in Chap. 12 have reported on the operational aspects of a wind turbine generator system considering grid connectivity issues.

There is no doubt that energy storage system (ESS) is an effective means to increase the wind penetration level to the power system. A lot of energy storage technology is available that can be successfully incorporated with an individual wind turbine generator unit or with the entire wind farm. However, in this regard the first question that comes to our mind is the huge investment cost of ESS, especially when we consider the megawatt class wind power integration to the power system. For autonomous islands the pump storage system is an attractive and viable solution and in Chap. 13 the authors present an excellent work on this technology where focus is given on hybrid wind-hydro power stations. The outline of the scheme, operating policy for island system, economic evaluation, detailed modelling and control strategy, and finally a real-case scenario from Ikaria Island, Greece are demonstrated in the chapter.

In Chap. 14, frequency fluctuation minimization of a grid-connected wind farm is discussed. To minimize wind power fluctuation, a superconducting magnetic energy storage (SMES) system is used as the energy storage device, which is considered to be connected at the terminal of the wind farm. A realistic power system model is used in the simulation and the detailed switching model is considered for SMES modelling. The optimum power and energy storage capacity required for SMES is focused and finally, it is shown that SMES is a good tool to mitigate the frequency fluctuation of a grid connected wind farm. The similar control strategy can be adopted with other energy storage systems such as energy capacitor system, battery energy storage system, etc.

Are we moving to fixed type of offshore wind farms in the future! The answer might be YES. The authors of Chap. 15 have given tremendous efforts in the past and are continuing the works with the support from the National Aeronautics and Space Administration (NASA). The authors are preliminary focused on developing wind mapping in sea based on spaceborne scatterometer. Using scattorometer data average wind in sea, frequency of strong wind, power density, etc. can be obtained. It is possible to get the global distribution of wind strength and the authors have also reported on that. The dependence of wind strength on height and stability is examined as well. The authors are working on identifying near-shore locations of strong wind. In general it can be said that the technology presented in Chap. 15 may explore many branches in offshore wind power extraction research.

Chapter 16 is focused on offshore wind farm grid interconnection using a line-commutated HVDC scheme. Offshore wind farm composed of squirrel cage wind generator might be a good choice due to robustness, simplicity and lower cost of induction generator. In that case, the issue of reactive power requirements of induction generators for their magnetizations should be handled with proper care, otherwise the bus voltage will be affected widely. In this chapter, the authors have emphasised on damping enhancement and mitigation of wind power fluctuation of parallel operated wind farms through line-commutated HVDC link with a modal-control designed PID rectifier current regulator (RCR). The authors have presented both frequency and time domain analyses to demonstrate the validity of the proposed scheme.

Fault ride-through is equally important in the case of HVDC connected offshore wind farms. A major challenge might be to make coordination in bulk power transfer between offshore and onshore stations.The authors of Chap. 17 present three schemes to resolve this problem. In this chapter voltage source converter-based HVDC scheme is used for grid interconnection of offshore wind farms. The control scheme is applicable for multi-terminal HVDC topology and can successfully be used to augment the fault ride-through capability of offshore wind farms.

The details of the control scheme for HVDC-based offshore wind farms can be obtained in Chap. 18. The authors explain variable speed wind generator control including front-end converter current control scheme, diode-based HVDC control, modelling of individual components such as wind turbine, transformer, back to back converter scheme and distributed voltage and frequency control of offshore wind farms connected with a diode-based HVDC link. The fault ride-through capability augmentation is covered as well. Islanded operation, self-starting and voltage-dependent current order limit for thyristor-based HVDC system are also explained.

In Chap. 19, a coordinated control scheme for offshore wind farms using line commutated converter (LCC) is discussed focusing on the load frequency control. A frequency drop characteristic is considered in the control loop of HVDC rectifier to share the wind farm active power with the power grid efficiently taking into account change of generation and load. The method demonstrated in this chapter is a simple but effective one where active power flow through HVDC link is ramped down or up when the grid frequency is too high or low respectively. Inertial response and blade angle control are discussed as well.

Tidal power conversion came into the picture in the recent years and there might be a bright future for electricity generation from offshore tidal and wind turbine generating systems placed in the same region. The authors in Chap. 20 have worked a lot towards this technology and a prototype of the proposed scheme is given in this chapter. Offshore wind and hybrid systems are explained in detail and the power conversion scheme is also presented. The authors provide both the simulation and experimental results of the proposed scheme. There is scope to work further on the topological scheme, system layout, and bulk power transmission system.

DC-based wind farms will become a popular trend in the near future. For large-scale DC wind farms, high voltage of the sending end is the key issue as the conventional DC–DC boost converter has limitations on its output voltage regulation. A full-bridge DC–DC converter can be a key component of a DC-based wind farm that can resolve this problem. In Chap. 21, the authors propose a coordinated control scheme of a DC-based offshore wind farm that transmits power to the onshore grid through high voltage DC cables, utilizing full-bridge DC–DC converter at the offshore HVDC station. This might be a cost-effective solution compared to the existing technology and the scope for future research is open considering the voltage level of the wind farm including its topology, soft switching full-bridge DC–DC converter, line length of DC-cable, Power transmission capacity, etc.

1.5 Conclusions

At present, wind power is playing a major role in the world renewable energy market and from its growth it can be said that wind power will certainly hold the leading position in the next few decades. Due to its technological maturity and cost-effectiveness compared to the other renewable sources, it is expected that wind energy will also make a significant contribution to the world energy market which will reduce the carbon emission to a large extent. A few of the technological challenges such as adopting variability of wind power, power quality issues, etc., are yet to be solved, however, the combined efforts from researchers and scientists from different disciplines will ensure its fastest growth I believe.

References

1. GWEC Publications (2010) Global wind 2010 report. The Global Wind Energy Council
2. GWEC Publications (2010) Global wind energy outlook 2010. The Global Wind Energy Council
3. AWEA Resources (2010) U.S. Wind industry year-end 2010 market report. The American Wind Energy Association
4. EWEA Resources (2010) The European offshore wind industry key trends and statistics 2010. The European Wind Energy Association (EWEA) report
5. McGugan M et al. (2008) Fundamentals for remote condition monitoring of offshore wind turbines. Risø-R-1639(EN), Risø National Laboratory for Sustainable Energy, Denmark
6. Pullen A, Sawyer S, Teske S, Aubrey C (2008) Global wind energy outlook. The Global Wind Energy council and Greenpeace International

Part I
Wind Energy Conversion Systems

Chapter 2
Calculation Method of Losses and Efficiency of Wind Generators

Junji Tamura

Abstract In the recent years, many wind turbine generation systems (WTGS) have been installed in many countries. However the electric power obtained from wind generators is not constant due to wind speed variations. The generated electric power and the loss in WTGS change corresponding to the wind speed variations, and consequently the efficiency and the capacity factor of the system also change. In this chapter, methods to evaluate the losses and output power of wind generator systems with Squirrel-Cage Induction Generator (IG), Permanent Magnet Synchronous Generator (PMSG), and Doubly-fed Induction Generator (DFIG) are explained. By using the presented methods, it is possible to calculate the generated power, the losses, total energy efficiency, and capacity factor of wind farms quickly.

2.1 Introduction

Wind energy is a clean and renewable energy source. In the recent years, many wind turbine generation systems (WTGS) have been installed in many countries from the viewpoints of global warming and depletion of fossil fuels. In addition, WTGS is of low cost in comparison with other generation systems using renewable energy. However the electric power obtained from wind generators (WG) is not constant due to wind speed variations. The generated electric power and the loss in WTGS change corresponding to the wind speed variations, and consequently the efficiency and the capacity factor of the system also change. In addition, the wind characteristic of each

J. Tamura (✉)
Department of Electrical and Electronic Engineering, Kitami Institute of Technology, 165 Koen-Cho, Kitami 090-8507, Japan
e-mail: tamuraj@mail.kitami-it.ac.jp

S. M. Muyeen (ed.), *Wind Energy Conversion Systems*,
Green Energy and Technology, DOI: 10.1007/978-1-4471-2201-2_2,

area is different and thus the optimal WTGS for each area is different. Therefore, it is necessary to analyze the optimal WTGS in each area. In the determination of optimal WTGS, annual energy production and capacity factor are very important factors.

In order to capture more energy from wind, it is essential to analyze the loss characteristics of WG, which can be determined from wind speed. Furthermore, since many non-linear losses occur in WG, making prediction profit by using average wind speed may cause many errors. This chapter presents a method to represent various losses in WG as a function of wind speed, which is based on the steady-state analysis. By using the presented method, wind turbine power, generated power, copper loss, iron loss, stray load loss, mechanical losses, converter loss, and energy efficiency can be calculated quickly.

First, a calculation method of the efficiency for constant speed WGs using Squirrel-Cage Induction Generator (IG) is presented, in which, using the wind turbine characteristics and IG steady-state equivalent circuit, wind turbine output, generator output, and various losses in the system can be calculated. Next, a calculation method of the efficiency for variable speed WGs using permanent magnet synchronous generator (PMSG) is presented. PMSG has some advantages over constant speed IG; i.e., PMSG can operate at the speed corresponding to the maximum power coefficient of wind turbine; noise can be decreased because PMSG WG does not need slip ring, brush, and gear system. However, since it needs power electronics devices for being connected to the power grid, loss evaluation of the power electronics devices is also needed in order to calculate the total efficiency of the wind generation system. Finally, a method to calculate loss, power, and efficiency of WTGS with Doubly-fed Induction Generator (DFIG) is presented. In recent years, the number of wind farms with large size DFIGs has increased all over the world. This type of system has power converters in the rotor circuit, and thus it can be operated at variable speed. The power rating of the power converter can be lower in this system than those in other types of systems, for example, WTGS with a synchronous generator with a field winding or permanent magnet. Thus, the power converter cost becomes lower than those of other systems.

In the methods presented in this chapter, wind speed is used as the input data, and then all state variables and conditions of the WG system, for example, wind turbine output, generator output, output power to the power grid, and various losses in the system etc., can be obtained. Generator state variables are calculated using the d-q axis equivalent circuit.

As one application of the presented methods, annual energy production and capacity factor of the wind farm can easily be evaluated by using wind speed characteristics expressed by Weibull distribution function. Weibull distribution function is commonly used to express the annual wind speed characteristics. Coefficients of Weibull distribution function can be determined by the geography and climate data of each area. Using the data of Weibull distribution function of different areas, capacity factor is calculated and compared among three types of WTGS, i.e., Squirrel-Cage IG, PMSG, and DFIG.

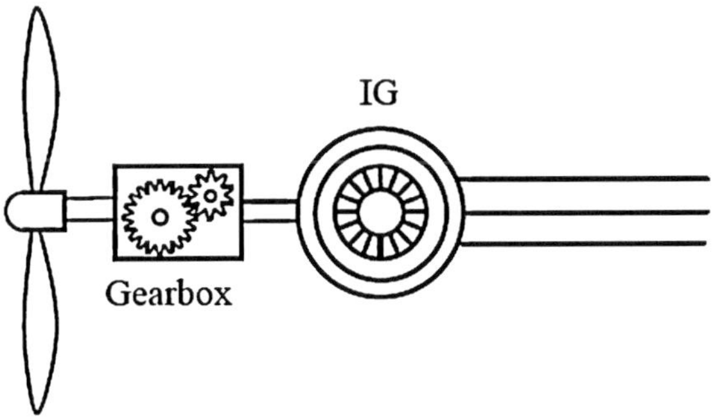

Fig. 2.1 System configuration with IG

Table 2.1 Losses of wind generator

Mechanical loss	Gear box losses
	Windage loss
	Ball bearing loss
Copper loss	Primary winding copper loss
	Secondary winding copper loss
Iron loss	Eddy current loss
	Hysteresis loss
Stray load loss	

2.2 Calculation Method for Squirrel-Cage Induction Generator

2.2.1 Outline of the Calculation Method

Induction generator is widely used as WG due to its low cost, low maintenance, and direct grid connection. However, there are several problems regarding the induction generator as given below.

- Usually the input, output, and loss conditions of induction generator can be determined from rotational speed (slip). However, it is difficult to determine slip from wind turbine input torque.
- Generator input torque is reduced by mechanical losses, but mechanical losses are a function of rotational speed (slip). It is difficult to determine mechanical losses and slip at the same time.
- It is hard to measure stray load loss and iron loss.
- It is difficult to evaluate gear loss analytically as a function of rotational speed.

In this section, a method of calculating the efficiency of WG correctly is presented, taking into account the points mentioned above. Figure 2.1 shows the system configuration for the analysis in this section. Table 2.1 shows the losses of this type of WG. The equivalent circuit of the induction generator used in the

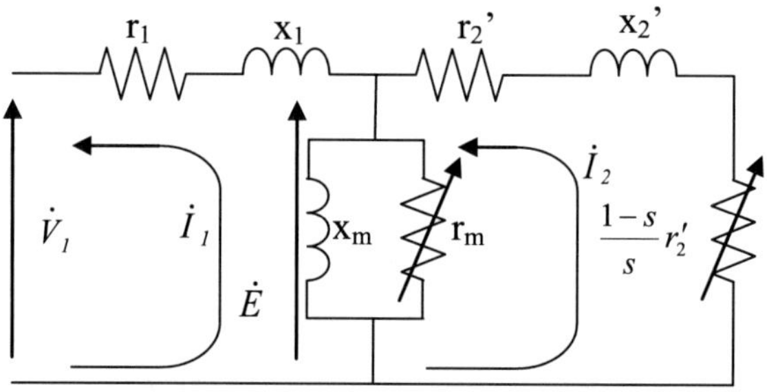

Fig. 2.2 Equivalent circuit of induction generator
r_1 = stator resistance,
r_2' = rotor resistance,
x_1 = stator leakage reactance,
x_2' = rotor leakage reactance,
r_m = iron loss resistance,
x_m = magnetizing reactance,
s(slip) = (Ns-N)/Ns, N = rotor speed, Ns = synchronous speed

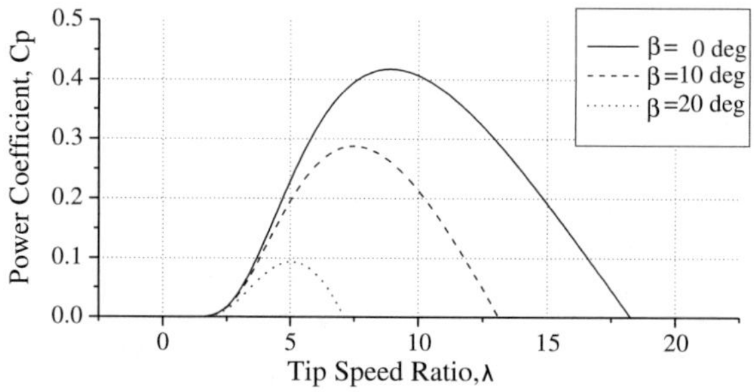

Fig. 2.3 Power coefficient versus tip speed ratio characteristics

method is shown in Fig. 2.2. The input torque and copper losses are calculated by solving the circuit Eq. 2.1.

$$\left.\begin{aligned}\dot{V}_1 &= -\left(r_1 + jx_1 + \frac{jr_m x_m}{r_m + jx_m}\right)\dot{I}_1 + \frac{jr_m x_m}{r_m + jx_m}\dot{I}_2 \\ 0 &= -\frac{jr_m x_m}{r_m + jx_m}\dot{I}_1 + \left(\frac{jr_m x_m}{r_m + jx_m} + \frac{r_2'}{s} + jx_2\right)\dot{I}_2\end{aligned}\right\} \quad (2.1)$$

2.2.2 *Models and Equations Necessary in the Calculations*

2.2.2.1 Wind Turbine Power

The MOD-2 [1] model is used as a wind turbine model in this chapter. The power captured from the wind can be expressed as Eq. 2.2, tip speed ratio as Eq. 2.3, and power coefficient C_P as Eq. 2.4. As shown in Fig. 2.3, this turbine characteristic is non-linear, and it has a characteristic similar to those of actual wind turbines.

$$P_{wtb} = \frac{1}{2}\rho C_p(\lambda, \beta)\pi R^2 V_w^3 (\mathrm{W}) \quad (2.2)$$

$$\lambda = \frac{\omega_{\mathrm{wtb}} R}{Vw} \tag{2.3}$$

$$C_{\mathrm{P}}(\lambda, \beta) = 0.5(\Gamma - 0.022\beta^2 - 5.6)e^{-0.17\Gamma}\left(\Gamma = \frac{R}{\lambda} \cdot \frac{3600}{1609}\right) \tag{2.4}$$

In Eqs. 2.2–2.4, P_{wtb} = turbine output power (W), ρ = air density (kg/m^3), C_{p} = Power coefficient, λ = Tip speed ratio, R = Radius of the blade (m), V_{w} = wind speed (m/s), ω_{wtb} = Wind turbine angular speed (rad/s), and β = blade pitch angle (deg).

2.2.2.2 Several Losses in the Generator System

Generator input power can be calculated from the equivalent circuit of Fig. 2.2 as shown below:

$$I_2^2\left(\frac{1-s}{s} \times r_2'\right) \quad (\mathrm{W}) \tag{2.5}$$

Copper losses are resistance losses occurring in the winding coils and can be calculated using the equivalent circuit resistances r_1 and r_2' as

$$w_{\mathrm{copper}} = r_1 \times I_1^2 + r_2' \times I_2^2(\mathrm{W}) \tag{2.6}$$

Generally, iron loss is expressed by the parallel resistance in the equivalent circuit. However, iron loss is the loss produced by the flux change, and it consists of eddy current loss and hysteresis loss. In the calculation method here, the iron loss per unit volume, $w_{\mathrm{f},}$ is calculated first using the flux density, as shown below [5].

$$w_{\mathrm{f}} = B^2\left\{\sigma_{\mathrm{H}}\left(\frac{f}{100}\right) + \sigma_{\mathrm{E}} d^2\left(\frac{f}{100}\right)^2\right\} \quad (\mathrm{W/kg}) \tag{2.7}$$

where B: flux density (T), σ_{H}: hysteresis loss coefficient, σ_{E}: eddy current loss coefficient, f: frequency (Hz), and d: thickness of iron core steel plate (mm). Generally, flux φ and the internal voltage E can be related to Eq. 2.8. Therefore, if the number of turns of a coil is fixed, proportionality holds between the flux density and the internal voltage as shown in Eq. 2.9.

$$E = 4.44 \times f \times k_{\mathrm{w}} \times w \times \phi(\mathrm{V}) \tag{2.8}$$

$$B = B_0 \times \frac{E}{E_0}(\mathrm{T}) \tag{2.9}$$

where k_{w} : winding coefficient, w : number of turns, φ : flux, E_0 : nominal internal voltage. Then the iron loss resistance can be obtained with respect to the internal voltage E determined by the flux density as shown below, where W_{f} is the total iron loss which is determined using Eq. 2.7 and the iron core weight.

$$r_{\mathrm{m}} = \frac{E^2}{W_{\mathrm{f}}/3} \tag{2.10}$$

Bearing loss is a mechanical friction loss due to the rotation of the rotor, which can be expressed as below.

$$W_{\mathrm{b}} = K_{\mathrm{B}}\omega_{\mathrm{m}}\,(\mathrm{W}) \tag{2.11}$$

where K_{B} is a parameter concerning the rotor weight, the diameter of an axis, and the rotational speed of the axis.

Windage loss is a friction loss that occurs between the rotor and the air, and is expressed as follows.

$$W_{\mathrm{m}} = K_{\mathrm{W}}\omega_{\mathrm{m}}^2\,(\mathrm{W}) \tag{2.12}$$

where K_{w} is a parameter determined by the rotor shape, its length, and the rotational speed. Stray load loss is expressed as follows.

$$W_{\mathrm{s}} = 0.005\frac{P^2}{P_{\mathrm{n}}}(\mathrm{W}) \tag{2.13}$$

where P is generated power (W) and P_{n} is the rated power (W).

Gear box losses [6, 2], are primarily due to tooth contact losses and viscous oil losses. In general, these losses are difficult to predict. However, tooth contact losses are very small compared with viscous losses, and at fixed rotational speed, viscous losses do not vary strongly with transmitted torque. Therefore, simple approximation of gearbox efficiency can be obtained by neglecting the tooth losses and assuming that the viscous losses are constant (a fixed percentage of the rated power). A viscous loss of 1% of rated power per step is a reasonable assumption. Thus the efficiency of a gearbox with "q" steps can be computed using Eq. 2.14. Generally, the maximum gear ratio per step is approximately 6:1, hence two or three steps of gears are typically required.

$$\eta_{\mathrm{gear}} = \frac{P_{\mathrm{t}}}{P_{\mathrm{m}}} = \frac{P_{\mathrm{m}} - (0.01)qP_{\mathrm{mR}}}{P_{\mathrm{m}}} \times 100(\%) \tag{2.14}$$

where P_{t} is gear box output power, P_{m} is turbine power, and P_{mR} is the rated turbine power. Figure 2.4 shows the gear box efficiency for three gear steps. In this chapter, three steps are assumed, according to a large-sized WG in recent years.

2.2.2.3 Calculation Method

The efficiency of a generator is determined using the loss expressions described above. The input, output, and loss conditions of induction generator can be determined from rotational speed (slip). However, it is difficult to determine slip from wind turbine input torque. Therefore, an iterative process is needed to obtain

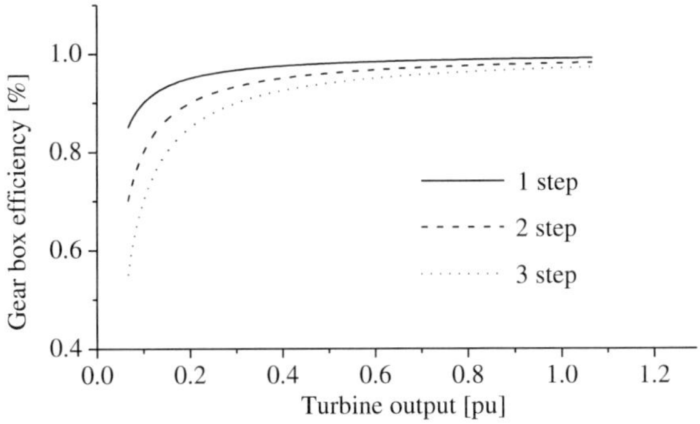

Fig. 2.4 Gear box efficiency

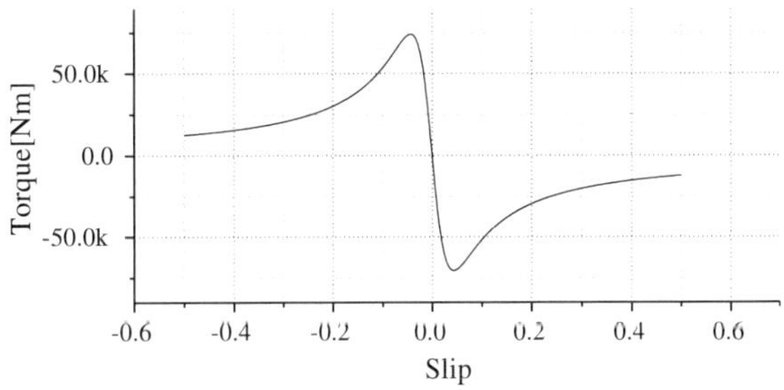

Fig. 2.5 Slip-torque curve

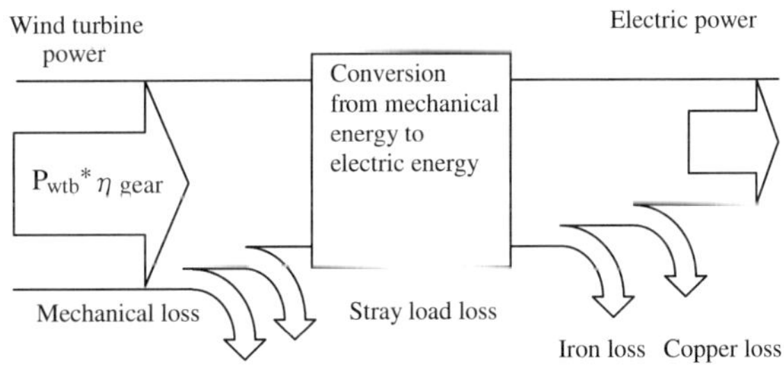

Fig. 2.6 Expression of power flow in the proposed method

a slip, which produces torque equal to the wind turbine torque, from a slip-torque curve as shown in Fig. 2.5. Furthermore, it is difficult to determine mechanical losses and slip at the same time, because mechanical losses are a function of rotational speed (slip). Mechanical loss can also be obtained in the iterative calculation. The power transfer relation in the WG is shown in Fig. 2.6.

Since mechanical losses and stray load loss cannot be expressed in a generator equivalent circuit, they are deducted from the wind turbine output. Figure 2.7 shows the flowchart of the calculation method, which is described below.

1. Wind velocity is taken as the input value, and from this wind velocity all states of WG are calculated.

Fig. 2.7 Flowchart of the proposed method

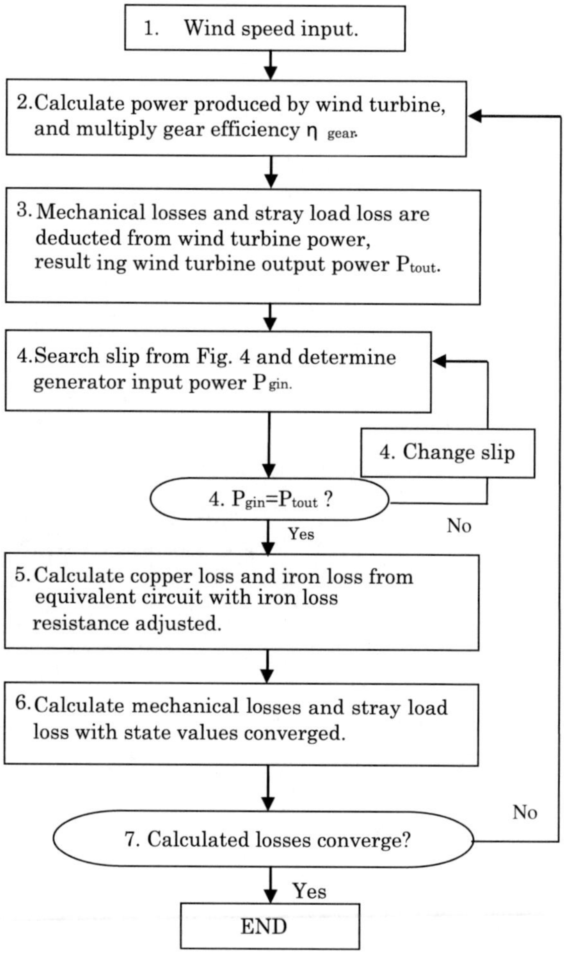

2. Wind turbine output is calculated from Eq. 2.2. The synchronous angular velocity is taken as the initial value of the angular velocity and wind turbine power is multiplied by the gear efficiency, η_{gear}.
3. Ball bearing loss and windage loss which are mechanical losses are deducted from the wind turbine output calculated in step 2, and stray load loss is also deducted. These losses are assumed to be zero in the initial calculation.
4. At this step the slip is changed using the characteristic of Fig. 2.5 until giving the same generated power as the power calculated in step 3.
5. By using the slip calculated in step 4 and using Eq. 2.1, the currents in the equivalent circuit can be determined, and consequently the output power, copper loss, and iron loss can be calculated. Next, loss W_f is calculated from the flux density using the iron loss calculation method mentioned above, and the iron loss resistance, r_m, which produces the same loss as W_f, is also determined.

Table 2.2 Induction generator parameters

Rated power	5 MVA	Rated voltage	6,600 V
Rated frequency	60 Hz	Pole number	6
Stator resistance	0.0052 pu	Stator leakage reactance	0.089 pu
Rotor resistance	0.0092 pu	Rotor leakage reactance	0.13 pu
Iron loss resistance (Initial value)	135 pu	Magnetizing reactance	4.8 pu

6. Ball bearing loss and windage loss are calculated by using Eqs. 2.11 and 2.12, and the rotational slip of the generator determined in step 5. And stray load loss is calculated from Eq. 2.13.
7. If the calculated losses converge, the calculation will stop, otherwise it will return to step 2.

2.2.3 Calculated Results

The parameters of the WG used in this section are shown in Table 2.2. A 5 MVA induction generator is used. The cut-in and rated wind speeds are 5.8 and 12.0 m/s respectively. Moreover, it is assumed that the generated power of the induction generator is controlled by pitch controller when the wind speed is over the rated wind speed. Figure 2.8 shows the calculated results of power and various losses of the generator, in which the curves for the windage loss, bearing loss, and iron loss are enlarged for clear and easy understanding. From the Figures, it is clear that all losses are non-linear with respect to the wind speed. Iron loss decreases with the increase of wind speed. When wind speed increases, the generator real power increases, and thus the generator draws more reactive power and the internal voltage of the generator decreases. As a result, flux density and iron loss decrease.

2.3 Calculation Method for Permanent Magnet Synchronous Generator

2.3.1 System Configuration

In this section, a calculation method of the efficiency for variable speed WGs using PMSG is explained. In the method, wind speed is used as the input data in a similar way as in the previous section, and then all state variables and conditions of the WG system, for example, wind turbine output, generator output, output power to the power grid, and various losses in the system etc. can be obtained.

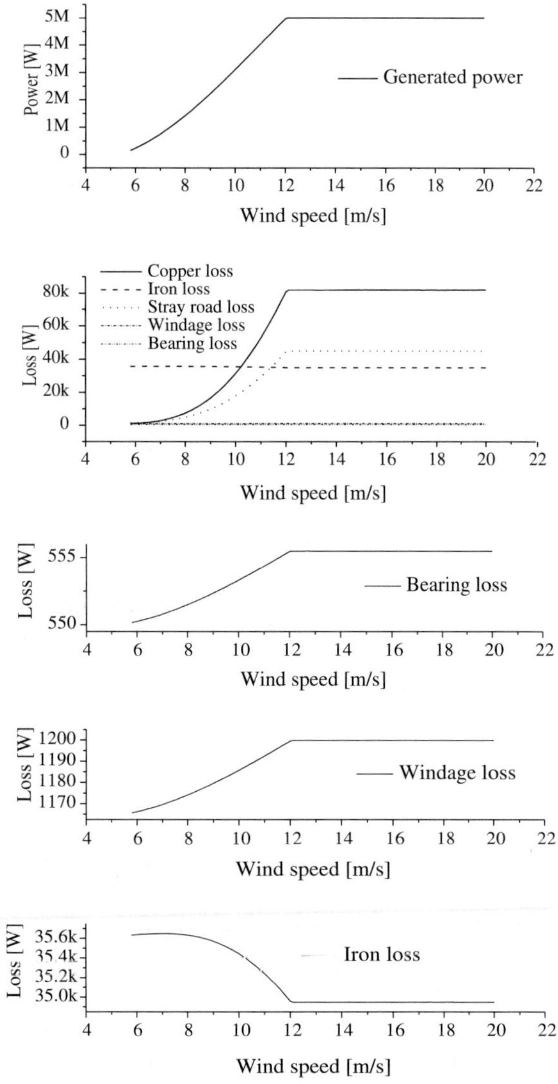

Fig. 2.8 Power and various losses of induction generator

Figure 2.9 shows the system configuration for the analysis in this section. The same model (MOD-2) as shown in Eqs. 2.2–2.4 is used as a wind turbine model. Figure 2.10 shows the wind turbine characteristic in a different manner from Fig. 2.3. Because this system can be operated in variable speed condition with the range of 0.4–1.0 pu where 1 pu is the synchronous speed, the turbine power can follow the maximum power point tracking (MPPT) line as shown in the figure. The rotor speed is controlled by the pitch controller in the high wind speed area and then kept at the rated level.

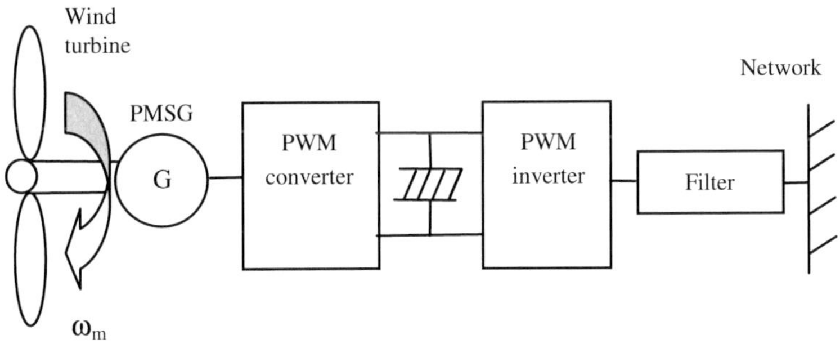

Fig. 2.9 System configuration with PMSG

Fig. 2.10 Wind turbine characteristics

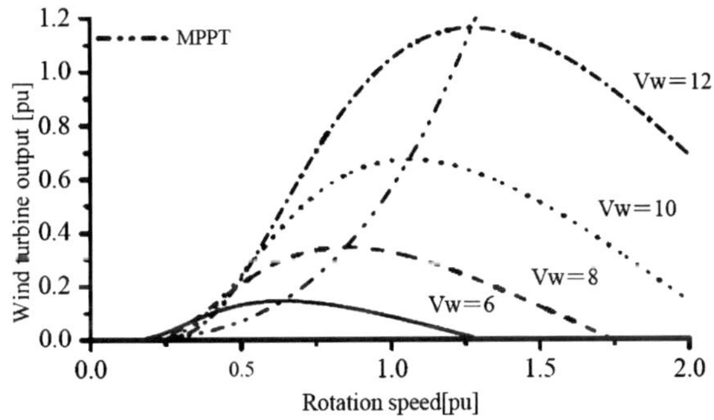

2.3.2 *Models and Equations Necessary in the Calculations*

2.3.2.1 Several Losses in the Generator System

Table 2.3 shows the various losses occurring in PMSG WG. Wind turbine output power is calculated by using the model equations presented above, and then generator input power can be calculated using the d-q axis equivalent circuit of Fig. 2.11 and Eqs. 2.15–2.22, where reactive power output of the generator is assumed to be controlled to zero.

$$v_\mathrm{d} = -r_\mathrm{a} i_\mathrm{d} + r_\mathrm{m} i_\mathrm{di} \tag{2.15}$$

$$0 = r_\mathrm{m} i_\mathrm{di} + \omega_\mathrm{m} \Phi_\mathrm{q} \tag{2.16}$$

$$v_\mathrm{q} = -r_\mathrm{a} i_\mathrm{q} + r_\mathrm{m} i_\mathrm{qi} \tag{2.17}$$

$$0 = r_\mathrm{m} i_\mathrm{qi} + \omega_\mathrm{m} \Phi_\mathrm{d} \tag{2.18}$$

$$\Phi_\mathrm{d} = -L_\mathrm{d}(i_\mathrm{d} + i_\mathrm{di}) + \Phi_\mathrm{f0} \tag{2.19}$$

Table 2.3 Losses of permanent magnet synchronous generator

Mechanical loss	Windage loss
	Ball bearing loss
Stray load loss	
Copper loss	
Iron loss	
Converter loss	
Inverter loss	
Filter loss	

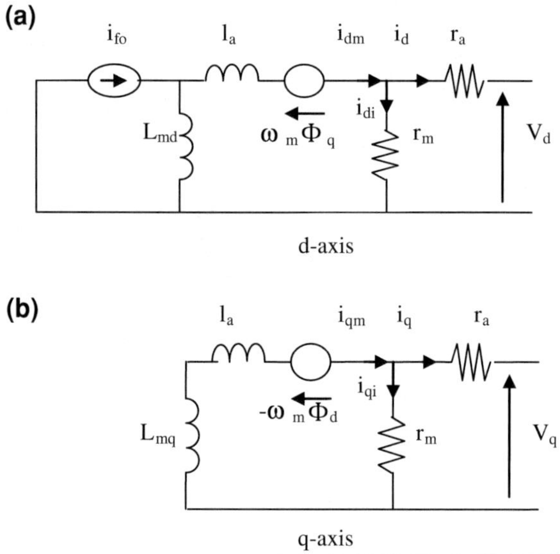

Fig. 2.11 d-q axis equivalent circuit. **a** d axis. **b** q axis r_a = Stator winding resistance, r_m = Iron loss resistance, l_a = Leakage inductance, L_{md} = d axis magnetizing inductance, L_{mq} = q axis magnetizing inductance, Φ_d = d axis flux linkage, Φ_q = q axis flux linkage, ω_m = Mechanical angular speed

$$\Phi_q = -L_q(i_q + i_{qi}) \tag{2.20}$$

$$\Phi_{f0} = L_{md} i_{f0} \tag{2.21}$$

$$P_{MG} = -\omega_m (L_d - L_q) i_d i_q + \omega_m \Phi_{f0} i_q \tag{2.22}$$

where, $L_d = l_a + L_{md}, L_q = l_a + L_{mq}$; L_d: d axis inductance; L_q: q axis inductance; P_{MG}: internal active power (W). Copper losses occur in the stator coil, and are calculated using stator winding resistance, r_a, in the equivalent circuit as below.

$$W_c = r_a(i_d^2 + i_q^2)(\mathrm{W}) \tag{2.23}$$

Mechanical losses, ball bearing loss W_b and windage loss W_m, are friction losses due to the rotation of the rotor. In general, bearing has two types, that is, plain bearing and ball-and-roller bearing. The bearing loss can be, in general, expressed as Eq. 2.24, where K_B is a parameter concerning the rotor weight, the diameter of the axis, and the rotational speed of the axis. Windage loss is a friction loss that occurs between the rotor and the air. Since it is difficult to calculate windage loss correctly, it is approximately expressed as Eq. 2.25 in this section, where K_w is a parameter determined by the rotor shape, its length, and the rotational speed. In general, bearing loss and windage loss in the case of PMSG WG are very small because its rotational speed is very low.

$$W_b = K_B \omega_m (W) \tag{2.24}$$

$$W_w = K_w \omega_m^2 (W) \tag{2.25}$$

Stray load loss is the electric machine loss produced under loading condition, and it is difficult to calculate accurately. The main factors for the stray load loss are the eddy current losses in conductors, iron core, and adjoining metallic parts produced by leakage flux. Stray load loss can be expressed approximately as Eq. 2.26 due to IEEE standard expression.

$$W_s = 0.005 \times \frac{P^2}{P_n} (W) \tag{2.26}$$

where, P: generated power (W); P_n: rated output (W).

Power electronic converter/inverter devices are necessary to connect PMSG WG with the power grid. Since the converter/inverter circuits include switching operations of IGBT devices, in general, it is difficult to calculate the losses in the devices accurately. In this section, power electronics device (PED) loss Eqs. 2.27 and 2.28 are used which is obtained from the semiconductor device catalogs [5]. PED loss is calculated by the combination of Eqs. 2.27 and 2.28.

$$P_{IGBT} = \sqrt{2} \cdot \frac{I_0}{\pi} \cdot (k_{ton} + k_{toff}) \cdot f_c + D \cdot \left(b \cdot \frac{I_0^2}{2} + \frac{\sqrt{2}}{\pi} \cdot a \cdot I_0 \right) \tag{2.27}$$

$$P_{FWD} = \sqrt{2} \cdot \frac{I_0}{\pi} \cdot k_{rr} \cdot f_c + (1 - D) \cdot \left(d \cdot \frac{I_0^2}{2} + \frac{\sqrt{2}}{\pi} \cdot c \cdot I_0 \right) \tag{2.28}$$

where, I_0: Phase current (A); f_c: Carrier frequency; D: IGBT duty ratio; k_{ton}: IGBT turn on switching energy (mJ/A); k_{toff}: IGBT turn off switching energy (mJ/A); k_{rr}: FWD recovery switching energy (mJ/A); a,b: IGBT on voltage approximation coefficient; $V_{CE} = a + b*I$; c,d: FWD forward voltage approximation coefficient; $V_F = c + d*I$. Figure 2.12 shows an example of loss characteristics of IGBT and FWD [7].

Filter is, in general, used to reduce the high harmonic components resulted from the inverter. Filter efficiency is assumed to be 98% here.

Iron loss mainly occurs in the stator iron core. Iron loss is expressed by using the varying iron loss resistance, r_m, in the equivalent circuit as shown in Fig. 2.11. However, real iron loss varies depending on the magnetic flux density in the core

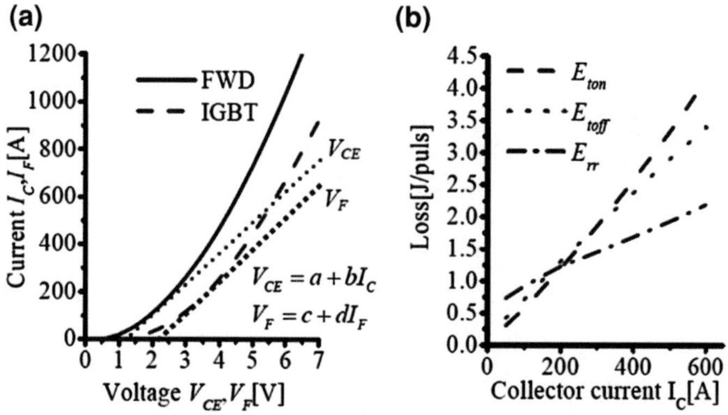

Fig. 2.12 Loss characteristics of IGBT and FWD. **a** Output characteristics. **b** Dependence of switching loss on I_C

which varies depending on the load condition. Therefore, if the iron loss is calculated using constant iron loss resistance, the result can have some error. Moreover, it should be noted that magnetic flux densities in heel piece and teeth in the stator core differ from each other.

Generally, iron loss consists of eddy current loss and hysteresis loss, both of which are proportional to the square of the magnetic flux density. In addition, eddy current loss is proportional to the square of the frequency and hysteresis loss is proportional to the frequency of alternating magnetic flux. In this section, iron loss is expressed as Eq. 2.29 in the same way as Eq. 2.7 for each of heel piece and teeth, which denotes the loss per 1 kg core. Therefore, the total iron loss W_f for each of heel piece and teeth is obtained by multiplying Eq. 2.29 by the core weight of each part. Then, the value of iron loss resistance, r_m, in the equivalent circuit is changed in order for the iron loss calculated from the equivalent circuit, W_r, to be equal to the iron loss, W_f.

$$w_f = B^2\left\{\sigma_H\left(\frac{f}{100}\right) + \sigma_E d^2\left(\frac{f}{100}\right)^2\right\}(\mathrm{W/kg}) \tag{2.29}$$

where, B: Magnetic flux density (T); σ_H: Hysteresis loss coefficient; σ_E: Eddy current loss coefficient; f: Frequency (Hz); d: Thickness of iron core steel plate (mm).

Calculation method of the iron loss is described below. Generally, magnetic flux and internal voltage can be related to each other as Eq. 2.30. Therefore, the magnetic flux density can be calculated from Eq. 2.31 and then the iron loss is calculated from Eq. 2.29, where magnetic flux density and internal voltage for the rated operating condition are expressed as nominal values, B_0 and E_0.

$$E = 4.44 f k_w w \phi (\mathrm{W}) \tag{2.30}$$

where, k_w: Winding coefficient; w: Number of turns; f: frequency.

$$B = B_0 \frac{E}{E_0} (\mathrm{T}) \tag{2.31}$$

where, E_0: Nominal internal voltage. Magnetic flux density is determined using the above equation, and then the total iron loss W_f is calculated. The iron loss resistance r_m can be obtained with respect to the internal voltage E and the total iron loss W_f as follows.

$$r_m = \frac{E^2}{W_f/3} \tag{2.32}$$

By continuing the above calculations until W_f to be equal to W_r, converged results can be obtained for the iron loss and iron loss resistance in the equivalent circuit. For the calculation of iron loss, initial value of iron loss is assumed to be 2.5% of the rated power.

2.3.2.2 Calculation Method

Figure 2.13 shows the flowchart of the calculation method, which is described below.

1. Wind speed V_w m/s is taken as the input value, and then all state variables of WG will be calculated.
2. Wind turbine output power is calculated from Eq. 2.2. Then, MPP(Maximum Power Point) produced by wind turbine is searched, resulting in the maximum wind turbine output power and the corresponding rotor speed. However, if the obtained power is greater than the rated power, the power is changed to 1 pu.
3. Bearing loss, windage loss, and also stray load loss are deducted from the wind turbine power calculated in step 2, yielding the input power to the generator. However, generator rotor speed is the value calculated in step 2.
4. Assuming generator reactive power to be 0, d and q axis currents are calculated from the d-q axis equivalent circuits, and then internal active power is calculated from Eq. 2.22.
5. Comparing the generator input power with the internal active power, if they are not equal to each other, calculation returns to step 4 with changing d-q axis currents, which will be continued until the generator input power is equal to the internal active power.
6. Using generator frequency and d-q axis currents calculated in step 5, internal voltage E, and then, W_f, W_r, and r_m, are calculated.

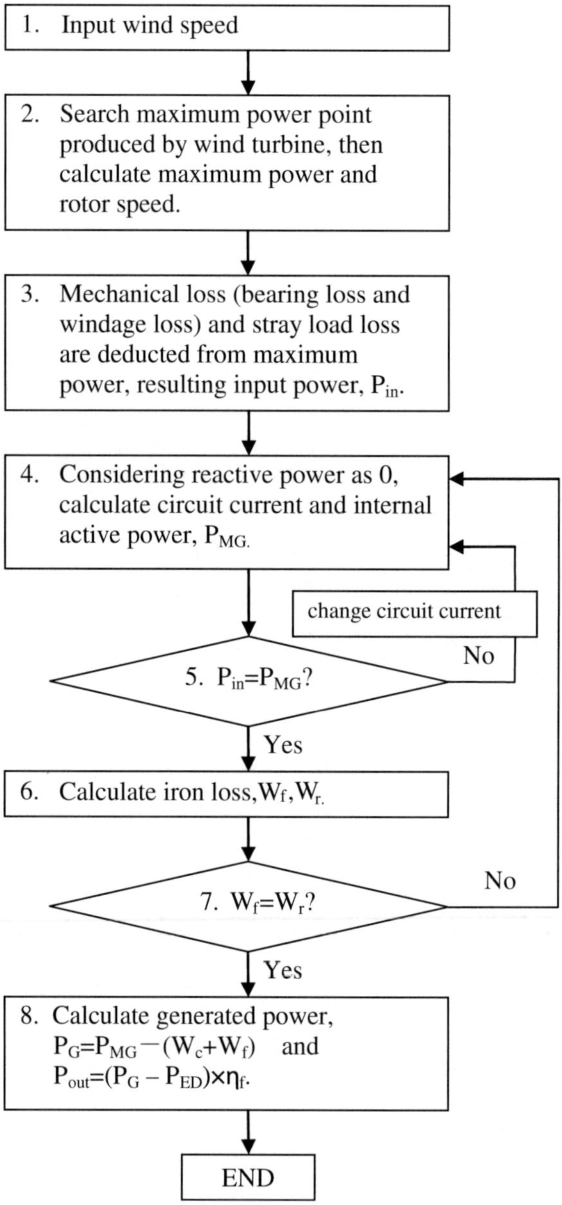

Fig. 2.13 Flowchart of calculation for PMSG wind generator

7. Comparing W_f with W_r, if W_f is not equal to W_r, calculation returns to step 4, with replacing the value of r_m by the new value calculated in step 6.
8. Generator active power P_G and AC/DC/AC converter loss, P_{ED}, are calculated. And then, deducting P_{ED} from P_G and multiplying the result by the filter efficiency, η_f, yields the final output power, P_{out}.

Table 2.4 PMSG wind generator parameters

Rated power	5 MVA	Rated voltage	6,600 pu
D axis reactance	0.88 pu	Q axis reactance	0.97 pu
Stator resistance	0.012 pu	Field flux	1.4 pu
Iron loss resistance (rated condition)	116 pu	Number of poles	96

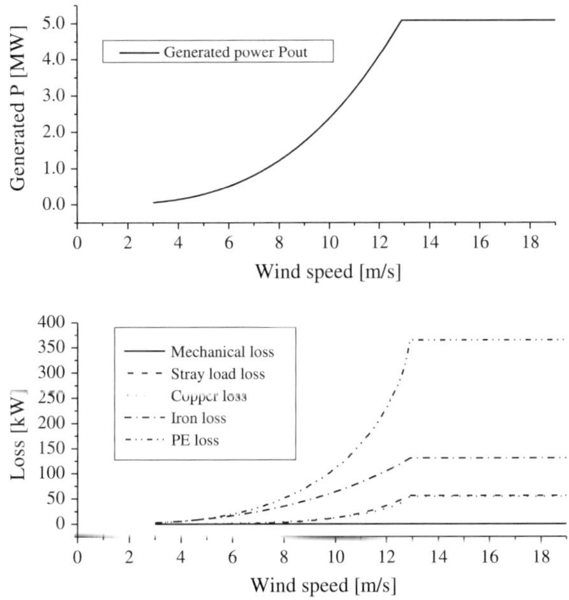

Fig. 2.14 Output and losses of PMSG wind generator

2.3.3 Calculated Results

The parameters of the PMSG WG used in the calculation are shown in Table 2.4.

Figure 2.14 shows the results of output power and various losses of PMSG. Bearing loss and windage loss characteristics are shown in Fig. 2.15. Figure 2.16 shows the characteristics of the iron loss, generator frequency, and the internal voltage. It is seen from Fig. 2.14 that output power and each loss increase with the increase in wind speed. Bearing loss and windage loss are small as shown in Fig. 2.15, because rotor speed of PMSG WG is very low. It is seen from Fig. 2.16 that the iron loss, the generator frequency, and the internal voltage increase with the wind speed. It is also seen from Fig. 2.17 that the iron loss resistance decreases with the wind speed, which can be thought to be due to the increases in generator frequency and internal voltage.

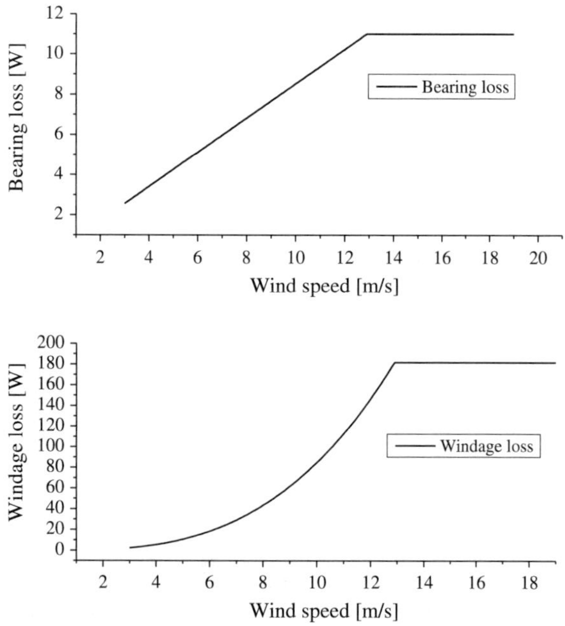

Fig. 2.15 Bearing loss and windage loss of PMSG wind generator

2.4 Calculation Method for Doubly-Fed Induction Generator

2.4.1 System Configuration

Figure 2.18 shows the system configuration of DFIG WG analyzed in this section. Wind speed is used as an input data, and then all state variables and conditions of the WG system can be obtained. This system can be operated in variable speed with the range of 0.7–1.3 pu, where 1 pu is the synchronous speed. Rated power of DFIG is set at 5 MVA.

The same model (MOD-2) as shown in Eqs. 2.2–2.4 is used as a wind turbine model. Because this system can also be operated in variable speed condition, the turbine power can follow the maximum power point tracking (MPPT) line as shown in Fig. 2.10. The rotor speed is controlled by the pitch controller within 1.3 pu.

2.4.2 Models and Equations Necessary in the Calculations

2.4.2.1 Several Losses in the Generator System

Various state values in the generator can be calculated using the equivalent circuit of Fig. 2.19, in which reactive power output of the generator is assumed to be controlled zero. Each internal voltage, E_1, E_2, is expressed as Eq. 2.33. Table 2.5 shows the various losses considered in DFIG WG.

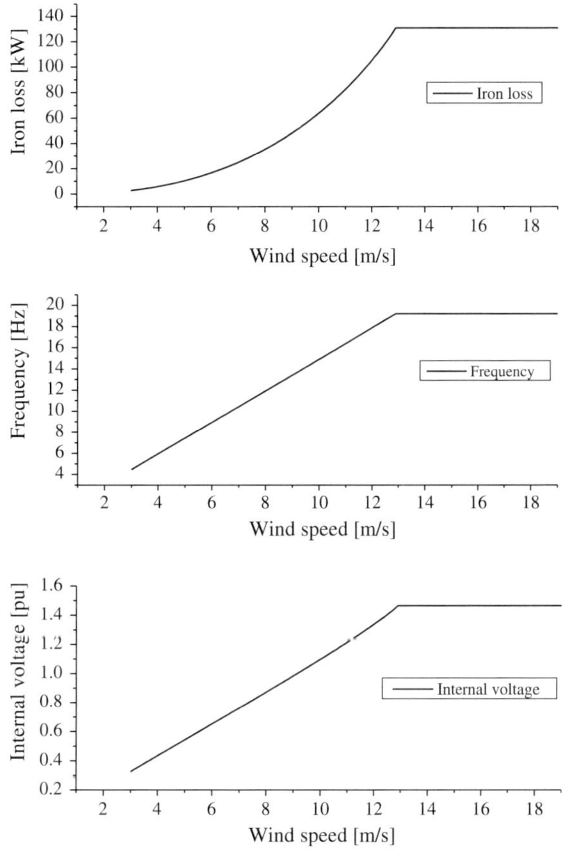

Fig. 2.16 Iron loss, frequency, and internal voltage of PMSG wind generator

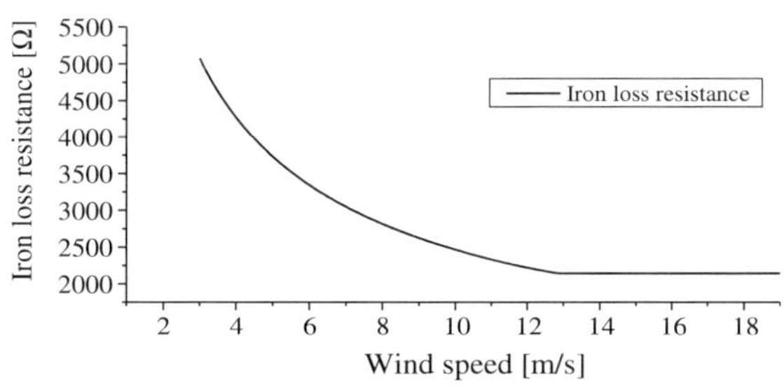

Fig. 2.17 Iron loss resistance of PMSG wind generator

$$\left.\begin{aligned}\dot{E}_1 &= jx_\mathrm{m}\dot{I}_{2\mathrm{a}}\\ \dot{E}_2 &= jsx_\mathrm{m}\dot{I}_{1\mathrm{a}}\end{aligned}\right\} \tag{2.33}$$

Gear loss, W_{gear}, among mechanical losses [3] is generated in speed-up gear between wind turbine and generator. Bearing loss, W_b, is a mechanical friction loss due to the rotation of the rotor. Windage loss, W_w, is a friction loss between the rotor surface and the surrounding air. Although it is difficult, in general, to

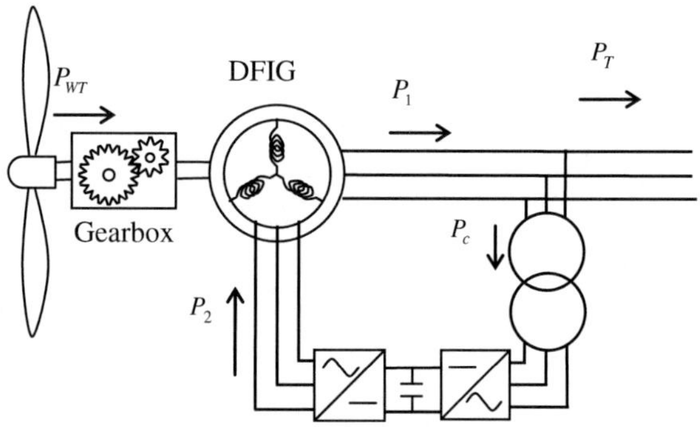

Fig. 2.18 System configuration with DFIG

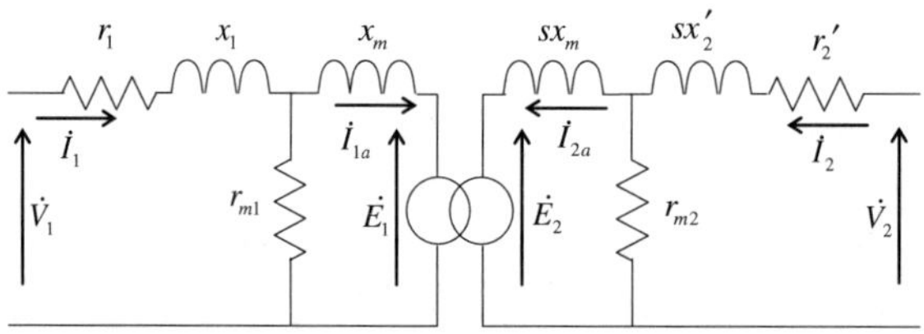

Fig. 2.19 Equivalent circuit of DFIG r_1 : stator winding resistance, r_2' : rotor winding resistance, x_1 : stator leakage reactance, x_2' : rotor leakage reactance, r_{m1} : stator iron loss resistance, r_{m2} : rotor iron loss resistance

Table 2.5 Losses in DFIG wind generator

Mechanical loss	Gear loss
	Bearing loss
	Windage loss
Iron loss	
Copper loss	
Stray load loss	
Power converter loss	
Transformer loss	

calculate these mechanical losses accurately, approximate expressions for the losses shown by Eqs. 2.34–2.36 are used in this section, where coefficients of bearing loss and windage loss, K_B, K_W, are determined by using generator structure and dimensions. Input power to the generator can be calculated by subtracting these mechanical losses from the wind turbine output.

$$W_{gear} = 0.01qP_{mR} \tag{2.34}$$

$$W_b = K_b\omega_m \tag{2.35}$$

$$W_w = K_w\omega_m^2 \tag{2.36}$$

In the equations above, P_{mR} is the rated turbine power and ω_m is angular speed.

Iron loss varies dependent on the flux density and frequency. In this section, iron loss is expressed by using variable iron loss resistances, r_{m1}, r_{m2}, as shown in Fig. 2.19. On the other hand, since the iron loss consists of hysteresis loss and eddy-current loss, it can be expressed as Eq. 2.37. Moreover, it should be noted that magnetic flux densities in yoke core and teeth core differ from each other. Therefore, iron loss of each core is calculated separately using Eq. 2.37 for each part of the stator and rotor, and then the total iron loss is obtained by summing them. Flux density can be expressed to vary in proportion to the internal voltage as shown in Eq. 2.38. Iron loss resistance in the equivalent circuit can be determined by Eq. 2.39. In these equations, K_1 and K_2 are coefficients of hysteresis and eddy-current losses, E_0 is the reference internal voltage, E' is the internal voltage, B_0 is the reference flux density, and W is iron core weight.

$$W_i = W_h + W_e = K_1 f B^{1.6} + K_2 f^2 B^2 (\mathrm{W/kg}) \tag{2.37}$$

$$B = B_0 \frac{E'}{E_0} \tag{2.38}$$

$$r_m = \frac{E'^2}{W_i \cdot W/3} \tag{2.39}$$

Copper losses in the stator coil P_{co1} and the rotor coil P_{co2} can be calculated using winding resistances, r_1 and r_2', in the equivalent circuit as follows.

$$P_{co1} = r_1 I_1^2 \tag{2.40}$$

$$P_{co2} = r_2' I_2^2 \tag{2.41}$$

Stray load loss can be expressed approximately as Eq. 2.42.

$$W_S = 0.005 \times \frac{P^2}{P_R} \tag{2.42}$$

where P_R is the rated power of the generator and P is generator output.

The Power converter is composed of IGBT and FWD. Therefore, power converter loss is calculated as a summation of IGBT switching loss, reverse recovery loss of FWD, and steady-state losses of IGBT and FWD [5]. It is expressed as Eq. 2.43.

$$P_{PC} = \frac{1}{2} DT \left[\frac{2\sqrt{2}}{\pi} I_0 a + I_0^2 b \right] + \frac{1}{2}(1 - DT) \left[\frac{2\sqrt{2}}{\pi} I_0 c + I_0^2 d \right] + \frac{1}{2} f_C (E_{on} + E_{off}) + \frac{1}{2} f_C E_r \tag{2.43}$$

where, I_0: Phase current (A); f_c: Carrier frequency (Hz); DT: IGBT duty ratio; k_{ton}: IGBT turn on switching energy (mJ/A); k_{toff}: IGBT turn off switching energy

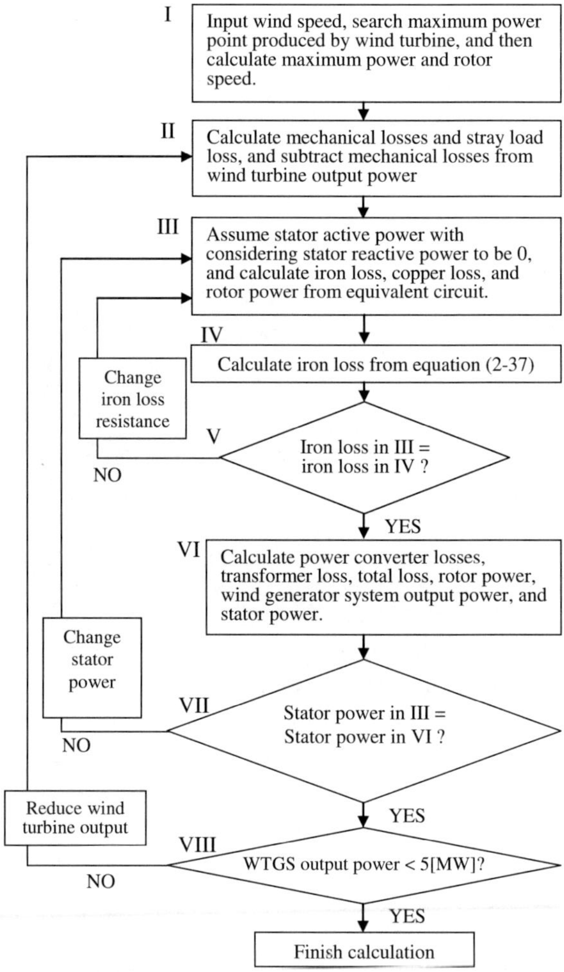

Fig. 2.20 Flowchart of calculation for DFIG wind generator

(mJ/A); k_{rr}: FWD recovery switching energy (mJ/A); a,b: IGBT on-voltage approximation coefficient, $V_{CE} = a + b*I$; c,d: FWD forward voltage approximation coefficient, $V_F = c + d*I$.

For simplicity, transformer is expressed by leakage impedance, and its loss is calculated as a resistance loss. The iron loss of the transformer is not considered.

2.4.2.2 Calculation Method and Results

The flowchart for the entire calculation using each loss expression explained above is shown in Fig. 2.20, which is described below.

Table 2.6 DFIG wind generator parameters

Rated output	5 MVA	Rated voltage	6,600 V
Frequency	60 Hz		
r_1	0.0053 pu	r_2	0.0052 pu
x_1	0.076 pu	x_2	0.14 pu
x_{m}	4.4 pu		
r_{m1}	287 pu	r_{m2}	166 pu

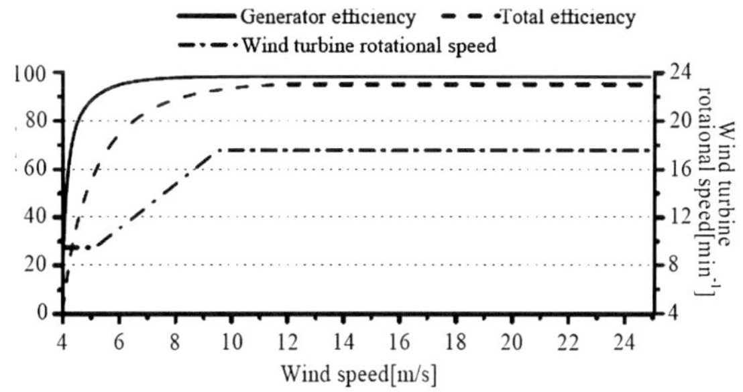

Fig. 2.21 Calculated results of efficiencies and rotational speed of DFIG wind generator

1. Wind speed V_{w} m/s is taken as the input value, and then all state variables of WG will be calculated. Wind turbine output power is calculated from Eq. 2.2. Then, MPPT(Maximum Power Point) power produced by wind turbine is searched, resulting the maximum wind turbine output power and the corresponding rotor speed.
2. Gear loss, bearing loss, windage loss, and also stray load loss are calculated and deducted from the wind turbine power calculated in step 1, yielding the input power to the generator. Generator rotor speed is the value calculated in step 1.
3. Assuming generator reactive power to be 0 and stator active power to be an appropriate value, iron loss, copper loss, and rotor power are calculated from the equivalent circuit, in which the stator voltage is set to be 1 pu.
4. Iron loss is calculated using Eq. 2.37 and iron core weight.
5. The above calculation is repeated until the iron loss in step 3 is equal to that in step 4 with changing the iron loss resistance.
6. Power converter losses, transformer loss, total loss, rotor power, WG system output power, and stator power are calculated.
7. Above calculation is repeated until the stator power in step 3 is equal to that in step 6 with changing the assumed stator power.
8. If the WG system output is greater than 1 pu, wind turbine output is reduced by the pitch controller and go to step 2.

The parameters of the DFIG WG used in the calculation are shown in Table 2.6. Figures 2.21, 2.22 and 2.23 show the obtained results of efficiency and losses of DFIG WG with respect to wind velocity. The cut-in wind speed, cut-out wind speed, and rated wind speed are 4.0 m/s, 25.0 m/s, and 12.1 m/s, respectively. Figure 2.21 shows that the generator efficiency becomes highest when wind

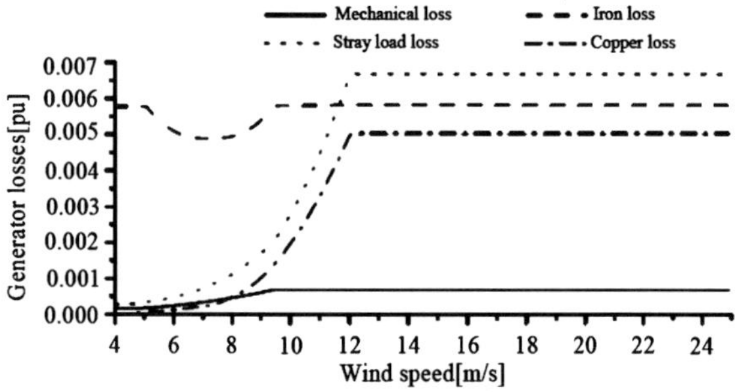

Fig. 2.22 Calculated results of losses in DFIG wind generator

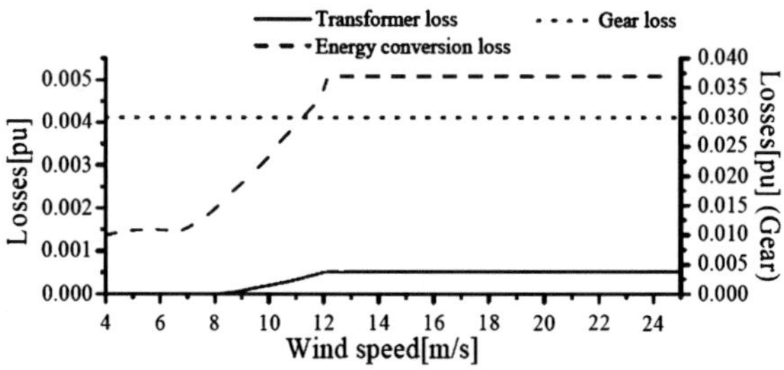

Fig. 2.23 Calculated results of other system losses in DFIG wind generator

speed is about 10 m/s but the total system efficiency becomes highest when the wind speed is over the rated speed. From Figs. 2.22 and 2.23, it can be understood that the stray load loss is greater than other generator losses and the gear loss is very large among all losses.

2.5 Comparative Study About Capacity Factor Among Three WGs (IG, PMSG and DFIG)

2.5.1 Weibull Distribution Function

If real wind speed data is available as a function of time, the efficiency calculation of WG can be precisely performed. However, it is difficult to calculate the annual generated energy and capacity factor by using the real wind data for one year expressed as a function of time. If the Weibull distribution function of wind speed for a specific area is available, it is possible to calculate the amount of annual generated energy and capacity factor for that area by using the method described above. The Weibull function can be expressed as Eq. 2.44, where k is shape factor and c is scale factor. $f(V_w)$ denotes a probability density distribution function that wind speed V_w appears. The annual generated energy can be calculated from Eq. 2.45, where $P_g(V_w)$ is generated power, E_{total} is annual energy production, V_{max} is cut-out wind speed (m/s), and V_{min} is cut-in wind speed (m/s). Capacity factor is calculated from Eq. 2.46.

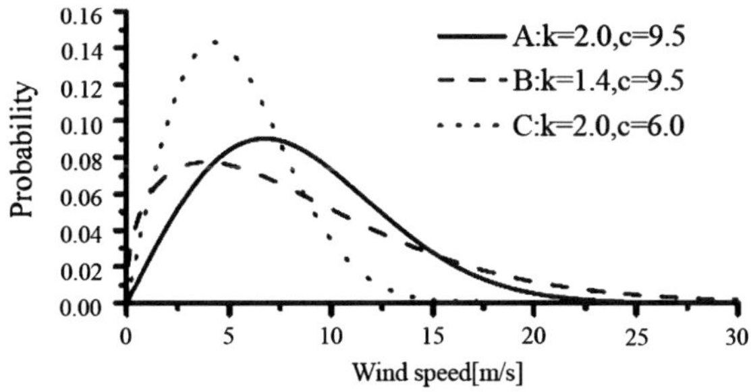

Fig. 2.24 Weibull distribution of wind speed

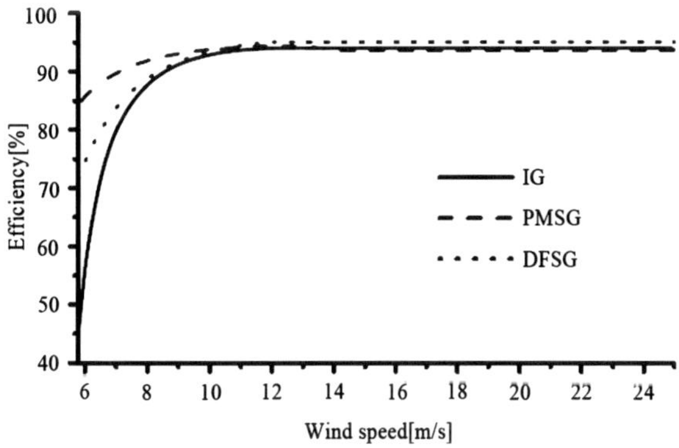

Fig. 2.25 Efficiency of each wind generator system

$$f(V_{\mathrm{w}}) = \frac{k}{c}\left(\frac{V_{\mathrm{w}}}{c}\right)^{\mathrm{k}-1} \exp\left[-\left(\frac{V_{\mathrm{w}}}{c}\right)^{\mathrm{k}}\right] \tag{2.44}$$

$$E_{\mathrm{total}} = \int_{V\,\mathrm{min}}^{V\,\mathrm{max}} P_{\mathrm{g}}(V_{\mathrm{w}}) \times f(V_{\mathrm{w}}) \times 8760 dv \tag{2.45}$$

$$\text{Capacity Factor} = \frac{E_{\mathrm{total}}}{P_{\mathrm{R}} \times 365 \times 24(\mathrm{h})} \tag{2.46}$$

2.5.2 Calculated Results of Capacity Factor

Capacity factor of WG systems with Squirrel-Cage Induction Generator (IG), PMSG, and DFIG is calculated and compared to each other. Although a power converter is needed, the latter two systems can be operated in variable speed condition. On the other hand, the first system is operated in fixed speed. In the comparison analysis, 5.8 m/s is used for the fixed speed WG (IG) as the cut-in wind speed, but both 5.8 and 4.0 m/s are used for the variable speed WGs (PMSG and DFIG). Coefficients of Weibull distribution function have been determined as shown in Fig. 2.24. Figure 2.25 shows the system efficiency of

Table 2.7 Calculated results of capacity factor

Area	IG capacity factor (%)	PMSG capacity factor (%)		DFIG capacity factor (%)	
		4.0 m/s	5.8 m/s	4.0 m/s	5.8 m/s
A	40.36	43.20	42.14	42.48	41.64
B	37.40	39.74	38.79	39.12	38.37
C	12.81	16.28	14.56	15.58	14.22

each WG system with respect to wind speed. Table 2.7 shows the result of capacity factor of each WTGS for each Weibull distribution function. It is clear that capacity factors of variable speed WGs (PMSG and DFIG) are higher than that of the fixed speed one (IG).

2.6 Conclusions

In this chapter, methods to evaluate the losses and output power of WG systems with Squirrel-Cage Induction Generator (IG), PMSG and DFIG are explained, in which values of losses and state variables in each system can be calculated with respect to wind speed. By using the presented methods, it is possible to calculate the generated power, losses, total energy efficiency and capacity factor of WG system quickly.

In addition, if the Weibull distribution function of annual wind speed condition at a certain area is available,the annual generated energy and capacity factor of WG system for that area can easily be obtained. Using the method, capacity factors of three WG systems (IG, PMSG, and DFIG) for three wind conditions expressed by Weibull distribution function data have been evaluated, and then, it has been clearly shown that capacity factors of variable speed WGs (PMSG and DFIG) are higher than that of the fixed speed one (IG). The presented method can be used effectively for improving WG design, construction planning, and economic conditions of wind farms for specific areas.

References

1. Anderson PM, Bose A (1983) Stability simulation of wind turbine systems. IEEE Trans Power Apparatus Syst PAS-102(12):3791–3795
2. Cotrell J (2002) A preliminary evaluation of a multiple-generator drivetrain configuration for wind turbines. National renewable energy laboratory, NREL/CP-500-31178
3. Hitachi, ltd. Power and industrial systems power semiconductor dept. power and industrial systems div., IGBT module application manual, http://www.pi.hitachi.co.jp/pse/images/pdf/igbt-aple.pdf [online]
4. Hitachi, ltd.: High-voltage high-power IGBT, http://www.pi.hitachi.co.jp/pse/product/igbt/2006222_14590.html

5. I.E.E. Japan (1988) Electric machinery design. Ohm Publisher, Scottsdale
6. Johnson GL Wind energy systems. Electronic edition, http://www.rpc.com.au/products/windturbines/wind-book/WindTOC.html [online]
7. Polinder H, Van der Pijl FFA, De Vilder GJ, Tavner PJ (2006) Comparison of direct-drive and geared generator concepts for wind turbines. IEEE Trans Energy Convers 21(3):725–733

Chapter 3
Superconducting Direct Drive Wind Turbine Generators: Advantages and Challenges

Asger Bech Abrahamsen and Bogi Bech Jensen

Abstract This chapter contains a discussion of the advantages and challenges of introducing superconducting generators in future wind turbines. A special focus is given to the European offshore wind turbine marked, because this is the most mature and because the European Union (EU) has decided on a 20% renewable energy share of the electricity by 2020. Thus there are already scenarios of how the offshore wind power capacity is expected to develop in EU over the next two decades and this is used as the framework for a discussion of the advancements needed to make the superconducting drive trains feasible. The text is organized in a section first outlining the EU offshore plans; a section on the different drive trains; a section on the materials used to produce and shape the magnetic field in the generators and finally a section on the superconducting, vacuum, cryostat and cooling challenges of the superconducting direct drive technology.

3.1 Introduction

Wind power has developed into a mature renewable energy technology providing 160 GW of the world's demand for energy by 2010 as illustrated in Fig. 3.1. The development was initiated in the late seventies after the oil crisis and focus was on

A. B. Abrahamsen (✉)
Materials Research Division, Risø National Laboratory for sustainable energy,
Technical University of Denmark, Building 229, Room S47, Frederiksborgvej 399,
4000 Roskilde, Denmark
e-mail: asab@risoe.dtu.dk

B. B. Jensen
Department of Electrical Engineering, Technical University of Denmark,
2800 Lyngby, Denmark

S. M. Muyeen (ed.), *Wind Energy Conversion Systems*,
Green Energy and Technology, DOI: 10.1007/978-1-4471-2201-2_3,

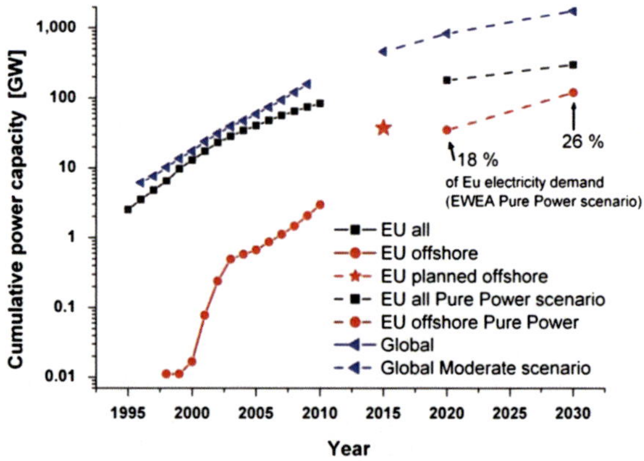

Fig. 3.1 Total installed wind power capacity in EU (*black*), globally (*blue*) and the EU offshore fraction (*red*) by 2010 [1]. The planned offshore capacity is marked with a star and the expected developments until 2030 are marked with *dashed lines* [2]

land-based turbines with a power range less than 100 kW. However, the size of the turbines has grown to the 2–3 MW range for onshore turbines, which constitute the majority of the wind market today. The areas suitable for onshore wind farms near the major cities of Europe are limited and offshore installation is seen as an option even though the price of the produced electricity is higher than for onshore turbines. Figure 3.1 shows that the offshore market of EU first showed a visible increase around the year 2000 and the growth is expected to be two orders of magnitude from the present 3 GW and to 120 GW by 2030 [1, 2]. Thus offshore wind power should not be considered mature yet and the offshore turbine technology is expected to divert from the onshore technology by aiming at bigger and more reliable machines.

A central question is what the optimal size of offshore turbines should be in order to be economically feasible, but the answer will depend heavily on the initial assumptions made for such an investigation. First of all one would need to decide whether the renewable energy sources should be competing directly with the fossil fuel sources, such as coal power plants, which will remain cheap for decades if the impact on the environment is not included in the economic evaluation. Secondly one will need to consider a growing global demand for energy and that fossil fuel resources are fundamentally finite causing a peak in the production. This peak in the oil production has already been passed by many industrialized countries, which rely on imports from foreign states. We can only guess at how the above scenarios will develop, but the issues of climate change and energy security are likely to become more important in the future, whereby the scenario outlined in Fig. 3.1 would call for more installed capacity. In this context it should be said that several EU countries are discussing a goal of being independent of fossil fuel in 2050.

3.2 Upscaling Offshore Turbines

The basic properties of future offshore turbines can be estimated by considering the amount of kinetic energy density in the wind, which can be converted into kinetic energy of the turbine shaft. The power density of the wind is simply given by

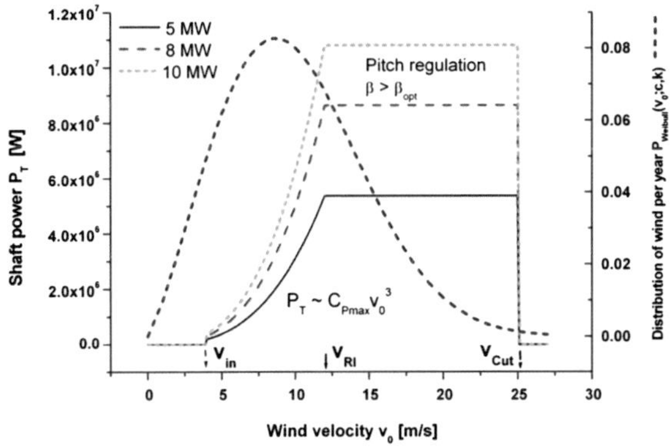

Fig. 3.2 Power curve of a Multibrid 5 MW turbine scaled to 8 and 10 MW by increasing the swept area of the rotor. The shaft power scales with the wind speed by v^3 up until the rated wind speed, where pitching of the turbine blades is used to limit the shaft power at the rated production. The blue curve represents the Weibull distribution of the wind power in the North Sea and the annual energy production of the turbine is determined by the integral of the product of the Weibull distribution and power curve [3]. Reproduced with permission from Institute Of Physics

$$P_{\text{wind}} = \frac{1}{2}\rho A v^3 \tag{3.1}$$

where ρ is the density of the air, A is the swept area of the turbine rotor with a diameter D giving $A = \pi(D/2)^2$ and v is the velocity of the wind passing the rotor. The efficiency by which the turbine blades can convert the wind power into kinetic energy of the shaft is quantified in the power coefficient C_{p} and is determined by the aerodynamic properties of the blades. The power coefficient depends on the ratio between the tip speed v_{Tip} of the blades and the wind speed v, as well as the pitch angle β of the rotor blades

$$\lambda = \frac{v_{\text{Tip}}}{\text{v}} = \frac{R\omega}{\text{v}} \tag{3.2}$$

where R is the radius of the turbine rotor and ω is the rotation speed. Thus the power on the turbine shaft is then given by

$$P_{\text{Shaft}} = P_{\text{wind}} C_{\text{P}}(\lambda, \beta) \tag{3.3}$$

where the power coefficient can be changed from a maximum value at an optimum pitch angle and λ to zero by pitching the blades. This is used to control the power of the turbine shaft as illustrated in Fig. 3.2, which shows the ideal power curve of a 5 MW Multibrid M5000 turbine. Between the cut-in wind speed $v_{\text{in}} = 4$ m/s and the rated ideal wind speed $v_{\text{RI}} = 12$ m/s the controller of the turbine will regulate the blades to maximize the power coefficient, whereby the shaft power will scale with the wind speed to v^3. The controller will pitch the blades to get a constant rated shaft power for wind speeds above $v_{\text{RI}} = 12$ m/s and finally stop the turbine above the cut-out wind speed $v_{\text{Cut}} = 25$ m/s [3].

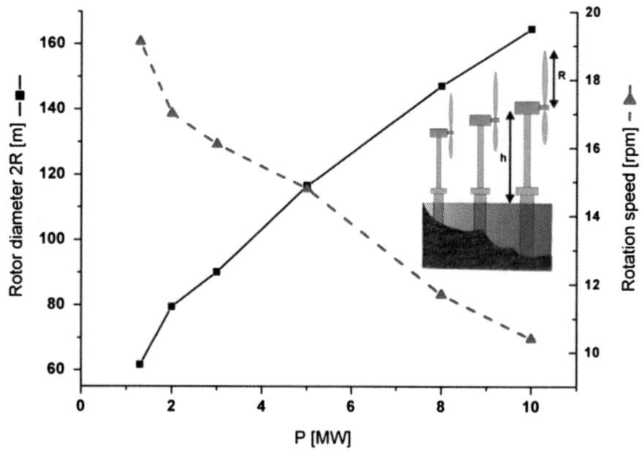

Fig. 3.3 Turbine rotor diameter as a function of the power rating of a series of existing turbines ($P \leq 5$ MW) and the resulting rotation speed, which is decreasing in order to limit the tip speed of the blades. The values for $P > 5$ MW have been obtained by scaling the properties of the 5 MW Multibrid turbine [3]

The conservative up-scaling of a turbine can be done by increasing the area of the rotor A in Eq. 3.1, while limiting the tip speed to less than $v_{tip} < 90$ m/s in order to limit the noise and the load on the rotor blades. This choice results in a decreasing rotation speed according to Eq. 3.2 as the turbines become bigger as illustrated in Fig. 3.3.

3.3 Drive Trains

There have been many suggestions on drive trains for wind turbines, but the majority are based on a gearbox inserted between the turbine shaft and a generator supplied with a suitable power converter connected to the collection grid of a wind farm [4, 5]. Figure 3.4 illustrates the mechanical configurations, where the gearing ratio changes from approximately 1:100 with a generator rotating at approximately 1,500 rpm, into a gear ratio of 1:10 connected to a medium speed generator rotating at approximately 150 rpm and finally into the direct drive option where the gearbox is completely omitted. The consequence of the simplification of the gearbox of the different drive train choices in Fig. 3.4 is that the generators must provide an increasing torque T at lower rotational speeds ω_g as the direct drive is approached, since the power P produced by the generator is

$$P = T\omega_g \tag{3.4}$$

On general grounds one can formulate the torque of a generator by only considering the fundamental harmonic of the airgap flux density B_g and of the stator electric loading I_S

$$B_g(\theta) = \hat{B}_g \cos(p\theta) \tag{3.5}$$

$$I_S(\theta) = \sqrt{2}A_S \cos(p(\theta - \gamma)) \tag{3.6}$$

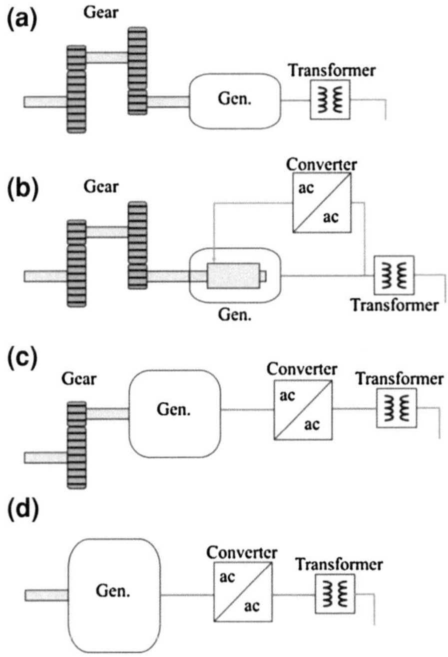

Fig. 3.4 Typical drive train configurations for wind turbines based on **a** a multi-stage gearbox and a squirrel-cage induction generator rotating at 1,500 rpm connected directly to the grid via a transformer **b** a multi-stage gearbox and a double fed induction generator (*DFIG*) which has a converter connected to the rotor windings via slip rings. **c** Hybrid setup with a 1:10 ratio gearbox and a generator rotating at medium speed of 150 rpm and **d** the direct drive with the generator rotating at the speed of the rotor blades. The frequency of the produced electricity will not match the grid frequency in the last two cases and a full power converter must be inserted between the generators and the grid transformer

where θ is the angle around the circumference of the airgap, γ is the angular displacement between the magnetic field and the stator current distribution and p is the number of pole pairs. The peak airgap flux density is denoted $\hat{B}_g$ and the rms (root mean square) value of the electric loading is denoted A_S.

The total torque T of the generator is now found by considering that the airgap flux density B will cause a Lorentz force F_L on a stator wire carrying a current I, $F_L \sim I \times B$. This force results in a torque on the wire, $T \sim r \times F_L$. By performing an integration of Eqs. 3.5 and 3.6 one obtains

$$T = \sqrt{2} A_S B_g V \cos(p\gamma) \tag{3.7}$$

$$V = \pi r^2 L \tag{3.8}$$

where V is the airgap volume of the generator with an airgap radius r and active length L.

3.4 Generator Types

In the previous sections it was illustrated that the rotation speed of the turbine blades is expected to decrease as the turbines are up-scaled to power ratings of 10 MW and furthermore the transition to the direct drive will call for generators capable of delivering torques in the order of $T = 10^6$ Nm at a speed of 10 rpm. From Eq. 3.7 it is seen that in order to obtain such an increase of the torque one must either increase the electric loading A_S, the airgap flux density B_g or finally the volume of the generator. The airgap flux density is limited by saturation and cannot be increased substantially in a more conventional machine. The electric loading is limited by stator cooling. This can be increased to a certain extent by improving the thermal design, but would not be expected to increase by several factors. It is therefore obvious from Eq. 3.7 that if the torque is to increase, then the generator volume must increase. From Eq. 3.8 increasing the radius of the generator is more efficient, since the torque scales as r^2, compared to a linear scaling with the length L of the generator.

Alternative to just increasing the volume of the generator it is interesting to investigate whether it is possible to increase the airgap flux density B_g. Direct drive generators can be characterized by the method by which the magnetic field is created and shaped.

3.5 First Generation: Copper and Steel

The simplest version of the direct drive generator is based on copper (Cu) wires wound around magnetic steel making up the rotor structure, which is magnetized when a current is passed through the field windings. The stator is constructed in a similar way where copper wires are inserted between the teeth of the magnetic steel of the stator. This is illustrated in Fig. 3.5, where the green circles represent the Cu wires in one of the rotor coils and the red circles represent one of the stator windings of a 3-phase stator. The blue arrows illustrate the magnetic flux path. It should be noticed that each stator winding will be distributed in several slots in a realistic machine.

The advantage of this kind of machine is the choice of the cheap raw materials Cu and Steel, and the ability to control the excitation. However, some limitations of the parameter entering Eq. 3.7 are imposed by the physical properties of the materials, which will be discussed below. Enercon has been using the wound direct drive generator in its wind turbines since the early nineties, which have a characteristic nacelle shape, because they must hold a large diameter ring generator [6]. The weight of the generator for the E-112 turbine of 6 MW power rating and having a rotor diameter of 114 m is 212 t and the total nacelle weight is 500 t [7].

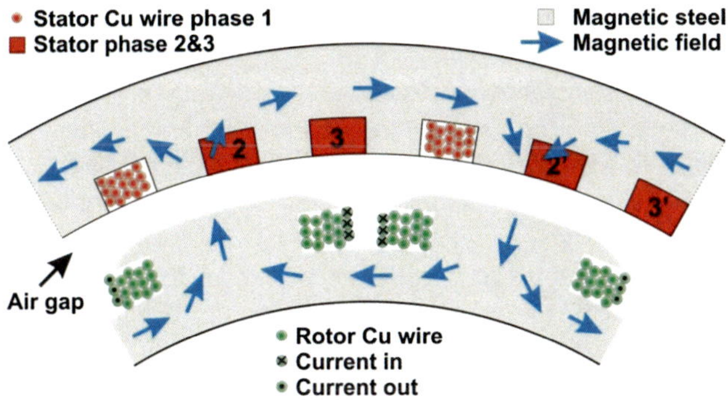

Fig. 3.5 Simplified illustration of the first generation direct drive generator based on copper wires (*green circles*) wound into a coil placed on the magnetic steel structure of the rotor. The stator is constructed in a similar way by winding copper wires (*red*) in the slots of the stator magnetic steel. A 3-phase stator is made by adding two more coils marked by index 2 and 3. The generator is magnetized by passing a current through the rotor windings and the shape of the magnetic field is schematically illustrated by the arrows (*blue*)

3.5.1 Ohm's Law and Heat Generation

The resistivity ρ of the wires making the rotor winding will cause a total resistance of

$$R_{\mathrm{Coil}} = \rho \frac{l_{\mathrm{wire}}}{A_{\mathrm{wire}}} \tag{3.9}$$

where l_{wire} is the length of the wire and A_{wire} is the cross-sectional area of the wire. The rotor is magnetized by sending a current I through the rotor winding, whereby the magnetic flux Φ produced is proportional to the magneto motive force (mmf = NI) and inversely proportional to the reluctance (R) of the magnetic path.

$$\Phi = \frac{NI}{R} \tag{3.10}$$

N is the number of turns in the coil. The power dissipation P_{R} in the wire will however scale with the square of the current

$$P_{\mathrm{R}} = RI^2 \tag{3.11}$$

and cause Joule heating, which must be managed by the cooling of the rotor to prevent a thermal run away. Thus the second thing to consider is the thermal conductivity κ, the specific heat c_{p} and the mass density p_{m} of the wires and materials surrounding the coil windings. One can then determine the final operating temperature of the rotor windings by solving the general heat equation

$$\nabla^2 T = \frac{\rho_{\mathrm{m}} c_{\mathrm{p}}}{\kappa} \frac{\partial T}{\partial t} \tag{3.12}$$

where T is the local temperature, ∇ is the Laplace operator and $\partial/\partial t$ is the time derivative.

The above equations illustrate that the current passing through the copper windings of both the rotor and stator will be limited by the cooling of the generator. This is why the current density of the copper wires of air cooled machines is usually quite limited and forced air or liquid cooling of the windings is needed to achieve higher current densities. From Eq. 3.10 it is seen that the flux will scale linearly with the current until the current density limit is reached. This is however only true as long as the iron does not saturate. When the iron saturates the reluctance will start to increase and hence the flux will not remain proportional to the current.

3.5.2 Magnetic Steel and Flux Circuits

Magnetic steel is primarily iron holding some percentage of silicon and additional traces of other elements to increase the electrical resistivity compared to pure iron. Iron is a ferromagnetic material, because the conduction electrons with spin up and down are shifted due to the exchange interaction between the electrons. This imbalance results in an effective magnetic moment of each of the Fe atoms with neighbouring moments pointing in the same direction. Long range alignment of the magnetic moments is however prevented by the formation of magnetic domains, which are reducing the stray fields of the ferromagnetic material. These domains can however rather easily be aligned along an external applied magnetic field, but the associated magnetization will eventually saturate when all domains point in the same direction [8]. This is illustrated in Fig. 3.6 which shows a typical magnetization curve of magnetic construction steel (0.42% Mn, 0.17% Si and 0.08% C). Thus the combination of a field coil and soft iron is used to construct a magnetic flux circuit, where the flux created by the coil is mediated by the alignment of the magnetic domains in the soft iron in order to maximize the air gap flux density B_g entering the torque Eq. 3.7. This technique is however only effective as long as the iron does not saturate too much, since the magnetic flux will leak out of the circuit above the saturation magnetization density $\mu_0 M_s = 1.5–1.8$ T.

3.6 Second Generation: $Nd_2Fe_{14}B$, Copper and Steel

The second generation direct drive generator illustrated in Fig. 3.7 utilizes permanent magnets to establish the magnetic field instead of using rotor coils, whereas the stator is constructed similar to the first generation. The advantage is that there is no need for connections of the rotor magnetization current and there

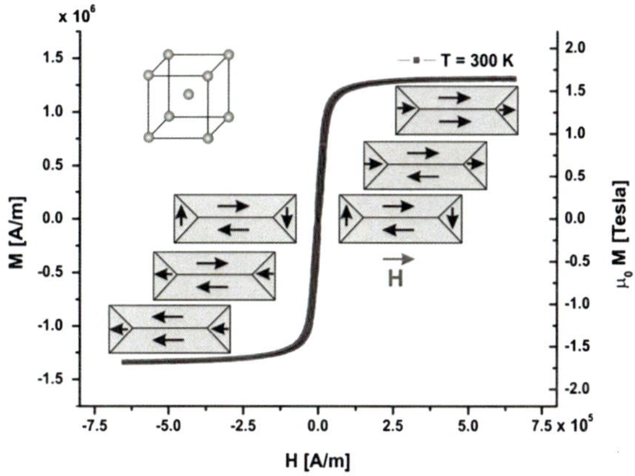

Fig. 3.6 Magnetization curve of soft magnetic iron, which is saturated when all the ferromagnetic domains are aligned with the applied magnetic field. Inset: Simple cubic crystal structure of iron

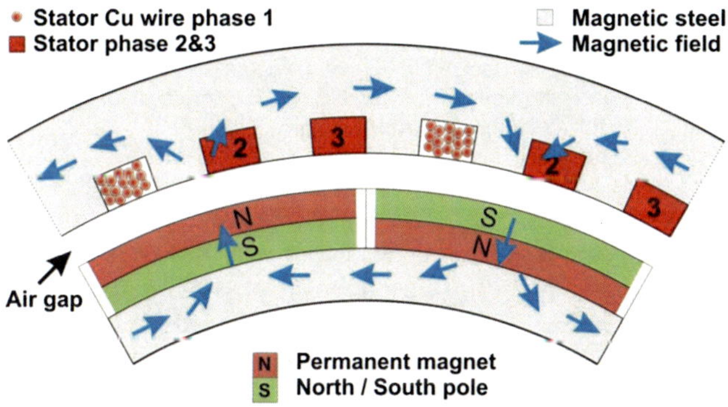

Fig. 3.7 Second generation direct drive generator based on permanent magnets mounted on the surface of the rotor, whereas the stator is similar to the first generation generator with cu wires (*red*) wound in the slots of the magnetic stator steel

are no resistive losses associated with the rotor coils. Thus the mechanical and electrical simplification of the rotor is considerable and is believed to improve the reliability, because fewer parts can fail in the generator.

A prerequisite for the construction of such a generator is however the availability of the strong $Nd_2Fe_{14}B$ type of permanent magnet. They primarily consist of iron atoms, but the additional neodymium (Nd) atoms locks the magnetization direction of the iron moments to the long axis of the crystal structure as shown in the inset of Fig. 3.8. This directional locking of the magnetic moments makes it possible to align small powder grains of $Nd_2Fe_{14}B$ by applying a magnetic field and then pressing the powder into blocks, which are finally sintered together in a furnace. Small ferromagnetic domains are formed as the blocks are cooled down below the magnetic ordering temperature $T_{\mathrm{Curie}} = 585$ K, but the ordering is random and the magnet does not cause any stray fields outside the block.

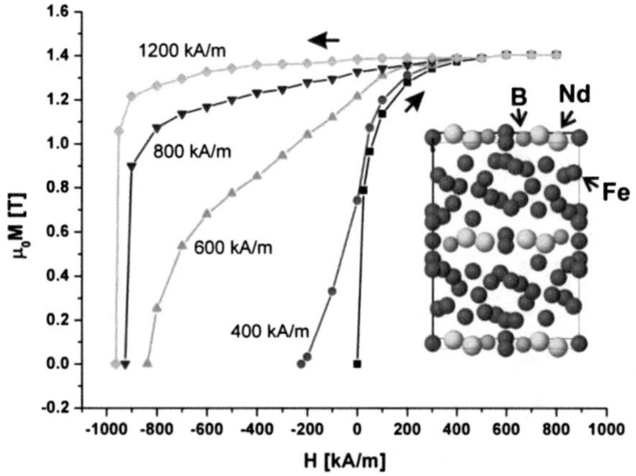

Fig. 3.8 Magnetization curves of sintered $Nd_{15}B_8Fe_{77}$ permanent magnets after cooling the sample below the magnetic ordering temperature $T_C = 585$ K. The saturation magnetization is reached by increasing the applied field and the coercivity field is observed when the magnetization becomes zero as the applied field is reversed (Reproduced from [9]). Inset: Illustrates the tetragonal crystal structure of the $Nd_2Fe_{14}B$ permanent magnet holding layers of Nd and B elements separated by thicker layers of iron

Figure 3.8 shows the original magnetization curves of sintered $Nd_{15}B_8Fe_{77}$ obtained by M. Sagawa and illustrates that it is possible to align all the magnetic domains, but the locking of the moment to the crystal lattice and pinning of magnetic domain walls prevent an alignment change even when the applied field is pointing in the opposite direction [9]. Such a permanent magnet is characterised by the saturation magnetization M_S and by the coercivity H_C field needed to force the magnetization to zero.

The PM direct drive generators have been demonstrated commercially by Goldwind in a 1.5 and 2.5 MW turbine [10]. Siemens wind power has recently also introduced the SWT 3.0-101 turbine holding a 3 MW PM direct drive generator [11] and GE have announced the 4.0-110 offshore wind turbine [12] holding a 4 MW PM direct drive generator based on the technology obtained from Scanwind, who originally collaborated with the Switch [13].

The size and weight of a PM direct drive have been analyzed extensively in the literature for power ratings from 1 to 20 MW [14, 15]. It is generally concluded that the active mass (Cu + Iron + PM) of the generator will be approximately equal to the structure supporting the active materiel up to power ranges of 5 MW, but above this level the support structure is likely to become a factor 3–5 times higher than the active mass at 10 MW.

Table 3.1 shows the expected values for PM direct drive generators. It should be noticed that the usage of permanent magnets is of the order 600–700 kg/MW for the turbines below.

Table 3.1 Typical size and weight estimates of PM direct drive generators for wind turbines

Power (MW)	3[a]	5[b]	10[c]
Airgap diameter(m)	5	6.1	10
Active length (m)	1.2	1.1	1.6
Weight (tons)	48	98	325
Weight of PM (kg)	1.7	3	6

[a] [14]
[b] [16]
[c] [17]

3.7 Third Generation: Superconductors, Copper and Steel

The third generation direct drive generator utilizes the ability of superconductors to conduct an electric direct current with a vanishing small resistance ($R \sim 0\ \Omega$), whereby superconducting field coils can produce the flux of Eq. 3.10 without the Joule heating of Eq. 3.11 found for the 1G direct drive generator. The superconducting state of zero resistance is however limited by a critical temperature T_C, an upper critical magnetic field B_{C2} and by a critical current density J_C(T,B), which depends on both the temperature and the applied magnetic field. The critical engineering current density of superconductors is 20–300 A/mm^2 in magnetic fields up to several Tesla and thereby orders of magnitude larger than the typical current density of copper in the 1G generator.

A consequence of the vanishing electric loss of superconducting field coils is that flux densities exceeding the saturation magnetization M_S of iron as illustrated in Fig. 3.6 can be considered. In this case the iron becomes ineffective and should be removed from the central generator parts in order to allow more space for the superconducting windings. Figure 3.9 illustrates the concept of an airgap wound superconducting direct drive generator, where the superconducting field coils are enclosed in a cryostat, which is providing the thermal insulation to maintain an operating temperature sufficiently below T_C. The cryostat consists of outer stainless steel walls, which can sustain a vacuum to prevent heat conduction by gas convection, and multi layer super insulation wound around the rotor support structure to prevent the heat inflow from radiation. Thus the flux density B_g entering the torque Eq. 3.7 can in principle be increased to several Tesla compared to the typical flux density of less than 1 T produced by the 1G and 2G generators. This opens up for the possibility to increase the torque of superconducting direct drive generators without increasing the size of the machine.

From the torque Eq. 3.7 it is natural to ask whether the stator could also be made superconducting in order to increase the stator current loading A_S. This is however a little more complicated as the superconductor will suffer higher losses when exposed to an AC current or magnetic field. The associated heat will have to be removed at low temperature, where the cooling machines are inefficient. Thus the most likely application is based on having the copper windings of the stator at room temperature outside the rotor cryostat as illustrated in Fig. 3.9. The stator is also of the airgap wound type, since any magnetic teeth would be completely

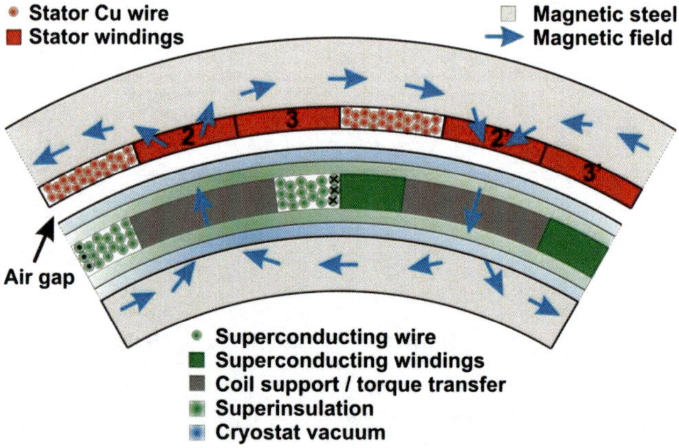

Fig. 3.9 Third generation direct drive generator based on superconducting rotor windings (*green*) embedded in multilayer thermal insulation material inside an evacuated cryostat. The cryostat is lying on the magnetic steel of the rotor, but the torque on the superconducting coils is transferred by a coil support structure inside the cryostat. The 3-phase stator winding is outside the cryostat and is kept at room temperature in order to be able to remove the heat from the copper windings. There are no magnetic steel teeth between the stator windings, because the air gap magnetic field is higher than the saturation limit of the steel. There would however typically be non-magnetic thin structural teeth to support the airgap stator winding. The outer steel ring is confining the flux in the generator and the losses will remain relatively low due to the low rotation speed

saturated during the AC cycle and high losses would be expected. The AC losses of the superconductors also dictate that the generator should be operated as a synchronous generator, so that the superconductors are exposed to a constant field from the stator during steady-state operation.

3.7.1 Superconductivity

The superconducting state is caused by the pairing of conduction electrons of certain materials into a quantum mechanically correlated condensate, which can move collectively through the material without electrical resistance ($R \equiv 0\ \Omega$). This is however only possible if the thermal fluctuations in the material are lower than the binding energy of the electron pairs and superconductivity therefore only exists below the critical temperature T_C.

Many of the electrical properties of practical superconductors are determined by the creation of circulating supercurrents inside the material when exposed to an applied magnetic field. A magnetic field line tends to create a rotational flow in the condensate, but this flow will be quantized due to the quantum mechanical nature of the condensate. This causes the magnetic flux associated to such a flow pattern

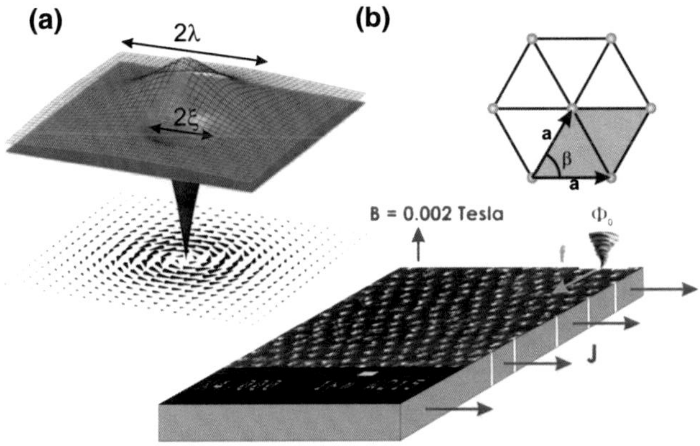

Fig. 3.10 **a** A flux line in a superconductor is a tube-shaped region where a circulating super current (*arrows*) are confining a quantum of magnetic flux Φ_0. The magnetic field distribution (*mesh*) and the supercurrent are decaying exponentially with the penetration depth λ. The density of the superconducting condensate (*surface*) is suppressed at the centre of the flux line in the normal core having a size given by the coherence length ξ. **b** Bitter decoration image of flux line lattice formed in a $TmNi_2B_2C$ superconductor exposed to an applied field of B = 2 mT. It should be noticed that the distance between the flux lines is about a = 1 μm, which is large compared to $\lambda = 80$ and $\xi = 15$ nm [18]. A force f perpendicular to the flux line Φ_0 results if a transport current density J is applied to the superconductor and dissipation is caused if the flux lines moves

to be given by the flux quantum Φ_0, which is determined by the ratio between Planck's constant and the charge of the electron pair $q = 2e$

$$\Phi_0 = \frac{h}{2e} = 2.07 \cdot 10^{-15}\,\mathrm{T\,m^2} \tag{3.13}$$

The magnetic field is strongest at the centre of the flow pattern and is confined by the circulating supercurrent, which is decaying exponentially away from the centre with a characteristic length scale called the penetration depth λ. The magnetic field is also decaying exponentially with the penetration depth away from the centre, but it is finite at the centre because the condensate is broken down completely over a short length scale called the coherence length ξ. The tube-shaped region holding a flux quantum in a superconductor is called a flux line and the characteristic length scales are shown in Fig. 3.10a.

Flux lines are created at the edge of the superconductor and pushed into the material by the magnetic pressure of the applied field. Thus the number of flux lines present in the superconductor will increase with the applied magnetic field. Eventually the field profile of the different flux lines will start to overlap and a close packed lattice of circular symmetric objects is formed due to a repulsive interaction. A hexagonal flux line lattice is formed and the spacing between the flux lines a is directly related to the average magnetic flux density B_{av} in the superconductor. This is easily seen because the flux in the unit cell of the lattice Φ must be exactly one flux quantum and the area of the unit cell is $A_{unit} = a^2 \sin(\beta)$, where β is the opening of the two vectors spanning the unit cell ($\beta = 60^\circ$ for a hexagonal lattice).

$$\Phi = \int B dS = B_{av} a^2 \sin(\beta) = \Phi_0 \tag{3.14}$$

Figure 3.10b shows a scanning electron microscopy image of a superconductor, which has been decorated by iron particles at $T = 4.2$ K and in an applied field of $B = 2$ mT [18]. The flux line lattice is revealed by small clusters of iron particles collected at the centre of each flux line and the hexagonal lattice is observed. From Eq. 3.14 it can now be estimated how close the flux lines will be pushed together when the applied field is increased and $a = 1.5$ μm, 49 nm and 22 nm is obtained for $B_{av} = 1$ mT, 1 T and 5 T respectively.

The existence of the flux line lattice imposes an upper critical field limit B_{C2} above which the superconducting state is completely suppressed. This limit is reached when the distance between the flux lines a becomes comparable to the size of the normal cores ξ and is qualitatively found by inserting $a \sim 2\xi$ into Eq. 3.14.

A second consequence of the existence of flux lines in a superconductor is that the zero resistance state $R \equiv 0\ \Omega$ can be violated. Figure 3.10b shows that a Lorentz-like force F_{FL} will act on each flux line with the magnetic field along the $\hat{z}$-axis and depend on the local current density J in the material

$$\bar{F}_{FL} = \bar{J} \times \Phi_0 \hat{z} \tag{3.15}$$

The unit of F_{FL} is force per length of the flux line [N m^{-1}].

Now if the flux line is moving then there is a work performed on the system and the dissipation is observed as a flux flow resistance. Thus any practical superconductor will contain pinning centres, which can prevent the flux lines from moving when exposed to both a magnetic field and a transport current. Pinning centres appear naturally due to defects and impurities on the superconductors, but they are also created by nano-engineering, where nano particles of a size equivalent to the coherence length are incorporated in the superconductors. The concept of a critical current density J_C can therefore be understood in the following oversimplified picture: a flux line will start to move when the Lorentz-like force F_{FL} exceeds the pinning force f_{pin} of the pinning site where the flux line is trapped

$$F_{FL} = f_{pin} \Rightarrow J_C = \frac{f_{pin}}{\Phi_0} \tag{3.16}$$

Again f_{pin} is in units of force per unit length of the flux line [N m^{-1}]. A more realistic picture includes the non-local competition between energy gain of including pinning sites in the flux lines and the energy cost of elastic deformation of the flux line lattice. Additionally this will contain distributions of the strength of pinning sites and a more gradual transition from pinned and into moving flux lines is observed. Figure 3.11 shows the critical surface of a practical superconductor, which is limited by the critical temperature T_C, the irreversibility flux density B_{Irr} and the critical current density $J_C(B,T)$. Thus a successful operation of superconducting field coils as illustrated in Fig. 3.9 will depend on the cryostat and

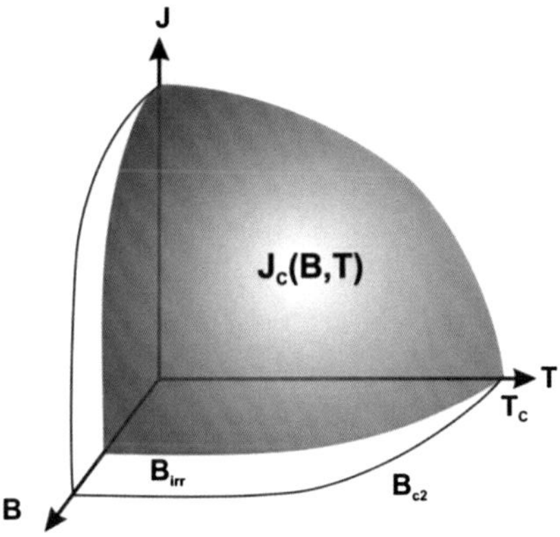

Fig. 3.11 Surface of the critical current density of a superconductor spanned by the temperature, the magnetic field and the current density. The irreversibility flux density B_{Irr} is lower than the upper critical flux density B_{C2} and indicates where the pinning sites become ineffective ($J_C \rightarrow 0$). The flux line lattice will melt into a liquid when $B_{Irr} < B < B_{C2}$

cryogenic cooling systems ability to maintain the operation temperature sufficiently below the critical surface.

3.7.2 High Temperature Superconducting Tapes

The discovery of high temperature superconductors (HTS) took place in 1986 and caused an increased optimism that superconductors could be utilized widely in power applications. However, the new materials were brittle ceramics and the manufacturing of practical conductors was not easy, because the ceramic power needed to be supported by metallic sheath materials in order to give it mechanical strength and also provide an additional current path in case the superconducting state was transformed into the normal state. The first material to be commercialized as an HTS tape was $Bi_2Sr_2Ca_2Cu_3O_{10+x}$ (Bi-2223) ceramic filaments enclosed in a silver sheath and laminated by stainless steel. The advantage is that the powder-in-tube (PIT) manufacturing process, where ceramic powder is filled into silver tubes, which are then drawn into tapes, can be applied, but the disadvantage is that silver is an expensive raw material [19]. Thus the price of the Bi-2223 tape will have a lower limit dictated by the silver price.

The second material to be commercialized in the form of tapes was $YBa_2Cu_3O_{6+x}$ (YBCO), because the superconducting properties are better at temperatures T = 40–77 K compared to the Bi-2223 material. YBCO tapes however cannot be made by the PIT methods, because the crystalline grains must be aligned better than a few degrees, before a supercurrent can cross the grain boundary. Thus an entire new production method has been developed which is based on coating of ceramic layers onto metallic substrates. The substrates are subsequently laminated by copper or steel and the final tape is called a coated conductor [20].

Figure 3.12 illustrates the architecture of the coated conductor tapes, which has the advantage that the substrate materials used are nickel alloys, which are cheaper than silver. The Yttrium of the YBCO material can be substituted by any of the

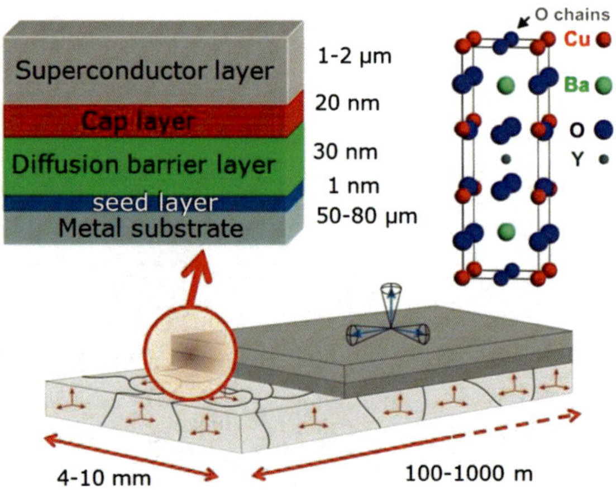

Fig. 3.12 Architecture of coated conductor tape where several ceramic layers are coated onto a metal substrate. The inset shows the crystal structure of the high temperature superconductor $REBa_2Cu_3O_{6+x}$ (RE = Rare Earth elements or Y), which is finally grown as a bi-axially textured layer, where the mis-alignment between the crystal planes in the grains of the superconductor material is only a few degrees. This ensures that a supercurrent can flow across the grain boundaries along the 100–1,000 m length of the tape

Rare Earth elements (La-Lu) and mixtures are often used to optimize the pinning sites of coated conductors [21, 22]. The usage of the Rare Earth elements is however small, which can be illustrated by calculating the volume of the superconducting layer in a km of tape:

$$V_{\mathrm{YBCO}} = twL = 2 \cdot 10^{-6}\ \mathrm{m} \cdot 4 \cdot 10^{-3}\ \mathrm{m} \cdot 10^{3}\ \mathrm{m} = 8 \cdot 10^{-6}\ \mathrm{m}^{3} \tag{3.17}$$

where t is the thickness of the YBCO layer, w the width of the tape and L the length of the tape as shown in Fig. 3.12. The mass density of YBCO is $\rho_{\mathrm{YBCO}} = 6.35\ \mathrm{g/cm^3}$ and the resulting mass is

$$m_{\mathrm{YBCO}} = \rho_{\mathrm{YBCO}} V_{\mathrm{YBCO}} = 6.35 \frac{\mathrm{g}}{\mathrm{cm}^3} \cdot 8\ \mathrm{cm}^3 = 51\ \mathrm{g} \tag{3.18}$$

The disadvantage of the present coated conductor technology is the usage of advanced high vacuum deposition techniques, which are slow and expensive. Large research efforts are however made to introduce the all chemical method of making the ceramic layers by simple dip coating [23].

3.7.3 Race Track Coils and Generator Layout

The interconnection between the ceramic grains of the HTS tapes can easily be broken if the tapes are bent more than a critical bending diameter of the order 5 cm for the Bi-2223 tapes and down to 1 cm for the coated conductors. Thus the

Fig. 3.13 High temperature superconducting race track coil holding 32 m of coated conductor from American Superconductor. The HTS tape is wound on an inner stainless steel frame with fiberglass tape inserted as insulation. Two copper blocks are used as current contacts and an outer stainless steel frame provides mechanical support for the winding. The coil is vacuum impregnated by epoxy after the winding. The dimensions of the coil are 240 × 145 × 5 mm and it was produced as part of the Superwind project [25]

winding of HTS tapes into coils is often done in the race-track geometry, where the corners of the former must have bending diameters bigger than the critical bending diameter. Figure 3.13 shows a small-scale example of a race track coil based on coated conductor tape from American Superconductors [24]. The coil was produced as part of the Superwind project at the Danish University of Technology with the purpose to examine the challenges of manufacturing large-scale race track coils for future multi MW wind turbines [25].

One of the issues of the coil fabrication is the obtained engineering critical current density $J_{C,e}$, which is defined as the critical current in the superconductor tape I_C divided by the cross-sectional area of the tape A_{tape} and also the insulation $A_{insulation}$

$$J_{C,e} = \frac{I_C}{A_{tape} + A_{insulation}} \tag{3.19}$$

This was determined for several types of the state-of-the-art high temperature superconducting tapes using fibreglass tape as well as a thin layer of epoxy applied to the tape as electrical insulation. Table 3.1 shows the typical values obtained from the critical current supplied by the manufacturers and then divided with the cross sectional area from the coils. Once the tape is wound into a coil the critical current is suppressed by the field created, since more flux lines are pushed into the superconductor. Figure 3.14 shows the electric field across race track coils based on a coated conductor and a Bi-2223 tape as the current of the coils is increased [26]. The noise limit of the measuring circuit is dominating for current less than about 60 A and then the electric field is given by a power law

$$E(I) = E_0 \left(\frac{I}{I_C(B,T)} \right)^{n(B,T)} \tag{3.20}$$

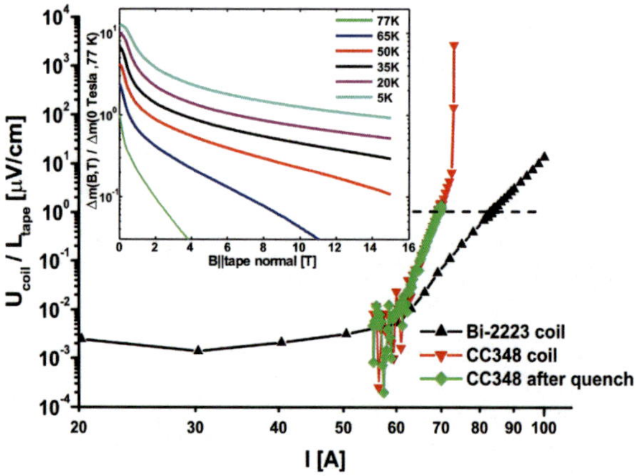

Fig. 3.14 Electric field across the coated conductor coil (CC348) in Fig. 3.13 and a similar coil based on Bi-2223 tape as function of the current in the coils when cooled in liquid nitrogen (LN_2) at T = 77 K. The dashed line indicates when the average electric field is crossing the criteria of the tape critical current E_0 = 1 μV/cm and is used to define the critical current of the coil. The abrupt increase of the electric field across the CC348 coil at I = 73 A is caused by a quench, where the power dissipation exceeds the cooling power of the LN_2 and the superconducting state is suppressed by a propagating heat wave. Inset: Scaling of the magnetization hysteresis of the tape, which is proportional to the critical current density of the tape. Thus the critical current of the tape at T = 77 K and in zero applied field must be scaled by this factor to obtain value for low temperature and high field [26]. Reproduced with permission from Elsevier

where the pre-factor E_0 = 1 μV/cm is the electric field observed when the current I in the superconductor reaches the critical current I_C (B,T). The n-value indicates how abruptly the electric field will increase as the current changes in the vicinity of I_C. The inset of Fig. 3.14 shows how the hysteretic magnetization of the coated conductor changes with applied field and temperature as determined from magnetization measurements. Since the critical current density of the superconductor is proportional to the hysteresis of the magnetization this shows how the critical current will increase as the superconductor is cooled further below T_C and how the critical current will decrease as the applied field causes more flux lines to enter the superconductor. This scaling can then be used to extrapolate the critical current values of Table 3.2 to the operation magnetic field and the temperature of a possible generator.

The race track coil geometry causes some limitation to the design of a direct drive superconducting machine and Fig. 3.15 illustrates how a number of race track coils can be placed on a rotor support to create a multi pole machine. The advantages and challenges of realizing a superconducting drive train will be discussed in the following section. This will be done by illustrating the possible properties of a 5 MW direct drive generator and by a discussion of the technical challenges, which must be solved.

Table 3.2 Critical current I_C of several commercial high temperature superconductor tapes measured at T = 77 K and in the self field. Engineering critical current density is determined from the cross-sectional area of the tape and the insulation. The two numbers of the thickness indicate the tape and insulation thickness respectively. The first two tapes are of the coated conductor type, whereas the last is based on Bi-2223

Tape type	I_C(77 K, sf) (A)	Width (mm)	Thickness (mm)	$J_{C,e}$ (A/mm^2)
CC348[a]	95	4.8	0.22 + 0.10	61.8
SP4050[b]	125	4.2	0.1 + 0.06	186.0
Bi-2223 HS[a]	145	4.3	0.28 + 0.1	88.7

[a] American Superconductor [24]
[b] Super power inc [27]

3.8 Advantages of a Direct Drive Superconducting Generator

The primary advantage of the superconducting direct drive generator is that the operational airgap flux density in principle can be increased above the limit dictated by the saturation of iron and of $Nd_2Fe_{14}B$ permanent magnets. However, a generator consisting of only superconductors and copper will become very expensive, because the coated conductors are not yet produced for a mass market. Thus the first realizations of a superconducting direct drive generator will contain magnetic steel to confine the magnetic field inside the machine.

A higher air-gap flux density would allow for more compact generators and possibly also for a smaller generator than the 2G direct drive at large power ratings. This is illustrated by the example below, where a superconducting direct drive generator, which can fit into the nacelle of the 5 MW Repower turbine 5 M, is considered. A ring-shaped geometry is chosen for the direct drive generator and aiming at an active diameter and active length of

$$D_{\text{active}} = 4.2 \text{ m} \tag{3.21}$$

$$L_{\text{active}} = 1.5 \text{ m} \tag{3.22}$$

The target air-gap flux density was chosen as $B_g = 2.4$ T and the electric loading of the stator was chosen as $A_S = 90$ kA/m with a targeted efficiency of 95% for 5 MW power rating obtained at a rotation speed of 12 rpm. The flux density limit in the magnetic steel enclosing the generator was chosen as 2.5 T. Two current densities of the superconducting coils of $J_{e1} = 70$ A/mm^2 and $J_{e2} = 300$ A/mm^2 were considered for the AmSC CC348 and the Superpower 4050 tape respectively. These current densities can be obtained by cooling the sample to $T = 30$–40 K and aiming at B = 4 T on the superconducting tape as seen from the inset of Fig. 3.14. The thickness of the cryostat vacuum is assumed to be 4 cm on both sides of the superconducting coils in order obtain the above operation temperature.

It was then examined how the mass of the direct drive generator changes with the number of poles and Fig. 3.16 shows a finite element simulation of a magnetic field distribution of a 24 pole generator. Figure 3.17 shows that the usage of coated

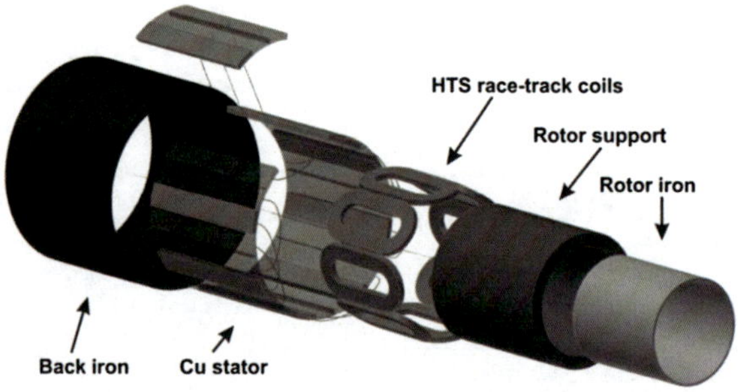

Fig. 3.15 Illustration of superconducting direct drive generator based on high temperature superconducting race track coils mounted on the rotor support structure. The stator has airgap windings without any teeth of magnetic steel as shown in Fig. 3.9. Both an inner and outer magnetic steel support is used to confine the magnetic flux inside the generator [3]. Reproduced with permission from Institute Of Physics

Fig. 3.16 Magnetic field distribution of a 24 pole superconducting direct drive generator. The components from the left are: rotor iron, cryostat vacuum, superconductor coils and support, cryostat vacuum, mechanical air gap, 3 phase stator and stator iron. Reproduced from [26] with permission from Elsevier

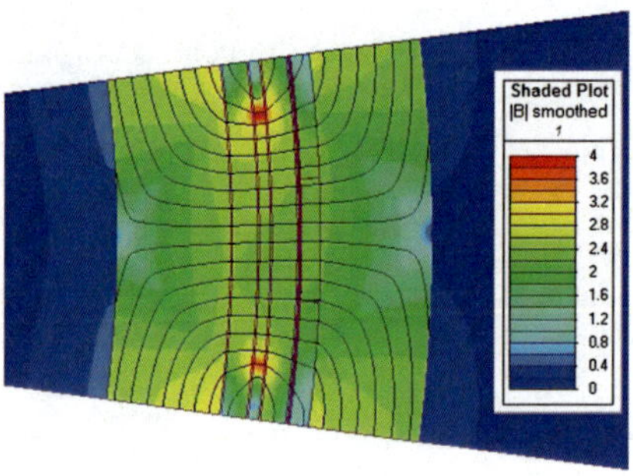

conductor increases as the number of poles is increased and the total weight of the machine becomes smaller.

Reasonable limits on the mass and usage of HTS material was estimated from the expected installation price of offshore wind power in 2020, which is approximately 2 M Euro/MW. Thus the installation price of a 5 MW offshore turbine must be in the order of 10 M Euro. With about 15% being the cost of the drive train it is reasonable to set an upper limit of 1.5 M Euro for the superconducting direct drive generator. Now the present price of the tape is about 30–45 Euro per meter. By assuming this price to fall by a factor of 4 to approximately 8 Euro per meter by 2020 one can estimate the amount of superconducting tape to be used for a direct drive generator. The length is about

$$L_{\text{Tape}} = \frac{1.5\text{MEuro}}{8\text{Euro/m}} = 188\ \text{km} \tag{3.23}$$

Thus the generator should be based on less than 200 km of superconducting tape. Now the weight of the superconductor including insulation depend on the

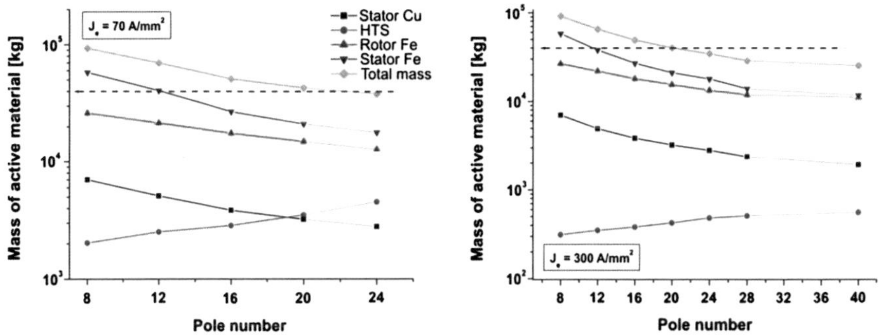

Fig. 3.17 Active mass of the superconducting direct drive generator for a superconducting coil with a current density of **a** 70 A/mm^2 and **b** 300 A/mm^2 as a function of the number of poles in the machine [26]. Reproduced with permission from Elsevier

type of tape used and one gets 0.1 km/kg and 0.3 km/kg for the CC348 and 4050 tape respectively in Table 3.2. A reasonable mass limit is imposed by maintaining the active mass at half of the drive train mass (gear + generator) in the Repower turbine, which amounts to approximately 80 t. Thus the dashed line in Fig. 3.17 indicates the $m_{active} = 40$ t limit, which is fulfilled for a 24 and 20 pole machine for the J_{e1} and J_{e2} respectively. However the usage of HTS tape in the first case is too high with $m_{tape} = 4$ t ($\sim$400 km), but improves to $m_{tape} = 425$ kg ($\sim$130 km). Thus the price of the illustrated generator does become smaller than the cost limit, but further reduction of the cost would be desirable to also allow funding for the cryogenic cooling system. Finally it should be noted that the kilo price for the coated conductor in the last case will be

$$C = \frac{130\text{ km} \cdot 8\text{kEuro/km}}{425\text{ kg}} = 2.5\frac{\text{kEuro}}{\text{kg}} \tag{3.24}$$

An advantage of using the coated conductor is that the amount of superconductor in the 5 MW generator outlined above is very small

$$m_{\text{YBCO,5 MW}} = 51\frac{\text{g}}{\text{km}} \cdot 130\text{ km} = 6.6\text{ kg} \tag{3.25}$$

Furthermore, it is interesting to determine the power loss of the superconductor due to the current in the field coils. This can be done as follows

$$\begin{aligned} P &= E_0 I L\left(\frac{I}{I_c(B,T)}\right)^{n(B,T)} \\ &= 10^{-4}\frac{\text{V}}{\text{m}} \cdot 70\text{ A} \cdot 130\text{ km} \cdot (0.90)^{40} = 13.5\text{ W} \end{aligned} \tag{3.26}$$

since the current in the tape is I = 70 A and it is assumed to be 10% below the critical current. The n-value is of the order n $\sim$ 40 and the length of the tape is assumed to be L = 130 km.

3.9 Technical Challenges

The biggest challenge of realizing the superconducting direct drive based on coated conductors is primarily the price of the tape, which must be reduced by a factor of 10 while the worldwide production capacity is increased by a factor of 40 in order to cover 10% of the EU offshore marked by 2030 as indicated in Fig. 3.1 [3]. This scenario for the price reduction in terms of Euro per meter is not considered easy by the manufacturers, but it is generally believed that the current density J_e of the tapes can be improved by increasing the thickness of the YBCO layer in Fig. 3.12 and also optimizing the pinning [28]. Thus the tape will probably become better, whereby less space is needed for the superconductors and the effective airgap can be decreased causing a more cost effective machine.

Besides the challenge of reducing the price of the tape there are also a number of technical challenges connected to the construction of the cryostat providing the thermal insulation of the superconducting coils. This can be illustrated on a general basis by considering that the efficiency of the superconducting drive train should be in the order of 93–95%. Since the losses of the generator are in the order of 5% and the loss of the power electronics in the order of 2% there is about 1% reserve for maintaining the cold environment of the superconductors. Thus the design criteria of the generator is to minimize the heat flow into the cold parts through the cryostat walls and also through the torque transfer element supporting the superconducting coils in order to be able to remove the heat inflow by suitable cryocooler machines.

The efficiency of cryocoolers are in the order of 1–2% at temperatures T = 30–40 K [29]. Thus if the 5 MW generator is used as an example, the cryogenic cooling loss should be in the order of $P_{\rm cryo} = 50$ kW and the cold loss should therefore be in the order of $P_{\rm cold} = 500$ W.

First of all the power dissipation of the superconductor as determined in Eq. 3.26 must be removed by the cryocoolers. These can be placed around the circumference of the generator and if we assume that each cryocooler provides approximately 6 W at $T = 30$ K on the second stage and 65–110 W at $T = 77$ K on the first stage then there is a need of approximately 5–7 cryocoolers per 5 MW machine (see the CH-208R and CH-210 from Sumitomo [30]). Additionally one must consider whether the heat from the superconductors can be mediated to the cryocoolers by conduction in the support metal or whether a gas circulation circuit must be established.

Secondly the position of the cryostat vacuum region between the superconducting coils and the stator as illustrated in Fig. 3.9 contributes to the effective air-gap of the machine. This will have a big influence on the performance of the generator. Thus the success of the superconducting drive trains will be determined by the optimization of this electromagnetic-thermal–mechanical problem and also by proving that the reliability of the combined system can comply with the load conditions imposed by the wind and wave conditions of the wind turbine.

3.9.1 Minimizing the Cryostat Thickness

The heat conduction through the cryostat is mainly mediated by conduction in the gas of the vacuum region and by the radiation from the cryostat wall. The gas heat conduction between two parallel walls in units of [Wm^{-2}] is approximately given by

$$\dot{Q} \sim c_1 a P (T_2 - T_1) \tag{3.27}$$

where the constant $c_1 \sim 1.2$ for air, a is the accommodation coefficients for the gas molecules colliding with the surface ($a \sim 0.5$ is often used in cryogenic design), P is the pressure of the gas, whereas T_2 and T_1 are the temperature of the hot and cold wall [31]. Vacuum levels of $P_1 = 10^{-3}$ m bar are quite easily obtained using rotary pumps, whereas the more demanding $P_2 = 10^{-6}$ m bar often used for cryostats needs turbo pumps. If the cold temperature of $T_1 = 40$ K and $T_2 = 295$ K is inserted one will obtain a heat conduction of the order

$$\dot{Q} \sim 0.015 - 15\ \mathrm{Wm}^{-2} \tag{3.28}$$

Since the cryostat wall area is $A_{\mathrm{cryowall}} = 2\ \pi\ \mathrm{r}\ \mathrm{L} = 2\ \pi\ 2.1\ \mathrm{m} \cdot 1.5\ \mathrm{m} = 20\ \mathrm{m}^2$, it is seen that a vacuum level close to P_2 is needed and the system would have to rely on turbo pumps. It should be said that the cold parts will provide a substantial cryopumping capability once the operation temperature is reached and the turbo pumps might actually be turned off. However the operation of turbo pumps in a rotating frame and exposed to mechanical vibrations must be proven or alternatively one would have to pump out the cryostat and cool down the superconducting coils with the turbine rotor at a standstill. Then the turbo pumps could be turned off and operation of the turbine could be started. Today many research cryostats maintain the vacuum for several years and ideally the vacuum of the generator should only be pumped during planned maintenance intervals of 2 years, whereby the turbo pumps would only need to be connected as part of the maintenance.

The radiation heat conduction is determined by the Stefan-Boltzmann law which can be formulated for two parallel plates

$$\dot{Q} = \frac{\sigma \left(T_1^4 - T_2^4\right) \varepsilon}{2} \tag{3.29}$$

where $\sigma = 5.67 \cdot 10^{-8}\ \mathrm{Wm}^{-2}\mathrm{K}^{-4}$ is the Stefan's constant, T_1 and T_2 are the temperatures of the two plates and ε is the emissivity of the surfaces ($\varepsilon \sim 0.05$ for polished metal surfaces) [31]. Again if a cold temperature of $T_1 = 40$ K and $T_2 = 295$ K is used, the radiation heat is

$$\dot{Q} \sim 11\ \mathrm{Wm}^{-2} \tag{3.30}$$

which seems problematic at first, but this can be reduced by inserting Multi Layer Insulation (MLI) sheet in the vacuum space as indicated in Fig. 3.9. MLI consist of thin plastic layers coated with a metal layer, which reflects ½ of the radiation heat back to the hot wall. N layers of MLI can thereby reduce the heat load by a factor 1/(N−1). Thus somewhere between $N = 10$–100 would be desirable in the case of the superconducting generator. Typical MLI sheets are produced in 10 or 30 layers with the latter being approximately 1 cm thick [32]. Thus with a 30 layer MLI the radiation heat load can be reduced to $Q = 0.4\ \text{W/m}^2$ taking up 1 cm thickness of the air gap.

3.9.2 *Torque Transfer Tube*

The mechanical transfer of the torque from the wind turbine blades to the superconducting rotor of the generator must be provided by a tube of either stainless steel or some glass/carbon fibre material. The thermal conductivity should be minimized by increasing the length of the tube, but keeping the torsion deformation at some fraction of the angle spanned by a rotor field winding. Additionally, the possibility of resonance frequencies of the torque tube combined with the superconducting coils should be compared with the frequency spectrum of the excitations due to the fluctuating wind and water waves. Coincidence should be avoided and appropriate damping must be provided.

3.9.3 *State of the Art and Alternatives*

Superconducting machines have been investigated since the first fabrication of the low temperature superconducting NbTi wires, but the need for liquid helium as coolant has made the commercial utilization difficult. The discovery of the high temperature superconductors in 1986 caused a revisit of the technology and opened up the possibility to use cryocoolers to operate the superconducting machines [33, 34]. The main focus of the machine development has been on electrical propulsion motors for ships [35], but the direct drive generators for wind turbines is often considered as a further development of the medium speed marine motor with rotation speeds of 100 rpm into high torque generators with a rotation speed approaching 10 rpm as illustrated in Fig. 3.3.

There are several projects focused on the development of superconducting direct drive wind turbine generator and the most visible is probably the 10 MW superconducting SeaTitan turbine proposed by American Superconductor [36]. It is interesting to notice that the diameter of the SeaTitan generator is $D_{\text{SeaTitan}} = 4.5$–$5$ m and that weight is estimated to be $m_{\text{SeaTitan}} = 150$ t, which should be compared to the numbers stated in Table 3.1. Thus it is concluded that the superconducting direct drive does offer a size and weight, which are approximately 50% of the expected values for the PM direct drive for a $P = 10$ MW wind turbine. However, both the reliability and a reasonable price of the SeaTitan generator remains to be verified. The second project is conducted by Converteam and Zenergy

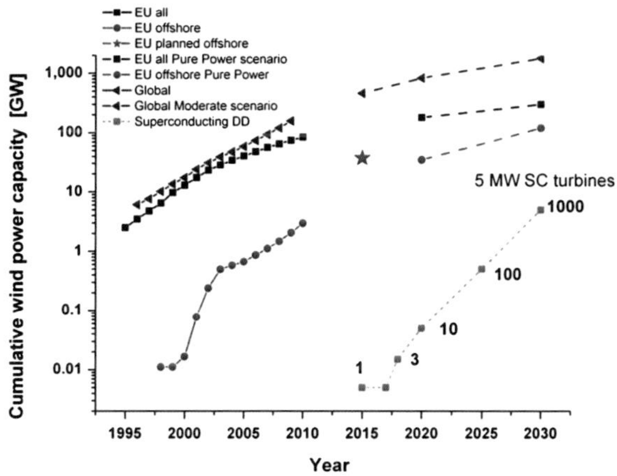

Fig. 3.18 Installed and expected wind power capacity from Fig. 3.1 along with a scenario for introducing superconducting direct drive generators. The numbers indicate the possible installed numbers of 5 MW superconducting turbines

with the focus on an 8 MW direct drive generator [37]. Several alternative generator topologies have also been suggested and the Double Helix winding technology from Advanced Magnet Lab might be based on a fully superconducting machine with both rotor and stator winding made of superconductors. A multifilament wire with very small filaments will however be needed to minimize the AC losses caused by movement of flux lines in the stator winding [38].

3.9.4 Road Map

The major question whether superconducting direct drive generators will ever become economically feasible will probably depend on the success of increasing the production volume of superconducting tapes or wires while simultaneously reducing the production price. This "chicken and egg" problem can be solved by either waiting to see whether the superconductors become good enough or by demonstrating the technical feasibility of the generators and thereby encourage further development of the superconductors. There are a number of supply chain issues, which should be mentioned in this context. In order to discuss them Fig. 3.18 shows a possible scenario of introducing the superconducting direct drive into the EU market. This scenario contains a demonstration period with the installation of just one 5 MW turbine in 2015, 2 turbines in 2018 and 8 additional in 2020. This should give sufficient experience with the technology and provide reliability data to ensure successful large-scale implementation, where a total superconducting wind turbine capacity of 5 GW (1,000 5 MW turbines) is obtained in 2030. These 5 GW are however only about 4% of the expected EU offshore capacity and could be expanded more. The following challenges must be considered:

- HTS tape: Demand in the order of 100,000 km up to 2030, whereas the current annual production is in the order of 1,000 km/year [21, 39]. Thus the demonstration phase can be based on the current production capacity, but it will then have to be increased.
- Cryocoolers: Demand of 5–7,000, which is comparable to the 7,000 cryocoolers installed in Magnetic Resonance Imaging (MRI) devices since 1995 [40]. The cryocoolers of the wind turbine must comply with the marine environment.
- Cryostats: Demand 1,000, which could be compared to the 22,000 cryostats of the currently installed MRI systems [40].

Finally, by assuming the PM direct drive as the major competitor to the superconducting generator from 2020, the demand for Rare Earth element-based permanent magnets must be considered. From Table 3.1 it is seen that the needed amount of $Nd_2Fe_{14}B$ PM material is approximately 600–700 kg/MW. Thus if it is assumed that the increased offshore capacity in EU from 2020–2030 is all covered by PM direct drive then the total PM material needed is

$$\mathrm{m}_{\mathrm{Nd_2Fe_{14}B}} = 90\ \mathrm{GW} \cdot 0.7 \frac{\mathrm{ton}}{\mathrm{MW}} = 63000\ \mathrm{t} \tag{3.31}$$

The weight fraction of the Nd elements is 0.27, which gives $m_{\mathrm{Nd}} = 17{,}000$ t. This is of the same order of magnitude as the annual worldwide production of Nd in 2006 as stated in the recent report “Critical raw materials for the EU” [41]. Thus the introduction of the PM direct drive will cause a considerable increase in the demand for the Rare Earth elements Nd. Additionally, it should be said that if the global scenario in taken into account then the demand will become even larger and supply shortages and increasing prices are likely to become an issue.

The superconducting direct drive will be able to lift this dependency of a worldwide increase in the supply of Rare Earth element, since the usage is about 400 times smaller as shown in Eq. 3.25. Thus this argument might become dominating when the old turbines, which are installed today, must be replaced in 2030.

3.10 Conclusion

The superconducting direct drive generator does offer weight and volume reduction compared to the expected permanent magnet direct drives in the power ranges from 5 to 10 MW. This can be accomplished by the utilization of the high current density of $YBa_2Cu_3O_{6+x}$ high temperature superconducting layer in coated conductors. There is no doubt that the price of the coated conductors must be reduced before the superconducting option becomes economical feasible, but the reliability of the technology must additionally also be proven. A road map including a technology demonstration of 10 turbines from 2015 to 2020 is suggested to determine the reliability and economical feasibility before the production capacity must be increased by several orders to make an impact on the wind power marked in 2030. Several technical challenges should be addressed in the demonstration period. The performance of the coated conductors are expected to improve over the next 5–10 years with higher engineering current densities and it will be important to optimize the electromagnetic-thermal–mechanical problem of the superconducting generator to take advantage of the better performance. In this context it will be important to evaluate whether the reduced weight of the generator can allow for a higher generator installation cost. Secondly, the reliable usage of cryocoolers, vacuum equipment and cryostats in a rotating frame must be mastered and documented.

Finally, the usage of the superconducting direct drive generators will cause a 400 times smaller and more flexible usage of Rare Earth elements compared to the $Nd_2Fe_{14}B$ based permanent magnet direct drive generators, which are expected to cause a considerable increase in the worldwide demand for Nd.

Acknowledgment The funding of the Superwind project was provided by the Technical University of Denmark from the globalization funds and the support of Henrik Bindslev is acknowledged. Also the valuable discussions with colleges and co-workers are acknowledged. This work is dedicated to the memory of Steen Tronæs Frandsen.

References

1. EWEA (2010) Wind statistics. EWEA, Brussels
2. Zervos A, Kjaer C (2009) Pure power—wind energy targets for 2020 and 2030. European Wind Energy Association, Brussel
3. Abrahamsen AB et al (2010) Superconducting wind turbine generators. Supercond Sci Technol 23(034019):1–8
4. Gardner P et al. (2009) Wind energy—the facts. ISBN 9781844077106. Brussel: Earthscan
5. Poore R, Lettermaier T (2003) Alternative design study report: WindPACT advanced wind turbine drive train designs study. National Renewable Energy Laboratory. NREL/SR-500-33196
6. Enercon. www.enercon.com. [Online]
7. Wobben A (2006) Majesties in the wind. Windblatt. February
8. Blundell S (2003) Magnetism in condensed matter. New York: Oxford university press. ISBN 0 19 850591 5
9. Sagawa M et al (1984) New material for permanent magnets on a base of Nd and Fe (Invited). J Appl Phys 55(6):2083–2087
10. Goldwind www.goldwindglobal.com. [Online]
11. Siemens www.energy.siemens.com. [Online]
12. GE. www.gepower.com. [Online]
13. Switch. www.theswitch.com. [Online] The Switch
14. Polinder H et al (2006) Comparison of direct drive amd geared generator concepts for wind turbines. IEEE trans energy convers 21(3):725
15. Shrestha G, Polinder H, Ferreira JA (2009) Scaling laws for direct drive generators in wind turbines. IEMDC proceedings. p 797
16. Bang D et al. (2008) Comparative design of radial and transverse flux PM generators for direct-drive wind turbines. International conference on electrical machines, proceedings, vol. Paper ID 1325, p 1

17. Polinder H et al (2007) 10 MW wind turbine direct-drive generator design with pitch or active speed stall control. IEMDC proceedings, p 1390
18. Eskildsen MR et al (1998) Intertwined symmetry of the magnetic modulation and the flux-line lattice in the superconducting state of TmNi2B2C. Nature 393:242
19. Maeda H, Togano K (ed) (1996) Bismuth-based high temperature superconductors. Marcel dekker, ISBN 0-8247-9690-X
20. Foltyn SR et al (2007) Materials science challenges for high-temperature superconducting wire. Nature Mater 6:631
21. Li X et al (2009) The development of second generation hts wire at american superconductor. IEEE Trans Appl Supercond 19(3):3231
22. Hazelton DW, Selvamanickam V (2009) SuperPower's YBCO coated high-temperature superconducting (HTS) wire and magnet applications. Proceedings of the IEEE, vol 97, 11, p 1831
23. Vlad VR et al (2009) Growth of chemical solution deposited TFAYBCO/MOD(Ce, Zr)O2)/ ABADYSZ/SS coated conductors. IEEE Trans Appl Supercond 19(3):3212
24. www.amsc.com. [Online] American Superconductor
25. www.superwind.dk. [Online]
26. Abrahamsen AB et al. (2011) Feasibility study of 5 MW superconducting wind turbine generator. Physica C. vol. DOI: 10.1016/j.physc.2011.05.217
27. www.superpower.com. [Online]
28. Selvamanickam V et al (2010) Enhanced and uniform in-field performance in long (Gd, Y)–Ba–Cu–O tapes with zirconium doping fabricated by metal–organic chemical vapor deposition. Supercond Sci Technol 23(014014):1–6
29. ter Brake HJM, Wiegerinck GFM (2002) Low-power cryocooler survey. Cryogenics 42:705
30. SHICryogenics www.SHIcryogenics.com. [Online]
31. White GK, Meeson PJ (2002) Experiental techniques in low-temperature physics. 4. Oxford University press, ISBN 0 19 851427 1
32. RUAG. www.ruag.com/space/ch. [Online]
33. Kalsi SS et al (2004) Development status of rotating machines employing superconducting field windings. Proc IEEE 92:1688
34. Barnes PN, Sumption MD, Rhoads GL (2005) Review of high power density superconducting generators: present state and prospects for incorporating YBCO windings. Cryogenics 45:670–686
35. Nick W et al (2010) Development and construction of an HTS rotor for ship propulsion application. J Phys Conf Ser 234(032040):1–9
36. Snitchler G (2010) Progress on high temperature superconductor propulsion motors and direct drive wind generators. International Power Electronics Conference—ECCE Asia -, IPEC. 2010, pp 5–10
37. Lewis C, Muller J (2007) A direct drive wind turbine HTS generator. IEEE Power Engineering Society General Meeting. pp 1–8
38. AML Advanced Magnet Lab. www.magnetlab.com. [Online]
39. Selvamanickam V et al (2009) High performance 2G wires: from R&D to pilot-scale manufacturing. IEEE Trans Appl Supercond 19(3):3225
40. Radebaugh R (2009) Cryocoolers: the state of the art and recent developments. J Phys: Condens Matter 21(164219):1–9
41. EU (2010) Critical raw materials for the EU (Report). Brussels: European Commision, June

Chapter 4
Potential Applications and Impact of Most-Recent Silicon Carbide Power Electronics in Wind Turbine Systems

Hui Zhang and Haiwen Liu

Abstract Power electronics is an enabling technology found in most renewable energy generation systems. In a wind turbine system, it plays an important role in system integration, power quality, and reliability control. Moreover, the fast growth of wind energy poses the increasing need for high-power, low-loss, and fast-switching power electronic devices in order to reduce the system complexity and cost, and improve reliability and compactness. Among the technologies addressing this need, silicon carbide (SiC) power electronics as the most-recent technology stands out because of its superior voltage blocking capabilities and fast switching speeds. As the research samples of SiC power switches become available, it is possible to discuss the design of a wind turbine system using SiC devices and estimate its performance based on the characteristics of practical devices. Therefore, considering the high-power density and high voltage capability of SiC power devices and the recent trend on wind turbine converters, this chapter focuses on the studies of the application of SiC power devices in a full-scale wind turbine converter. First, the characteristics of the most recent devices are obtained through tests. Then, wind turbine system modeling including models for the major electrical components such as generator, power converter, etc., is discussed in detail. Next, the potential benefits from the use of SiC devices in a wind turbine system are explored by a comparison study of the SiC converter and its Si counterpart. Results are presented and analyzed at different wind speeds, temperatures, and switching frequencies. The conclusions drawn from these studies verify that the

H. Zhang (✉)
Department of electrical engineering,
Tuskegee University, Tuskegee, AL 36088, USA
e-mail: hzhang18@ieee.org

H. Liu
Intertek Inc, Cortland, NY 13045, USA
e-mail: haiwenliu@gmail.com

S. M. Muyeen (ed.), *Wind Energy Conversion Systems*,
Green Energy and Technology, DOI: 10.1007/978-1-4471-2201-2_4,

application of SiC converters in wind turbine systems can improve the wind system power conversion efficiency and reduce system size and cost due to the low-loss, high-frequency, and high-temperature properties of SiC devices even for one-for-one replacement for Si devices. It is also pointed out that the application of SiC devices may enable medium converter technology for wind turbine applications when such devices become available. In this way, substantial improvement can be achieved. The chapter is organized as follows: Section 4.1 introduces the present status of wind energy and power electronics. It briefly reviews the electrical technologies used in wind turbine systems such as generator, power converter technology, and power electronics suitable for wind turbine applications. It also summarizes the future trends on wind turbine systems. Section 4.2 focuses on studies of the application of SiC power devices in a full-scale wind turbine converter, including discussions on the present SiC device characteristics, system modeling, simulations of two wind turbine systems with the same components expected for the power converters (One is with SiC power converter, and the other is with a Si power converter). Section 4.3 draws the conclusions and discusses the future work.

4.1 Introduction to Wind Energy and Power Electronics

In the United States and globally, renewable energy capacity has increased rapidly, although it is a relatively small portion of the total energy supply. According to the key findings of the 2009 Renewable Energy Data Book, it has more than tripled in the last 10 years. At the same time, wind energy, as the fastest growing renewable technology, has its installed capacity increased by a factor of 9 worldwide, and by a factor of 14 in the United States. In 2009, the new installed nameplate wind capacity in the United States was 9,922 MW, which accounted for 92% of the annual installed renewable electricity capacity and 51% of all new electrical installations. The major contribution to the fast growth of wind energy is the fact that wind energy technology is the most cost-effective renewable technology able to provide electricity at a relatively low price (6–13 cents/kWh) [1]. Therefore, wind energy will play an important role in the future energy supply.

Modern wind turbines are used to generate electricity for individual use or for contributions to utility grids. The ratings of wind turbines range from tens of kilowatts to multi-megawatts. The early developed wind turbine systems are based on squirrel-cage induction generators, which are typically connected directly to the grid and operate at a fixed speed. Consequently, the perturbations in the drive train are almost transferred to the grid. Variable speed capability allows a wind turbine to operate at speeds which produce the greatest amount of power and minimizes torque perturbations in the drive train [2–6]. This capability tends to decrease the overall cost of energy because the amount of energy generated is increased, and the cost of the drive train and its maintenance are reduced. Since the voltage and frequency of the generated power vary with turbine speed, a power converter is required to reconcile the output with the fixed voltage and frequency of the grid [7–11]. Even for a fixed-speed system, power electronics are required as a soft starter [12]. For systems with energy storage, power electronics also serves as integration interfaces to decouple the generation from demand so that electricity can be stored at times of low demand or low-generation cost and released at times of high demand or high-generation times [13]. In recent years, due to the increasing attention on wind farms' performance in power systems, power electronic technologies are also needed to deal with power quality issues (reactive power control, harmonics, etc.) and system stability issues (different types of faults in the network, such as tripping of transmission lines, loss of production capacity, and short circuits, etc.) [12]. Thus, power electronics are the key components in wind energy systems and play important roles in system integration, power qualities, and stability controls.

Two major factors have enabled the development of cost-effective and control-friendly power electronics for power systems. One factor is the development of fast semiconductor switches that can handle high power at relatively high switching speeds. The other is the introduction of real-time microprocessors that can implement advanced and complex control algorithms [13]. Today, the dominant semiconductors are silicon (Si) material based. They are widely utilized in the industry and in traction, generation, transmission and distribution, etc. With more than half a century's efforts since the first BJT was invented in the 1950s, Si semiconductor technology has been so highly developed that it is hard to achieve any breakthrough. After the invention of IGBT, the main research efforts have been integrating and perfecting the features and characteristics of the existing devices. A substantial reason behind this is that Si power electronic devices are reaching the fundamental limits imposed by the Si material due to its small bandgap. While SiC semiconductor technology, as the representation of wide bandgap semiconductor technologies, has been recognized as the substitutes for Si semiconductor technology in high-power, high-temperature, and high-frequency applications. Several technical and market reports [14–17] have recognized SiC power electronics as a potential technology for wind turbine power converters. The primary benefits of SiC-based power devices include low losses, high temperature tolerance, and fast switching. These can be exploited to reduce generation losses and increase net energy production. The low losses, along with high temperature tolerance, can also be used to improve the reliability of the converter and reduce the thermal management requirements. Moreover, fast switching has the potential to reduce the filtering passive component size and cost, and thus the total cost of the system. More discussions about the state-of-the-art power electronics for wind energy systems and SiC power electronics and their potential applications in wind energy systems will be presented in the following two sections.

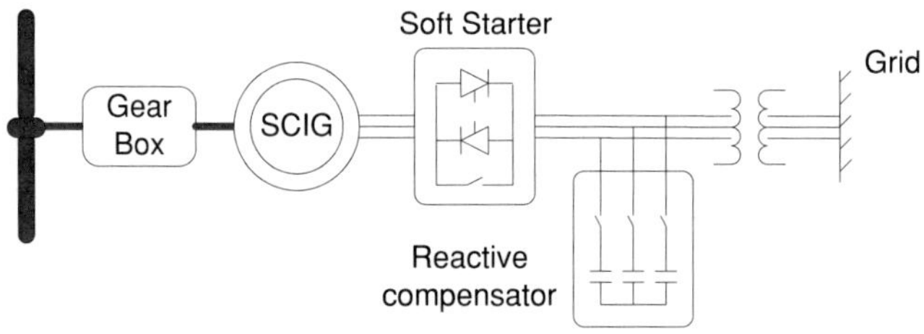

Fig. 4.1 SCIG fixed-speed wind turbine systems with soft starter

4.1.1 State-of-the-Art Power Electronics for Wind Turbine Systems

Wind energy systems can be categorized into standalone systems, hybrid systems, and grid-connected systems. In the former two types of systems, the roles of power electronics are similar to those in grid-connected systems, except when used as an energy storage interface. Thus, this chapter mainly focuses on grid-connected systems.

4.1.1.1 Introduction to Wind Energy Conversion Systems

Although a fixed-speed grid-connected system requires less power electronics, usually a thyristor soft starter is connected between a squirrel-cage induction machine (SCIG) and grid to limit the short-duration inrush current to prevent the disturbances to both grid and drive train, as shown in Fig. 4.1. Stall control or pitch control is used to control the power. Typically, the generator slip change is within 2–3%. Simplicity and cheapness are the major advantages of this kind of a system. If a wound rotor IG is used instead of SCIG, a limited generator speed control (slip change is upto 10%) can be done by variable rotor resistance via slip rings on rotor.

In a grid-connected variable-speed wind generation system shown in Fig. 4.2, a wind turbine captures the kinetic energy in wind, and converts it into rotating mechanical energy to drive a generator. The mechanical energy converted at high wind speeds must be well limited for system safety, which can be implemented by stall control or pitch control. Typically, the mechanical energy has low speed which can be adapted to the high speed of the generator energy by a gearbox. For multi-pole generator systems, gearbox is not necessary. The generator converts mechanical energy into electricity which is connected to the grid through power converters, a reactive compensator (optional), and a transformer with circuit breakers [12].

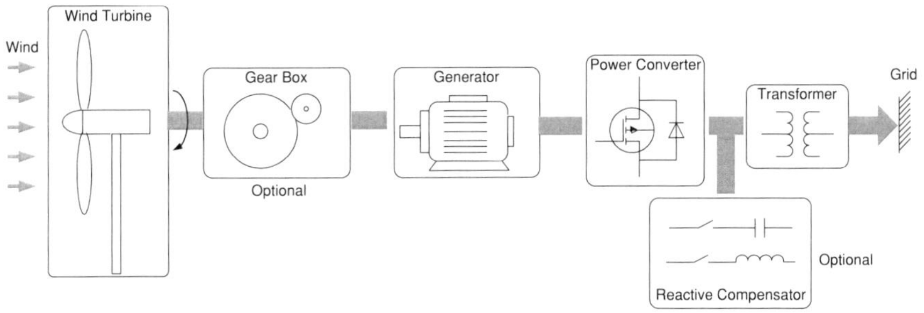

Fig. 4.2 System configuration for grid-connected wind turbines

4.1.1.2 Generators

There are four types of commonly used generators in wind turbine systems, namely Permanent Magnet Synchronous Generator (PMSG), Induction Generator (IG), Doubly Fed Induction generator (DFIG), Synchronous Generator (SG). Among the four types of generators, PMSG is favored by small-scale wind turbines (in kilowatts) due to the high power density and resultant compactness, low loss and simple excitation and cooling design on the rotor side, possibility to eliminate gearbox, and low maintenance cost. However, its power capability and initial cost are limited by the high price of magnets. IG has relative low capital cost and is robust, but the requirement of a full-scale power converter increases the total system cost and losses. Unlike IG, DFIG allows a converter connected to the rotor which typically processes about 30% of the total system power, and consequently the cost and loss of the converter are reduced. The major drawbacks of the kind of generators are control complexity and high maintenance cost. SG is a self-excited machine and its real power and reactive power can be controlled independently. It costs more than IG but less than PMSG. Therefore, both DFIG and SG are favored for high power systems [18].

4.1.1.3 Power Converters

There are mainly two types of converters for variable-speed wind turbine systems. They are full-scale converters and partial-scale converters. The system configurations associated to the two types of converters are shown in Fig. 4.3. In type A configuration, the generator can only be wound rotor IG, and a partial-scale converter is connected to rotor side, which is the so-called DFIG. The partial-scale converter can also be a matrix converter, which is excellent for the elimination of the DC link or energy storage, and simple control for one converter. DFIG has the ability to deliver power more than its rated power without overheat, and its converter only needs to process 25–30% of the total power. Thus, DFIG is excellent for high power applications in MW range [18]. Currently, it dominates

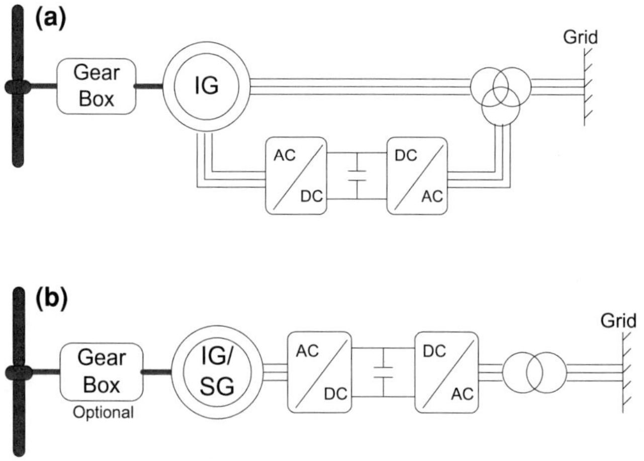

Fig. 4.3 Wind turbine system configurations. **a** Type A: with partial-scale converters. **b** Type B: with full-scale converters

the market for variable-speed wind turbine systems. In type B configuration, the generator can be IG, SG, and PMSG. The full-scale converters can be bidirectional or unidirectional back-to-back converters. Bidirectional back-to-back converters are composed of a controlled rectifier at generator side and a voltage source inverter (VSI) at grid side. For PMSM and SM, a diode rectifier can be used to substitute the controlled rectifier, that is, a unidirectional back-to-back converter. Full-scale configuration provides an opportunity to eliminate gearbox since the input power is always converted into power at grid frequency. With type B configuration, it may be more effective and less complicate to deal with grid-related issues, such as reactive power control, possibility for active grid support, and the potential to operate wind farms as power plants [12]. Thus, wind turbine systems with full-scale converters will become more attractive.

Besides the power electronics for system integration, wind systems also depend on power electronics to improve power quality and system stability. For more information about these applications, please refer to [12].

The performance and reliability of power converters in wind conversion systems are directly related to the availability and the electrical and thermal characteristics of power electronics devices. IGBT is now the main power electronic device for wind turbine applications. The rating of the recently developed high voltage IGBT (HVIGBT) is upto 3,300 V/1,200 A [19]. IGCT is another recent technology used in wind turbine converters, which is the combination of a low-inductance gate drive and an improved GTO structure. Thus the IGCT has higher power handling capability (upto 4,500 V/600 A) [20]. But it has relatively low switching speed, and is susceptible to the heating and cooling cycle experienced in wind turbine converters [18]. Therefore, advanced power electronics with

Table 4.1 Advantages and application scope of SiC devices

Device characteristics	System benefits	Application scope
High breakdown voltage	Large power capacity	Military: combat vehicles, weapons, electric ships
High current density	High reliability, compactness	Aerospace: spacecraft and satellite applications
High operational temperature	Less cooling requirements	Energy: power transmission and distribution
High switching frequency	Reduced passive components	Industry: deep earth drilling for energy exploring
Low power losses	High efficiency	Future: domestic automobiles, motor drives

higher power capability, better characteristics, and reliability are always the need of wind turbine systems.

4.1.2 SiC Power Electronics and Their Potential Applications in Wind Energy Systems

Using SiC material to substitute Si material is expected to bring the substantial improvement to the performance and reliability of power electronic devices. SiC, as a wide bandgap material, has larger electrical breakdown field, reasonably high electron mobility, higher thermal conductivity, and larger saturated electron conductivity compared to Si. Theoretically, devices made of SiC have almost a 10 times improvement when compared to Si devices for such properties as thermal conductivity, voltage blocking capability, and reverse recovery characteristics. Consequently, systems composed of SiC devices have substantial improvements in efficiency, reliability, size, and weight even in harsh environments. Therefore, they are attractive especially for high-voltage, high-temperature, high-efficiency applications. Table 4.1 summarizes the advantages and applications scope of SiC devices. Much work has been conducted in the areas of military and aerospace [21–25], where cost is not the main concern. More applications are expected for vehicles, motor drives, industry, and utilities when technology is mature and the price drops [26–29]. Also, as mentioned previously, wind turbine systems have been recognized as one of such potential applications.

4.1.2.1 Present Status and Future Trends of SiC Power Devices

SiC Schottky diodes were commercialized in 2001. These devices have relatively high blocking voltage (upto more than 1,700 V [30]), and their switching losses are significantly reduced due to the negligible reverse recovery. These features were successfully explored in power factor correction (PFC) circuits [31]. Besides, increased efforts have been made on their use as freewheeling diodes for Si switching devices. As compensation for high-switching-loss switches, SiC Schottky diodes help to reduce the total system switching loss and enable high-frequency operation so that higher power density can be achieved. For an example, they are used in a 1,200 V/300 A Si IGBT single phase module (Cree) [32] and a 55 kW Si IGBT inverter [33].

The research efforts on switching devices mainly focus on unipolar devices. The major types of devices investigated are JFETs and MOSFETs, which are available as research samples in a voltage range from 600 V to several kVs. The technology of JFETs is relatively mature and are ready to enter the commercial market. The development of MOSFETs gained new attention in the recent years due to improvements in channel mobility [30]. Both SiC JFETs and SiC MOSFETs offer superior static and dynamic characteristics over Si switching devices, as well as high temperature capability with proper packaging. These features can be explored to achieve a system of smaller size, less weight, and/or with higher efficiency, and/or high temperature capability with less cooling. For a higher voltage blocking range, the development of bipolar devices like IGBTs and BJTs are inevitable.

Currently, most interests on SiC devices are in high power applications. However, the availability of such SiC devices prevents them from virtual applications. Therefore, many research efforts are being made to improve power ratings of SiC devices by either improving single device rating [34, 35] or developing high power modules [36–41]. So far, the single SiC diode and DMOSFET voltage rating is upto 10 kV with die area exceeding 1.5 cm^2 and 0.64 cm^2, respectively [34]. The largest SiC power modules reported in the literature is 10 kV/100 A [38] or 1.2 kV/880 A [35]. These are SiC devices which could be used in future wind turbine systems.

4.1.2.2 Characteristics of Present SiC Power Devices

In this section, the characteristics of SiC devices are discussed in general. Sample characteristics for major device types are presented.

First, SiC Schottky diodes are superior to Si diodes for the negligible reverse recovery. Figure 4.4a clearly demonstrates the reverse recovery waveform of a SiC Schottky diode versus. a Si diode with similar rating. Morever, unlike Si diodes, the reverse recovery phenomenon in SiC Schottky diode is not influenced by the changing of junction temperature as shown in Fig. 4.4b. In addition, the reverse recovery current is more dependent on voltage across than current through the diode since SiC Schottky diodes are majority devices.

Second, the static characteristics of a SiC JFET and a SiC MOSFET are shown in Figs. 4.5 and 4.6. The I–V characteristics of the SiC switching devices are close to linear, and their on-state resistances increase slightly with higher temperature. The transfer characteristics of these devices are pretty steady at different temperatures as shown in Figs. 4.5b and 4.6b, especially for the SiC JFET. Their switching characteristics are represented by the energy losses shown in Fig. 4.7, which is independent of temperature but increased as current increases.

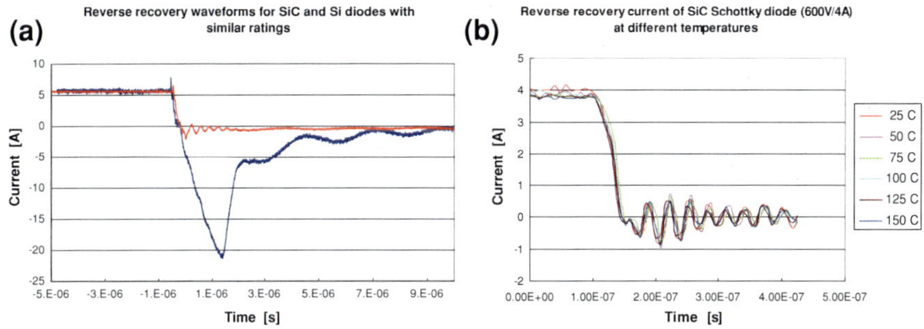

Fig. 4.4 Reverse recovery waveforms. **a** SiC (*red*) Versus. Si (*blue*). **b** SiC at different temperatures

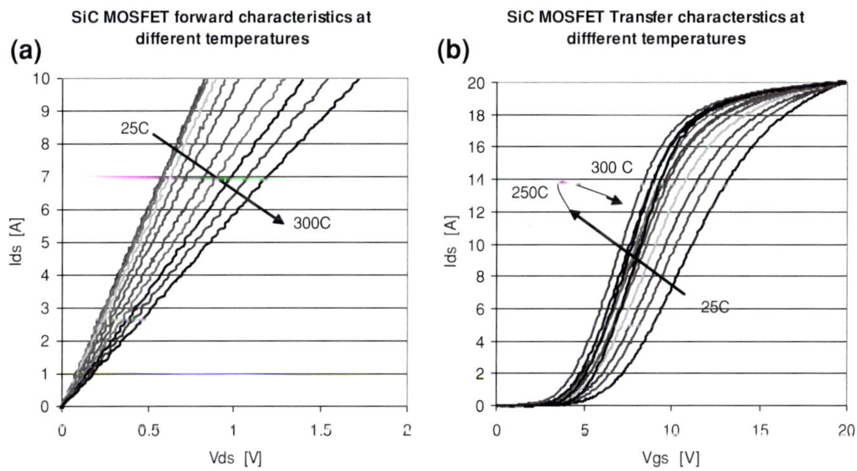

Fig. 4.5 Static characteristics of the SiC MOSFET (1,200 V/10 A). **a** forward characteristics. **b** transfer characteristics

Note that, the JFET shown in Fig. 4.6 is a normally-on device, and so a negative gate voltage is required to turn off the device. To demonstrate the advantages of SiC switches over Si ones, the primary properties of several SiC and Si devices are compared in Table 4.2. As a result of their superior characteristics, SiC switching devices have less condition losses and switching losses, and this benefit becomes more dramatic as the temperature increases.

Finally, high power modules with high-temperature package are needed to take the full advantages of SiC devices. Table 4.3 presents the characteristics of such SiC modules developed by APEI Inc. versus those of Si IGBT modules from major manufacturers. As these kinds of SiC modules become available, wind converters can be designed using SiC devices to achieve high efficiency and compactness at

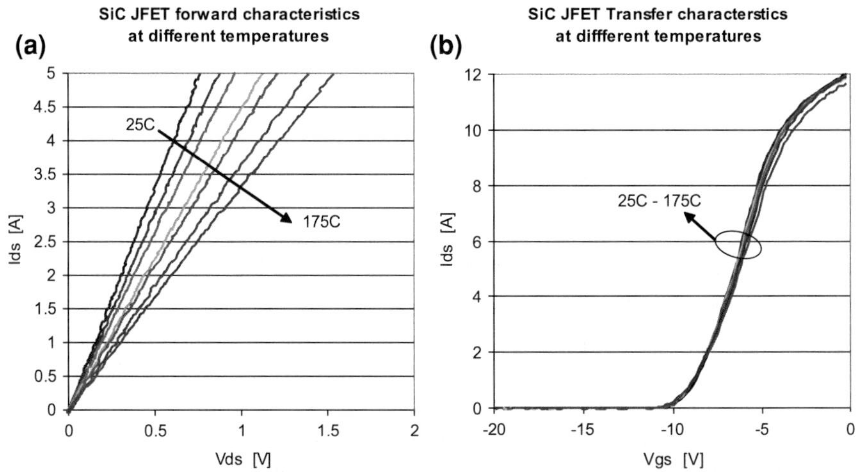

Fig. 4.6 Static characteristics of the SiC JFET. **a** forward characteristics. **b** transfer characteristics

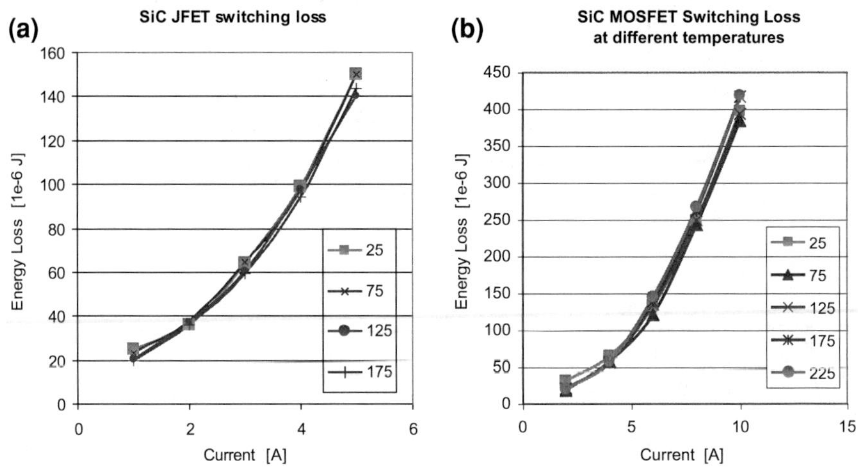

Fig. 4.7 Switching energy loss of the SiC devices. **a** SiC JFET (1,200 V/5 A). **b** SiC MOSFET(1,200 V/10 A)

the high power level. In addition, as the finding in [42], medium voltage (1 ~ 5 kV) wind turbine converters are superior to low voltage converters which currently dominate at all power levels. Typically, the parallel connection of low power converter modules are needed at power levels in excess of 500 kVA. This requires more space and weight support from the mechanical systems which complicate the whole system design. Instead, the using of medium voltage converters can reduce the system components and costs due to the decreased

Table 4.2 Characteristics of some 1,200 V SiC and Si single switching devices

Characteristics	SiC MOSFET cree	SiC JFET (normally-on) SiCED	SiC JFET (normally-off) SemiSouth	Si IGBT Toshiba GT10Q301	Si MOSFET APT, APT 1201R5BVR
Current	10 A	5 A	5 A	10 A	10 A
Voltage drop @rated current	0.84 V @25°C	0.79 V @25°C	0.59 V @25°C	2.1@25°C	15 V @25°C
	1.04 V @150°C	1.42 V @150°C	1.41 V@150°C		
Switching energy loss	402 μJ @200 V	146 μJ @200 V	244 μJ @200 V	1250 μJ @200 V/ 25°C	

Table 4.3 Characteristics of some SiC and Si half-bridge power modules [43]

Characteristics	SiC MOSFET APEI	Si IGBT PowerEx CMD150DY-12NF	SiC JFET APEI	Si IGBT IXYS MII-45-12A3
Voltage	600 V	600 V	1,200 V	1,200 V
Current	180 A @ 250°C	150 A @ 25°C	100 A @ 250°C	160 A @ 25°C
Maximum junction temperature	150°C	250°C	250°C	150°C
On-state resistance	12.75 mΩ @ 25°C	15 mΩ @ 25°C	15.5 mΩ @ 25°C	25 mΩ @ 25°C
Gate charge	480 nC	600 nC	160 nC	600 nC
Turn-off delay	300 ns	300 ns	< 100 ns	600 ns
Figure of Merit (FOM) switching speed[a]	6.1 nΩ s	9 nΩ s	2.5 nΩ s	15 nΩ s

[a] FOM directly relates to switching speed capability (low FOM = fast switching)

current load, and improved reliability, especially for high-power levels. Thus, medium voltage converters become competitive, but are dependent on the availability of high voltage power electronics devices. Since SiC devices have 10 times higher breakdown voltage than Si devices theoretically, using SiC devices can be a solution for medium wind turbine converters in the future.

In the following section, the performance of SiC full-scale wind turbine converter is studied by simulations in order to validate the applications of SiC devices in wind turbine systems and illustrate the benefits qualitatively and quantitatively.

4.2 Studies on a SiC Full-Scale Wind Turbine Converter

This section focuses on studies of the application of SiC power devices in a full-scale wind turbine converter. Modeling of wind turbine components including generator, power converter, thermal system, and electrical system are discussed in

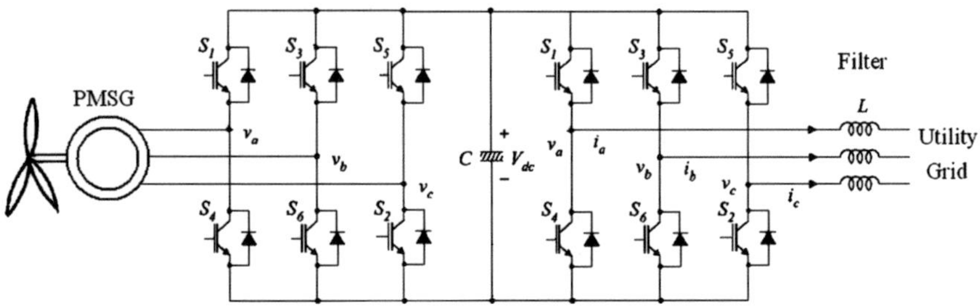

Fig. 4.8 Wind generation system structure with full-scale converter

Fig. 4.9 System modeling diagram

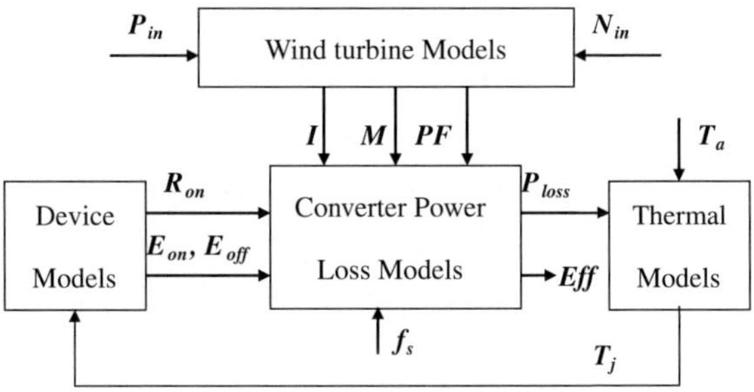

detail. Simulations are developed for two wind turbine systems with the same components expected for the power converters. One is with a power converter based on Cree SiC MOSFETs, and the other is with a power converter based on modern Si IGBT products. The performance of these two systems is presented in quantities at different wind speeds, temperatures, and switching frequencies.

4.2.1 Wind Energy System Components and Modeling

A wind turbine system is designed based on a National Renewable Energy Laboratory (NREL) baseline wind turbine described by [44]. It is composed of a 1.5 MW wind turbine, a permanent magnet (PMSG) generator rated at 690 V, a bi-directional full scale frequency converter comprising two back-to-back inverters, and a utility filter (simplified as a single inductance), as shown in Fig. 4.8 [45]. The wind energy is converted into electricity by the PMSG, and then transferred to the utility. In this process, the converter plays two roles, one is to control the generator to capture as much wind energy as possible, and the other is to deliver the energy to the utility. In both roles, the power loss in the converter is of utmost concern, which not only determines the design of the converter but also affects other components in the system, such as the filter. The converter efficiency will be

Table 4.4 Wind generation system parameters

Generator parameters		Others	
Rated power, MW	1.5	DC link voltage V_{dc}, V	1,100
Nominal voltage, V	690	Grid voltage V_{ll}, V	690
Rated speed n_0, rpm	164	Grid power factor, cos ϕ	0.95
Back emf at n_0, Emf_0, V	150	filter loss constant, k	0.0097
Base machine pole number p	56		
Stator phase resistance R, Ω	1.23e-2		
Stator phase inductance L, H	6.62e-4		
Eddy loss at n_0, P_{le0}, kW	4.284		
Hysteresis loss at n_0, P_{lh0}, kW	1.848		

demonstrated by simulation results, and the effect of applying SiC devices on such a system is also analyzed. The associated modeling work is presented next.

As shown in Fig. 4.9, the system modeling is composed of four models: wind turbine models, device models, converter power loss models, and thermal models. The system inputs are input power and speed of the generator, ambient temperature, and the switching frequency of the converter. The output will be the system efficiency. The four groups of models are connected by parameters. For example, wind turbine models and device models provide parameters for the converter models, and the output of the converter power loss models is fed to the thermal models. Then, the junction temperature obtained from the thermal models is provided to the device models, which starts the next calculation step based on the new temperature information. Each group of models is explained in detail in the following sections.

4.2.1.1 Wind Turbine Models

The wind turbine model is mainly a PMSG model, whose function is to calculate the current, power factor, and modulation index which are needed by the converter model based on the input power and the speed of the generator. The electrical parameters of the generator studied in this work are listed in Table 4.4. Assume the generator back emf is in phase with the generator current, then a simplified PMSG model can be developed using d - q rotating coordinate system analysis. With the generator model, the output voltage and current of the generator can be obtained. They are also the ac side current and voltage of the rectifier in the back-to-back converter. Then based on the relationship between ac side and dc voltage in an inverter controlled by SPWM technique, the modulation of the converter can be calculated. The list of equations is as follows:

Generator frequency, f_g:

$$f_g = \frac{p \cdot n}{120} \tag{4.1}$$

Back Emf (line-neutral, peak), *Emf* (subscript '0' denotes the parameter's value at rated speed n_0)

$$Emf = Emf_0 \cdot \frac{f_g}{f_{g0}} \tag{4.2}$$

Core losses, P_{lc}, eddy loss, P_{le}, and hysteresis loss, P_{lh}:

$$P_{le} = P_{le0} \cdot \frac{f_g}{f_{g0}} \tag{4.3}$$

$$P_{lh} = P_{lh0} \cdot \frac{f_g}{f_{g0}} \tag{4.4}$$

$$P_{lc} = P_{le} + P_{lh} \tag{4.5}$$

Phase current on q-axis (peak), I_{sq}:

$$I_{sq} = \frac{1,000 \cdot P_{in} \cdot 2}{3 \cdot Emf} \tag{4.6}$$

Output voltage without considering the core losses, E_c:

$$E_c = \sqrt{(Emf - I_{sq}R)^2 + (I_{sq}X)^2} \quad X = \omega_g L \tag{4.7}$$

Effective current due to core losses (peak), I_c, I_{cq} (q-axis component), I_{cd} (d-axis component):

$$I_c = \frac{1,000 \cdot P_{lc} \cdot 2}{3 \cdot E_c} \tag{4.8}$$

$$I_{cq} = I_c \cdot \cos \phi_c \cdot \tag{4.9}$$

$$I_{cd} = I_c \cdot \sin \phi_c \; \phi_c = \arccos\left(\frac{E_c}{Emf}\right) \tag{4.10}$$

Total phase current (peak), I_s:

$$I_s = \sqrt{I_{stq}^2 + I_{std}^2} \tag{4.11}$$

$$I_{stq} = I_{sq} + I_{cq} \tag{4.12}$$

$$I_{std} = I_{sd} + I_{cd} \tag{4.13}$$

Output voltage (peak), U_{gout}:

$$U_{gout} = \sqrt{U_{sq}^2 + U_{sd}^2} \tag{4.14}$$

$$U_{sq} = Emf - I_{std}X - I_{stq}R \tag{4.15}$$

$$U_{sd} = I_{stq}X - I_{std}R \tag{4.16}$$

Power factor, $\cos\phi$:

$$\cos\phi = \cos(\alpha_I - \alpha_U) \tag{4.17}$$

$$\alpha_I = \arctan\left(\frac{I_{stq}}{I_{std}}\right). \tag{4.18}$$

$$\alpha_U = \arctan\left(\frac{U_{sq}}{U_{sd}}\right) \tag{4.19}$$

Copper loss, P_{lcu}:

$$P_{lcu} = \frac{3 \cdot I_s^2 \cdot R}{2 \cdot 1000} \tag{4.20}$$

Generator efficiency, η_g:

$$\eta_g = \frac{P_{in} - P_{lcu} - P_{lc}}{P_{in}} \times 100\% \tag{4.21}$$

Rectifier modulation index, M_r:

$$M_r = \frac{U_{gout}}{V_{dc}/2} \tag{4.22}$$

For the grid inverter, if the power loss of the DC link capacitor is neglected, the input power of the grid inverter is equal to the output power of the generator rectifier. Then, for a certain grid output voltage V_{ll} (line–line, rms) and power factor $\cos\phi$, the phase current (peak) I_{grid}, is

$$I_{\text{grid}} = \frac{\sqrt{2}P_{in}\eta}{\sqrt{3}V_{ll}\cos\phi} \tag{4.23}$$

and its modulation index M_i is

$$M_i = \frac{2\sqrt{2}\cdot V_{ll}}{\sqrt{3}\cdot V_{dc}} \tag{4.24}$$

Also, the filter loss P_{lf} is considered when calculating the system efficiency.

$$P_{lf} = k \cdot I_{\text{grid}}^2 \tag{4.25}$$

where k is filter loss constant.

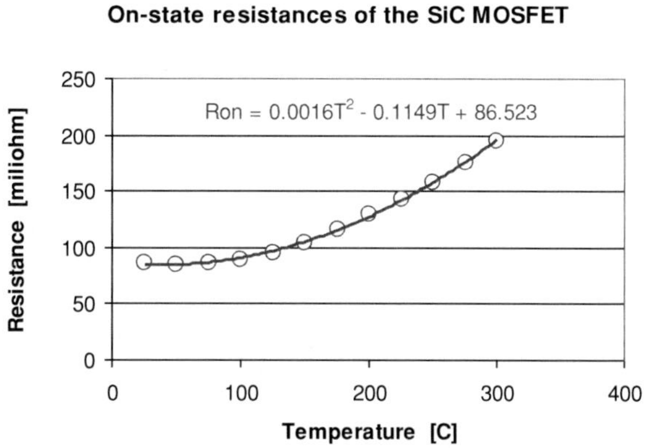

Fig. 4.10 On-state resistance of the SiC MOSFET

4.2.1.2 SiC Power Device Models

Device models describe the device characteristics related to the operation losses. In this work, they are look-up tables or polynomial functions based on the curve fitting of test results. Most recent SiC MOSFET prototypes were obtained and tested for both static and dynamic characteristics, which were then used for curve-fitting.

The prototype SiC MOSFETs listed in Table 4.5 were experimentally tested with a curve tracer at different ambient temperatures from 25 to 300°C with an increment of 25°C. As expected, the on-state resistances of the MOSFETs increase with temperature but remain constant with respect to current, as shown in Fig. 4.5a. After scaling the rating of the SiC MOSFETs to 1,700 V and 1,200 A, which is the rating of the Si IGBTs used in this work (see Table 4.5), not only is the on-state resistance (0.76 mΩ @ RT) of the SiC MOSFETs is smaller than that of the Si IGBTs (0.84 mΩ @ RT), but also the change rate of the resistances with temperature is smaller. The on-resistance of the SiC MOSFETs increases by 11.8% when temperature changes from room temperature to 125°C, compared to 41.8% for the Si IGBTs. Thus, the SiC devices are more efficient in terms of having lower conduction loss especially at higher temperatures. The transfer characteristics of the MOSFETs in Fig. 4.5b change slightly with increase in temperatures above 150°C. This indicates that the change in switching losses of the MOSFETs with temperature will be small. This is confirmed by the switching tests (Fig. 4.7b).

To count the influence of variable temperature, the SiC MOSFET is modeled as a function of the device junction temperature. By letting it be a quadratic function, the on-state resistance can be represented as

$$R_{M,\,on} = 0.0016T^2 + 0.1149T + 86.523 \ \ \text{m}\Omega, \tag{4.26}$$

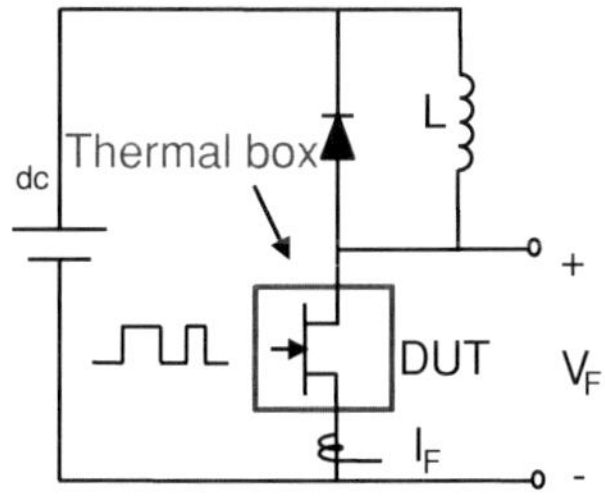

Fig. 4.11 Switching test circuit

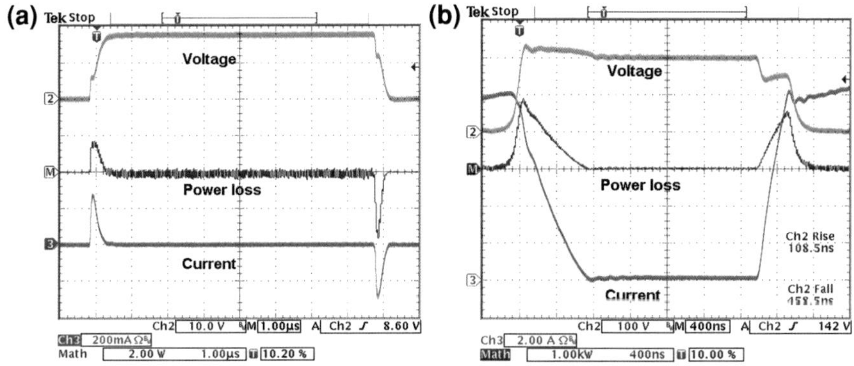

Fig. 4.12 Test waveforms of the SiC MOSFET (Cree) at 225°C. **a** gate signal waveforms. **b** switching waveforms

where the coefficients are obtained by performing curve fitting on test results as shown in Fig. 4.10. Similar equations can be established for on-state resistance of SiC Schottky diodes based on the manufacturer's data as (4.27).

$$R_{D,on} = 0.0016T^2 + 0.516T + 37.4\,\text{m}\Omega \tag{4.27}$$

The circuit shown in Fig. 4.11 is used to obtain the switching characteristics in Figs. 4.12 and 4.7b. The ambient temperature of the SiC MOSFETs is varied from 25 to 225°C with the increment of 50°C. With a pure inductive load, the current in the switches can be controlled by adjusting the duty ratio of the first pulse when applying a double-pulse control signal. Commercial gate driver IC HCNW3120 is selected to drive the SiC MOSFETs. It generates a voltage of about 20 V at turn on and 0 V at turn off. The gate drive board is separate from the MOSFET, and its ambient temperature is room temperature. As shown in Fig. 4.12, with a gate resistance of 10 Ω, the MOSFET works functionally upto 225°C ambient at a power level of 200 V and 10 A. The peak gate current at this condition is 280 mA. The switching losses, including both turn-on loss and turn-off loss, are calculated at each test condition. They are plotted versus the drain current at each tested

Table 4.5 Devices used in the converters

Item	Voltage rating	Current rating	Part number
SiC MOSFETs	1,200 V	10 A × 20 × 10	CREE, Prototype
SiC Schottky diodes	1,200 V	10 A × 20 × 10	CREE, C2D10120
Si IGBT/diode modules	1,700 V	1,200 A × 2	DYNEX, DIM1200FSM17

temperature in Fig. 4.7b. The switching losses increase as the current increases, but change very slightly when ambient temperature changes, while the increase of the switching losses of Si devices is much more obvious. For the Si IGBT used in the work, its switching loss increases by 56.3% when case temperature changes from 25 to 125°C. Thus, the substitution of SiC devices for Si devices will improve the system efficiency, and with higher temperature and higher switching frequency, the system will gain even more benefits compared to the Si-based system.

Since the switching loss of the SiC MOSFET increases with increasing current and is nearly constant with increasing temperature, it can be modeled as a polynomial function of current as (4.28).

$$E_{J,sw}(i) = 0.1059i^3 + 2.6733i^2 + 3.0064i(\mu\text{J}) \tag{4.28}$$

Similar to (4.26), the coefficients in (4.28) are also obtained by performing curve fitting on the test results shown in Fig. 4.7b.

In addition to the dependency of switching losses on current as shown in (4.28), the switching losses in SiC MOSFETs are also affected by applied voltage. This effect can be accounted for by introducing a factor corresponding to the ratio between the applied voltage and the test voltage at which the switching losses are measured [46]; for this case, the test voltage is 200 V. This ratio is reflected in (4.33) in the following section.

Similarly, the change of switching losses with voltage in SiC Schottky diodes is also accounted for by introducing such a factor (see (4.34)). Moreover, this change is the only one needed to be considered for a SiC Schottky diode operating under different conditions because its switching losses (mainly reverse recovery loss) are more dependent on voltage than current and do not vary with temperature [47].

4.2.1.3 Converter Power Loss Models and Thermal Models

A 1.5 MW converter is required to provide full power conditioning for the full output of the generator. Because no SiC devices are presently available at this rating, the converter is assumed to be composed of 10 SiC-based converters that each use twenty 10 A SiC MOSFETs for a power rating of 150 kW each in the simulation, which are based on the devices listed in Table 4.5. The multiples in the current column mean the number of devices in parallel and the number of converters in parallel.

Similarly, Si IGBT/diode modules with maximum available ratings are selected to form a converter representing today's Si power electronic techniques. The Si modules are single switch modules with two single IGBT and two single diodes rated at 1,200 A in parallel. To meet the power requirement for this application, two such Si converters will be needed in parallel.

As high current rated SiC power devices and modules mature in their development, future SiC inverters will be comparable to the present Si IGBT inverter in terms of difficulty of paralleling devices and complexity resulting from the additional components such as gate drivers and connectors. This will enable the practical application of SiC inverters in such high power systems as the wind turbine system discussed in this chapter.

Because of the unavailability of such high power SiC modules at present, the best prediction of what will be achievable when using SiC devices in the near future can be extracted from the characteristics of the presently available devices. Two assumptions are made for this prediction: (1) future SiC devices will have the same performance as today's SiC devices, and (2) module packaging technology does not introduce parasitics or issues other than those found in single device packaging. The first assumption is a conservative one, while the latter is overly optimistic. Thus, the actual performance will vary from the predicted performance in future SiC inverters and depends on the continued future device development and packaging technologies for large current rated SiC power electronic modules.

Power Loss Models

A widely used averaging technique [48–51] is employed to study the inverter power loss. This technique takes a sample in each switching cycle, and then uses these values to find the effective value at the fundamental cycle of the output. The resultant equations are presented in the following paragraphs. In these equations, where ± appears in the equation, the upper sign should be used for inverter calculations, and the bottom one is for rectifier calculations [52, 53].

For SiC MOSFETs, the conduction loss is caused by on-state resistance, R_{on}.

It is calculated by

$$P_{M,\,cond} = I^2 \cdot R_{on,M}\left(\frac{1}{8} \pm \frac{1}{3\pi} M \cos\phi\right) \tag{4.29}$$

where M is modulation index, I is the peak of phase current, and $\phi$is phase angle of the current with respect to voltage. For the SiC Schottky diode and Si IGBT, the voltage drop is not zero when current is zero. So there is additional loss associated with this voltage drop, V_O. Conductive loss equations are shown as (4.30) for the diode and (4.31) for the IGBT.

$$P_{D,\,cond} = I^2 \cdot R_{on,D}\left(\frac{1}{8} \mp \frac{1}{3\pi} M\cos\phi\right) + I \cdot V_0 \cdot \left(\frac{1}{2\pi} \mp \frac{M\cos\phi}{8}\right) \tag{4.30}$$

$$P_{I,\,cond} = I^2 \cdot R_{on,I}\left(\frac{1}{8} \pm \frac{1}{3\pi} M\cos\phi\right) + I \cdot V_0 \cdot \left(\frac{1}{2\pi} \pm \frac{M\cos\phi}{8}\right) \tag{4.31}$$

The extraction of R_{on} of SiC devices have been presented in Sect. 4.2.1.2. Similar mathematical expressions can be obtained for the Si diodes and IGBT using manufacturer values [54, 55]. Because of the limited data points, the linear models based on the data at two different temperatures are used for Si devices.

Effective switching power loss for the SiC MOSFETs in an SPWM controlled inverter can be derived from the instantaneous energy loss expression (4.28) using equation (4.33).

$$P_{sw} = f_{sw} \cdot \frac{1}{2\pi} \int_{\phi}^{2\pi+\phi} E[i(\theta)]d\theta \tag{4.32}$$

Thus, the switching power loss of the SiC MOSFET is represented by

$$P_{M,\,sw} = f_{sw} \cdot \left(\frac{V}{V_0}\right)^{\frac{3}{2}} \cdot \left(0.1797I^3 + 2.6733I^2 + 1.9139I\right) \times 10^{-6}. \tag{4.33}$$

The same method can be applied to Si diodes and IGBTs. The switching loss of the SiC diode can be calculated by [47, 52]

$$P_{D,sw} = f_s \frac{V}{4S} \sqrt{\frac{V}{V_0}} \left(\frac{dI}{dt}\right) \left(\frac{St_{rr}}{S+1}\right)^2. \tag{4.34}$$

The total power loss of the SiC inverter is the sum of the conduction loss (4.29) and switching loss (4.33) of the MOSFETs and the conduction loss (4.30) and switching loss (4.34) of the diodes. Since there are six each of the MOSFETs and diodes in an inverter and they are assumed to be identical, the total power loss of an inverter is

$$P_{Inv,\,SiC} = 6 \times \left(P_{M,\,cond} + P_{M,\,sw} + P_{D,\,cond} + P_{D,\,sw}\right) \tag{4.35}$$

Thermal Models

The equivalent thermal circuit shown in Fig. 4.13 is used to analyze the thermal response of the converter [49, 50]. Power losses at the current temperature are fed to this model to calculate new device junction temperatures for the next computation iteration in order to update power losses based on the new temperature. In this way, the self-heating of the devices is counted dynamically. As shown in the figure, the power sources of the circuit are the power losses (P_m and P_d) from

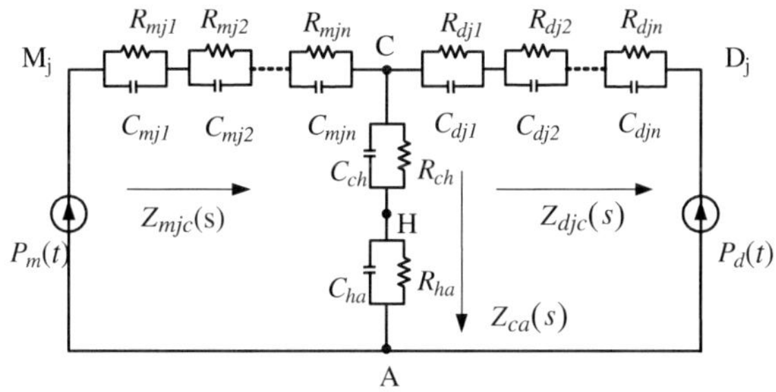

Fig. 4.13 Thermal equivalent circuit of an inverter

switches and diodes (assumed to be on the same heatsink), and the interfaces between these devices, heatsink, and thermal grease in between are modeled as a series of *RC* pairs [46, 53, 56].This circuit can be solved using transfer functions (4.36) and (4.37) in the frequency domain [57]. *R* is the thermal resistance, and τ is the thermal time constant. The parameters used in this model are direct or extracted from manufacturer data.

$$Z_{jjc}(s) = \frac{R_{jj1}}{1 + s\tau_{jj1}} + \frac{R_{jj2}}{1 + s\tau_{jj2}} + \cdots + \frac{R_{jjn}}{1 + s\tau_{jjn}} \tag{4.36}$$

$$Z_{djc}(s) = \frac{R_{dj1}}{1 + s\tau_{dj1}} + \frac{R_{dj2}}{1 + s\tau_{dj2}} + \cdots + \frac{R_{djn}}{1 + s\tau_{djn}} \tag{4.37}$$

$$Z_{ca}(s) = \frac{R_{ch}}{1 + s\tau_{ch}} + \frac{R_{ha}}{1 + s\tau_{ha}} \tag{4.38}$$

4.2.2 Simulations and Discussions

The simulations of wind generation systems are done for the wind speed range from 6 m/s to 11 m/s in which the wind energy density is the best for modern wind turbines. Different switching frequencies are also studied. The junction temperature limit for both Si and SiC systems is assumed to be 25°C for Case A (switching frequency of 3 kHz) and Case B (switching frequency varies from 1 to 50 kHz). In order to do a fair comparison, the power losses of SiC MOSFET are scaled corresponding to the rated voltage ratio of the SiC MOSFET and Si IGBT (see Table 4.5). In other words, the on-state resistance and switching loss used in the model assumes that 1,700 V devices were used for the SiC MOSFETs and Si IGBTs.

4.2.2.1 At Switching Frequency of 3 kHz

Currently, most commercial wind turbine converters are switched at frequencies around 3 kHz. The generator is designed to work at rated power at the wind speed of 11 m/s. By the simulation, from 6 to 11 m/s, the converter efficiency (including

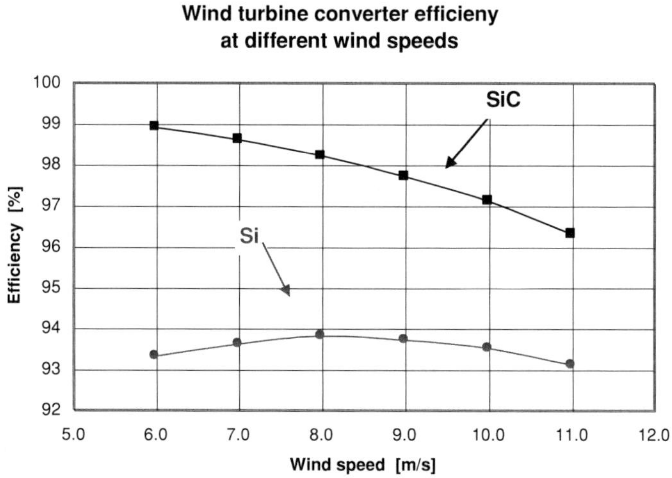

Fig. 4.14 Efficiency of SiC and Si wind turbine converters at different wind speeds and 3 kHz switching frequency

the rectifier, inverter, and the power loss of the filter shown in Fig. 4.8) is shown in Fig. 4.14. At the whole speed range, the efficiency of the Si converter is lower than that of the SiC converter. More specifically, the average efficiency of the SiC converter is 97.8% compared to 93.5% of the Si converter. If power loss saving of a SiC device compared to a Si device is defined by the difference of the power losses in the two devices over the power loss of the Si device, the average power loss saved by the SiC converter in this application is about 58.4%. At the worst case (wind speed 11 m/s), it is about 43.2 kW.

The average power loss of each component in the two wind power generation systems is shown in Fig. 4.15. The efficiencies of the generators in the two systems are the same (96.4% on average) whether the Si or the SiC converter is used. However, for the Si based system, the converter loss accounts for the most loss, which is as large as 54.3% (compared to 32.9% of the SiC-based system). Thus, it is necessary to reduce the loss in the converter in order to improve the generation system efficiency. The SiC converter is a good alternative.

4.2.2.2 At Frequency upto 50 kHz

The power loss of the filter also accounts for a substantial portion in the total system loss for both converters (12.1% for Si vs. 18.3% for the SiC-based system in Fig. 4.15). The size and loss of the filter is related to the switching frequency of the inverter. Increasing the switching frequency of the inverter can reduce the size and loss of the filter, and thus the cost. Because high switching frequency capability is one of the merits of SiC devices, an analysis was made to consider the option of increasing the switching frequency of the converter in this application.

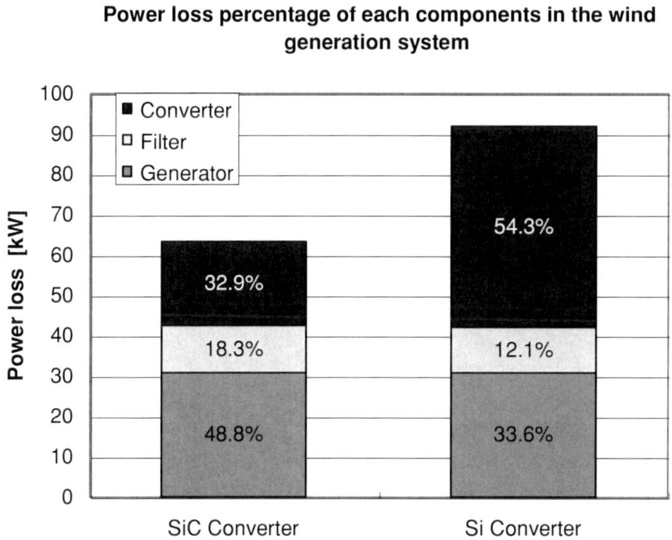

Fig. 4.15 Average power losses of SiC and Si wind turbine systems operating at full power rating and 3 kHz switching frequency

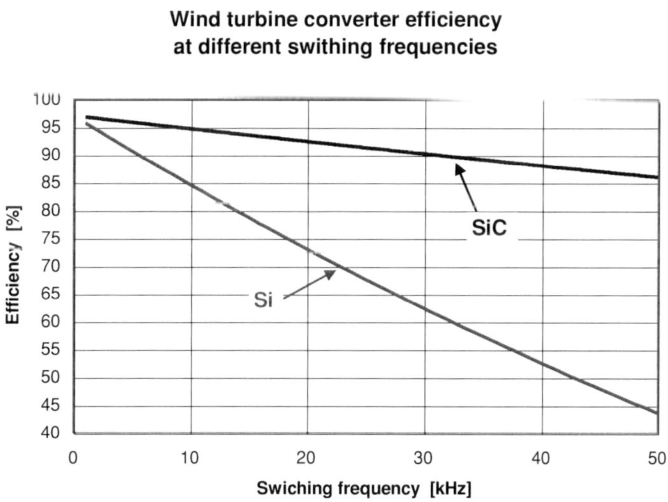

Fig. 4.16 Efficiency of SiC and Si wind turbine converters at full power rating and different switching frequencies

As shown in Fig. 4.16, as frequency increases from 1 to 50 kHz, the efficiency of the SiC converter (not including filter loss) at rated power and 25°C ambient linearly decreases at a rate of 1.1% per 5 kHz, and that of the Si converter decreases much more quickly at a rate of about 4.9% per 5 kHz.

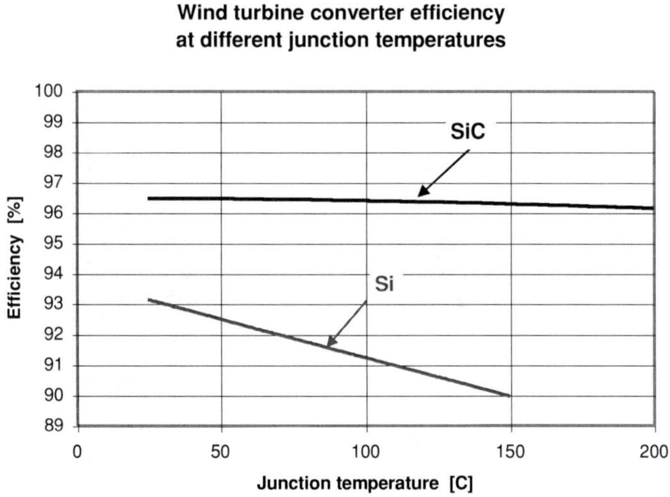

Fig. 4.17 Efficiency of SiC and Si wind turbine converters at full power rating and different temperatures

In practice, the switching frequency of a large power electronic Si IGBT cannot exceed a few kHz because of the large amount of loss. For this case, the efficiency of the Si converter at 20 kHz is 73.1%, which is not acceptable. While for the SiC converter, it has a relatively high efficiency of 85.9% even at 50 kHz. Thus, it is possible to improve the efficiency and reduce the cost of the system at the same time by using the SiC converter.

For example, increasing the switching frequency of the SiC converter to 6 kHz, its efficiency is 95.6%, which is the same as the efficiency of the Si converter switching at only 3 kHz, and at the same time, the size, loss, and cost of the filter are reduced for the SiC-based converter because of the 2 × higher switching frequency.

4.2.2.3 High Temperature Capability of the SiC Converter

When operating at full power rating and using a switching frequency of 3 kHz, the efficiency of the SiC converter decreases by only 0.2% from 25 to 150°C as shown in Fig. 4.17. However, the efficiency of the Si converter is lowered by 3.2% from 25 to 150°C (see Fig. 4.17). Moreover, the efficiency of the SiC converter is higher for the tested temperature range, and the efficiency difference is more at higher temperature. For example, the efficiency of the SiC converter at 6.3% points higher than that of the Si converter at 150°C.

Therefore, the cooling requirement of the SiC converter can be less than that of the Si converter even with the same junction temperature limit. For example, the junction temperature limit for all devices is chosen to be 150°C. As calculated by

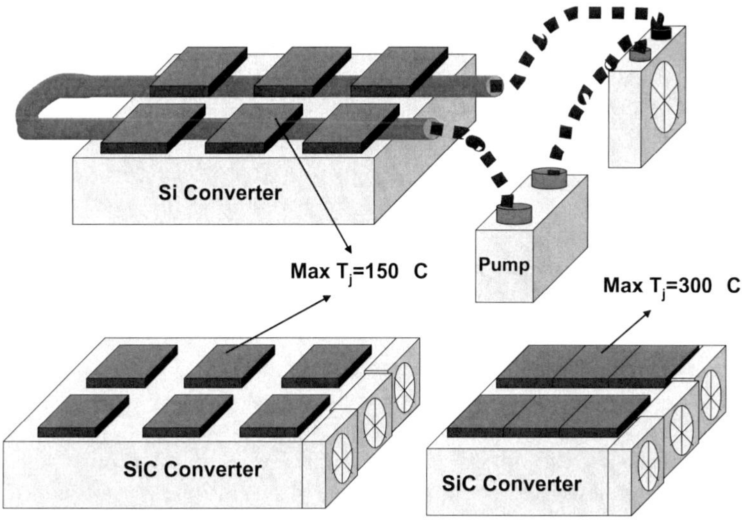

Fig. 4.18 Cooling system of Si and SiC converters

the simulation, the thermal resistance of the heatsinks required by the SiC converter is 0.11 K/W assuming that all the devices of the 10 SiC back-to-back converters are evenly distributed on two heatsinks with the same thermal resistance value. The value of 0.11 K/W can be realized by a natural or forced convection heatsink. To illustrate the size of heatsink, a commercial heatsink product MF18-1515 from Conrad Engineering [58] is selected but it may not be the best. The required thermal resistance can be achieved by forced cooling at an airflow rate of 27 cfm (12.7 l/s). Then the approximate volume of the heatsinks not including cooling fans is 23,480 cm^3.

Similarly, the thermal resistance of the heatsinks required by the Si converter is 0.0035 K/W assuming that all the devices of the 2 Si back-to-back converters are evenly distributed on two heatsinks with the same thermal resistance value. Liquid cooling is needed to achieve such low values. If Hi-Contact liquid cold plates from Aavid Thermalloy, LLC [59] are used, the volume of the heatsinks will be about 24,278 cm^3 (and this does not include any accessories such as pumps, tubing, radiator, etc.). If the space occupied by the accessories is considered, the total volume of the Si converter would be much larger than that of the SiC converter. Besides, the cooling management of the SiC converter is much simpler as shown in Fig. 4.18 and less expensive. Based on the current market prices for the two kinds of heaksinks used here, the cost of the heatsinks for the SiC converter is about 1/8 of that of the Si converter.

Furthermore, SiC devices can work at higher temperatures (at least 300°C junction temperature) with proper packaging. This can further reduce the size of heatsinks. By the simulation, the required thermal resistance of the SiC converter can be increased to 0.18 K/W under the same assumptions. If using the same series of heatsink products and force cooling conditions, the thermal result values can be realized by MF18-75, which has shorter length compared to MF18-1515. The approximate volume of the heatsinks not including cooling fans will be 11,624 cm^3 (about 49.5% of that with 150°C temperature limit). The SiC converter with a smaller heatsink is also drawn in proportion to the other designs in Fig. 4.18.

4.3 Conclusions and Future work

The full-scale converter concept is gaining more attention due to the possibility to simplify the system design and control, and effectiveness in dealing with grid-related problems. But the application of full-scale converters is limited by the relatively high cost and losses, also the availability of high-power and high-efficiency power devices. The simulations in Sect. 4.2 lead to a conclusion that the application of a SiC converter in the wind generation system will improve the system efficiency, provide more output power, and reduce system size and cost due to the low-loss, high-frequency, and high-temperature properties of SiC devices even for one-for-one replacement. Furthermore, since SiC devices inherently have high power capability and power density, this technology could promote the use of full-scale converters when it is mature.

Applications of SiC technology in wind turbine systems may result in the evolution of wind turbine system design, because more benefits can be obtained by elevating the rated voltage of the system in order to take advantage of the high voltage capability of SiC devices. Medium voltage converter technology is believed to have benefits such as less requirements on current handling capability and so less system components, as well as improved reliability.

For any of the above benefits to appear in wind generation systems, however, will require that manufacturers of SiC switching devices are able to produce the devices with sufficient power rating and quantities at costs that can show an overall system cost savings in installed cost and/or operating costs of the wind turbines.

References

1. DOE (2020) 2009 renewable energy data book. [Online] Available: http://www1.eere.energy.gov/maps_data/pdfs/eere_databook.pdf
2. Echenique E, Dixon J, Cardenas R, Pena R (2009) Sensorless control for a switched reluctance wind generator, based on current slopes and neural network. IEEE Trans Ind Electron 56(3):817–825
3. Ni B, Sourkounis C (2008) Investigations on control methods for variable speed wind energy converters at strongly fluctuating wind power. In Proceedings IEEE Ind. Electron. Conference: pp 24 1–246
4. Mirecki A, Roboam X, Richardeau F (2007) Architecture complexity and energy efficiency of small wind turbines. IEEE Trans Ind Electron 54(1):660–670

5. Abo-Khalil AG, Lee DC (2008) MPPT control of wind generation systems based on estimated wind speed using SVR. IEEE Trans Ind Electron 55(3):1489–1490
6. Cimuca GO, Saudemont C, Robyns B, Radulescu MM (2006) Control and performance evaluation of a flywheel energy-storage system associated to a variable-speed wind generator. IEEE Trans Ind Electron 53(4):1074–1085
7. Li H, Steurer M, Shi KL, Woodruff S, Zhang D (2006) Development of a unified design, test, and research platform for wind energy systems based on hardware-in-the-loop real-time simulation. IEEE Trans Ind Electron 53(4):1144–1151
8. Karimi S, Gaillard A, Poure P, Saadate S (2008) FPGA-based real-time power converter failure diagnosis for wind energy conversion systems. IEEE Trans Ind Electron 55(12):4299–4308
9. Portillo RC, Prats MM, Leon JI, Sanchez JA, Carasco JM, Galvan E, Franquelo LG (2006) Modeling strategy for back-to-back three-level converters applied to high-power wind turbines. IEEE Trans Ind Electron 53(5):1483–1491
10. Zhang S, Tseng KJ, Nguyen TD (2009) Modeling of AC–AC matrix converter for wind energy conversion system. In Proceedings. IEEE conference on industrial electronics and applications: pp. 184–191
11. Dessel MV, Deconinck G (2008) Power electronic grid connection of PM synchronous generator for wind turbines. In Proceedings. IEEE ind. electron. conf.: pp. 2200–2205
12. Chen Z, Guerrero M, Blaabjerg F (2009) A review of the state of the art of power electronics for wind turbines. IEEE Trans Power Electron 24(8):1859–1875
13. Carrasco JM, Franquelo LG, Bialasiewicz JT, Galvan E et al (2006) Power electric systems for the grid integration of renewable energy sources: a survey. IEEE Trans Ind. Electron 53(4):1002–1016
14. Casady JB (2008) SiC power device development for clean energy application [Online]. Available:http://www.sandia.gov/ess/Publications/Conferences/2008/PR08_Presentations/ritenour_semisouth.pdf
15. Wolk RH (2008) Proceedings of high Megawatt power converter technology R&D roadmap workshop.[Online].Available: http://www.nist.gov/eeel/high_megawatt/upload/Roadadmap workshopProceedingsFinal8-clean-al-1.pdf
16. Zhang H, Tolbert LM (2008) SiC's potential impact on the design of wind generation system. In Proceedings IEEE ind. electron. Conf.: pp. 2231–2235
17. Schugart P (2008) Progress in power converters enlarges wind turbine markets. Windtech Int 4(8):1–4
18. Baroudi JA, Dinavahi V, Knight AM (2007) A review of power converter topologies for wind generators. Renew Energy 32(14):2369–2385
19. Motto ER, Yamamoto M (1998) New high power semiconductors: high voltage IGBTs and GCTs. In proceedings power electronic conference, Santa Clara, CA: pp. 296–302
20. Wang X, Caiafa A, Hudgins J, Santi E (2003) Temperature effects on IGCT performance. In Proceedings IEEE industry applications society annual meeting: pp. 1006–1011
21. Neudeck P.G., Matus L.G., (1992) An overview of silicon carbide device technology. ninth symposium on space nuclear power systems, Albuquerque, New Mexico
22. Ericsen T.,(2002) Future navy application of wide bandgap power semiconductor devices. Proceedings of the IEEE, Vol. 90(6): pp. 1077–1082
23. Burke T, Xie K, Singh H, Podlesak T, Flemish J, Carter J, Schneider S, Zhao JH (1997) Silicon carbide thyristors for electric guns. IEEE Trans Magn 33(1):432–437 part 1
24. News, "SiC modules set to cut ship transformer size by half," Compound semiconductor, August 9,2005.[Online]Available:http://compoundsemiconductor.net/csc/news details.php?id = 22834&name = SiC%20modules%20set%20to%20cut%20ship%20transformer%20 size%20by%20half
25. Cheng L, Sankin I, Merrett JN, Bondarenko V, Kelley R, Purohit S, Koshka Y, Casady JRB, Casady JB, Mazzola MS (2005) Cryogenic and high temperature performance of 4H-SiC vertical junction field effect transistors (VJFETs) for space applications. Proceedings of the 17 international symposium on power semiconductor devices & IC's, Santa Barbara, CA

26. Tolbert L.M., Ozpineci B., Islam S.K., Chinthavali M., (2003) Wide bandgap semiconductors for utility applications. IASTED International Conference on Power and Energy Systems (PES 2003), Palm Springs, California: pp. 317–321
27. Chang HR, Hanna E, Zhang Q, Gomez M (2004) 1500 V and 10 A SiC motor drive inverter module. Proceedings of the 16th International Symposium on Power Semiconductor Devices and ICs, ISPSD 04: pp. 351–354
28. Kelley R, Mazzola MS, Bondarenko V (2006) A scalable SiC device for DC/DC converters in future hybrid electric vehicles. Twenty-First Annual IEEE Applied Power Electronics Conference and Exposition, APEC '06: pp. 460–463
29. Dreike PL, Fleetwood DM et al (1994) An overview of high-temperature electronic device technologies and potential applications. IEEE Trans Compon Packag Manuf Technol 17(4):594–609
30. Friedrichs P., Rupp R., (2006) Silicon carbide power devices—current developments and potential applications. Power electronics and applications, 2005 European Conference on
31. Wang C, Xu M, Lee FC, Luo Z (2008) Light load efficiency improvement for multi-channel PFC. In Proc. IEEE Power Electron. Specialists Conf.: pp. 4080–4085
32. Zhang H., Tolbert L.M., Ozpineci B., (2006) System Modeling and Characterization of SiC Schottky Power Diodes. IEEE Workshop on Computers in Power Electronics: Troy, New York
33. Ozpineci B., Chinthavali M., Kashyap A., Tolbert L.M., Mantooth A., (2006) A 55 kW three-phase inverter with Si IGBTs and SiC Schottky diodes. IEEE applied power electronics conference, Dallas, Texas: pp. 448–454
34. Friedrichs P., (2007) Silicon carbide power devices—status and upcoming challenges. European conference on power electronics and applications, Aalborg, Denmark: pp.1–11
35. Callanan R.J., Agarwal A., Burk A.,Das M., Hull B., Husna F., Powell A., Richmond J., Ryu S., Zhang Q., (2008) Recent progress in SiC DMOSFETs and JBS diodes at Cree. annual conference of IEEE on industrial electronics, Orlando, Florida: pp. 2885–2890
36. Sugawara Y, Takayama D, Asano K, Ryu S, Miyauchi A, Ogata S, Hayashi T (2003) 4H-SiC high power SIJFET module. in Proceedings IEEE 15th international symposium on power semiconductor devices and ICs: pp. 127–130
37. Salem TE, Urciuoli DP, Green R, Ovrebo GK (2009) High-temperature high-power operation of a 100 A SiC DMOSFET module. in Proc. IEEE applied power electronics conference and exposition: pp. 653–657
38. Ortiz-Rodriguez JM, Hernandez-Mora M, Duong TH, Leslie SG, Hefner AR, (2008) Thermal network component models for 10 kV SiC power module packages. In Proceedings IEEE power electron. Specialists Conf.: pp. 4770–4775
39. Durham NC (2009) Cree and Powerex develop new SiC power switches for next generation military systems. Press release of Cree, Inc., [Online]. Available: http://www.cree.com/press/press_detail.asp?i=1234879464387
40. Nezu T (2009) Rohm Exhibits New SiC Power Module. Tech-on Newsletter, [Online]. Available: http://techon.nikkeibp.co.jp/english/NEWS_EN/20091016/176491/
41. Richmond J., Leslie S., Hull B., Das M., Agarwal A., Palmour J., (2009) Roadmap for megawatt class power switch modules utilizing large area silicon carbide MOSFETs and JBS diodes. IEEE energy conversion congress and exposition, San Jose, California: pp 106–111
42. Maibach P, Faulstich A, Eichler M, Dewar S (2009) Full-scale medium-voltage converters for wind power generators upto 7 MVA. [Online] Available: http://www05.abb.com/global/scot/scot232.nsf/veritydisplay/984712acf892432c125740f003c3d65/
43. "SiC power modules—the world's first 250c operating power module with SiC devices," Sandia National Laboratory, R&D 100 winner 2009, [Online] : http://www.sandia.gov/mission/ste/r&d100/…/SIC_Power_Module.pdf
44. Bywaters G, John V, Lynch J, Mattila P, Norton G, Stowell J, Salata M, Labath O, Chertok A, Hablanian D (2005) Northern power systems WindPACT drive train. Subcontractor Report, NREL/SR-500-35524, [Online]. Available: http://www.nrel.gov/wind/pdfs/35524.pdf

45. Marckx DA (2005) Breakthrough in power electronics from SiC. Subcontractor Report, NREL/SR-500-38515, [Online]. Available: http://www.nrel.gov/wind/pdfs/38515.pdf
46. Zhang H., Tolbert L.M., Ozpineci B., Chinthavali M., (2005) Power losses and thermal modeling of 4H-SiC VJFET inverter. IEEE industry applications society annual meeting, Hong Kong, China: pp. 2630–2634
47. Zhang H, Tolbert LM, Ozpineci B (2006) System modeling and characterization of SiC Schottky power diodes. IEEE workshop on computers in power electronics, Troy, New York: pp. 199–204
48. Lai JS, Young RW, Ott GW, McKeever JW (1995) Efficiency modeling and evaluation of a resonant snubber based soft-switching inverter for motor drive applications. In Proceedings IEEE power electron. specialists conf.: pp. 943–949
49. Bierhoff MH, Fuchs FW (2004) Semiconductor losses in voltage source and current source IGBT converters based on analytical derivation. In Proceedings IEEE power electron. specialists conf.: pp. 2836–2842
50. Blaabjerg F, Jaeger U, Munk-Nielsen S (1995) Power losses in PWM-VSI inverter using NPT or PT IGBT devices. IEEE Trans Power Electron 10(3):358–367
51. Ozpineci B, Tolbert LM, Islam SK, Hasanuzzaman M (2001) Effects of silicon carbide (SiC) power devices on PWM inverter losses. In Proceedings IEEE ind. electron. conf.: pp. 1061–1066
52. Zhang H (2007) Electro-thermal modeling of SiC power electronic systems. Ph.D. Dissertation, The University of Tennessee
53. Zhang H, Tolbert LM, Ozpineci B, Chinthavali M (2006) A SiC-based converter as a utility interface for a battery system. in Proceedings IEEE industry applications society annual meeting: pp. 346–350
54. Datasheet of silicon carbide Schottky diode-C2D10120. [Online]. Available: http://www.cree.com/products/pdf/C2D10120.pdf
55. Datasheet of silicon IGBT module—DIM1600FSM17. [Online]. Available: http://www.dynexsemi.com/assets/IGBT_Modules/Datasheets/DNX_DIM1600FSM17-A000.pdf
56. Hopkins T., Cognetti C., Tiziani R., (1988) Designing with thermal impedance. Fourth Annual IEEE semiconductor thermal and temperature measurement symposium, San Diego, CA: pp. 55–61
57. Blasko V., Lukaszewski R., Sladky R., (1999) On line thermal model and thermal management strategy of a three phase voltage source inverter. IEEE industry application society annual meeting, Phoenix, Arizona: pp.1423–1431
58. Conrad Heatsink 2005 catalog, http://www.conradheatsinks.com
59. Manufaturer page of Hi-Contact liquid cold plates from Aavid Thermalloy, LLC. [Online]. Available: http://www.aavidthermalloy.com/products/liquid/hi-contact.pdf

Chapter 5
A New Interconnecting Method for Wind Turbine/Generators in a Wind Farm

Shoji Nishikata and Fujio Tatsuta

Abstract In this chapter a new interconnecting method for a cluster of wind turbine/generators is discussed, and some examples of the basic characteristics of the integrated system are shown. This method can be achieved with a wind turbine generating system using a shaft generator system. A group of wind turbine generators can be interconnected easily with the proposed method, and high reliability and electric output power with high quality are also expected. Moreover, since this method enables transmission of the generated power through a long-distance DC transmission line, the optimum site for wind turbines can be selected so as to acquire the maximum wind energy.

5.1 Introduction

There is a need to construct a large-scale wind farm for generating a large amount of electric power by kinetic energy of wind, and such a site should be far distant from urban areas or off shore. In addition, the interconnecting method for the wind turbine/generators is essential in forming the wind farms.

We already proposed a DC-link type wind turbine generating system using a shaft generator system [1, 2], which is widely used for power sources in a ship [3], and revealed the usefulness of the system. In this type of generating system, a large-scale smoothing capacitor is unnecessary because current-source inverter is

S. Nishikata (✉)
Department of Electrical and Electronic Engineering, Tokyo Denki University,
2-2 Kanda-Nishikicho, Chiyoda-ku, Tokyo 101-8457, Japan
e-mail: west@cck.dendai.ac.jp

F. Tatsuta
Department of Information Systems and Multimedia Design, Tokyo Denki University,
2-2 Kanda-Nishikicho, Chiyoda-ku, Tokyo 101-8457, Japan

S. M. Muyeen (ed.), *Wind Energy Conversion Systems*,
Green Energy and Technology, DOI: 10.1007/978-1-4471-2201-2_5,

adopted. Moreover, since this type of inverter uses, in general, thyristors as switching devices, the system scale can be easily enlarged, and system reliability is improved greatly when compared with insulated-gate bipolar transistors in the voltage-source inverters. Furthermore, as in [1, 2], output electricity with high quality is obtained with this generating system.

On the basis of the aforementioned system, a new interconnecting method for a group of wind turbine/generators was proposed in [4]. In this method, the outputs of each ac generator coupled with the wind turbine are rectified, and these rectified outputs are connected in series and integrated in the DC link. The resultant output DC power is converted again to ac power with the thyristor inverter. Thus, only one inverter is enough, making the system very simple, and the inverter can be placed everywhere without restraint. Thereby, we can select the optimum site for wind turbine/generators with a long-distance DC transmission line.

In addition to the aforementioned advantages, the proposed system not only has an ability of standalone operation without the existence of utility but also it cooperates with the utility. Since this system is provided with a synchronous compensator on the output side as an essential component, the low voltage ride-through characteristic of the system is improved greatly when compared with the conventional system.

In this chapter, the basic idea of control method for the proposed system given in [4] is introduced. Then, simulated steady-state waveforms of the voltages and currents in the system with two wind turbine/generators are shown, and some examples of system dynamic performances are discussed in some detail as well.

5.2 Basic Equations of the System

Let us first show the basic equations for a wind turbine/generator which are necessary for the following discussion. As a wind turbine generator, a permanent magnet synchronous generator (PMSG) is used here. In Fig. 5.1, mechanical energy is acquired from kinetic energy possessed by wind through a wind turbine, and the PMSG converts it to electrical energy. The output of PMSG is converted to DC power through a thyristor rectifier. The output power of the wind turbine P_t, which is equal to converted DC power if the losses in the generator and rectifier are neglected, is given by

$$P_t = \frac{1}{2} C_p(\lambda, \beta) \rho A_W V_{wind}^3 = V_d I_d \tag{5.1}$$

where, ρ, A_w, V_{wind}, V_d, and I_d are air density, rotor swept area, wind velocity flowing into the wind turbine, output voltage, and current of the rectifier, respectively. In addition, C_p is the performance coefficient which is the conversion ratio of turbine output to wind power [5].

Here, it should be noted that C_p is expressed as a function of tip speed ratio λ and blade pitch angle β. The tip speed ratio λ is defined as

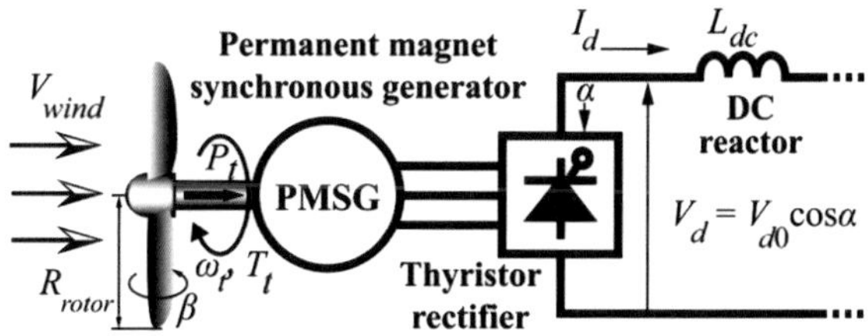

Fig. 5.1 Wind turbine generator and thyristor rectifier

$$\lambda = \frac{\omega_t R_{rotor}}{V_{wind}} \tag{5.2}$$

where, ω_t is the angular velocity of the wind turbine and R_{rotor} is blade radius.

In general, the wind turbine should be driven to keep λ a constant value to obtain C_p as large as possible. Thus, λ and C_p are assumed to be constant for a given pitch angle β in this discussion for the sake of simplicity. In this case, the angular velocity ω_t is proportional to the wind velocity V_{wind}. On the other hand, it should be recognized that the non-control output voltage of the rectifier V_{d0} is almost proportional to ω_t since PMSG is used as the wind turbine generator.

When the coefficient of proportion between V_{d0} and ω_t is assumed to be K_d, we have (5.3) for controlled output voltage of the rectifier.

$$V_d = V_{d0}\cos\alpha = K_d\omega_t\cos\alpha = \frac{K_d\lambda}{R_{rotor}}V_{wind}\cos\alpha \tag{5.3}$$

where, α is control angle of rectifier and $V_{d0} = K_d\ \omega_t$.

As to the torque of the wind turbine T_t, we obtain the following Eq. 5.1–5.3.

$$T_t = \frac{P_t}{\omega_t} = \frac{C_p\rho A_W R_{rotor} V_{wind}^2}{2\lambda} = I_d K_d \cos\alpha \tag{5.4}$$

From (5.4), it is recognized that the wind turbine torque T_t and hence, the operating point of the system can be controlled with the control angle α. The increase in α, however, results in an increase in reactive power of the system, and hence, α should be kept small to use the PMSG as effectively as possible.

5.3 System Configuration

Figure 5.2 shows the configuration of the proposed wind turbine generating system composed of two or more sets of wind turbine and PMSG. As in the figure, the output of each thyristor rectifier is connected in series, and the unified DC output is fed to the current-source thyristor inverter through a DC transmission line. It should be noticed that the whole system seems likely to come to a halt when one of the wind turbines is lost for some reason, including the lack of wind. However, we can prevent such a system shutdown by short-circuiting the corresponding

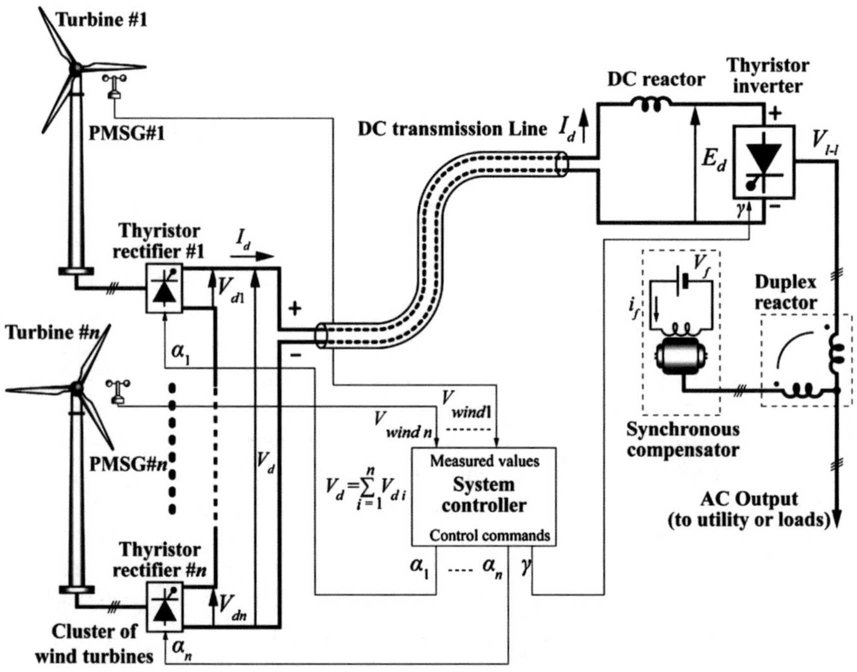

Fig. 5.2 Proposed interconnecting method for wind turbine/generators and system configuration

rectifier by means of giving signals to all the gates for the thyristors in the rectifier. Another means for protecting the system against the shutdown is to connect a diode in parallel with each thyristor rectifier in Fig. 5.2 [6]. That is, once one of the output voltages of the rectifiers is lost for some reason, the corresponding parallel-connected diode turns into ON-state immediately, continuing DC-link current I_d to flow and enabling the whole system to operate without interruption. We have to derive another set of equations introduced in this chapter for the system with the parallel diodes, and in the following, the system without parallel-connected diodes is discussed. Hence, we can say that system reliability is improved greatly with one of these ideas.

The synchronous compensator connected to the inverter through a duplex reactor provides reactive power needed for commutation of the inverter thyristors and that required in the ac output as well [1, 2]. Since the presence of this synchronous machine (compensator) allows the system output voltage to control, the proposed system can be operated as either an isolated generating system or an incorporated generating system into the utility. Hence, a good low-voltage ride-through characteristic can be obtained with this system [1, 2]. Moreover, when we drive the synchronous machine by a prime mover to generate active power as well, a new hybrid-type wind turbine generating system can be realized [7].

It should be also reminded that the output voltage distortion in the inverter caused by the commutation of thyristors is completely compensated with the well-designed duplex reactor so as to cancel the subtransient inductance of the synchronous compensator [3].

The system controller shown in the figure collects observed data on the wind velocities $V_{\text{wind }i}$ $(i = 1, 2,\ldots, n)$ at each wind turbine. In this controller, DC-link current I_{d} and then the leading angle for commutation of the inverter γ are calculated based on $V_{\text{wind }i}$ (i.e., total output power gained by the wind farm $\sum P_{\text{t }i}$), and the control angles of rectifier α_i $(i = 1, 2,\ldots, n)$ for the individual rectifier [output voltage $V_{\text{d }i}$ $(i = 1, 2,\ldots, n)$] are also determined.

When there exists only one wind turbine, it is the case that we reported in [1, 2]. In the following, we discuss how to control the system.

5.4 Operating Method for System Consisting of Arbitrary Number of Wind Turbine Generators

5.4.1 Loads Connected Through Thyristor Inverter

We discuss here the way of controlling α_i for individual thyristor rectifier when the system loads are connected in the DC link through the thyristor inverter as in Fig. 5.2.

For this case, let $V_{\text{wind }1} \sim V_{\text{wind }n}$ be the wind velocities flowing into #1 ~ #n wind turbines, respectively, and let $V_{\text{w max}}$ be the maximum value among these wind velocities. That is,

$$V_{\text{w max}} = \max(V_{\text{wind}\,1}, V_{\text{wind}\,2}, \ldots, V_{\text{wind}\,n}). \tag{5.5}$$

As previously mentioned, the control angle for the rectifiers α should be controlled as small as possible in order to reduce the reactive power of the system. Consequently, we set the control angle for the rectifier at which wind velocity is $V_{\text{w max}}$ as 0. Then, the output power and output DC voltage for this wind turbine become maximum as

$$P_{\text{t max}} = \frac{1}{2} C_{\text{p}} \rho A_{\text{w}} V_{\text{w max}}^3 \tag{5.6}$$

$$V_{\text{d max}} = \frac{K_{\text{d}}\lambda}{R_{\text{rotor}}} V_{\text{w max}}. \tag{5.7}$$

As a result, the DC-link current turns into

$$I_{\text{d}} = \frac{P_{\text{t max}}}{V_{\text{d max}}} = \frac{C_{\text{p}} \rho A_{\text{w}} R_{\text{rotor}} V_{\text{w max}}^2}{2 K_{\text{d}} \lambda}. \tag{5.8}$$

Aside from the wind turbine with the maximum wind velocity, the output power $P_{\text{t}i}$ and output DC voltage $V_{\text{d}i}$ for #i wind turbine are given as

$$P_{ti} = \frac{1}{2} C_p \rho A_w V_{\mathrm{wind}\,i}^3 = V_{di} I_d \tag{5.9}$$

$$V_{di} = \frac{K_d \lambda}{R_{\mathrm{rotor}}} V_{\mathrm{wind}\,i} \cos \alpha_i. \tag{5.10}$$

Since the DC-link current I_d is the same for all the rectifiers because of DC link, the following relationships are obtained:

$$\frac{P_{ti}}{P_{t\max}} = \frac{V_{di}}{V_{d\max}} = \frac{V_{\mathrm{wind}\,i}^3}{V_{w\max}^3} = \frac{V_{\mathrm{wind}\,i} \cos \alpha_i}{V_{w\max}}. \tag{5.11}$$

Hence, control angle α_i for rectifier #i should be controlled as

$$\begin{aligned} \cos \alpha_i &= \frac{V_{\mathrm{wind}\,i}^2}{V_{w\max}^2} \\ \alpha_i &= \cos^{-1} \frac{V_{\mathrm{wind}\,i}^2}{V_{w\max}^2} \end{aligned}. \tag{5.12}$$

Based on the control strategy introduced here, we can acquire the maximum wind power from the whole system, in which the individual wind turbine can be operated most effectively.

The total DC-link voltage V_d and the total output power P_{tTotal}, which are the input voltage and input power, respectively, to the inverter, become

$$V_d = \frac{K_d \lambda}{R_{\mathrm{rotor}}} \sum_{j=1}^{n} \left(V_{\mathrm{wind}\,j} \cos \alpha_j \right) \tag{5.13}$$

$$P_{\mathrm{tTotal}} = \frac{1}{2} C_p \rho A_w \sum_{j=1}^{n} \left(V_{\mathrm{wind}\,j}^3 \right). \tag{5.14}$$

It is recognized here that the DC-link current I_d given in (5.8) should be governed by controlling the leading angle of commutation for inverter thyristors γ.

If the angle of overlap in the inverter is neglected, the equation for inverter DC side voltage E_d is given as follows:

$$E_d = V_d - R_d I_d = \frac{3\sqrt{2}}{\pi} V_{l-l} \cos \gamma \tag{5.15}$$

where, R_d is total resistance in the DC link and V_{l-l} is the line-to-line rms voltage of the inverter output.

Hence, we have the equation for γ as

$$\gamma = \cos^{-1} \left\{ \frac{\pi (V_d - I_d R_d)}{3\sqrt{2} V_{l-l}} \right\}. \tag{5.16}$$

It is also noted that γ should be controlled so that the margin angle for commutation $\gamma - u > 0$ (u: overlapping angle of commutation) to secure a stable operation of the inverter.

5.4.2 Resistive Load Connected in DC Link

As another primitive investigation, we discuss the method of controlling the output voltage for each rectifier for the case of a load of constant resistance connected in the DC link instead of the thyristor inverter shown in Fig. 5.2.

The total power derived from all the wind turbines P_{tTotal} is given with (5.14), and the DC-link voltage V_{d} applied to the load resistance turns into

$$V_{\mathrm{d}} = \sqrt{R_{\mathrm{L}} \cdot P_{\mathrm{tTotal}}} = \sqrt{\frac{1}{2} R_{\mathrm{L}} C_{\mathrm{p}} \rho A_{\mathrm{w}} \sum_{j=1}^{n} \left(V_{\mathrm{wind}j}^{3} \right)}. \tag{5.17}$$

Since the same DC current I_{d} flows in all of the rectifiers because of DC link, the output DC voltage for the individual rectifiers has to be controlled, depending on the wind conditions for the turbines in order to provide a stable operation of the system. That is, the control angles should be determined based on the relationships between the power obtained by the wind and the consumption power.

Hence, the contribution of each wind turbine to the total power is assigned as

$$\frac{P_{\mathrm{t}i}}{P_{\mathrm{tTotal}}} = \frac{V_{\mathrm{wind}\,i}^{3}}{\sum_{j=1}^{n} \left(V_{\mathrm{wind}j}^{3} \right)}. \tag{5.18}$$

Thus, the output voltage for #i wind turbine rectifier should be

$$\begin{aligned} V_{\mathrm{d}i} &= V_{\mathrm{d}} \frac{V_{\mathrm{wind}\,i}^{3}}{\sum_{j=1}^{n} \left(V_{\mathrm{wind}j}^{3} \right)} = \sqrt{\frac{R_{\mathrm{L}} C_{\mathrm{p}} \rho A_{\mathrm{w}}}{2 \sum_{j=1}^{n} \left(V_{\mathrm{wind}\,j}^{3} \right)}} \, V_{\mathrm{wind}\,i}^{3} \\ &= \frac{K_{\mathrm{d}} \lambda}{R_{\mathrm{rotor}}} V_{\mathrm{wind}\,i} \cos \alpha_{i}. \end{aligned} \tag{5.19}$$

As a result, the control angle for #i rectifier can be calculated as

$$\alpha_{i} = \cos^{-1} \left(\frac{R_{\mathrm{rotor}} \cdot V_{\mathrm{wind}\,i}^{2}}{K_{\mathrm{d}} \lambda} \cdot \sqrt{\frac{R_{\mathrm{L}} C_{\mathrm{p}} \rho A_{\mathrm{w}}}{2 \sum_{j=1}^{n} \left(V_{\mathrm{wind}j}^{3} \right)}} \right). \tag{5.20}$$

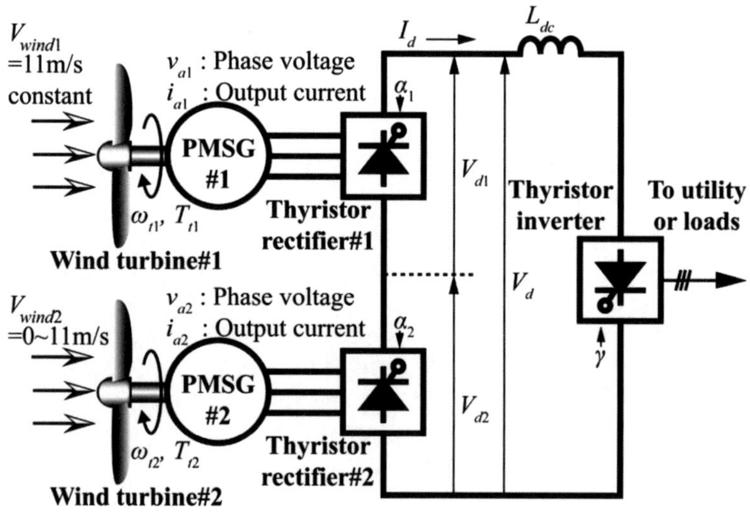

Fig. 5.3 Two wind turbine/generators connecting an inverter load

5.5 Basic Characteristics for Case of Two Wind Turbine Generators

On the basis of the system equations derived earlier, we explore here the system characteristics for the case of two sets of wind turbine/generator as a basic investigation.

In this case, $n = 2$, and the system is given by Fig. 5.3 for the inverter load. The whole load is shared between two generators, and the control angles of each rectifier should be properly controlled depending on the wind velocity. It should be remembered here that the output voltage and frequency of the inverter can be kept constant by controlling the field current of the synchronous compensator shown in Fig. 5.2 as well as the leading angle of commutation of inverter for the case when the system is used in the standalone operation [1, 2], while the voltage and frequency of the system when connected with the utility depend solely on those of the grid.

When $V_{\text{wind 1}} > V_{\text{wind 2}}$, then α_1 is set to be zero according to foregoing discussion and α_2 is calculated with (5.12). As an example of steady-state characteristics for the system shown in Fig. 5.3, those for "SUBARU 15/40" wind turbine/generator [8], which was constructed in our university [2], are investigated here. A photograph of the wind turbine is shown in Fig. 5.4, and the parameters used in the calculation are given in Table 5.1 [2].

Figures 5.5 and 5.6 show examples of calculated results for the cases when α_1 is fixed to be zero and $V_{\text{wind 1}} = 11$ m/s = constant; meanwhile, $V_{\text{wind 2}}$ changes in the range of 0 – 11 m/s.

Fig. 5.4 SUBARU15/40 wind turbine/generator

Table 5.1 Parameters used in the calculations (common in both #1 and #2 turbine/generator)

Blade radius	R_{rotor}	7.5 m
Tip speed ratio	λ	4.0 = constant
Performance coefficient	C_p	0.3 = constant
Air density	ρ	1.225 kg/m^3
Generator coefficient	K_d	40.0 V s/rad
Rated wind velocity	–	11 m/s
Rated output power	–	43 kW

In Fig. 5.5, the characteristics of angular velocities of the wind turbines ω_{t1}, ω_{t2}, and α_2 versus $V_{wind\ 2}$ are shown. It is clarified that, although ω_{t2} increases with $V_{wind\ 2}$, ω_{t1} is kept constant because of a constant tip speed ratio λ and that control angle α_2 decreases with an increase in $V_{wind\ 2}$ according to (5.12).

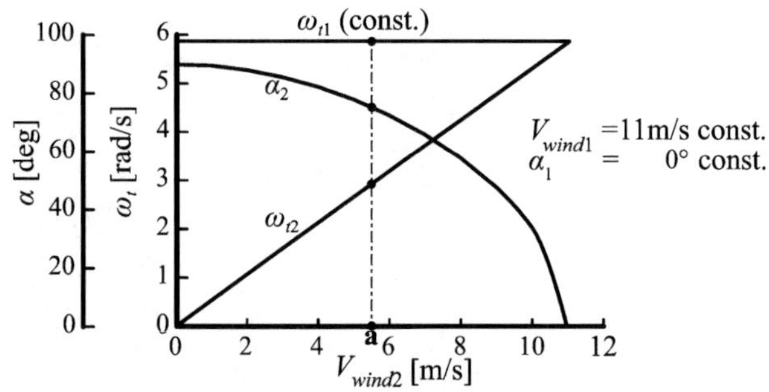

Fig. 5.5 Angular velocities of wind turbines ω_{t1}, ω_{t2} and control angle of rectifier α_2 versus $V_{wind\ 2}$

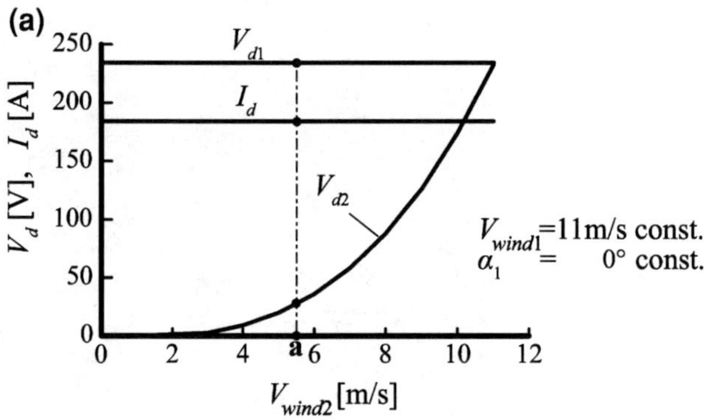

I_d, V_{d1}, V_{d2} versus V_{wind2}.

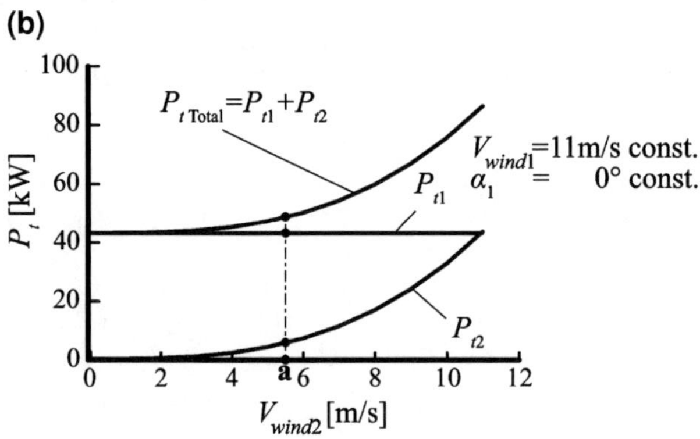

P_{t1}, P_{t2} versus V_{wind2},

Fig. 5.6 Characteristics of DC-link voltages and current, and output powers versus wind velocity at turbine #2 $V_{wind\ 2}$. **a** I_d, V_{d1}, V_{d2} versus $V_{wind\ 2}$. **b** P_{t1}, P_{t2} versus $V_{wind\ 2}$

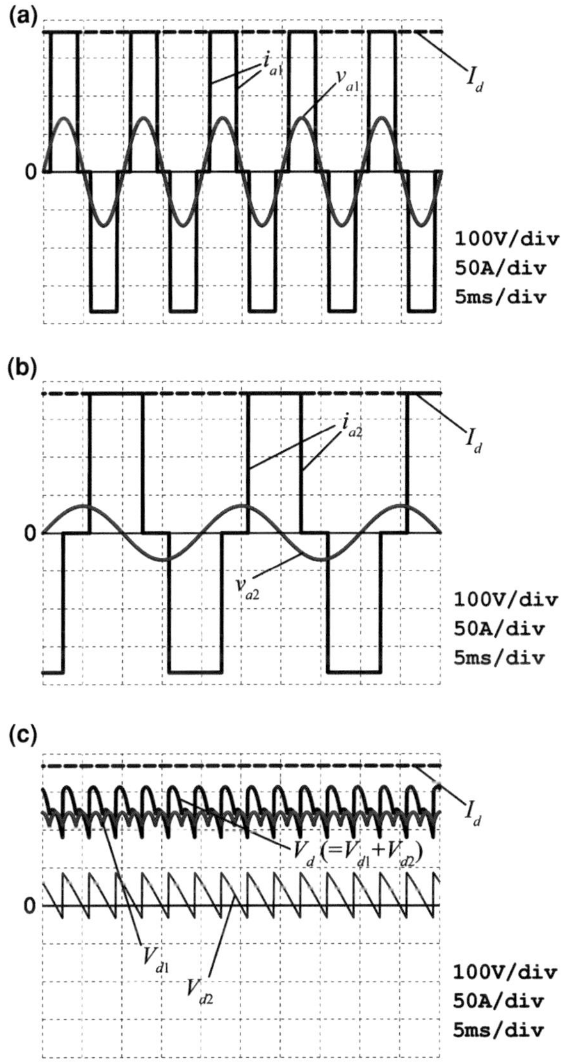

Fig. 5.7 Simulated waveforms of voltages and currents (At point **a** in Figs. 5.5, 5.6). **a** Phase voltage and output current (PMSG#1). **b** Phase voltage and output current (PMSG#2). **c** DC-link voltages and current

Figure 5.6 shows the characteristics of DC-link current I_d, DC voltages of the rectifiers V_{d1} and V_{d2} (in Fig. 5.6a), and system output P_{tTotal}, which is equal to $P_{t1} + P_{t2}$ (in Fig. 5.6b), when $V_{wind\ 2}$ changes. It can be seen from this figure that DC voltage V_{d2} and output power P_{t2} increase with an increase in $V_{wind\ 2}$, while V_{d1} and P_{t1}, as well as I_d, are kept to be constant independently of $V_{wind\ 2}$, since $V_{wind\ 1}$ and α_1 are constant in this case.

Examples of simulated waveforms of voltages and currents in PMSG outputs and DC link are shown in Fig. 5.7 for the steady-state operation. The operating points are indicated with '**a**' in Figs. 5.5 and 5.6 and given in Table 5.2, in which

Table 5.2 Steady-state operating points used for simulation

Wind turbine/generators		
	PMSG#1	PMSG#2
Wind velocity (m/s)	11.0	5.5
Output power (kW)	43.2	5.4
Phase voltage (rms) (V)	100.3	50.2
Output current (rms) (A)	150.4	150.4
Control angle (deg)	0.0	75.5
DC link		
DC-link current I_d (A)		184.2
Output voltage of thyristor inverter #1 V_{d1} (V)		234.7
Output voltage of thyristor inverter #2 V_{d2} (V)		29.3
DC-link voltage V_d (V)		264.0

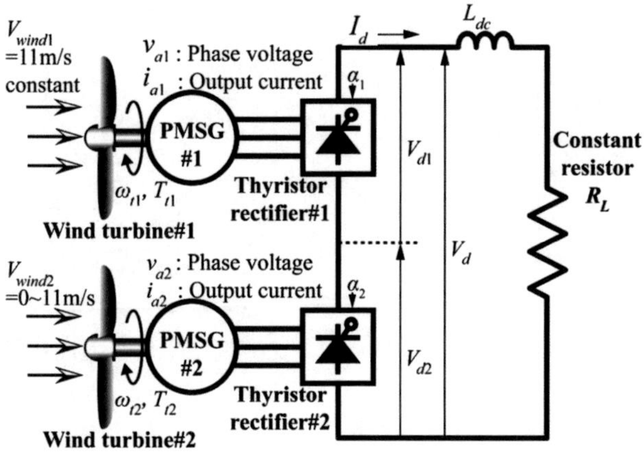

Fig. 5.8 Operation with the resistor load

the ratio of the wind velocities of each turbine is 2:1 (11:5.5 m/s). Because of constant tip speed ratio, the ratio of output phase voltages of PMSGs is also 2:1, as in Table 5.2. On the other hand, it should be recognized from the table that the ratio of output power of each PMSG becomes 8:1 because wind energy is proportional to cubic of the wind velocity.

It should be noticed that the rms values for output currents i_{a1} and i_{a2} are the same for the individual PMSG as shown in the table because the outputs of rectifiers are connected in series. It is also noted that there is a phase difference between phase voltage and output current for PMSG#2 (in Fig. 5.7b), whereas that for PMSG#1 is zero (in Fig. 5.7a). The phase difference in PMSG#2 is based on control angle α_2. As a result, the waveforms of the output DC voltages for individual rectifier V_{d1} and V_{d2}, and that of DC-link voltage V_d are shown as in Fig. 5.7c.

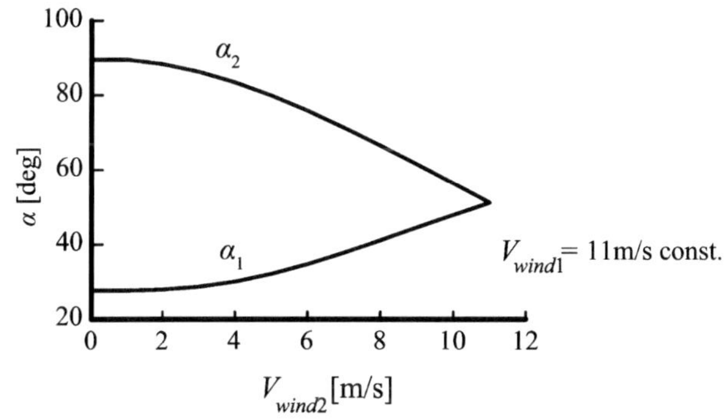

Fig. 5.9 Control angles α_1, α_2 versus $V_{\mathrm{wind}\ 2}$ for the case with constant load resistance

Next, the steady-state characteristics of the system are discussed for the case of a load of constant resistance connected in the DC link as in Fig. 5.8.

Based on (5.17)–(5.20) the characteristics of α_1 and α_2 can be calculated for the case of constant DC-link resistance load ($R_{\mathrm{L}} = 1\Omega$, in this case), and those of DC-link voltages and current can also be clarified.

Figure 5.9 shows the characteristics of α_1 and α_2 for a change in $V_{\mathrm{wind}\ 2}$ when $V_{\mathrm{wind}\ 1} = 11$ m/s = constant. In this calculation, the parameters shown in Table 5.1 are used again. It is noted that the control angles have to be controlled to realize a stable operation of the system.

Figure 5.10 shows steady-state characteristics of P_{tTotal}, V_{d1}, V_{d2}, V_{d} (=$V_{\mathrm{d1}} + V_{\mathrm{d2}}$), and I_{d} for the case when $V_{\mathrm{wind}\ 2}$ changes in the range of from 0 to 11 m/s. It is noticed that I_{d} is directly proportional to V_{d} because the load resistance is constant in this case, while for the system in Fig. 5.3, I_{d} is kept constant as in Fig. 5.6a.

5.6 Dynamic Performances

5.6.1 Dynamic Model of the System

Since the velocity of the natural wind fluctuates with respect to time, it is essential to investigate the system performances based on a dynamic model for the wind turbine generating system. In [9] we obtained such a model to predict the dynamic performances of the system for a single wind turbine. In Fig. 5.1, the torque equations of the wind turbine and generator (PMSG) are given by

$$\begin{cases} T_{\mathrm{t}} = J_{\mathrm{t}} \cdot \dfrac{\mathrm{d}\omega_{\mathrm{t}}}{\mathrm{d}t} + R_{\omega} \cdot \omega_{\mathrm{t}} + T_{\mathrm{g}} \\ T_{\mathrm{g}} = \dfrac{V_{\mathrm{d}} \cdot I_{\mathrm{d}}}{\omega_{\mathrm{t}}} = K_{\mathrm{d}} \cos\alpha \cdot I_{\mathrm{d}} \end{cases} \tag{5.21}$$

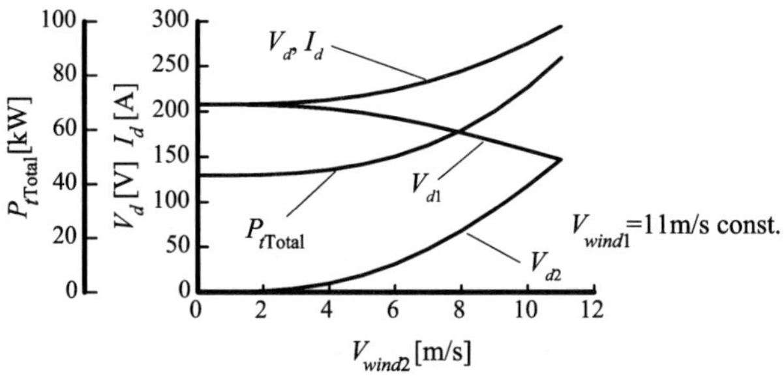

Fig. 5.10 Characteristics of output power P_{tTotal}, DC-link current I_{d}, and DC voltages V_{d} ($= V_{\mathrm{d1}} + V_{\mathrm{d2}}$), versus wind velocity $V_{\mathrm{wind\ 2}}$ for the case with constant load resistance $R_{\mathrm{L}} = 1\ \Omega$

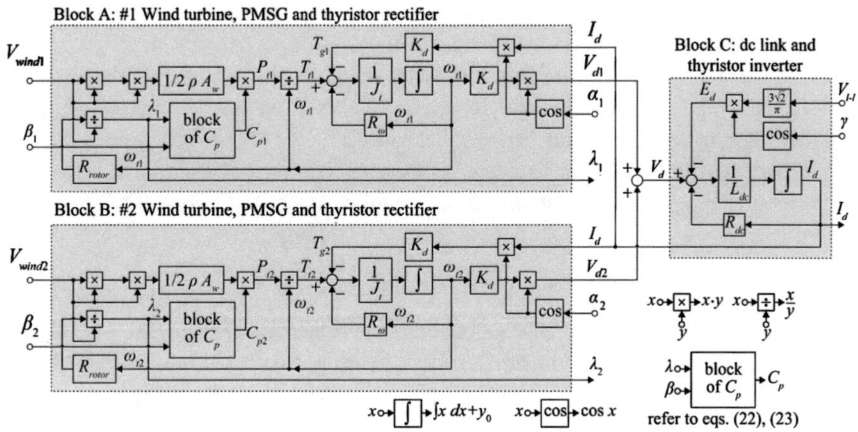

Fig. 5.11 Simulation block diagram for two wind turbine/generators and inverter

where, J_{t}, R_{ω} are the inertia moment and the braking friction coefficient of the mechanical system including wind turbine, and T_{g}, K_{d} are the load torque of the generator and the generator coefficient [see (5.3)], respectively.

The dynamic model for a set of the wind turbine generating system including rectifier can be derived from (5.1) to (5.4) and (5.21), and this model is shown in Fig. 5.11 as Block A or B. The whole dynamic model for the wind turbine generating system in Fig. 5.3, hence, is given by Fig. 5.11, where a model for DC link and thyristor inverter is included [10].

The performance coefficient C_{p} (block C_{p} in Fig. 5.11) is calculated based on (5.22) [5], which is a function of tip speed ratio of wind turbine λ and pitch angle of turbine blade β,

$$C_{\mathrm{p}} = c_1 \left(c_2 - c_3 \beta - c_4 \beta^2 - c_5 \right) \varepsilon^{-c_6} \tag{5.22}$$

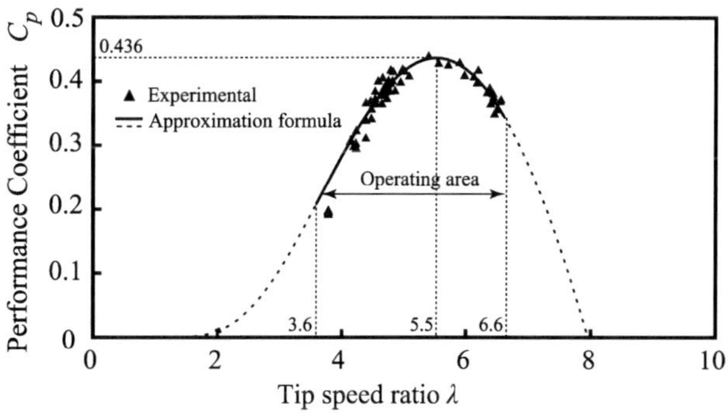

Fig. 5.12 C_p versus λ for the wind turbine

$$\left.\begin{aligned} c_1 = 0.749, \quad c_2 = 116/\lambda_i, \quad c_3 = 0.01 \\ c_4 = 0, \quad c_5 = 7.9, \quad c_6 = 21/\lambda_i \\ \frac{1}{\lambda_i} = \frac{1}{\lambda + 0.008\beta} - \frac{0.035}{\beta^3 + 1} \end{aligned}\right\} \tag{5.23}$$

It is noted that $c_1 \sim c_6$ in (5.22) should be determined experimentally. In Fig. 5.12, the relationships between performance coefficient C_p and tip speed ratio λ are shown for our wind turbine [2]. Plotted data were obtained through the experiments when $\beta = 0°$. Based on these data we decide $c_1 \sim c_6$ as (5.23), and the solid curve in Fig. 5.12 is used for the following discussion.

5.6.2 Control System for Constant Tip Speed Ratios

The tip speed ratios of the individual wind turbines should be kept constant as much as possible in order to obtain a large performance coefficient.

The block diagram of a closed-loop control system for constant tip speed ratio λ for each wind turbine/generators is shown in Fig. 5.13. Here, $G_{\gamma1}$, $G_{\alpha1}$, $G_{\gamma2}$, $G_{\alpha2}$ are the PI controllers.

There are two control systems shown in Fig. 5.13. These are applied depending on the wind condition. That is, Fig. 5.13 (a) is used for the case of $V_{\text{wind 1}} > V_{\text{wind 2}}$, while (b) for the case of $V_{\text{wind 1}} < V_{\text{wind 2}}$.

When $V_{\text{wind 1}} > V_{\text{wind 2}}$, for example, the control system (a) is applied. In this case, tip speed ratio λ_1 for #1 wind turbine is controlled to be constant with DC-link current I_d, which is controlled by leading angle of commutation γ_1 as in the figure, and simultaneously, tip speed ratio λ_2 for #2 turbine is controlled with the control angle α_2 of the rectifier #2. On the contrary, for the case when $V_{\text{wind 1}} < V_{\text{wind 2}}$, the control variables are replaced to obtain constant tip speed ratios as in (b).

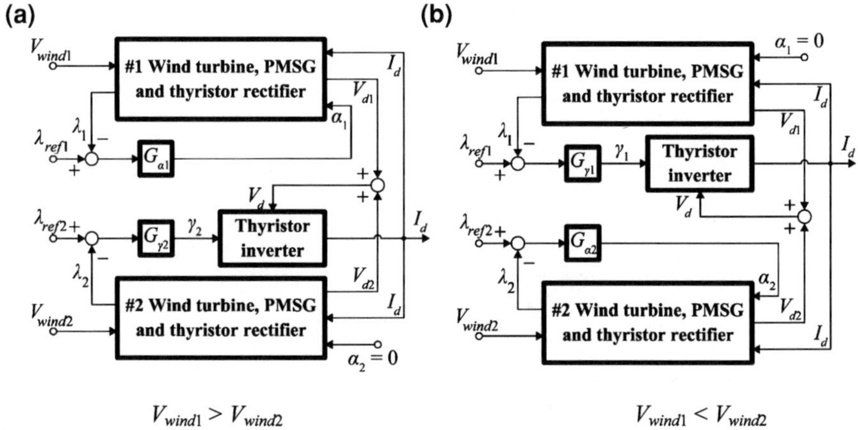

Fig. 5.13 Closed-loop control system for constant tip speed ratio (refer to Fig. 5.11) (λ_{ref1}, λ_{ref2}: Tip speed ratio references, $G_{\gamma 1}$, $G_{\alpha 1}$, $G_{\gamma 2}$, $G_{\alpha 2}$: PI controllers). **a** $V_{\text{wind 1}} > V_{\text{wind 2}}$, **b** $V_{\text{wind 1}} < V_{\text{wind 2}}$

5.6.3 Dynamic Responses when Wind Turbines are Driven by Natural Wind

Figure 5.14 shows an example of simulated dynamic responses of the system when wind velocities are changed as shown in (a), which are used as the inputs of the simulation. These wind data were acquired based on the natural wind observed at our wind turbine. In this figure it is assumed that the wind with the same pattern blows in each turbine with a constant time lag of 53 s. Here, the reference value of the tip speed ratio is set at 5.5 according to Fig. 5.12. Also, pitch angles of both turbine blades β are assumed to be zero. These responses are obtained through the closed-control system of Fig. 5.13, in which the simulation block in Fig. 5.11 is incorporated. The simulations were executed with MATLAB/Simulink (The MathWorks, Inc.).

It is shown in (d) that the tip speed ratios are almost kept constant since the DC-link current and the control angles of each rectifier are adjusted as in (b) and (c).

The responses of DC-link voltages and output power of the wind turbines are also shown in (e) and (f), respectively. It is clarified that both DC-link voltages (also powers) derived from wind turbine generators are integrated successfully at all times with almost constant tip speed ratios, confirming the usefulness of the proposed system. In (g) the output power of wind turbines [the same as P_{t} in (f)] and that of DC link (= $V_{\text{d}} I_{\text{d}}$) are shown. Although the output fluctuation in DC link is larger than that in wind turbines, it is recognized that the average powers are almost the same (DC link: 48.5 kW and wind turbine: 48.7 kW). This means that the energy of the wind can be always converted to electric power at its maximum.

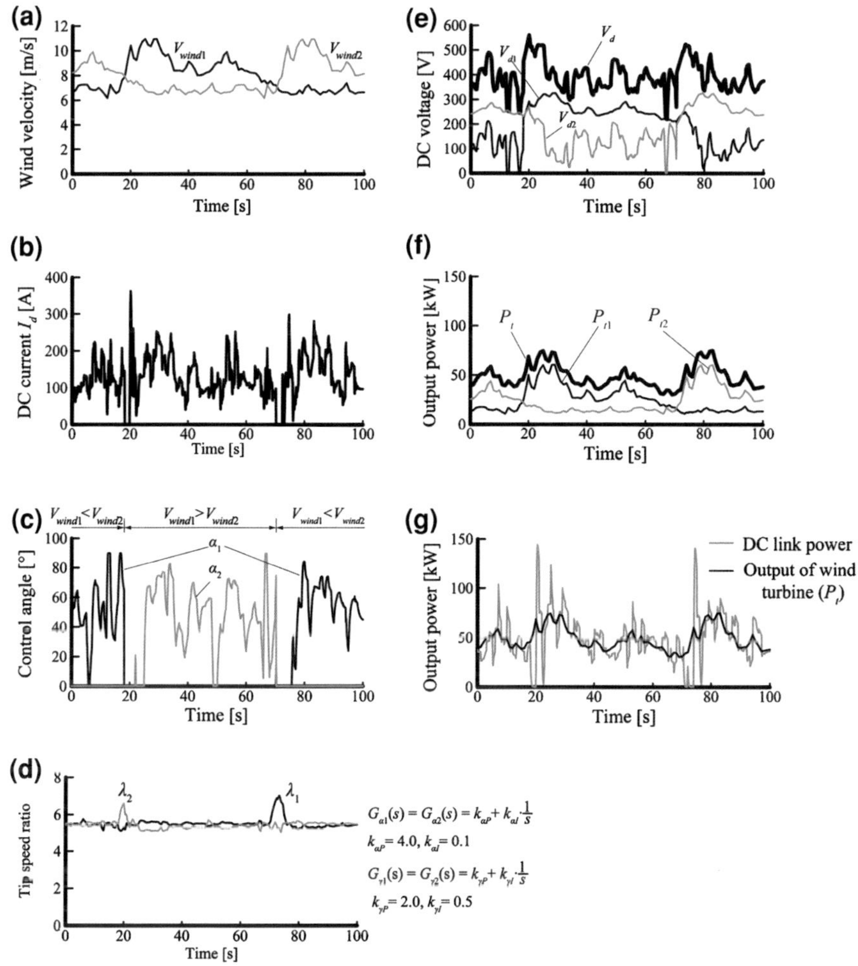

Fig. 5.14 Dynamic responses of the system. **a** Wind velocities. **b** DC-link current. **c** Control angles of the rectifier. **d** Tip speed ratios. **e** DC-link voltages. **f** Output powers of wind turbines. **g** Output power of wind turbines and DC link

5.7 Dynamic Responses for More Wind Turbines (In the Case of Four Wind Turbines)

In the foregoing sections, the wind turbine generating system composed of two generators has been discussed. However, it is necessary to clarify the applicability of the system that includes more than two generators. Here, a wind farm consisting of four wind generators is studied to confirm the usefulness of the proposed system.

Figure 5.15 shows the configuration of the wind turbine generating system composed of four sets of wind turbine and PMSG. Figure 5.16 shows an example

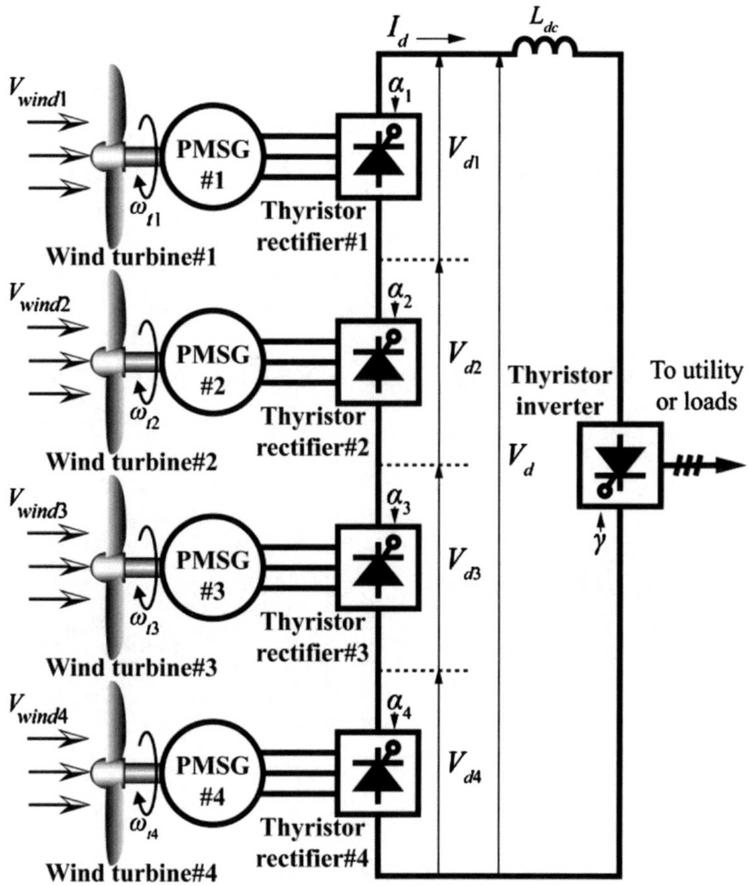

Fig. 5.15 System composed of four wind turbine/generators

of the simulated dynamic responses of the wind farm. (a) shows an example of wind velocities, which are used as the inputs of the simulation. It is shown in (b) that the tip speed ratios are almost kept constant. The dynamic responses for the leading angle of commutation of inverter and the control angles of each rectifier are shown in (c). Furthermore, in (d) and (e), it is noted that the output powers of each turbine or generator $P_{t1} \sim P_{t4}$ are integrated properly into the system output power P_t, and the performance coefficients $C_{p1} \sim C_{p4}$ can be always controlled toward the maximum value (=0.436) in Fig. 5.12, confirming that each wind turbine is operated in the optimum condition.

From these results, we can say that there is no limit in the connected number of the wind turbine/generators from the control point of view. However, since the voltage level of the system is stepped up when the number of generators is increased, we should take the insulation level of the whole system into careful consideration.

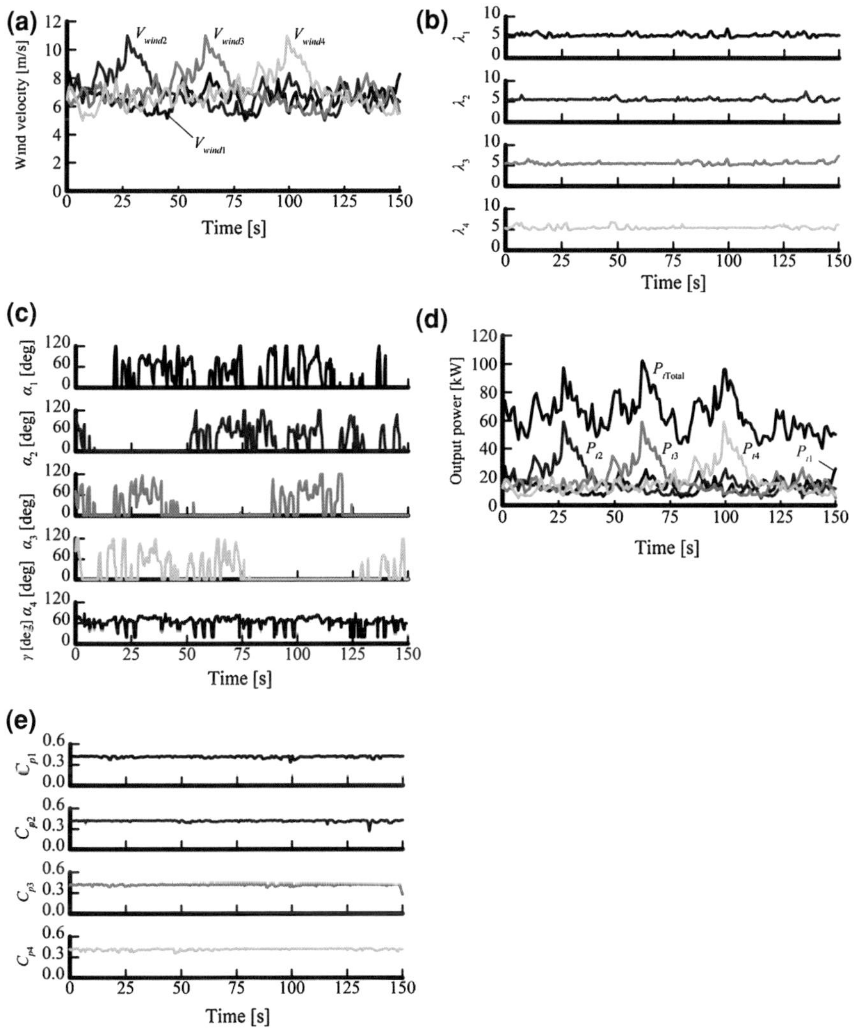

Fig. 5.16 Dynamic responses of the system (in case of four wind turbines). **a** Wind velocities. **b** Tip speed ratios. **c** Control angles and leading angle. **d** Output powers. **e** Performance coefficient

5.8 Conclusions

In this chapter, a new interconnecting method of two or more sets of wind turbine/generator used in a wind farm has been proposed, and basic characteristics of the integrated wind turbine generating system have been discussed.

In this type of system, only one externally commutated thyristor inverter is required for a cluster of wind turbines, and output voltage without distortion can be achieved with ease, realizing a very simple configuration of wind farm with high quality of output power as well as high reliability.

In addition to these advantages, only one DC link is used, and the optimum site for wind turbines, such as off shore, can be readily selected in order to obtain more power from wind because DC transmission system is entirely appropriate for the proposed system.

It should be recognized that, in general the DC-link voltage of the proposed system is changed fairly with the change in the wind velocity, so careful attention has to be made in designing the DC transmission system as for the insulation deterioration and losses. In addition, the voltage levels to the ground for the system components such as PMSGs and thyristor rectifiers increase considerably when compared with the case of single turbine/generator, and careful considerations should be made for the insulation class of the individual components. In addition, topologies in the rectifier to improve the power factor of PMSGs should be developed since the power factors for the proposed system depend on the wind conditions.

References

1. Tatsuta F, Nishikata S (2005) Studies on a wind turbine generator system using a shaft generator system. In: Proceeding of the 8th international conference on electrical machines and systems, Nanjing, China, pp 931–936
2. Tatsuta F, Nishikata S (2007) Performance characteristics of a practical scale wind turbine generating system using a shaft generator system. In: Proceedings of the 12th European conference on power electronics and applications, Aalborg, Denmark
3. Nishikata S, Koishikawa Y, Mita F, Kataoka T (1999) A shaft generator system without output voltage distortion. Trans IEE Jpn 119-D(12):1549–1555
4. Nishikata S, Tatsuta F (2010) A new interconnecting method for wind turbine/generators in a wind farm and basic performances of the integrated system. IEEE Trans Ind Electron 57(2):468–475
5. Heier S (2006) Grid integration of wind energy conversion systems, 2nd edn. Wiley, England, p 44
6. Tatsuta F, Nishikata S (2010) Dynamic performance analysis of a wind turbine generating system with series connected wind generators and bypass diodes using a current source thyristor inverter. In: Proceedings of the 2010 international power electronics conference–ECCE ASIA, Sapporo, Japan, 23H2-3
7. Fujii Y, Hasegawa C, Tatsuta F, Nishikata S (2008) Dynamic performance analysis of a hybrid wind turbine generator system. In: Proceedings of the 11th international conference on electrical machines and systems, Wuhan, China
8. http://www.subaru-windturbine.jp/SubaruWtsTop.htm (in Japanese)
9. Tatsuta F, Nishikata S (2008) Dynamic performance analysis of a wind turbine generating system using a current-source thyristor inverter. In: Proceedings of the 11th international conference on electrical machine and systems, Wuhan, China
10. Tatsuta F, Nishikata S (2009) Dynamic performance analysis of a wind turbine generating system with series connected wind generators using a current source thyristor inverter. In: Proceedings of the 13th European conference on power electronics and applications, Barcelona, Spain

Chapter 6
Grid Connection Scheme of a Variable Speed Wind Turbine Driven Switched Reluctance Generator

Hany M. Hasanien and Ahmed Aldurra

Abstract The variable speed wind turbine generator system (WTGS) has recently become more popular than the fixed speed system. In 2004, the worldwide market share of variable speed WTGS was more than 60%. Doubly fed induction generator (DFIG), wound field synchronous generator (WFSG), and permanent magnet synchronous generator (PMSG) are currently being used as variable speed wind generators. Besides the aforementioned classical machines used in variable speed operation of WTGS, the switched reluctance machine (SRM) has some superior characteristics suitable for wind power application. In this chapter, the construction and operation of switched reluctance generator (SRG) are presented. The static characteristics of SRG are involved. The power inverter circuits which can be used in SRG operation are presented. Furthermore, the control of a grid-connected variable speed wind turbine driving SRG is studied. Finally, the dynamic characteristics of variable speed wind turbine driving a switched reluctance generator are analyzed.

6.1 Introduction

A switched reluctance generator consists of a stator with exciting windings and a magnetic rotor, and both of them carry a set of salient poles. Rotor conductors or permanent magnets are not required because torque is produced by the tendency of

H. M. Hasanien (✉)
Faculty of Engineering, Electrical Power and Machines Department,
Ain Shams University, 1st Elsarayat Street, Abbasia, Cairo, 11517 Egypt
e-mail: hanyhasanien@ieee.org

A. Aldurra
Department of Electrical Engineering, The Petroleum Institute Abu-Dhabi,
Abu Dhabi, UAE
e-mail: aaldurra@pi.ac.ae

S. M. Muyeen (ed.), *Wind Energy Conversion Systems*,
Green Energy and Technology, DOI: 10.1007/978-1-4471-2201-2_6,

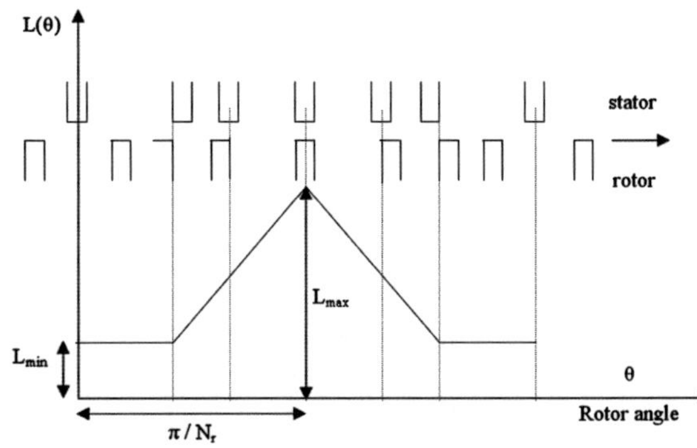

Fig. 6.1 The phase inductance related to the machine poles

the rotor poles to align with the excited stator poles in such a fashion as to maximize the stator flux linkages that result from a given applied stator current.

In order to produce torque from the switched reluctance machine it must be designed such that the stator winding inductance varies with the position of the rotor, because torque in this machine is directly proportional to the variation of the winding inductance with angular position, and the square of machine phase current.

$$\text{Torque } (T) \propto i^2 \frac{\mathrm{dL}}{\mathrm{d}\theta} \tag{6.1}$$

$$T \propto i^2 \frac{L_{\max} - L_{\min}}{\Delta\theta} \tag{6.2}$$

The variation of the phase inductance of SRM with rotor angle relative to the motor poles is shown in Fig. 6.1, where $L_{\max}$ is the maximum inductance of the aligned positions, which occurs when any pair of rotor poles is exactly aligned with the excited stator poles of a certain phase. $L_{\min}$ is the minimum inductance of the unaligned positions, which occurs when the interpolar axis of the rotor is aligned with the excited stator poles. The phase inductance is increased as the rotor pole enters under the stator pole until the aligned position where the inductance has its maximum value. As the rotor pole moves far from the aligned position, the phase inductance will decrease until reaching its minimum value at the unaligned position [1].

The produced torque in SRM is proportional to the square of the phase current which means that it depends on the magnitude of the phase current and not on its direction. Thus the machine drive which supplies the phase current can be uni-directional. It is important to note that no torque is produced at the aligned and unaligned positions, since ($\mathrm{d}L/\mathrm{d}\theta$) equal zero.

A switched reluctance machine is usually supplied from a DC power supply which is switched on and off while transferring among phases using electronic switches like transistors and thyristors. Therefore, the relationship between speed and fundamental switching frequency follows from the fact that if the poles are wound oppositely in pairs to form the phases, then each phase produces a pulse of torque on each passing rotor pole. The fundamental switching frequency in one phase is defined by:

$$f_1 = n.N_r \text{ Hz} \tag{6.3}$$

where n is the speed in rev/sec and N_r is the number of rotor poles.

The step angle (ε) is defined as the angle between two successive torque pulses, and is computed by:

$$\varepsilon = \frac{2\pi}{qN_r} \text{ rad} \tag{6.4}$$

where q is the number of phases, and thus there are qN_r steps per revolution.

6.2 SRG Construction

The switched reluctance generator is a doubly salient, singly excited generator. This means that it has unequal number of salient poles on both the rotor and the stator, but only one member (usually the stator) carries windings, and each two diametrically poles form usually one phase. The rotor has no winding, magnets, or cage winding and is built up from a stack of salient-pole laminations. So it is considered as a simple and robust construction machine. The switched reluctance generator is named by its pole numbers (the number of poles in the stator and in the rotor), so for example: 6/4 SRG means that the SRG contains six poles on stator and four poles on rotor. The classical form of a 3-ph 6/4 switched reluctance generator is shown in Fig. 6.2, also the constructional feature of a 4-ph 8/6 switched reluctance generator is shown in Fig. 6.3, other pole numbers are possible, including 4/2, 12/8, [2].

6.3 Torque Production

6.3.1 Principle of Operation

The operation of all types of reluctance machines (the synchronous and the switched reluctance machines), depends on the principle that a piece of soft magnetic iron will align itself with a magnetic field imposed upon it. In the long-established form of synchronous reluctance machine, salient poles on the rotor should be synchronized with the magnetic field produced by electric currents in windings on the stator. Such machines carry cage windings like induction motors on the rotor to enable the starting process to take place.

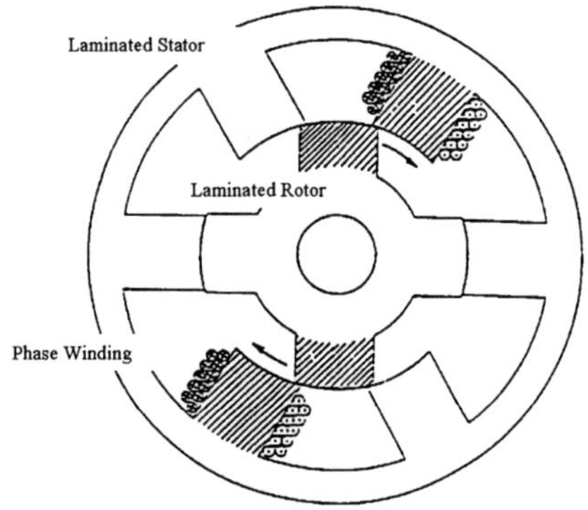

Fig. 6.2 A cross section of a 3-ph 6/4 SRG

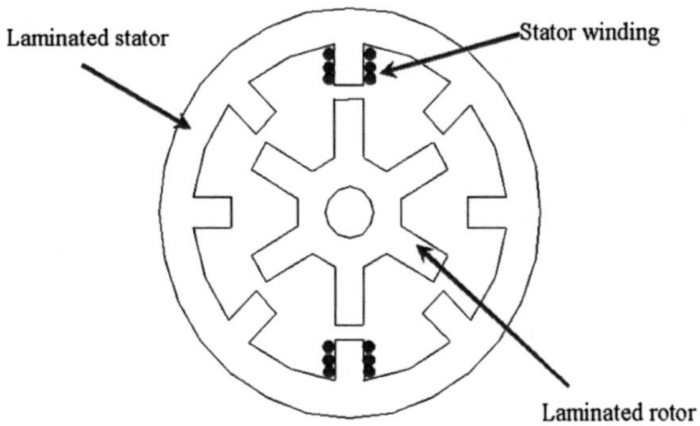

Fig. 6.3 A cross section of a 4-ph 8/6 SRG

Switched reluctance machines differ from these synchronous reluctance machines in two ways. First, they completely eliminate the need for any winding on the rotating member, and second, they employ salient poles on both the rotor and stator with the latter being the only ones to carry electrical windings [3].

SRM is similar to a variable reluctance stepping machine and they may appear identical except that the latter is designed as a low power positioning device which runs in synchronism with a square wave supply, whereas the SRM differs in its design properties to give an efficient and smooth variable speed power drive for which switching of the supply is dictated by the rotor position; hence the SRM has more affinity with the DC brushless drive.

6.3.2 Magnetization Curves

Magnetization curves of SRM represent the relationship between the flux-linkage and the current of a certain phase at different rotor positions, and their importance is because they include the machine capabilities. Thus, the generator behavior and characteristics (generator co-energy, torque, current, and speed) are calculated based on these curves. The most two important points in these curves are the unaligned and aligned positions.

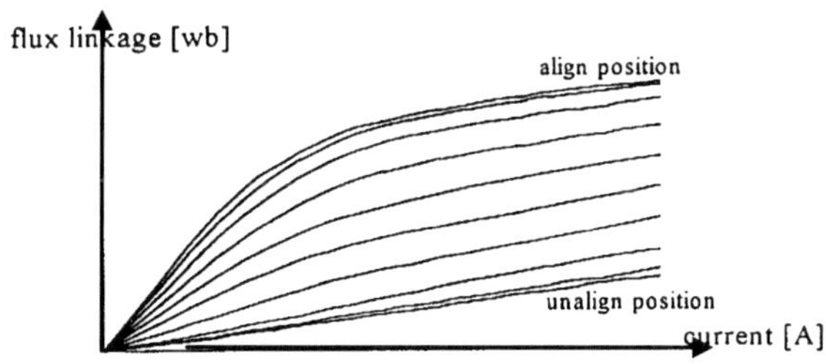

Fig. 6.4 The magnetization curves of SRM

Figure 6.4 shows the magnetization curves of an SRM where each curve belongs to a certain value of rotor angle that varies between the unaligned (the lowest curve) to the aligned position (the highest curve).

Figure 6.5 shows the idealized inductance curves of SRM against rotor position for different values of current. The region of increasing inductance is marked by motoring where a positive torque is produced through this region and the region of decreasing inductance is marked by generating or braking where a negative torque is produced through this period. As it is shown there is a small dwell at maximum inductance due to the difference between stator and rotor pole arcs, and another dwell is obtained at the minimum inductance due to the difference between the interpolar arcs of the rotor and the stator. In practical curves, the corners are smoother than those shown in the figure due to the fringing effect, which cannot be neglected as in the idealized curves.

Figure 6.6 shows the inductance of a 3-ph SRG. It can be noted from this figure that it is required to produce energy at all positions, that the entire 360° be covered by segments of falling inductance for different phases, and the phase currents must be commutated and sequenced to coincide with the appropriate segments.

The conduction period under idealized conditions (which is the difference between the on-angle and the off-angle of the phase) may exceed the step angle, this leads to an overlapping between phases (its length equal to the difference between the conduction angle and the step angle). This is desirable in a small amount because it results in minimizing the torque ripple; however, in a large amount it imposes transient or vibratory stresses on the shaft, coupling, and load.

The overlapping between the phases is not the same for all generator constructions. For example, with 6/4 3-ph SRG the step angle is 30°, the maximum conduction angle is 45°, and equal 1.5 times the step angle while the step angle with 8/6 4-ph SRG is 15°, the maximum conduction angle is 30° which is 2.0 times the step angle. Figure 6.7 shows the idealized phase inductance of a 4-ph 8/6 SRG.

6.3.3 Static Torque Curves

These curves represent the phase torque values as a function of rotor angle at different values of current. The curves are computed by integrating the

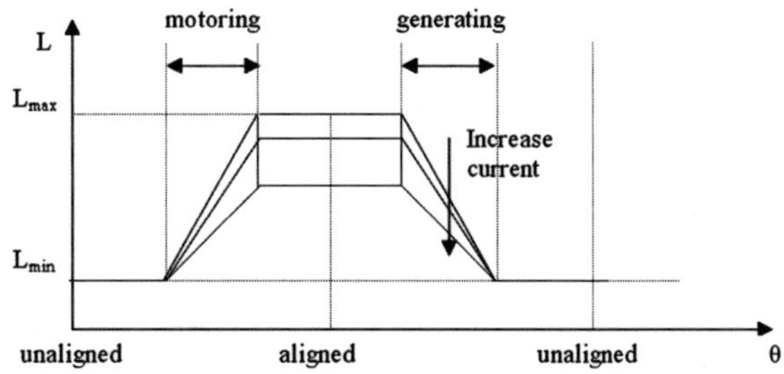

Fig. 6.5 The inductance curves of SRM

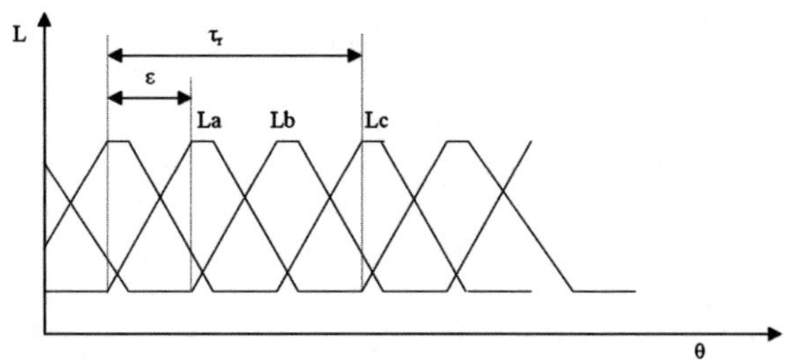

Fig. 6.6 The inductance of 3-ph 6/4 SRG

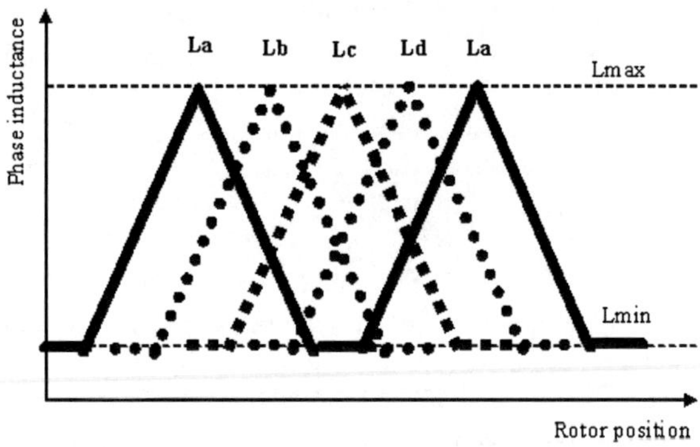

Fig. 6.7 The idealized phase inductance of a 4-ph 8/6 SRG

magnetization curves to obtain the co-energy curves as a function of rotor angle at different values of current; the curves are differentiated relative to the rotor angle at fixed values of current to get finally the static torque curves which have the form shown in Fig. 6.8 [4].

$$W' = \int_0^i \psi \, \mathrm{d}i|_{\theta=\text{Constant}} \tag{6.5}$$

Then

$$T = \frac{\partial W'}{\partial \theta}|_{i=\text{Constant}} \tag{6.6}$$

where T is the torque produced by one phase.

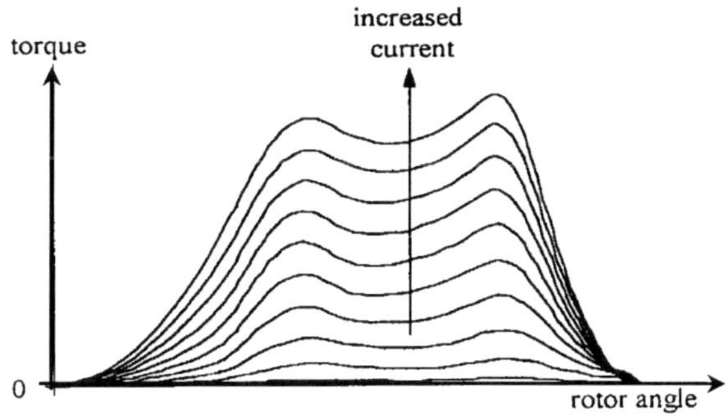

Fig. 6.8 The static torque curves of SRM

The importance of these curves comes from the fact that they imply all the possible capabilities of torque/phase produced on the generator shaft at any value of rotor angle and current.

6.4 Switched Reluctance Generator Converter System

Figure 6.9 illustrates a 4-ph SRG converter system with two controllable power semiconductor switches and two diodes per phase that source a resistive load. The filter capacitor C is chosen large enough to assure fairly constant DC-link voltage at each stroke.

Thus each phase has pulse nature parameters (current, flux-linkage, and torque). The torque is produced by the tendency of the rotor poles to align with the stator poles of the excited phase, and it is independent of the current phase direction.

The SRG possesses many inherent advantages such as simplicity, robustness, low manufacturing cost, high speed, and high efficiency. The SRG is under development for variable speed applications. To date, these applications include sourcing aerospace power systems, hybrid vehicles, and wind turbine applications. The aerospace and automotive applications are generally characterized by high speed operation. The wind energy application is characterized by low speed, high torque operation [5].

6.5 Switched Reluctance Generator Static Characteristics

The model of an SRG for dynamic analysis comprises the set of phase circuit and mechanical differential equations. In integerating these equations, the problem centers on handling the data (flux-linkage/angle/current) used to describe the magnetic nature of the switched reluctance machine. Different methods have been used for numerical integeration of the nonlinear differential equations of the SRG

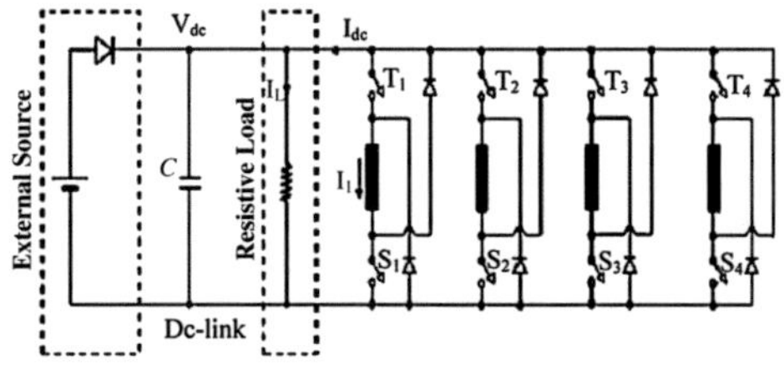

Fig. 6.9 Generator system with 4-ph SRG

with the magnetization data in the form of a look-up table $\psi(\theta, i)$. The integeration of these equations to obtain the waveforms of phase current and torque against time requires the definition of magnetic behavior of the switched reluctance machine in the form of look-up tables $i(\theta, \psi)$ and $T(\theta, i)$ to enable the values of current and torque of each phase to be updated after each step of numerical integeration.

6.6 Methods for Representing the Magnetic Curves of SRG

In this section, different methods used to represent the magnetization curves of SRG will be introduced. In the first method a mathematical function of the measured flux-linkage points versus rotor angle at fixed stator current values is obtained for a variable reluctance stepping motor, this function has the polynomial form:

$$\psi = f(\theta) = \mathrm{a}_0 + \mathrm{a}_1\theta + \mathrm{a}_2\theta^2 + \cdots + \mathrm{a}_k\theta^k \tag{6.7}$$

where $\mathrm{a}_o, \mathrm{a}_1, \mathrm{a}_2, \ldots, \mathrm{a}_k$ are the polynomial coefficients and they are determined using the least square error method, θ is the rotor angle in electrical degrees. For p values of current the following set of polynomials could be obtained as:

$$\psi_b = f_b(\theta) = \mathrm{a}_{0b} + \mathrm{a}_{1b}\theta + \mathrm{a}_{2b}\theta^2 + \cdots + \mathrm{a}_{kb}\theta^k \tag{6.8}$$

where $b = 1, 2, \ldots p$, and any one set of coefficient $\mathrm{a}_{ob}, \mathrm{a}_{1b}, \ldots \mathrm{a}_{kb}$ is independent of the chosen current. Equation 6.8 gives the flux-linkage at p discrete values of current and in order to represent the flux-linkage at any current the sets of coefficients $\mathrm{a}_{01}, \mathrm{a}_{02}, \ldots \mathrm{a}_{0p}$; $\mathrm{a}_{11}, \mathrm{a}_{12}, \ldots \mathrm{a}_{1p}$; $\mathrm{a}_{21}, \mathrm{a}_{22}, \ldots \mathrm{a}_{2p}$; $\mathrm{a}_{k1}, \mathrm{a}_{k2}, \ldots \mathrm{a}_{kp}$ should be replaced by polynomials of jth order in terms of current. Thus the general expression for the flux-linkage as a function of both current and rotor position is:

$$\psi = \sum_{b=0}^{k} \left(\mathrm{a}_{0b} + \mathrm{a}_{1b} \cdot i + \cdots + \mathrm{a}_{jb} \cdot i^j\right) \cdot \theta^b \tag{6.9}$$

The flux-linkage Ψ must be an even function of θ and an odd function of current, hence j is taken to be an odd integer number, and k is taken to be an even number.

In method 2 an exponential function is used to represent the magnetization curves of the SRG because these functions are a natural fit to typical magnetization curves. The used exponential function consists of three exponential terms in addition to a linear one.

$$\psi = \mathrm{a}_0.i + \mathrm{a}_1 \cdot (1 - e^{-\alpha_1.i}) + \mathrm{a}_2 \cdot (1 - e^{-\alpha_2.i}) + \mathrm{a}_3 \cdot (1 - e^{-\alpha_3.i}) \tag{6.10}$$

where $\mathrm{a}_0, \mathrm{a}_1, \mathrm{a}_2, \mathrm{a}_3$ are coefficients and $\alpha_1, \alpha_2, \alpha_3$ are constants. A nonlinear least square analysis is used to obtain each of the coefficients, then the Ψ/I curves are obtained at a constant measured rotor angle for each curve, and linear interpolation between coefficients and rotor angle is used to determine the intermediate curves of Ψ/I.

In method 3 the data defining the magnetic nature of the machine are stored as a look-up table to represent the function $\Psi(\theta, I)$ which has been formed from a set of measured curves, and a quadratic interpolation is used to get the intermediate values of flux-linkage for each curve because it has been found that this function is suitable to represent the saturated part. As it will be shown in this chapter, it is important to obtain the table of $I(\theta, \Psi)$ while integrating the motor equations, this table is obtained by inverting the table of $\Psi(\theta, I)$, and is formed at sufficient number of equally spaced angles and flux-linkages. The values of $I(\Psi)|_{\theta=\mathrm{constant}}$ at equally spaced flux-linkages are found using quadratic interpolation and $I(\Psi)|_{\theta=\mathrm{constant}}$ is represented by the equation:

$$I(\theta, \psi) = \mathrm{A}\psi^2 + \mathrm{B}\psi + \mathrm{C} \tag{6.11}$$

At least the parameters of three points are necessary to determine the coefficients (A,B,C) of Eq. 6.11.

Method 4 is used to represent the flux-linkage current curves of the SRG using the measured data. This method (which has been adopted in this chapter) is carried out using the Cubic Spline Interpolation technique, which is more accurate than the other methods and gives more smoothed representation of the magnetization curves.

After getting the magnetization curves of the SRG, the static torque characteristics can be obtained in order to get the whole static characteristics of the generator. The static torque characteristics are plotted against rotor angle for different values of current.

The static torque is computed by numerical differentiation of the co-energy which in turn is computed by numerical integration of the flux-linkage current curves. The method used in this study is based on some input curves obtained by measurement [6, 7].

6.7 Computation of the SRG Static Characteristics

The magnetization characteristics are extended using the cubic spline interpolation algorithm to cover the interval of rotor angles between the unaligned and the aligned positions as shown in Fig. 6.10. The co-energy curves are calculated from

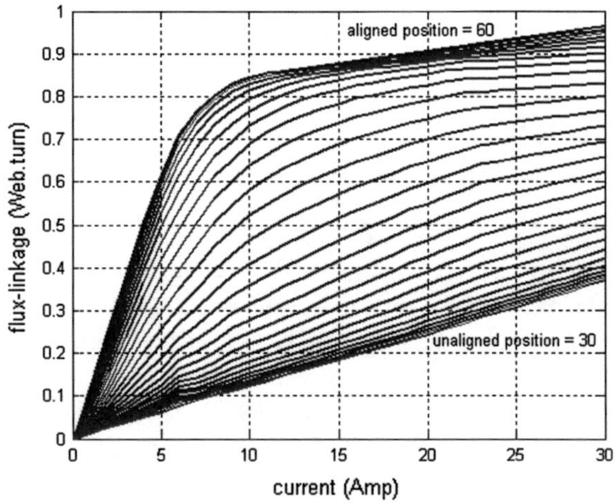

Fig. 6.10 The phase flux-linkage as a function of current and rotor position

Eq. 6.5 by applying the trapezoidal rule in numerical integration. The static torque curves of the SRG are computed according to Eq. 6.6 using numerical differentiation.

The previous characteristics data are stored as a look-up table. Thus there are two look-up tables for the flux-linkage and for the static torque characteristics available to use during the computation of the generator differential equations.

The computed torque as a function of the current and rotor angle is illustrated in Fig. 6.11. The computations are carried out at different rotor angles between the aligned position (60°) and the unaligned position (90°), which forms half the rotor pole-pitch [8].

6.8 Inverter Circuits of SRG

The phase windings of a doubly salient switched reluctance generator are fed from unipolar pulses of current from a suitable power inverter to control the speed and power of the generator. Ideally, a power inverter circuit for the switched reluctance generator should have:

1. Minimum number of switches to achieve the lowest possible losses to reduce the cost.
2. Complete flexibility in the number of phase windings to be used conveniently with various forms of SRM.
3. The full supply voltage applied to the generator phase windings.
4. Switching devices rated at values close to the generator voltage to reduce the inverter cost.

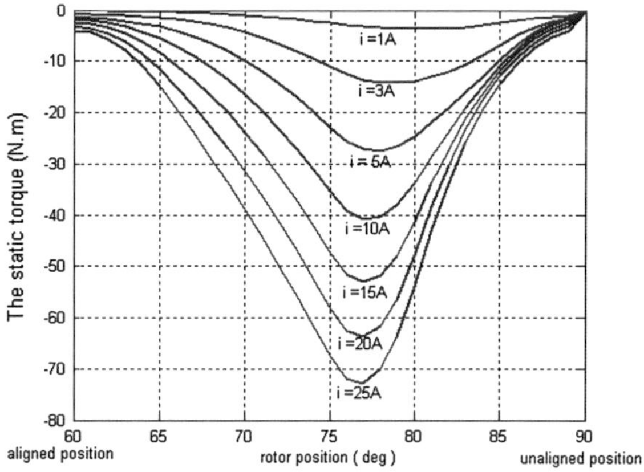

Fig. 6.11 The static torque curves of SRG

5. The ability to increase the phase current rapidly in order to bring the current waveform as close as possible to the ideal current waveform (rectangular waveform).
6. Effective means for current control by modulation of the switches.
7. Some methods to return the energy to the supply while the flux-linkage is decreasing after switching off the generator phase, and before reenergizing it for a next time.

In this part, some forms of SRG inverters will be outlined briefly:

6.8.1 Power Inverter with Asymmetric Half Bridge

The asymmetric half bridge inverter for a 3-ph SRG is shown in Fig. 6.12, where the switching devices and the freewheeling diodes must be rated to withstand the supply voltage plus any switching transients.

This asymmetric half bridge has three main modes of operation. The first, a positive voltage loop, occurs when both switching devices are turned on. The supply voltage is connected across the phase winding and the current in the phase winding increases rapidly, supplying energy to the generator. The second mode of operation is a zero voltage loop. This occurs if either of the two switching devices is turned off while current is flowing in the phase winding. In this case the current continues to flow through one switching device and one diode. Energy is neither taken from nor returned to the DC supply, minimizing the current ripple rating of the supply capacitor. The third mode of operation is a negative voltage loop, where both the switching devices are turned off. The current is forced to flow through both freewheeling diodes. The flux-linkage associated with the phase winding decreases rapidly as energy is returned from the generator to the supply.

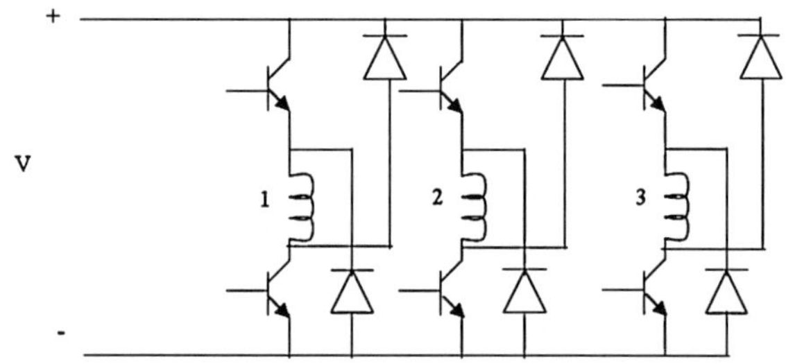

Fig. 6.12 An asymmetric half bridge inverter for a 3-ph SRG

The major advantage of this circuit is that all of the available supply voltage can be used to control the current in the phase windings. As each phase winding is connected to each own asymmetric half bridge, there is no restriction on the number of phase windings. However, as there are two switches per phase winding it is best suited to motors with few phase windings.

6.8.2 Power Inverter with Split DC Supply

A power inverter employing a split DC supply is introduced to provide the positive and negative voltages needed to increase and decrease the current in the phase winding. Figure 6.13 shows an inverter for 4-ph windings SRG.

The positions of the switching device and freewheeling diode are transposed for each phase winding to ensure that there is no power flow imbalance between the two supply capacitors. This arrangement means that this power inverter circuit is only suitable for generators that have an even number of phases. Each switching device and freewheeling diode must be rated to withstand the complete supply voltage plus any transient voltages due to the switching. However, only half of this voltage can be applied across the generator winding. If all of the available supply voltage could be used to increase and decrease the current in the phase winding, increased torque could be produced at higher speeds. Although this circuit requires only one switch per phase winding, this advantage is outweighed by under use of switch voltage, the need of extra capacitive components in the DC supply and the requirement for an even number of phase windings.

6.8.3 Power Inverter for SRG with Bifilar Windings

This type of SRG inverter is shown in Fig. 6.14 for a generator with 3-ph windings, where the bifilar winding is connected to a single switching device, and the other to a freewheeling diode.

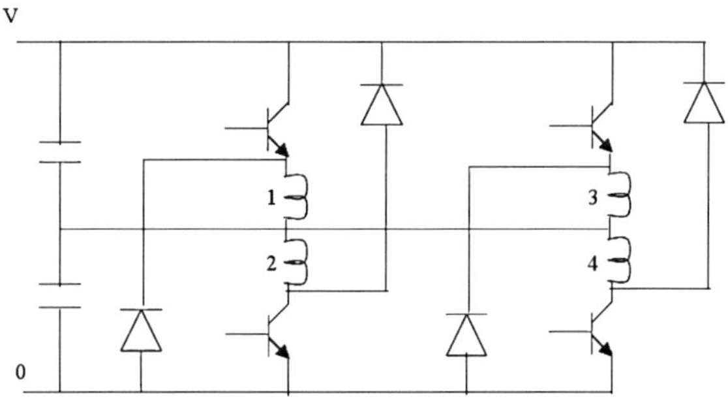

Fig. 6.13 Four-phase power inverter with split DC supply

Current is built up in the main winding when the switching device is turned on and transfers to the secondary winding when the switch is turned off.

Depending on the degree of coupling between the two windings and their turns ratio, the voltage across the switching device may rise to over twice the supply voltage at the instant of turn off. The switching device must be rated to withstand this. Although this power inverter has only one switch per phase, the voltage rating of that device must be at least twice the rating of the generator windings. A second disadvantage of this power inverter lies in the inefficient use of the copper in the generator since only one of the bifilar windings in each pair carries current at any time.

6.9 SRG for Wind Energy Applications

Recently, variable speed operation of SRG is becoming possible in many applications. To date, these applications include sourcing aerospace power systems [9], automotive applications [10, 11] hybrid vehicles [12], and wind turbine applications [13–15]. The aerospace and automotive applications are generally characterized by high speed operation. The wind energy application is characterized by low speed, high torque operation. In [13], the advantages of the SRG for wind energy application are discussed without any control strategy. In [14], the grid interfacing of wind energy conversion system is not considered. The authors reported the extension of [14] into [15] where the grid interfacing is also considered and buck converter based topologies were used in the generator side control of SRG.

This section, presents detailed modeling and control strategies for variable speed operation of SRG connected to the grid suitable for wind power application. For powering SRG, a voltage source topology is preferred in this study, which gives well-defined voltages over semiconductors and SRG-phases. The premise of a voltage source topology implies a unipolar DC-link voltage with a relatively large DC-link capacitor as energy buffer. The asymmetric half bridge converter based on hysteresis control is considered herein for the generator side control of

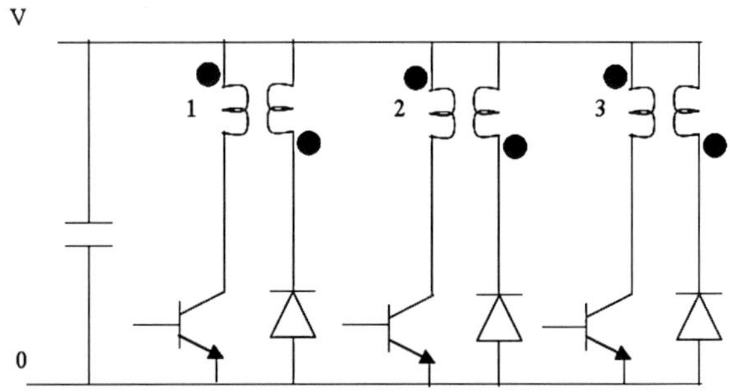

Fig. 6.14 Three-phase power inverter with bifilar winding

SRG. The switching on angle control of SRG is also taken into consideration to maintain optimum speed. For the grid side interfacing of VSWT-SRG, the well-known vector control based inverter system is considered in this study. To validate the effectiveness of the control strategy, real wind speed data is used in the simulation measured at Hokkaido Island, Japan. The simulation is carried out using PSCAD/EMTDC.

6.9.1 Wind Turbine Modeling

The mathematical relation for the mechanical power extraction from the wind can be expressed as follows [16]:

$$P_M = 0.5\rho C_p(\lambda, \beta)\pi R^2 V_W^3 \text{ (W)} \tag{6.12}$$

where, P_M is the extracted power from the wind, ρ is the air density (kg/m^3), R is the blade radius (m), V_w is the wind speed (m/s), and C_p is the power coefficient which is a function of both tip speed ratio, λ, and blade pitch angle, $\beta(°)$. C_p is expressed by the following equations [17].

$$C_p(\lambda, \beta) = 0.5(\Gamma - 0.02\beta^2 - 5.6)e^{-0.17\Gamma} \tag{6.13}$$

$$\lambda = \frac{\omega_m R}{V_W}, \Gamma = \frac{R}{\lambda} \cdot \frac{3600}{1609} \tag{6.14}$$

where ω_m is the rotational speed (rad/s).

In variable speed WTGS, the generated active power depends on the power coefficient, C_p, which is related to the proportion of power extracted from the wind. The optimum values of tip speed ratio and power coefficient are chosen constant values based on the turbine characteristics. For each instantaneous wind speed of VSWT, there is a specific turbine rotational speed, ω_r, which corresponds to the maximum active power, P_{max}, from the wind generator. In this study, power

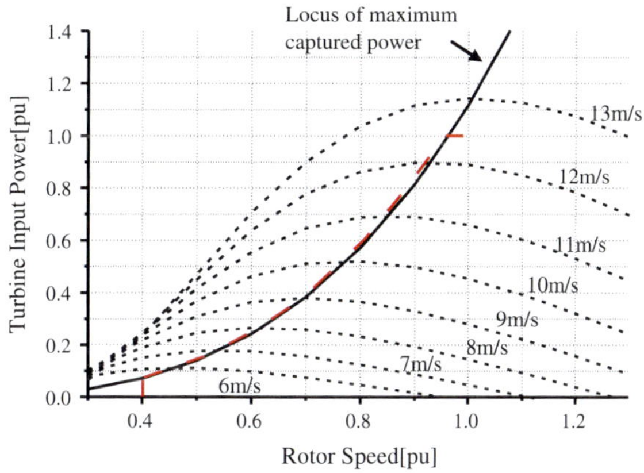

Fig. 6.15 Wind turbine characteristics for variable speed operation

coefficient curve with maximum power point tracking (MPPT) line is shown in Fig. 6.15. Since the precise measurement of the wind speed is difficult, it is better to calculate the maximum power, P_{max}, without the measurement of wind speed as shown below.

$$P_{\max} = 0.5\rho\pi R^2 \left(\frac{\omega_r R}{\lambda_{\text{opt}}}\right)^3 C_{\text{p\_opt}} \tag{6.15}$$

From Eq. 6.15, it is clear that the maximum generated power is proportional to the cube of rotational speed. The pitch converter works when the rotor speed exceeds the rated speed to control the mechanical torque of wind turbine [18].

6.9.2 SRG Modeling Include Converter

The SRG is considered as a simple and robust construction machine as explained previously. A 3-ph SRG converter system has two controllable power semiconductor switches and two diodes per phase, this circuit is called an asymmetric half bridge inverter for a 3-ph SRG. Thus each phase has pulse nature parameters (current, flux-linkage, and torque). The torque is produced by the tendency of the rotor poles to align with the stator poles of the excited phase, and it is independent of the current phase direction.

The model of the SRG for dynamic analysis comprises the set of phase circuit and mechanical differential equations. In integrating these equations, the problem centers on handling the data (flux-linkage/angle/current) used to describe the magnetic nature of the switched reluctance machine. Different methods have been used for numerical integration of the nonlinear differential equations of the SRG with the magnetization data in the form of a look-up table. In this work, the magnetization curves of the SRG are derived from the measured data. The cubic

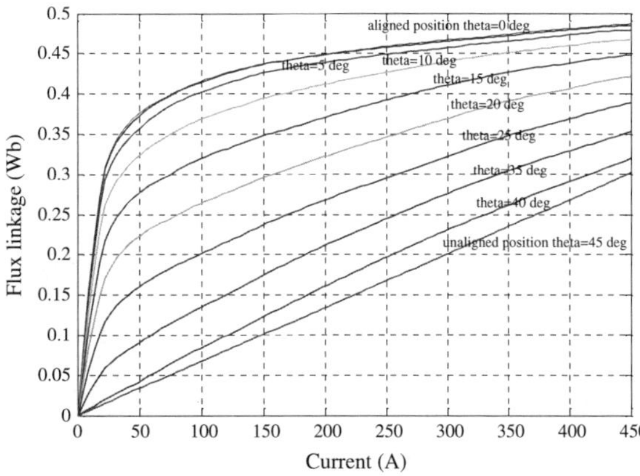

Fig. 6.16 The phase flux-linkage as a function of current and rotor position

spline interpolation technique is used, which is more accurate than many techniques and gives more smoothed representation of the magnetization curves. The magnetization characteristics are extended using the cubic spline interpolation algorithm to cover the interval of rotor angles between the unaligned and the aligned positions as shown in Fig. 6.16. The co-energy curves are calculated by applying the trapezoidal rule in numerical integration, and the static torque curves of the SRG are computed by numerical differentiation of the co-energy.

The previous characteristics data are stored as a look-up table and used in the laboratory standard power system simulator PSCAD/EMTDC. Thus there are two look-up tables for the flux-linkage and for the static torque characteristics available to use during the computation of the generator differential equations.

To predict the generator performance, it is necessary to solve the differential equations for the appropriate switched conditions, and an additional mechanical equation for a variable speed. It is valuable to note that the electromagnetic nature of the generator is reflected in the variation of the phase winding inductances with the rotor displacement, where the mutual inductances with other phase windings are often very small and arc neglected.

The phase equation of the generator has the general following formula:

$$\frac{\mathrm{d}\psi_{\mathrm{k}}(\theta_{\mathrm{k}}, \mathrm{i}_{\mathrm{k}})}{\mathrm{dt}} = \pm V - R\, i_{\mathrm{K}} \tag{6.16}$$

where R is the phase winding resistance, Ψ is the flux-linkage as a function of current and rotor angle, and V is the applied voltage which is positive during the conduction period. It is negative from the switch-off angle until the extinction angle, and otherwise equals zero.

Thus,

$$\begin{aligned} &V = E && \text{when} \quad \theta_{\mathrm{on}} < \theta < \theta_{\mathrm{off}} \\ &V = -E && \qquad\quad \theta_{\mathrm{off}} < \theta < \theta_{\mathrm{ext}} \\ &V = 0 && \qquad\qquad \theta > \theta_{\mathrm{ext}} \end{aligned} \tag{6.17}$$

Table 6.1 SRG specifications

The phase winding resistance	0.05 Ω
The DC supply voltage	240 V
The maximum phase current	200 A
The rated torque	152.79 N m
The rated speed	3,000 rpm
The rated power	48 KW
No. of motor phases	3
The rotor moment of inertia	0.05 kg m^2
The friction coefficient	0.02 m s

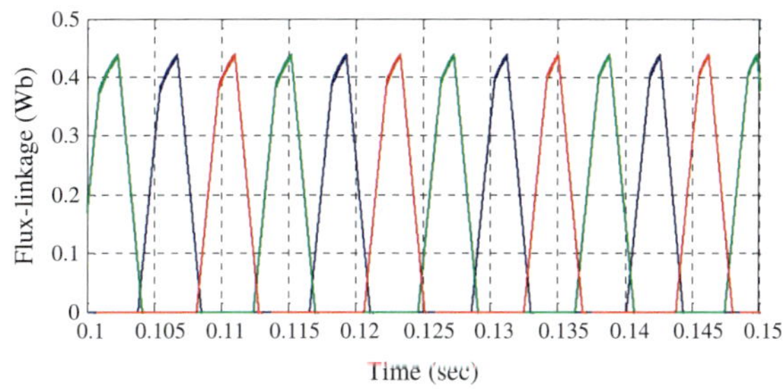

Fig. 6.17 The flux-linkage response of the SRG under $\theta_{\text{on}} = 10$ and $\theta_{\text{off}} = 30°$

where E is the dc supply voltage, θ_{on} is the switching on angle, θ_{off} is the switching-off angle and θ_{ext} is the extinction angle, which is the angle corresponding to zero current.

A switched reluctance machine operates in generating mode by retarding the fire-angles so that the conduction period comes after the aligned position, and where the phase inductance is decreasing, $dL/d\theta$ is negative. In such a case, the energy returned to the DC-link during the de-fluxing period P_o exceeds the excitation energy supplied during the dwell period P_{exc} and the difference is provided by the prime mover. The electrical output power can be written as follows:

$$P_{\text{out}} = P_o - P_{\text{exc}} \tag{6.18}$$

The generator under study is a 3-ph 6/4 SRG, the rated power is 48 kw at 3,000 rpm. The phase resistance is 0.05 Ω, the machine inertia is 0.05 kg m^2, and the supply voltage is 240 V; all the parameters are illustrated in Table 6.1.

Figures 6.17, 6.18 and 6.19 show the flux-linkage, the current, and torque curves of the SRG under normal operating conditions, switching-on angle $\theta_{\text{on}} = 10°$ and switching-off angle $\theta_{\text{off}} = 30°$. It can be predicted that the switching-on angle (θ_{on}) must be after the aligned position (i.e. 0°) to permit the current to increase rapidly with the effect of the back emf which appears at the start of the negative slope inductance.

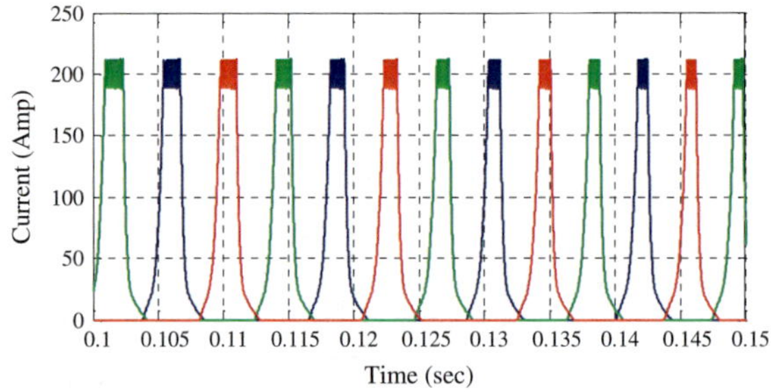

Fig. 6.18 The current response of the SRG under $\theta_{on} = 10°$ and $\theta_{off} = 30°$

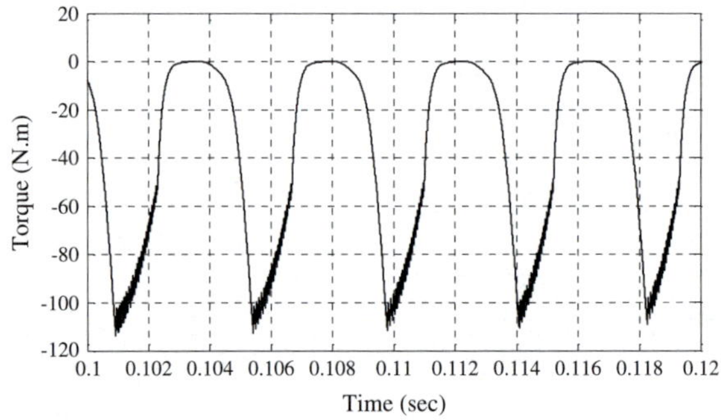

Fig. 6.19 The torque response of the SRG under $\theta_{on} = 10°$ and $\theta_{off} = 30°$

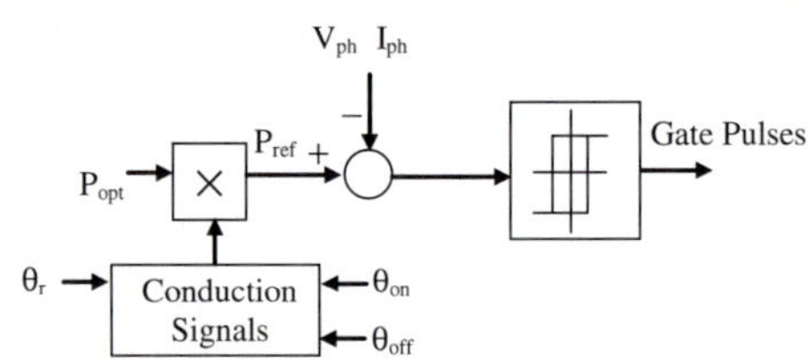

Fig. 6.20 Control block for SRG converter

The control block diagram for the asymmetric half bridge converter to generate gate pulse signals is shown in Fig. 6.20. The reference signal is determined from the maximum power point tracking (MPPT) algorithm. The conduction signals are generated according to the logic explained by Eq. 6.17. The SRG rotor angular position, θ_r is shifted by 30° in each phase for the 6/4 SRG. The hysteresis controller works well to generate the optimal firing angle for the converter in order to maximize the output power of the generator according to the reference signal. The optimum speed is maintained by controlling switching on angle, θ_{on}, as shown in Fig. 6.21, where ω_{r_opt} is the optimum rotational speed determined from MPPT.

6.9.3 Control of Grid Side Inverter

Control blocks for the grid side inverter are shown in Fig. 6.22, which is based on the cascaded control scheme. The dq quantities and three-phase electrical

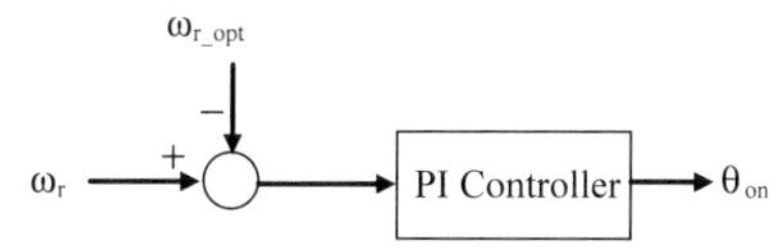

Fig. 6.21 Switching on angle control block

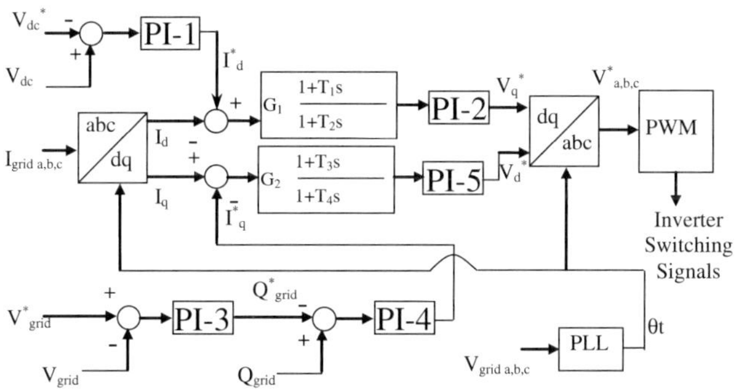

Fig. 6.22 Control block for grid side inverter

Fig. 6.23 Model system

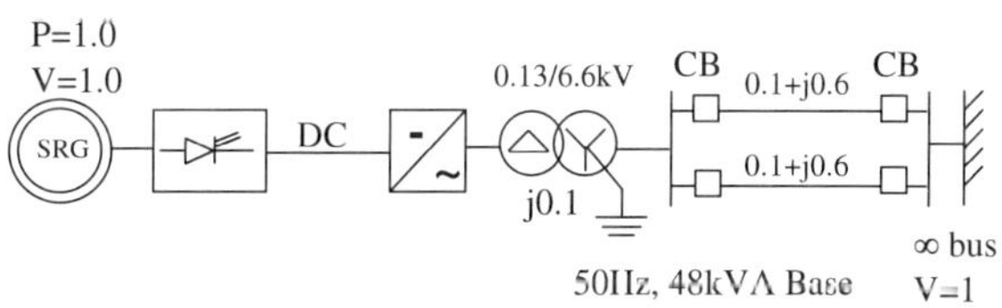

quantities are related to each other by reference frame transformation. The angle of the transformation is detected from the three-phase voltages (v_a,v_b,v_c) at the high voltage side of the grid side transformer. The DC-link voltage can be controlled by the d-axis current. On the other hand, the reactive power of grid side inverter can be controlled by the q-axis current. The reactive power reference is set in such a way that the terminal voltage at the high voltage side of the transformer remains constant. The triangular carrier signal is used as the carrier wave of PWM operation. The carrier frequency is chosen as 1,000 Hz.

6.9.4 Model System

The model system used for the dynamic analysis of VSWT-SRG is shown in Fig. 6.23. One SRG is connected to an infinite bus through the asymmetric half bridge converter, DC-link capacitor, grid side inverter, transformer, and double circuit transmission line. The system base is 48 kVA.

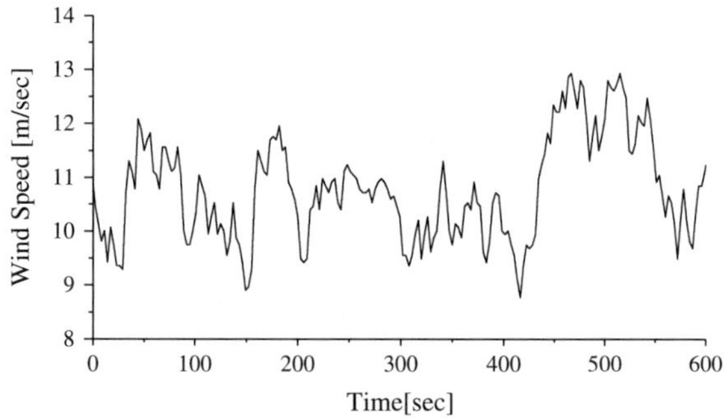

Fig. 6.24 Wind speed

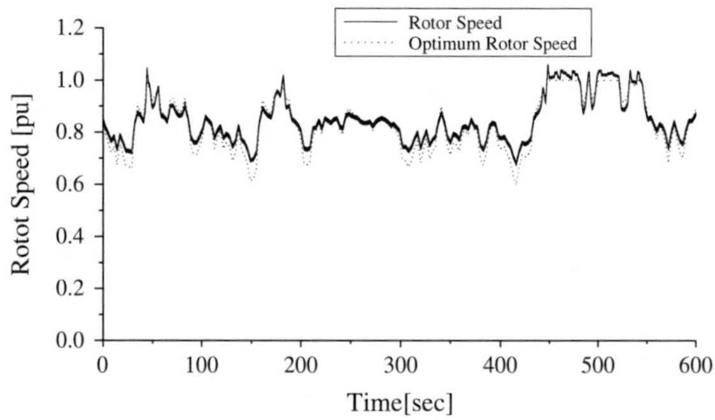

Fig. 6.25 Rotor speed of SRG

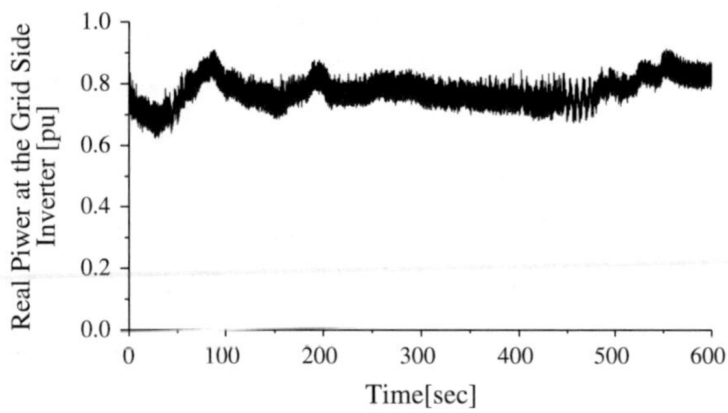

Fig. 6.26 Real power of grid side inverter

6.9.5 Simulation Results

The dynamic characteristic of VSWT-SRG is analyzed under wide range of wind speed variation shown in Fig. 6.24, which is a real data measured in Hokkaido Island of Japan. One of the control objectives is to maximize the wind power capture by adjusting the rotor speed of the wind turbine according to the wind speed variation, provided that the captured power should not exceed the rated power of SRG. Figure 6.25 shows the rotor speed of SRG and optimum rotor speed calculated from MPPT. The responses of real and reactive power at the grid side inverter are shown in Figs. 6.26 and 6.27, respectively. The terminal voltage at the grid side is maintained constant as shown in Fig. 6.28. The response of

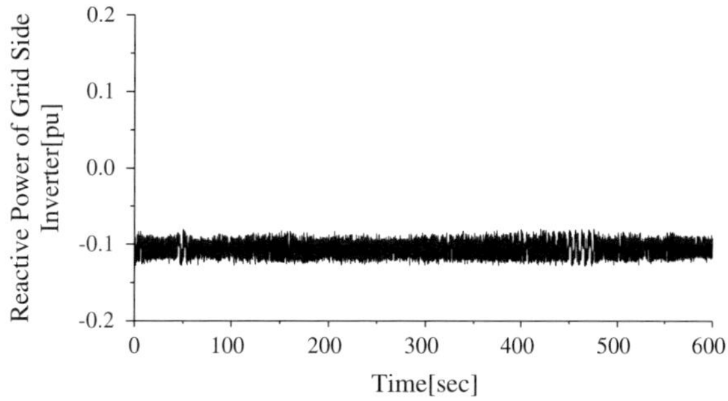

Fig. 6.27 Reactive power of grid side inverter

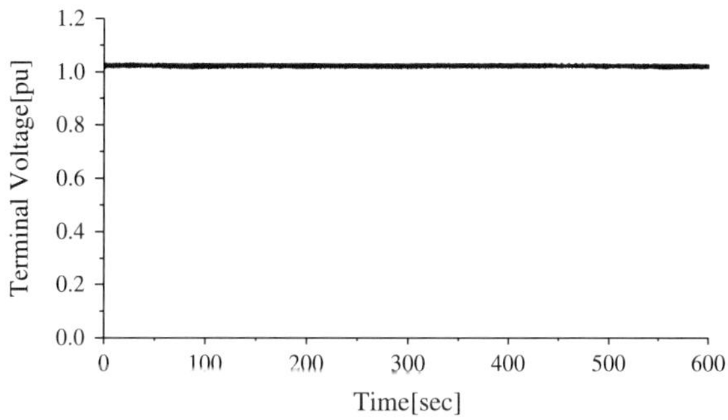

Fig. 6.28 Terminal voltage of the grid

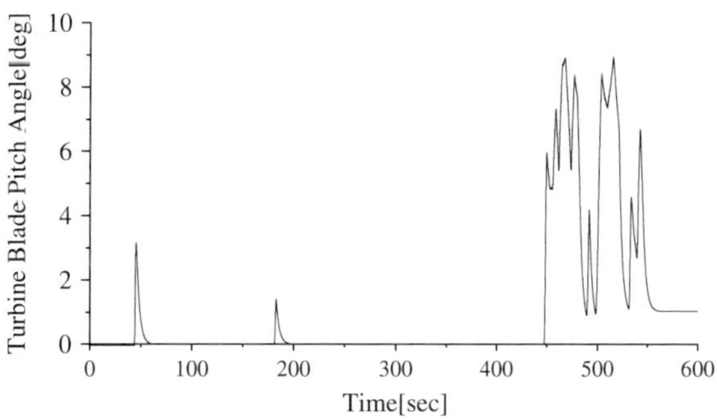

Fig. 6.29 Turbine blade pitch angle

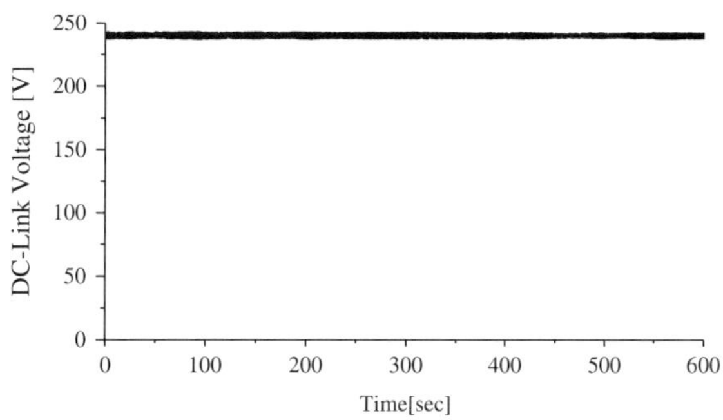

Fig. 6.30 DC-link circuit voltage

turbine blade pitch angle is shown in Fig. 6.29. The response of DC-link voltage is shown in Fig. 6.30. From the simulation responses it is seen that the proposed control system is well suited for wind power application.

6.10 Conclusion

This chapter presents a detailed study of the dynamic response of the variable speed wind turbine driving a switched reluctance generator. The modeling and control strategies for the variable speed operation of SRG including generator side asymmetric half bridge converter and grid side inverter have been presented. The proposed control strategy can make the wind generator to supply maximum power to the grid and also supply the necessary reactive power to maintain the terminal voltage of the grid constant. Finally, it has been concluded that the proposed control system is suitable enough for variable speed operation of switched reluctance generator under randomly fluctuating wind condition.

References

1. Miller TJE (1993) Switched reluctance motor and their control. Magna Physics Publishing and Clarendon Press, Oxford
2. Miller TJE (1988) Switched reluctance motor drives: a reference book of collected papers. Interec Communications Inc, Ventura, CA, USA, pp 1–69
3. Fitzgerard AE, Kingsley C Jr, Umans SD (1992) Electric machinery. McGraw-Hill, New York, pp 446–487
4. Stephenson JM, Corda J (1979) Computation of torque and current in a doubly salient reluctance motors from nonlinear magnetization data. IEE Proc 126(5):393–396
5. Hasanien HM, Muyeen SM, Tamura J (2010) Switched reluctance machine. Praise Worthy Prize, New York
6. Hasanien HM, Saad NH, Mostfa MA, Badr MA (2006) Steady state performance of axial laminations switched reluctance motor. Proceeding of the 11th international middle east power system conference (MEPCON), El Minia University, Egypt
7. Hasanien HM, Saad NH, Mostfa MA, Badr MA (2006) Speed control of axial lamination switched reluctance motor provided with digital pole placement controller. In proceeding of the 17th international conference on electrical machines (ICEM), Crete island, Greece
8. Hasanien HM (2008) Steady state performance of switched reluctance generator. J Electr Eng 8(1):53–60
9. Cameron DE, Lang JH (1993) The control of high-speed variable reluctance generators in electric power systems. IEEE Trans Ind Appl 29(6):1106–1109
10. Besbes M, Gasbi M, Hoang E, Lecrivian M, Grioni B, Plasse C (2000) SRM design for starter-alternator system. In proceedings international conference electric machines, pp 1931–1935
11. Ferreira C, Jones SR, Heglund W, Jones WD (1993) Detailed design of a 30-kw switched reluctance starter/generator system for a gas turbine engine. Ind Appl Soc Annu Meet 1:97–105 Torento, Canada

12. Kokernak JM, Torrey DA, Kaplan M (1999) A switched reluctance starter/alternator for hybrid electric vehicles. Power electronics proceedings (PCIM) conference, pp 74–80
13. Torrey DA (1993) Variable reluctance generators in wind-energy systems. In proceedings IEEE power electronics specialists conference, pp 561–567
14. Cardenas R, Ray WF, Asher GM (1995) Switched reluctance generators for wind energy applications. In proceedings IEEE power electronics specialists conference (PESC95), pp 559–564
15. Cardenas R, Pena R, Perez M, Clare J, Asher G, Wheeler P (2005) Control of a switched reluctance generator for variable-speed wind energy applications. IEEE Trans Energy Convers 20(4):781–791
16. Heier S (1998) Grid integration of wind energy conversion system. Wiley, Chicester, UK
17. Wasynczuk O (1981) Dynamic behavior of a class of wind turbine generators during random wind fluctuations. IEEE Trans Power Appar Syst 1 PAS-100(6):2837–2854
18. Muyeen SM, Tamura J, Murata T (2008) Stability augmentation of a grid-connected wind farm. Springer Verlag, London. ISBN 978-1-84800-315-6

Chapter 7
Dynamic Model and Control of a Wind-Turbine Generator

Ali Abedini

Abstract This chapter studies the dynamics and control methods of a wind turbine system. Although, there are various types of wind turbines, a general argument applicable to all types of wind turbines is attempted. Since the knowledge of the dynamics of each system is very essential to design a proper controller for that system, therefore, we first develop the dynamic equations of each part of a wind turbine system. The overall model will be the integration of all subsystem models. Next, the control algorithms of a wind turbine will be discussed. Since a wind turbine generator has several levels of control, these levels will be explained in detail. All subsystem control concept and implementation methods are explained after that.

7.1 Wind Turbine Structure

Wind turbines can be fixed speed or variable speed. Therefore, they can have various configurations. Since fixed speed wind turbines have less complicated structure, our argument is more about variable speed wind turbines. Figure 7.1 shows the structure of a wind turbine. The main components of a wind turbine are wind turbine (Blades, gearbox, shaft, hub and …), generator and converter. The converter can be placed in the rotor of the generator like a Doubly Fed Induction Generator (DFIG) Fig. 7.1a, or in the stator part like a Permanent Magnet Synchronous Generator (PMSG), Fig. 7.1b. Nevertheless, the following argument

A. Abedini (✉)
Electrical and Computer Engineering Department, Khaje Nasir Toosi University of Technology, P. O. Box. 16315-1355, 1431714191, Tehran, Iran (ISLAMIC REP.)
e-mail: abedini@eetd.kntu.ac.ir; d.aabedini@gmail.com

S. M. Muyeen (ed.), *Wind Energy Conversion Systems*,
Green Energy and Technology, DOI: 10.1007/978-1-4471-2201-2_7,

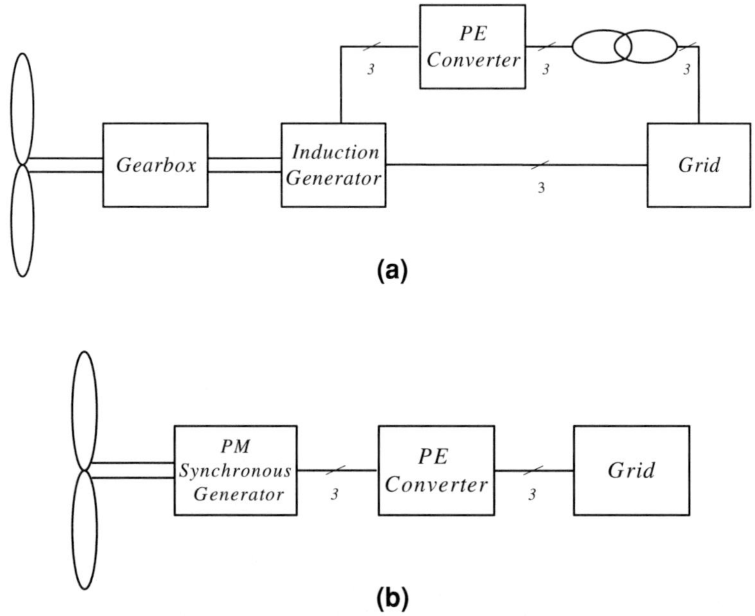

Fig. 7.1 Different configuration of wind turbines with different generators

can be applied for both of them. To find the dynamics of the wind turbine system, we model each section separately. The overall model is gained by integration of the model of each part.

7.2 Wind Turbine Model

In this section, wind turbine behavior will be investigated. Considering the turbine as a black box, its input is wind speed and its output is shaft speed. Since the shaft speed is determined by the control algorithm applied to wind turbines and wind speed, therefore, we first explain some of the control algorithms. The most applied algorithm which is Maximum Power Extraction (MPE) will be discussed in detail. The transfer function of the wind turbine system with MPE algorithm will be gained.

7.2.1 Wind Turbine Control Methods

The performance of the wind turbine not only depends on its hardware but also wind turbine control technique impacts on the performance of the wind turbines. Therefore, the wind turbine control technique can influence the output of the wind

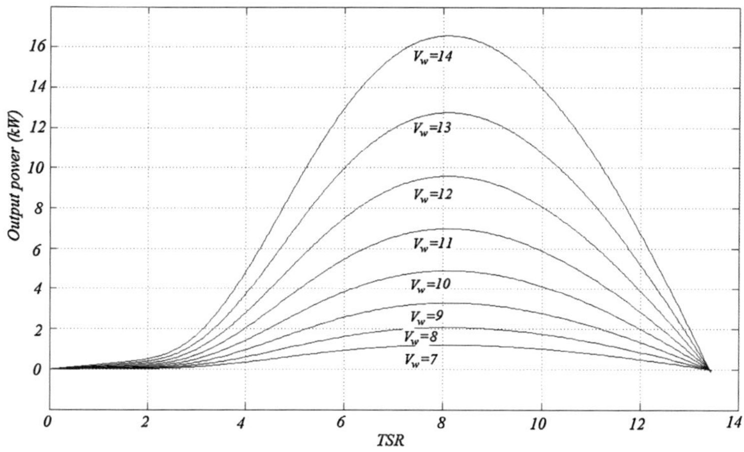

Fig. 7.2 Wind turbine output power versus Tip-Speed Ratio (TSR)

turbine. Fixed speed and variable speed control methods are two traditional control methods. Variable speed control method divides into several control methods.

Equation 7.1 shows the output power of a wind turbine.

$$P = c(\beta, \lambda)\frac{\rho A}{2} v_{\text{wind}}^3 \tag{7.1}$$

$$c(\beta, \lambda) = c_1\left(\frac{c_2}{\lambda_2} - c_3\beta - c_4\right)e^{-\frac{c_5}{\lambda_2}} + c_6\lambda \tag{7.2}$$

$$\frac{1}{\lambda_2} = \frac{1}{\lambda + 0.08\beta} - \frac{0.035}{\beta^3 + 1} \tag{7.3}$$

$$\lambda = \frac{R\omega}{v_{\text{wind}}} \tag{7.4}$$

where $\rho = 1.2\ \text{kg/m}^3$ is the air density, A is the swept area by the turbine blades, λ is the Tip-Speed-Ratio (TSR) given by (4), β is the pitch angle, c is performance coefficient of the turbine given in (7.2) and ω is the generator angular velocity. c_1–c_6 fare some coefficients which are dependent on structure of the wind turbine.

Figure 7.2 shows the output power of a wind turbine versus TSR. Output power changes with TSR. The TSR that corresponds to maximum output power is called optimal TSR. Figure 7.3 shows the output power versus wind turbine speed. Figure 7.3 compares wind turbine control methods. One of the lines shows the fixed speed control method and the other one shows the MPE method. In MPE algorithm the speed of the wind turbine is set so that the maximum power can be extracted from the blowing wind. This speed is called optimal speed. This control method is applicable to variable speed wind turbines and has better efficiency than fixed speed method.

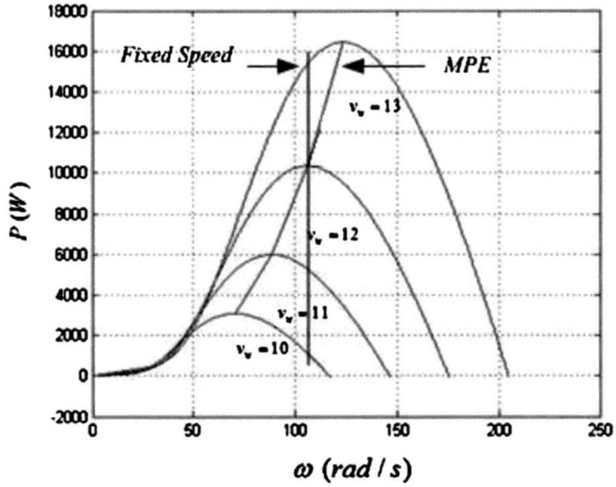

Fig. 7.3 Wind turbine output power versus rotor speed

Other control methods can be implemented in the wind turbines. The control algorithm is designed to achieve to certain criteria for the wind turbines. Therefore, the wind data and geographical characteristics of the wind site are important to design the wind turbine control algorithm.

7.2.2 Dynamics of the Wind Turbine with MPE Algorithm

Wind power variations are converted into kinetic energy and electrical power as follows:

$$\Delta P_w = J\omega_{t0}\frac{d\Delta\omega_t}{dt} + \Delta P_e \tag{7.5}$$

ΔP_w and ΔP_e show the variations in wind power and electrical power and ω_{t0} denotes the initial wind turbine speed. Using (7.1), the following equations are achieved:

$$\Delta P_w = P_{w0}\frac{c'_{p0}}{c_{p0}}\Delta\lambda + 3P_{w0}\frac{\Delta v_w}{v_0} \tag{7.6}$$

$$\Delta\bar{P}_w = \lambda_0\frac{c'_{p0}}{c_{p0}}(\Delta\bar{\omega}_t - \Delta\bar{v}_w) + 3\Delta\bar{v}_w \tag{7.7}$$

where $\Delta\bar{f} = \frac{\Delta f}{f_0}$ when f is P_w, v_w and ω_t. c'_p is the derivative of c_p.

$$\Delta\bar{P}_e = \left(\frac{\partial P_w}{\partial\omega_t}\right)\left(\frac{\omega_{t0}}{P_{e0}}\right)\Delta\bar{v}_w \tag{7.8}$$

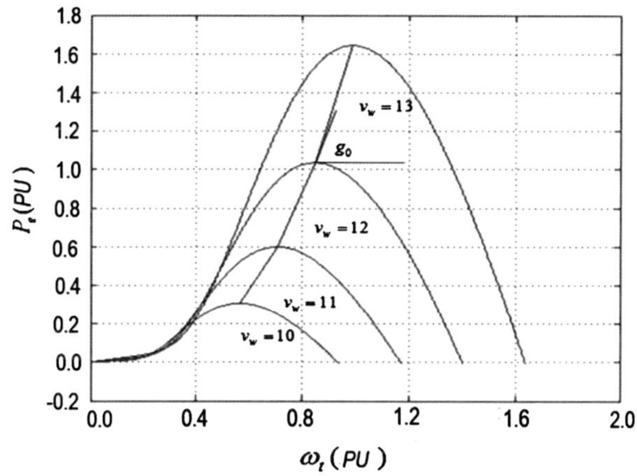

Fig. 7.4 Electrical power versus wind turbine speed

$\frac{\partial P_e}{\partial \omega_t}$ can be obtained from the control technique [1–8]. In MPE algorithm, the graph given below shows this ratio. This parameter is called g. P_{e0} is also equal to P_{w0} in the steady state (Fig. 7.4).

By substituting (7.6) and (7.8) in (7.5), we have:

$$\lambda_0 \frac{c'_{p0}}{c_{p0}}(\Delta\bar{\omega}_t - \Delta\bar{v}_w) + 3\Delta\bar{v}_w = J\omega_{t0}\frac{d\Delta\omega_t}{dt} + g_0\left(\frac{\omega_{t0}}{P_{w0}}\right)\Delta\bar{v}_w \tag{7.9}$$

By applying Laplace transform (7.10) is derived.

$$\frac{\Delta\bar{\omega}_t(s)}{\Delta\bar{v}_w(s)} = \frac{3 - \lambda_0\left(\frac{c'_{p0}}{c_{p0}}\right)}{\tau_0 s - \lambda_0\left(\frac{c'_{p0}}{c_{p0}}\right) + g_0\left(\frac{\omega_{t0}}{P_{w0}}\right)} \tag{7.10}$$

where $\tau_0 = J\frac{\omega_{t0}^2}{P_{w0}}$.

Combining (7.8) in (7.10), the transfer function is achieved as follows.

$$\frac{\Delta\bar{P}_e(s)}{\Delta\bar{v}_w(s)} = \frac{3 - \lambda_0\left(\frac{c'_{p0}}{c_{p0}}\right)g_0\left(\frac{\omega_{t0}}{P_{w0}}\right)}{\tau_0 s - \lambda_0\left(\frac{c'_{p0}}{c_{p0}}\right) + g_0\left(\frac{\omega_{t0}}{P_{w0}}\right)} \tag{7.11}$$

The magnitude of the transfer function of a typical system is depicted in Fig. 7.5.

7.3 Dynamics of the Generator

Depending on which type of generator is used in the wind turbine, the dynamics of the generator will be different. DFIG and PMSG are two types of generators which are used in the wind turbines more than any other generators. The first one is DFIG. Equation 7.12 expresses the state equation of the DFIG.

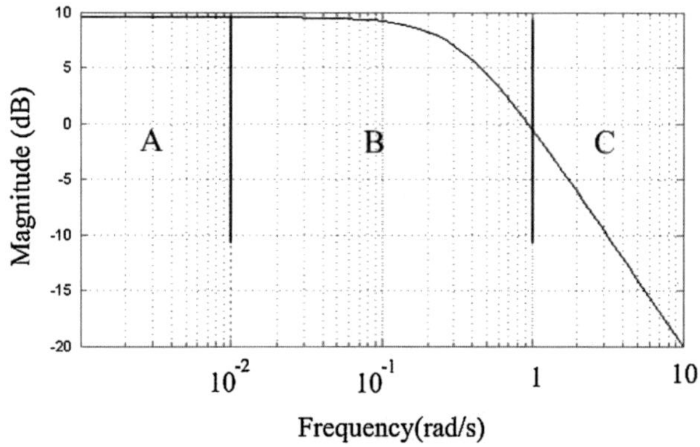

Fig. 7.5 Bode diagram of a wind turbine with MPE algorithm

$$\begin{bmatrix} E_{ds} \\ E_{qs} \\ S \end{bmatrix}^{\cdot} = \begin{bmatrix} -\frac{1}{T_0} & -S\omega_s & -SE_q \\ -S\omega_s & -\frac{1}{T_0} & -SE_d \\ -\frac{I_{ds}}{2J\omega_s} & -\frac{I_{ds}}{2J\omega_s} & 0 \end{bmatrix} \begin{bmatrix} E_{ds} \\ E_{qs} \\ S \end{bmatrix} + \begin{bmatrix} 0 & \frac{\omega_s L_m}{L_{rr}} & 0 \\ -\frac{\omega_s L_m}{L_{rr}} & 0 & 0 \\ 0 & 0 & \frac{1}{2J} \end{bmatrix} \begin{bmatrix} v_{ds} \\ v_{qs} \\ T_m \end{bmatrix} \quad (7.12)$$

where T_0 is time constant, L_m, L_{ss}, L_{rr} are mutual, stator and rotor inductance and S is slip.

The other type of the generator is PMSG. Equation 7.13 shows the dynamic of the generator.

$$\begin{bmatrix} i_{qs} \\ i_{ds} \\ \omega_g \end{bmatrix}^{\cdot} = \begin{bmatrix} \frac{R_s}{L_{qs}} & -\omega_s \frac{L_{ds}}{L_{qs}} & -\frac{\lambda_f}{L_{qs}} \\ \omega_s \frac{L_{qs}}{L_{ds}} & \frac{R_s}{L_{ds}} & 0 \\ -\frac{1.5P^2}{4J}\lambda_f & -\frac{1.5P^2}{4J}(L_{ds} - L_{qs})i_{qs} & -\frac{BP}{2J} \end{bmatrix} \begin{bmatrix} i_{qs} \\ i_{ds} \\ \omega_s \end{bmatrix} + \begin{bmatrix} \frac{-1}{L_{qs}} & 0 & 0 \\ 0 & \frac{-1}{L_{ds}} & 0 \\ 0 & 0 & \frac{P}{2J} \end{bmatrix} \begin{bmatrix} v_{qs} \\ v_{ds} \\ T_m \end{bmatrix} \quad (7.13)$$

where R_s is the stator resistance, L_{ds} and L_{qs} are the d–q axis inductances, P is number of poles, J is the rotor inertia, v_{ds} and v_{qs} are terminal voltage, i_{ds} and i_{qs} are d and q components of terminal currents, T_m is the input mechanical torque of the wind turbine, λ_m is the magnitude of the flux produced by the permanent magnet, $\omega_g = \frac{P}{2}\omega_m$ is the electrical rotor speed and ω_m is the mechanical rotor speed.

7.4 Dynamics of the Power Electronic Converter

Figure 7.6 shows a detailed representation of the back-to-back converter of Fig. 7.1. It includes a converter at the generator side which is connected to the DC-link of a voltage-sourced converter at the grid side. The generator-side

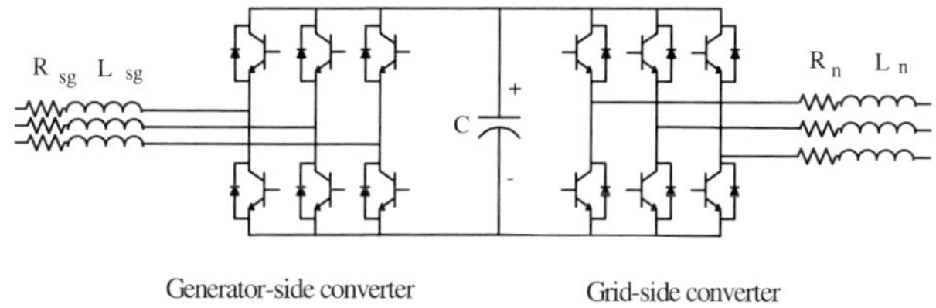

Fig. 7.6 The topology of the power electronic converter system

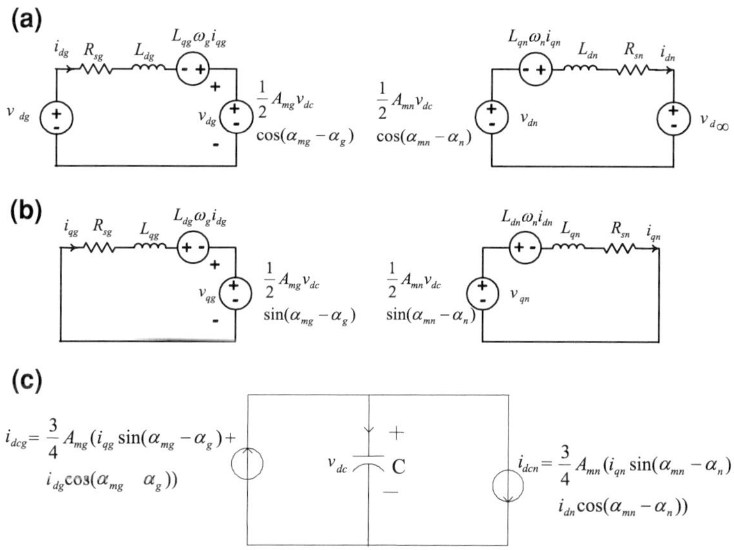

Fig. 7.7 Dynamic model of the power electronic converter

converter rectifies the generator output voltage to DC voltage which is then converted to ac three phase voltage by the grid-side converter. In the figure, l_{sg} is Generator synchronous inductance, L_n is Network inductance, R_{sg} is Generator stator resistance, R_n is Network resistance and C is the DC-link capacitance.

Figure 7.7 shows a dynamic model of the power electronic converter. The model includes dq representations of the generator side and the grid side. Details related to development of the dynamic model of Fig. 7.6 are presented in [2].

Based on the model of Fig. 7.7, the wind conversion system can be represented by the following mathematical equations.

$$\begin{bmatrix} v_{dg} \\ 0 \end{bmatrix} = \begin{bmatrix} R_{sg} + L_{dg}D & -\omega_g L_{qg} \\ \omega_g L_{dg} & R_{sg} + L_{qg}D \end{bmatrix} \begin{bmatrix} i_{dg} \\ i_{qg} \end{bmatrix} + \begin{bmatrix} \frac{1}{2} A_{mg} v_{dc} \cos(\alpha_{mg} - \alpha_g) \\ \frac{1}{2} A_{mg} v_{dc} \sin(\alpha_{mg} - \alpha_g) \end{bmatrix} \tag{7.14}$$

$$\begin{bmatrix} v_{d\infty} \\ 0 \end{bmatrix} = \begin{bmatrix} -R_{sn} - L_{dn}D & \omega_n L_{qn} \\ -\omega_n L_{dn} & -R_{sn} - L_{qn}D \end{bmatrix} \begin{bmatrix} i_{dn} \\ i_{qn} \end{bmatrix} + \begin{bmatrix} \frac{1}{2} A_{mn} v_{dc} \cos(\alpha_{mn} - \alpha_n) \\ \frac{1}{2} A_{mn} v_{dc} \sin(\alpha_{mn} - \alpha_n) \end{bmatrix} \tag{7.15}$$

$$\frac{dv_{dc}}{dt} = 1/C(i_{dcg} - i_{dcn}) = 1/C\left(\frac{p_g}{v_{dc}} - \frac{p_n}{v_{dc}}\right) \tag{7.16}$$

where A_{mx} is Amplitude of the voltage phasor x of the inverter output, i_{dcg} is DC-link current coming from the generator, i_{dcn} is DC-link current going to the network, i_{dg} is d component of the generator current, i_{dn} is d-axis of the network current, i_{qg} is q component of the generator current, i_{qn} is q-axis of the network current, v_{dc} is DC-link voltage, v_{dg} is d component of the generator voltage, $v_{d\infty}$ is d-axis of the network voltage, v_{qg} is q component of the generator voltage, $v_{q\infty}$ q-axis of the network voltage, L_{dg} is generator d-axis inductance, L_{qg} is generator q-axis inductance, L_{dn} is network d-axis inductance, L_{qn} is network q-axis inductance, P_g is generator power, P_n is Network power, α_{mx} is phase angle of voltage phasor x of the inverter output and D denotes $\frac{d}{dt}$.

7.5 Control of Wind Turbines

In previous sections we have investigated the dynamics of the various parts of the wind turbine system. This is very essential when we design controllers. In a wind turbine system, there are different levels of control which are overall control, system control and subsystem control. The highest level of control is overall control or supervisory control. This level of control determines turbine start and stop times, wind turbine reactions to wind speed variations, and also checks the health of the wind turbine. The second level of control is called system control or operational control which specifies how the wind turbine system must operate in each wind speed. Subsystem control is the control technique which is associated with each part of a wind turbine system such as generator, power electronic converter, yaw drive, pitch drive and other actuators of the wind turbine.

7.5.1 Yaw Control

Large wind turbines have an actuator which aligns the wind turbine axis with wind direction. If a motor is assigned for this purpose, this wind turbine is called active yaw control. Sometimes this control can be done by special shaping of the wind turbine which is called passive yaw control.

The optimal speed of the yaw control system is 1°/s.

7.5.2 Pitch Control

Usually a wind turbine has three blades. Therefore, there are three actuators which control the pitch of each blade separately. In order to get the maximum aerodynamic torque the pitch angle of the blade must be different at different positions.

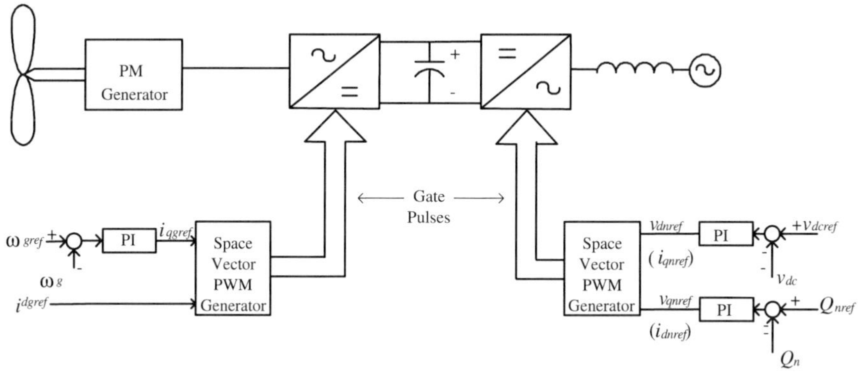

Fig. 7.8 Overall study system including the power and control sub-systems

The speed of the pitch actuator is different from 18°/s for a 600 kW wind turbine down to 8°/s for a 5 MW wind turbine. Furthermore, pitch controllers can limit the power by pushing the pitch angle of the blades to stall point. This will protect the wind turbine from the damage caused by high speed wind. The dynamic of the blade can be simplified down to a first-order system. And pitch controller cab designed based on the dynamics of the blades and desired speed.

7.5.3 Generator Control

The generator controller must adjust the speed of the wind turbine. Therefore, the electrical torque of the generator must be controlled. Depending on the type of the generator used in the wind turbine, the structure of the controller is different. One can implement different switching algorithms such as DTC, vector control and so on for different purposes.

Figure 7.8 shows the block diagram representation of a controller for the generator-side converter of Fig. 7.1. To capture a maximum power from the wind, the generator reference speed ω_{gref} is calculated based on the measured wind speed and the optimal tip speed ratio. The generator reference speed and the wind speed are related by $\omega_{\mathrm{gref}} = \frac{\lambda_{\mathrm{t}} v_{\mathrm{w1}}}{r_{\mathrm{t}}}$, where v_{w1} is the wind speed, λ_{t} is the optimal tip speed ratio and r_{t} is the wind turbine radius. The reference speed is compared with the measured generator speed and the error is given to a PI controller whose output is the generator q-axis current reference i_{qgref}. The generator d-axis reference current i_{dgref} is proportional to the generator output real power and is specified based on the available power from the wind. The generator d- and q-axis current references input a block that controls the generator-side converter by space vector modulation [3].

Similarly, the same control algorithm can be developed for DFIG.

7.5.4 Power Electronic Converter Control

In variable speed wind turbines, the power electronic converter is assigned for power conditioning. The grid-side converter regulates the active and reactive power delivered to the grid. In a wind conversion system, the DC-link voltage may have large fluctuations as a result of the wind speed changes and the resultant generator output power variations. Therefore, the control of grid-side converter must ensure a constant voltage at the DC-link in the presence of large generator voltage and power fluctuations. The grid-side converter must also regulate the exchange of reactive power with the grid to control the load voltage.

Figure 7.8 shows the block diagram representation of a control system for the grid-side converter for PMSG wind turbine system. The inputs are v_{dcref} which is the reference DC-link voltage and Q_{nref} which is the reference reactive power specified by the utility system. The converter terminal currents and line voltage are measured and used to estimate the reactive power Q. Comparing the measured reactive power and DC-link voltage with the corresponding references, the PI controllers provide references for the dq voltages. The current references are transformed to the corresponding abc current references. They are compared with the measured phase currents and the errors enter the switching signal generator.

The following provides key equations used to design the control system of Fig. 7.8. Equation 7.17 represents a dynamic model for the DC-side circuit of the converter system. The model is nonlinear since the state variable v_{dc} appears in the denominator. To design a controller for the grid-side converter, the model of (7.17) is linearized employing the feedback linearization technique [5, 6]. Let us define $\hat{i}_{dc}$ as

$$\hat{i}_{dc} = \frac{p_g}{v_{dc}} - \frac{p_n}{v_{dc}} \tag{7.17}$$

The linearized model of the DC-side circuit of converter from (7.15) is

$$C\frac{dv_{dc}}{dt} = \hat{i}_{dc} \tag{7.18}$$

The first-order dynamic model of (7.18) is used to design the controller which is effective for all operating conditions of the system [7]. In the controller of Fig. 7.8, the real and reactive power components at the grid are calculated from

$$P_n = \frac{3}{2} v_{d\infty} i_{dn} \tag{7.19}$$

$$Q_n = \frac{3}{2} v_{d\infty} i_{qn} \tag{7.20}$$

Note that P_n, and Q_n are proportional to i_{dn} and i_{qn}, respectively, since it is assumed that the q-axis component of network voltage ($v_{q\infty}$) is zero.

References

1. Abbey C, Joos G (2006) Attenuation of wind power fluctuations in wind turbine generators using a DC bus capacitor based filtering control scheme. Proc 41st IEEE Ind Appl Ann Meet 1:216–221
2. Nikkhajoei H, Iravani R (2007) Dynamic model and control of AC–DC–AC voltage-sourced converter system for distributed resources. IEEE Trans Power Delivery 22(2):1169–1178
3. Vas P (1994) Vector control of AC machines. Clarendon Press, Oxford
4. Abbey C, Joos G (2007) Supercapacitor energy storage for wind energy applications. IEEE Trans Ind Appl 43(3):769–776
5. Luo C, Ooi BT (2006) Frequency deviation of thermal power plants due to wind farms. IEEE Trans Energy Convers 21(3):708–716
6. Ackermann T (2005) Wind power in power systems. Wiley, New York
7. Lee TS (2003) Input-output linearization and zero-dynamics control of three-phase AC/DC voltage-source converters. IEEE Trans Power Electron 18(1):11–22
8. Abedini A, Nikkhajoei H (2011) Dynamic model and control of a wind-turbine generator with energy storage. IET Trans Renew Energy, pp 67–78

Part II
Prime Issues for Wind Industry

Chapter 8
Voltage Flicker Measurement in Wind Turbines

J. J. Gutierrez, P. Saiz, A. Lazkano, J. Ruiz, L. A. Leturiondo and I. Azkarate

Abstract In the last years the use of wind farms has drastically increased. The question is how those new generation systems will affect to the whole grid. In principle, wind energy can be considered a risky source in terms of power quality that must be certified on the basis of measurements performed according to international standards and guidelines. The IEC 61400-21 standard is the reference normative for the certification of the power quality of wind turbines. The first edition was published in 2001 and it specifies the main relevant features of power quality that should be measured in a wind turbine as well as the procedures for their measurement and assessment. According to it, measurements should be performed for harmonic content, flicker, voltage drops as well as active and reactive power, during normal and switching operations. Among those disturbances, flicker has the most complex and sensitive testing procedure.

Power fluctuations of the wind turbine generate voltage fluctuations causing changes in the luminous flux from light sources. Such changes may produce a very unpleasant visual sensation, known as flicker, leading to complaints from utility customers. The IEC 61000-4-15 standard describes the functional specifications of a flicker measurement device and provides a short-term indicator, P_{st}, to characterize the discomfort. This chapter demonstrates the sensitivity of the procedure defined by the IEC 61400-21 standard to asses the flicker by means of different signal processing techniques.

J. J. Gutierrez (✉) · P. Saiz · A. Lazkano · J. Ruiz · L. A. Leturiondo · I. Azkarate
Departamento de Electronica y Telecomunicaciones (UPV-EHU), Escuela Tecnica Superior de Ingenieria de Bilbao, Alda. Urquijo s/n, 48013 Bilbao, Bizkaia, Spain
e-mail: josejulio.gutierrez@ehu.es

S. M. Muyeen (ed.), *Wind Energy Conversion Systems*,
Green Energy and Technology, DOI: 10.1007/978-1-4471-2201-2_8,

8.1 Introduction

Since the early 1990s, wind power has been one of the fastest-growing energy sources in the world, and annual predictions are easily exceeded. According to a report released by the Global Wind Energy Council, the installed wind capacity of the world will increase from 158 GW at the end of 2009, to more than 400 GW by the end of 2013 [2]. Such numbers pose the challenge of integrating the increasing levels of wind power into the power system. One of the key aspects of this challenge is to ensure compliance with grid regulations and standards to meet power quality issues [15].

Grid-connected wind turbines may affect power quality. Since the end of the last century it has been confirmed that the injection of power from wind turbines can be considered as a risky source of interferences that compromise voltage quality [7]. The causes are mostly turbine-specific (for instance, the tower shadow effect, blade pitching, and yaw errors), but can also be attributed to meteorological or geographic conditions (for instance, variability in wind speed, turbulence, and wind shear). The International Electrotechnical Commission (IEC) set the definitions, rules, and procedures for the characterization of waveform quality in a wind turbine connected to the grid in the IEC 61400-21 standard [5]. The first edition of this standard was published in 2001 and was preceded by the research project *European Wind Turbine Testing Procedure Developments* [18], funded by the European Union. This standard was updated and reedited in 2008. It aims to define and specify the magnitudes that need to be determined for wave quality characterization, measurement procedures for quantifying the characteristics, and procedures for assessing compliance with power quality requirements. The power quality characteristics include wind turbine specifications, voltage quality (emission of flicker and harmonics), voltage drop response, power control (control of active and reactive power), grid protection, and reconnection time. The international Measuring Network of Wind Energy Institute (MEASNET) has defined some guidelines based on the above mentioned standard with the aim of adapting the procedures and hence the measurement results obtained by its members [12].

Among the power quality disturbances, flicker has the most complex testing procedure. Power fluctuations of the wind turbine cause voltage fluctuations. The typical sources of these fluctuations are the tower shadow, yaw errors, wind shears, wind turbulence, or fluctuations in the control system [1, 14]. Voltage fluctuations are defined as variations in voltage with amplitude below 10% of the nominal value. These fluctuations may cause changes in the illumination intensity of light sources. Such changes may produce a very unpleasant visual sensation, known as flicker, leading to complaints from utility customers. The IEC 61000-4-15 standard [6] provides a description of the structure and functional specifications of a flicker measurement device called a flickermeter. This measurement tool represents the relationship between voltage fluctuations and the human discomfort, providing a short-term indicator, P_{st}, and a long-term indicator, P_{lt}. The P_{st} is the flicker severity evaluated over a short period (10 min is used in practice) and the conventional threshold of irritability is set in $P_{st} = 1$.

This chapter aims to give an in-depth overview of flicker characterization in grid-connected wind turbines. It includes the description of a system designed for flicker measurement according to the IEC standard, detailing both hardware features and the digital signal processing module. The most critical point of the complete flicker measurement procedure is solving the fictitious grid proposed by the standard. Regarding this aspect, the error sensitivity of the parameters involved in flicker evaluation is analyzed thoroughly. Therefore, the main objective is to study the effects produced by different signal processing methods used for the implementation of the fictitious grid.

8.2 Test Procedure for Voltage Fluctuations in Wind Turbines

The IEC 61400-21 standard defines the procedure for measuring and assessing the flicker caused by grid-connected wind turbines [5]. Variable meteorological conditions and the constructive characteristics of the wind turbines are the main sources of power fluctuations during normal operation of wind turbines, which in turn produce flicker [10]. However, there is another source of flicker in wind turbines, namely the start, stop, and switching between generators or generator windings. Any one of these events may cause a variation in power production, which also produces flicker [9]. Therefore the IEC 61400-21 standard specifies that flicker caused by a wind turbine has to be characterized for two situations: continuous operation and switching operations. Continuous operation refers to the normal operation of the wind turbine excluding start-up and shutdown operations and switching between generators or generator windings. The operations that are excluded from normal operation are taken into account under what is called switching operations.

For continuous operation a wind turbine is characterized by the flicker coefficients, $c(\psi_k, v_a)$. These coefficients are non-site-specific, and are obtained as a result of the processing and subsequent statistical evaluation of a number of current and voltage time-series measured at the wind turbine terminals during continuous operation. Based on these coefficients, the flicker emission from a single wind turbine (or a group of wind turbines) in a specific site can be assessed.

Similarly, for switching operations a wind turbine is characterized by the maximum number of switching operations within a 10 min and a 2 h period, $N_{10\,m}$ and $N_{120\,m}$ respectively, the flicker step factors $k_f(\psi_k)$ and the voltage change factors $k_u(\psi_k)$, which are also non-site-specific. The maximum number of switching operations can be provided by the manufacturer, whereas the flicker step and voltage change factors have to be measured and calculated from a number of current and voltage time-series measured at the wind turbine terminals during switching operations. Based on these factors, the flicker emission and voltage changes caused by switching operations in a specific site can be estimated.

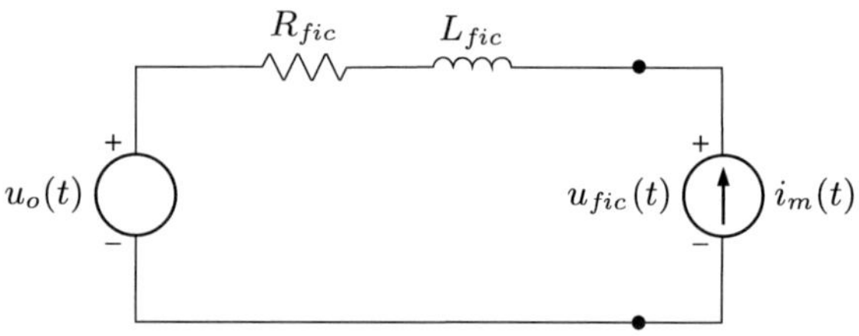

Fig. 8.1 Fictitious grid used for flicker assessment in wind turbines

In the preliminary stage of the standardization of power quality for wind turbines, power signal-based flicker evaluation was proposed. The working group of the technical committee in charge of the development of the standard compared this method with others based on current and voltage signals. Finally, IEC 61400-21 proposed that flicker evaluation be based on the current and voltage time-series measured at the wind turbine terminals. However, the voltage fluctuations of the grid at the point where flicker is measured are not caused only by the wind turbine itself. The wind turbine being tested is usually connected to a medium-voltage network that has other fluctuating loads that may cause significant voltage fluctuations. In addition, the voltage fluctuations imposed by the wind turbine depend on the grid conditions. To provide non-site-specific coefficients regardless of these facts, the standard establishes a model that causes results to be independent of the voltage fluctuations caused by the grid to which the wind turbine is connected. This model is based on a fictitious grid that enables evaluation of voltage fluctuations caused exclusively by the wind turbine.

8.2.1 Fictitious Grid

The standard establishes a specific test procedure with the aim of obtaining a normalized measure of the flicker emission. The phase–to–neutral voltage and the line current, $u_{\mathrm{m}}(t)$ and $i_{\mathrm{m}}(t)$, need to be acquired with a digital data acquisition system and processed following the model of the fictitious grid.

The fictitious grid is shown in Fig. 8.1. The turbine is represented by a current generator with a value of $i_{\mathrm{m}}(t)$ (the instantaneous phase current of the turbine), and the grid to which it is connected is represented by its Thevenin equivalent circuit. The grid impedance is given as a resistance R_{fic} in series with an inductance L_{fic}. Values for these elements are chosen to reflect an appropriate grid impedance phase angle ψ_{k} and short-circuit apparent power $S_{\mathrm{k,fic}}$ for the fictitious grid. The grid's ideal voltage source $u_0(t)$ models a grid free of fluctuations and is defined as:

$$u_0(t) = \sqrt{\frac{2}{3}} \cdot U_{\mathrm{n}} \sin \alpha_{\mathrm{m}}(t), \tag{8.1}$$

where U_n is the nominal voltage of the wind turbine.

The standard defines that this modeled signal can be generated in different ways, but always fulfilling these conditions:

1. $u_0(t)$ should be without any fluctuation, i.e. the flicker on this voltage should be zero;
2. $u_0(t)$ should have the same electrical angle $\alpha_m(t)$ as the fundamental of the measured voltage $u_m(t)$.

The electrical angle $\alpha_m(t)$ of the fundamental frequency can be described as:

$$\alpha_m(t) = 2\pi \int_0^t f(t)dt + \alpha_0 \tag{8.2}$$

where $f(t)$ is the fundamental frequency of $u_m(t)$, which may vary over time, and α_0 is the electrical angle of the fundamental frequency at $t = 0$.

After $u_0(t)$ is generated, the simple model can be solved and the fictitious voltage $u_{fic}(t)$ at the wind turbine terminals is calculated as:

$$u_{fic}(t) = u_0(t) + R_{fic} \cdot i_m(t) + L_{fic} \cdot \frac{di_m(t)}{dt} \tag{8.3}$$

The voltage $u_{fic}(t)$ will be then input to the processing systems specified in the subsequent sections.

8.2.2 Continuous Operation

As previously stated, the flicker coefficients that characterize flicker in continuous operation have to be estimated from the current and voltage time-series measured at the wind turbine terminals, $i_m(t)$ and $u_m(t)$. As flicker caused by a wind turbine also depends on the wind speed [10], the standard specifies that input current and voltage time-series should be measured at different wind speeds. More specifically, at least fifteen 10 min time-series of instantaneous voltage and currents (five tests, three phases) have to be recorded for each 1 m/s wind speed bin between the cut-in wind speed and 15 m/s.

The procedure defined by the standard to obtain the flicker coefficients and assess flicker is shown in Fig. 8.2. Each registered time-series pair $[i_m(t), u_m(t)]$ is input to the fictitious grid. In fact this is done four times, one for each of the network impedance phase angle values specified in the standard ($\psi_k = 30, 50, 70$ and $85°$) to represent different grid conditions, and the corresponding fictitious voltage $u_{fic}(t)$ is calculated for each of them. Using $u_{fic}(t)$ as the input to a IEC 61000-4-15 compliant flickermeter, a flicker emission value, $P_{st,fic}$, is obtained. Each $P_{st,fic}$ is then normalized to a flicker coefficient $c(\psi_k)$ by applying (8.4):

$$c(\psi_k) = P_{st,fic} \cdot \frac{S_{k,fic}}{S_n} \tag{8.4}$$

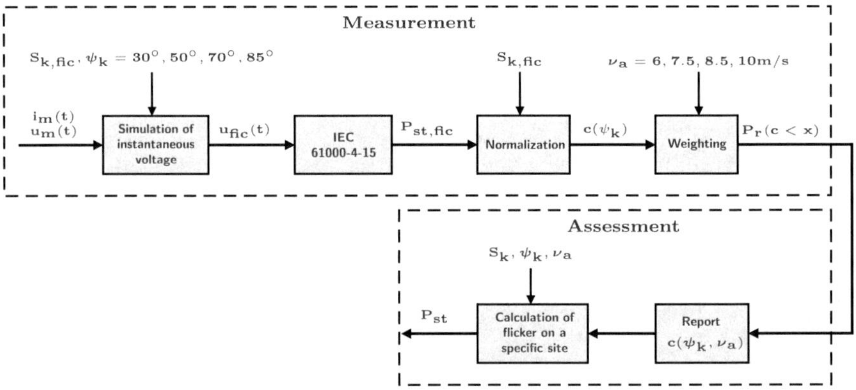

Fig. 8.2 Measurement and assessment procedures for flicker during continuous operation of the wind turbines in accordance with IEC 61400-21

where S_n is the rated apparent power of the wind turbine and $S_{k,fic}$ is the short-circuit apparent power of the fictitious grid.

The normalized coefficients obtained for each network impedance phase angle ψ_k are then weighted for four different wind speed distributions (Rayleigh distributions) with different annual average wind speeds ($v_a = 6,\ 7.5,\ 8.5$ and $10\ \text{m/s}$), as annual average wind speed varies from site to site. The weighted accumulated distribution of the measured flicker coefficients, $P_r(c<x)$ [the probability of $c(\psi_k)$ value being less or equal to the value x], is obtained for each wind speed distribution. For each weighted accumulated distribution, the 99% percentile, $c(\psi_k, v_a)$, is then reported. That is, flicker coefficients $c(\psi_k, v_a)$ are estimated for four different network impedance phase angles, $\psi_k = 30,\ 50,\ 70$ and $85°$, and for four different wind speed distributions with annual average wind speed $v_a = 6,\ 7.5,\ 8.5$ and $10\ \text{m/s}$.

The assessment procedure specifies how to use the reported flicker coefficients to estimate the flicker emission from a single wind turbine or a group of wind turbines in a specific location. For a single wind turbine, the following expression is used:

$$P_{st} = P_{lt} = c(\psi_k, v_a) \cdot \frac{S_n}{S_k} \tag{8.5}$$

where S_n is the rated apparent power of the wind turbine, S_k is the short-circuit apparent power at the Point of Common Coupling (PCC), and $c(\psi_k, v_a)$ is the flicker coefficient of the wind turbine for the actual ψ_k and v_a at the site, which can be obtained by interpolation of the 4×4 estimated flicker coefficients.

In case more wind turbines are connected to the PCC, the flicker emission due to the sum of them can be estimated as:

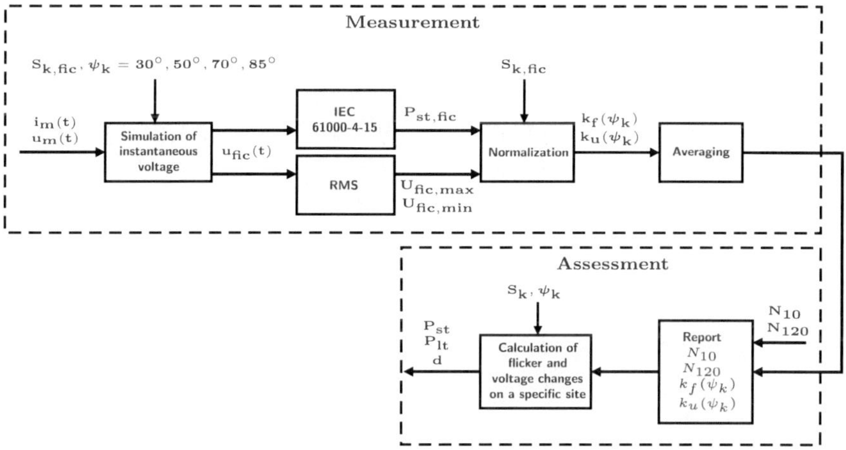

Fig. 8.3 Measurement and assessment procedures for flicker during switching operations of the wind turbines in accordance with IEC 61400-21

$$P_{\text{st}\Sigma} = P_{\text{lt}\Sigma} = \frac{1}{S_\text{k}} \cdot \sqrt{\sum_{i=1}^{N_\text{wt}} \left[c_\text{i}(\psi_\text{k}, v_\text{a}) S_{\text{n,i}}\right]^2} \tag{8.6}$$

where $c_i(\psi_\text{k}, v_\text{a})$ is the flicker coefficient of an individual wind turbine, $S_{\text{n,i}}$ is the rated apparent power of the individual wind turbine and N_wt is the number of wind turbines connected to the PCC.

Finally, the short and long-term flicker emission from the wind turbine installation must be compared with the short and long-term flicker emission limits for the relevant PCC.

8.2.3 Switching Operations

The standard establishes an alternative processing and statistical evaluation scheme during start-up or switching between generators (see Fig. 8.3).

The specification establishes a procedure of measurements and subsequent calculations to determine $k_\text{u}(\psi_\text{k})$ and $k_\text{f}(\psi_\text{k})$ for each type of switching operation. The phase–to–neutral voltage and the line current, $u_\text{m}(t)$ and $i_\text{m}(t)$, need to be processed for at least 15 registers of a period T_p long enough to pass the transient of the switching operation. As in the case of the continuous operation, the fictitious voltage, $u_\text{fic}(t)$, and the flicker emission values, $P_{\text{st,fic}}$, are calculated. Flicker step factor and voltage change factor can be obtained by applying (8.7) and (8.8) respectively, and finally they are determined as the average result of the 15 calculated values.

$$k_{\mathrm{f}}(\psi_{\mathrm{k}}) = \frac{1}{130}\frac{S_{\mathrm{k,fic}}}{S_{\mathrm{n}}} \cdot P_{\mathrm{st,fic}} \cdot T_{\mathrm{p}}^{0.31} \tag{8.7}$$

$$k_{\mathrm{u}}(\psi_{\mathrm{k}}) = \sqrt{3} \cdot \frac{U_{\mathrm{fic,max}} - U_{\mathrm{fic,min}}}{U_{\mathrm{n}}} \cdot \frac{S_{\mathrm{k,fic}}}{S_{\mathrm{n}}} \tag{8.8}$$

where $U_{\mathrm{fic,max}}$ and $U_{\mathrm{fic,min}}$ are the maximum and minimum one period root mean square (RMS) value of the voltage on the fictitious grid during the switching operation.

The assessment procedure for switching operations specifies how to estimate the flicker emission and voltage changes during switching operations on any specified site, using the reported flicker step factors and voltage change factors. For a single wind turbine, the following expressions are used:

$$P_{\mathrm{st}} = 18 \cdot N_{10\mathrm{m}}^{0.31} \cdot K_{\mathrm{f}}(\psi_{\mathrm{k}}) \cdot \frac{S_{\mathrm{n}}}{S_{\mathrm{k}}} \tag{8.9}$$

$$P_{\mathrm{lt}} = 8 \cdot N_{120\mathrm{m}}^{0.31} \cdot K_{\mathrm{f}}(\psi_{\mathrm{k}}) \cdot \frac{S_{\mathrm{n}}}{S_{\mathrm{k}}} \tag{8.10}$$

In the case that more wind turbines are connected to the PCC, the flicker emission from the sum of them can be estimated from (8.11) and (8.12):

$$P_{\mathrm{st}\Sigma} = \frac{18}{S_{\mathrm{k}}} \cdot \left(\sum_{i=1}^{N_{\mathrm{wt}}} N_{10_{\mathrm{m,i}}} \cdot \left[k_{\mathrm{f,i}}(\psi_{\mathrm{k}}) \cdot S_{\mathrm{n,i}}\right]^{3.2} \right)^{0.31} \tag{8.11}$$

$$P_{\mathrm{lt}\Sigma} = \frac{18}{S_{\mathrm{k}}} \cdot \left(\sum_{i=1}^{N_{\mathrm{wt}}} N_{120_{\mathrm{m,i}}} \cdot \left[k_{\mathrm{f,i}}(\psi_{\mathrm{k}}) \cdot S_{\mathrm{n,i}}\right]^{3.2} \right)^{0.31} \tag{8.12}$$

where $N_{10_{\mathrm{m,i}}}$ and $N_{120_{\mathrm{m,i}}}$ are the number of switching operations of the individual wind turbine within a 10 min and 2 h period respectively, and $k_{\mathrm{f,i}}(k)$ is the flicker step factor of the individual wind turbine.

Finally, the obtained short and long-term flicker parameters have to be compared with the established flicker emission limits.

8.3 A System for the Measurement of the Power Quality Characteristics

The assessment of the power quality characteristics of a wind turbine requires the obtaining of several voltage and current time-series for different wind speeds.

All that information must be processed to measure several parameters, which requires, in principle, the connection of different power quality analyzers. There are not so many commercial analyzers particularly designed to fulfill the requirements of the IEC 61400-21 standard. We developed a measurement system for the acquisition, storage and processing of the voltage, current and wind speed time-series required by the standard. In order to provide more flexibility to the measurement system, instead of using a commercial equipment, we chose the implementation of our own system. The main rationale of our system is to divide the whole measurement into independent processes:

1. The recording and storage of the wind, voltage and current time-series.
2. The off-line measurement and assessment of the power quality characteristics by post-processing the stored time-series.

This philosophy of measurement provides several advantages for the wind turbine certification process. Since the power quality standards may be changed, the off-line processing facilitates these changes to be accommodated by software modification. The post-processing allows the analysis of the waveforms producing a specific flicker characteristic without being necessary to retest the wind turbine. Finally, all parties involved in the certification process have access to the stored information.

To perform those processes we developed two interconnected sub-systems: a signal conditioning system (SAC-2) and a control and processing system (SARPE 2.1). The scheme of the overall measurement system is shown in Fig. 8.4. The conditioning system is a hardware module transforming the three-phase voltage and current, as well as the wind characteristics, to the appropriate levels for post-processing. That operation is managed by the control system. This is a software tool that controls the acquisition and stores the voltage, current and wind speed time-series. The control system also includes a post-processing module that works off-line by reading the recorded time-series and calculating the parameters of the power quality characteristics of the wind turbine.

8.3.1 Conditioning System SAC-2

This system converts the voltage, current and wind characteristics to the appropriate levels for the final measurement. Figure 8.5 shows a photograph of the developed hardware platform, SAC-2.

For a precise conditioning of the input levels, this hardware platform provides four voltage channels and four different scales per channel (see Table 8.1a).

There are also four current channels and four different types of current sensors (see Table 8.1b). There are two additional analog channels to register wind characteristics.

The system provides several clocks to be used by the control module as external sampling frequencies in the acquisition. The first set of sampling frequencies allows the use of a number of samples per cycle of 50 Hz always corresponding to a power of 2 ($f_s = 1,600,\ 3,200,\ 6,400,\ 12,800$ and $25,600$ S/s). A second set provides a number of samples per 10 cycles of 50 Hz or 12 cycles of 60 Hz

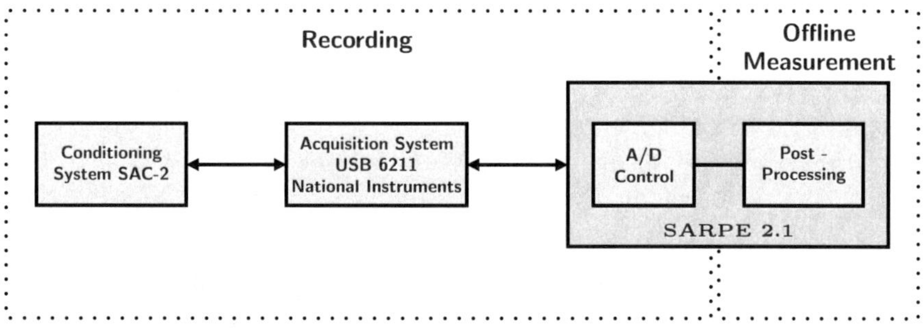

Fig. 8.4 Scheme of the measurement system

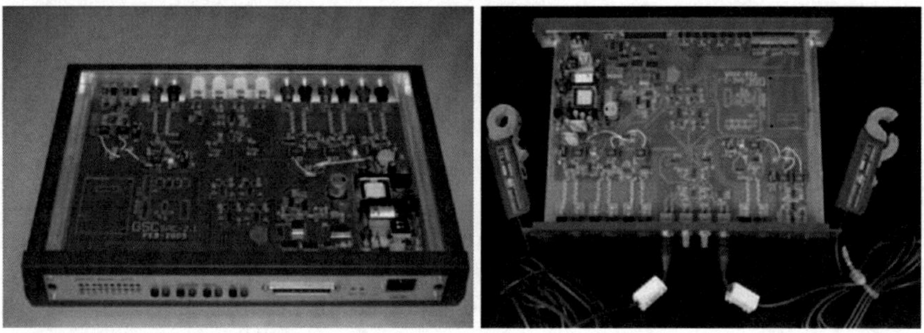

Fig. 8.5 Layout of the conditioning system, SAC-2

Table 8.1 Conditioning system working range (a) Working ranges for the voltage channels. (b) Working ranges for the current sensors

	(a)	
Scale (V_{rms})	Range (V_{rms})	
60	16–78	
150	40–195	
300	78–390	
600	156–780	
	(b)	
Sensitivity (mV/A)	Scale (A_{rms})	Range (A_{rms})
0.1	5,000	1,300–6,500
	500	130–650
1	500	130–650
	50	13–65
10	50	13–65
	5	1.3–6.5
100	50	1.3–6.5
	0.5	0.13–0.65

corresponding to a power of 2 ($f_s = 1,280,\ 2,560,\ 5,120,\ 10,240$ and 20, 480 $\frac{S}{s}$). This second group of sampling frequencies makes possible the implementation of the harmonics and interharmonics measurement method specified by the standard IEC 61400-21. Each channel includes a fifth order Butterworth anti-aliasing filter with adjustable cutoff frequency. A phase-locked loop (PLL) synchronizes the sampling rate to the first channel grid frequency, either 50 or 60 Hz.

The system also includes four digital inputs, activated by dry contacts, to trigger the start of the acquisition.

8.3.2 Control System SARPE 2.1

The control system is a MATLAB tool running on a laptop and consists of two modules: the acquisition module and the post-processing module.

The acquisition module manages both the acquisition and the recording operations. Its main functions are:

1. Acquisition of the conditioned signals using the USB 6211 card from National Instruments (16-bit resolution).
2. Selection of different parameters that configure the register:
   a. Internal clock from the acquisition card or external clock from the conditioning system.
   b. Sampling frequency.
   c. Use of the anti-aliasing filters.
   d. Activation of the PLL.
   e. Channels to be recorded.
   f. Scale of each selected voltage and current channels.
   g. Duration of the register.
3. Checking the functional status of the wind turbine to validate the storage of the corresponding time-series.
4. Selection of the trigger type:
   a. Delayed start.
   b. Digitally controlled start.
5. Communication by GSM/GPRS system to remotely control the status of the registering process.

8.3.3 Post-Processing Module for Flicker Measurement

Once the voltage, current and wind-speed time series are registered the post-processing module recovers the stored data and processes them according to the procedures specified by the IEC 61400-21 standard. According to it, measurements should be performed for harmonic content, flicker, voltage drops as well as active and reactive power. In the next paragraphs we will summarize the procedure for the measurement and assessment of the flicker characteristic of the wind turbine.

The flicker assessment procedure consists of two main tasks: the calculation of the flicker coefficients, $c(\psi_k)$, associated to every pair of 10 min registered voltage and current series in terms of four different network impedance phase angles ($\psi_k = 30, 50, 70$ and $85°$); and the statistical processing of those flicker coefficients, $c(\psi_k)$, in terms of the given annual average wind speed, v_a, to obtain the new flicker coefficients characterizing the flicker of the tested wind turbine, $c(\psi_k, v_a)$.

The statistical processing is perfectly specified in the standard. However, for the calculation of the $c(\psi_k)$ coefficients there are two relevant aspects that should be considered. First, the implementation of the IEC flickermeter, according to the functional specifications defined by the IEC 61000-4-15 standard. There are several works detailing the implementation of a discrete IEC flickermeter. Furthermore, we have analyzed in depth the IEC flickermeter in previous works and this should not be a risk for the flicker assessment [17, 16]. Second, the estimation of the fictitious voltage, $u_{fic}(t)$, obtained from the resolution of the fictitious grid specified by the IEC 61400-21 standard. This is the main processing task that could generate relevant errors when calculating the flicker produced by the wind turbine. Although the solution of the fictitious grid seems to be a simple task, a small error in the estimation of the phase of the fundamental component of $u_m(t)$ can generate important changes in $u_{fic}(t)$ that significantly affect the P_{st} value calculated.

In the next section we present different signal processing techniques to solve the fictitious grid and analyze the errors produced by every method in the P_{st} value of the fictitious grid output, $u_{fic}(t)$.

8.4 Analysis of the Fictitious Grid

As it has been described in previous sections, the IEC 61400-21 standard specifies a method that uses current and voltage time-series measured at the wind turbine terminals to simulate the voltage fluctuations on a fictitious grid with no source of voltage fluctuations other than the wind turbine, $u_{fic}(t)$ (see Fig. 8.1).

The signal processing needed to calculate $u_{fic}(t)$ is focused on the estimation of $u_0(t)$ and the differentiation of $i_m(t)$. The main errors affecting the calculation of the flicker coefficients could come from the estimation of $u_0(t)$. To obtain an accurate estimation of $u_0(t)$, the selection of a proper signal processing technique turns out to be a key factor. First, it is important to understand that $u_m(t)$ is a band-limited signal and most of its power is concentrated around its fundamental frequency, which is equal or very close to 50 Hz. Under this hypothesis, two classical techniques for obtaining the instantaneous phase of $u_m(t)$ are described next.

8.4.1 Calculation of $\mathbf{u_0(t)}$ from $\mathbf{u_m(t)}$

The voltage source $u_0(t)$, representing the power network, must be calculated using (8.1). The difficulty lies in the calculation of $\alpha_m(t)$, the phase of the fundamental component of the registered voltage $u_m(t)$. To solve this problem we have selected two signal processing techniques, namely the Short Time Fourier Transform (STFT) and the Zero Crossing Detection (ZCD). Each technique entails different problems which can put the accuracy of the calculation of $u_0(t)$ at risk.

The low-frequency disturbances that usually contaminate the voltage generated by the wind turbine are harmonic and interharmonic components [8]. The influence of the interharmonic components over the flicker measurement has been well described by several studies [13, 4, 21, 3, 19]. In the present work we have considered interharmonics only as disturbing terms which produce measurement errors in the calculation of $u_o(t)$, but not as the cause of perceptible flicker through $i_m(t)$. To test the method, $u_m(t)$ has been analytically formed as a summation of a fundamental component of frequency ω_0, and another harmonic/interharmonic component of frequency ω_i, from 1 to 100 Hz, and amplitude A_i, from 0 to 3% :

$$u_m(t) = \sqrt{2}U_n\left[\cos(\omega_0 t + \alpha_0) + A_i \cos\left(\omega_i t\right)\right]. \tag{8.13}$$

The goal of the estimation method should be the calculation of a voltage, $u_{0,est}(t)$, equal to the reference voltage, $u_{0,ref}(t)$:

$$u_{0,ref}(t) = \sqrt{\frac{2}{3}}U_n \cos(\omega_0 t + \alpha_0). \tag{8.14}$$

The parameters of the wind turbine affect the fictitious grid and the waveforms of the test signals. In this sense, all the tests were made using the wind turbine characteristic shown in Table 8.2.

8.4.1.1 Short-Time Fourier Transform

The Short-Time Fourier Transform (STFT) is normally used to determine the sinusoidal frequency and phase content of local sections of a signal as it changes over time. The input data is sectioned into time portions or windows, which usually overlap each other. Each portion is then Fourier transformed, and the complex result provides the magnitude and phase for each point in time and frequency.

The window length is an important parameter that must be carefully selected by considering different aspects such as computational load or spectral resolution. The minimum window length necessary to characterize the fundamental component is one cycle. To avoid discontinuities in the estimation of the signal, we chose a one-sample sliding rectangular window of one cycle, providing very accurate time

Table 8.2 Wind turbine characteristics for analytic tests

Parameter	Description	Value
U_n	Nominal voltage	690 V
S_n	Rated power	600 kVA
f_0	Fundamental frequency	50 Hz
$\frac{S_{k,fic}}{S_n}$	Ratio between power parameters	20
Ψ_k	Network impedance angle	85°

resolution. On one hand, there would be no spectral leakage due to the application of that rectangular window when $u_m(t)$ is polluted with harmonic components. Nevertheless, when interharmonic components are considered, the spectral leakage may disturb the output phase, depending on the frequency of the interharmonic and its amplitude.

We applied this technique to estimate $u_0(t)$ from $u_m(t)$ defined in (8.13), at a sampling rate $f_s = 3,200$ S/s. Figure 8.6 shows the relationship between the power of the deviation $(u_{0,est}(t) - u_{0,ref}(t))$ and the power of $u_{0,ref}(t)$ as a function of ω_i and A_i. The error grows to a maximum value of 0.035% when ω_i is near 50 Hz. Furthermore, the flicker produced by $u_{0,est}(t)$ is always negligible as it is demanded by the standard.

Those values should indicate a negligible influence over the P_{st} calculation of the fictitious voltage, $u_{fic}(t)$. However, the spectral leakage of the estimated $u_{0,est}(t)$ produces a periodic phase superimposed onto the phase of the fundamental component, quite similar to a typical phase modulation:

$$u_{0,est}(t) = \sqrt{\frac{2}{3}} U_n \cos(\omega_0 t + \alpha_0 + \Delta\phi(t)). \tag{8.15}$$

The result of these periodic phase variations, when they are in the range of 1–35 Hz, should be noticed in terms of flicker severity. Figure 8.7 represents the estimated flicker severity, $P_{st,est}$, of the fictitious voltage when the current $i_m(t)$ is a plain sinusoidal component of 50 Hz, and $u_m(t)$ has been formed as it was described in (8.13). The upper plot shows the 3D representation of the $P_{st,est}$ values in terms of the combination of the amplitude and frequency of the interharmonic. The lower plot shows the projection on the XZ plane of some level curves from the 3D representation.

In spite of the low estimation errors, when $u_{0,est}(t)$ is combined with $i_m(t)$ (8.3) the $P_{st,est}$ of $u_{fic}(t)$ reaches maximum values close to 1.75. The reference flicker severity value for the defined $i_m(t)$ and $u_m(t)$ signals should be $P_{st,ref} = 0.0086$.

The maximum deviations correspond to frequencies of 41 and 59 Hz. This is explained by means of Fig. 8.8 which shows the phase deviation, $\Delta\phi(t)$, of $u_{0,est}(t)$ for $u_m(t)$ formed as the fundamental frequency and an interharmonic of 41 Hz with $A_i = 1\%$. In this case, the interharmonic generates a periodic phase $\Delta\phi(t)$ with 9 and 91 Hz components. These components, superimposed as a phase modulation, produce a main 9 Hz fluctuation over the fundamental component. This point is quite close to the maximum sensitivity frequency of the IEC flickermeter (8.8 Hz).

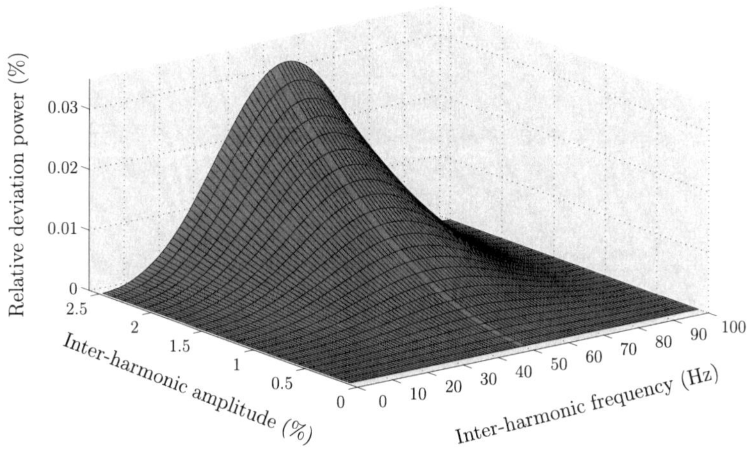

Fig. 8.6 Relative error power between the estimated voltage and the reference voltage in case of the STFT technique

The same effect occurs with all the interharmonics that are located near the fundamental component ($|\omega_{\mathrm{i}} - \omega_0| < 2\pi 35$). This is because the resulting components in $\Delta\phi(t)$ will appear inside the bandwidth of the weighing filter of the IEC flickermeter (from 0 to 35 Hz).

Summarizing, the STFT method does not behave properly when interharmonics of low amplitude are added to the fundamental component of $u_{\mathrm{m}}(t)$. The method is quite sensitive to small errors in the phase estimation of the fundamental component, providing quite anomalous values in the $P_{\mathrm{st,est}}$.

8.4.1.2 Zero Crossing Detection

The estimation of the frequency of the power system using the zero-crossing technique has been well known for a long time [11]. Given the frequency or period of each cycle of $u_{\mathrm{m}}(t)$, constructing the instantaneous phase of the signal $u_{\mathrm{m}}(t)$ is straightforward.

Working in the discrete domain, the algorithm searches for the positions of the contiguous samples of $u_{\mathrm{m}}(t)$ that mark a transition of values from positive to negative. To achieve a more precise approximation to the zero-crossing point, a linear interpolation between the points of the transition is used, as is shown in Fig. 8.9.

The fraction of the sampling period that places the zero-crossing can be obtained from the following expression:

$$\Delta_{\mathrm{n}} = \frac{Y_1}{Y_1 - Y_2} \tag{8.16}$$

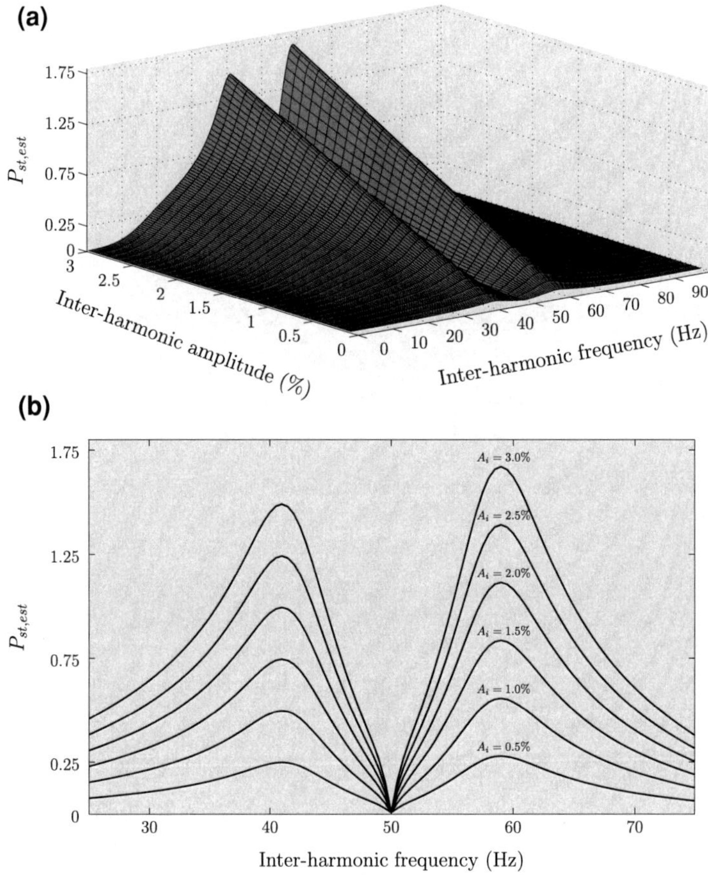

Fig. 8.7 P_{st} of the fictitious voltage in case of the STFT technique when errors in the estimation of $u_0(t)$ occur. **a** 3D representation. **b** Contour lines from $A_i = 0.5\%$ to $A_i = 3.0\%$

Knowing the number and the fraction of the samples that make up a period, reconstruction of the instantaneous phase of the fundamental component is done, dividing the 2π radians uniformly for each sampling instant.

Figure 8.10 represents the estimated flicker severity $P_{st,est}$ for ZCD method, using the same test signals as in case of STFT method, at a sampling rate $f_s = 3,200\ \frac{S}{s}$. The upper plot shows the 3D representation of the $P_{st,est}$ values in terms of the combination of the amplitude and frequency of the interharmonics. The lower plot shows the projection on the XZ plane of some level curves from the 3D representation.

The results are slightly better than in case of STFT method. However, the errors are still too relevant, reaching a $P_{st,est}$ value above 1.25. The maximum deviations also come from those interharmonic components that have influence on the estimation of $u_0(t)$ and generate errors leading to 9 Hz fluctuations over the

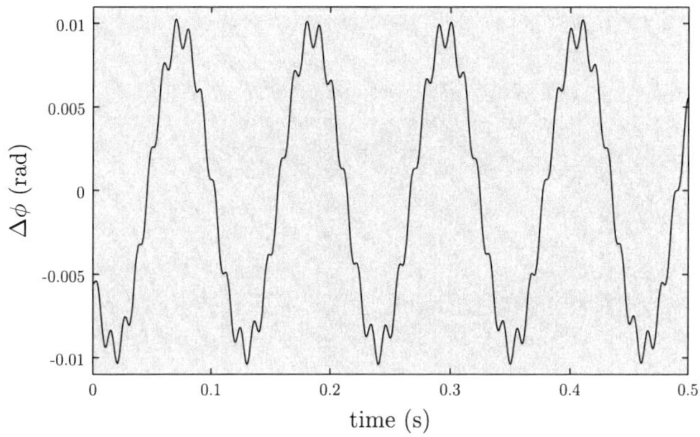

Fig. 8.8 Phase deviation in the estimation of $u_0(t)$ signal for an interharmonic of 41 Hz and $A_\mathrm{i} = 1\%$

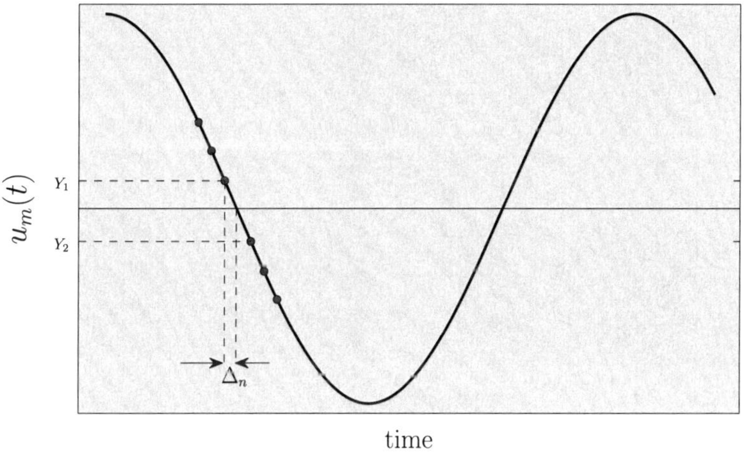

Fig. 8.9 The zero-crossing detection scheme

fundamental component. Those interharmonic components correspond to 9, 41, 59 and 91 Hz in case of ZCD method as it can be observed in Fig. 8.10.

In the light of the previous results it can be concluded that the obtaining of a precise estimation of $u_0(t)$ entails necessarily the removal of the interharmonic components and hence the band limitation of $u_\mathrm{m}(t)$. We propose the implementation of a narrow band-pass filter that will be described in the next section.

8.4.2 *Calculation of* $\mathbf{u_0(t)}$ *After Filtering* $\mathbf{u_m(t)}$

To extract the fundamental component from $u_\mathrm{m}(t)$ the use of a narrow band-pass filter around $f_0 = 50\,\mathrm{Hz}$ is needed. Since we process discrete signals, this section

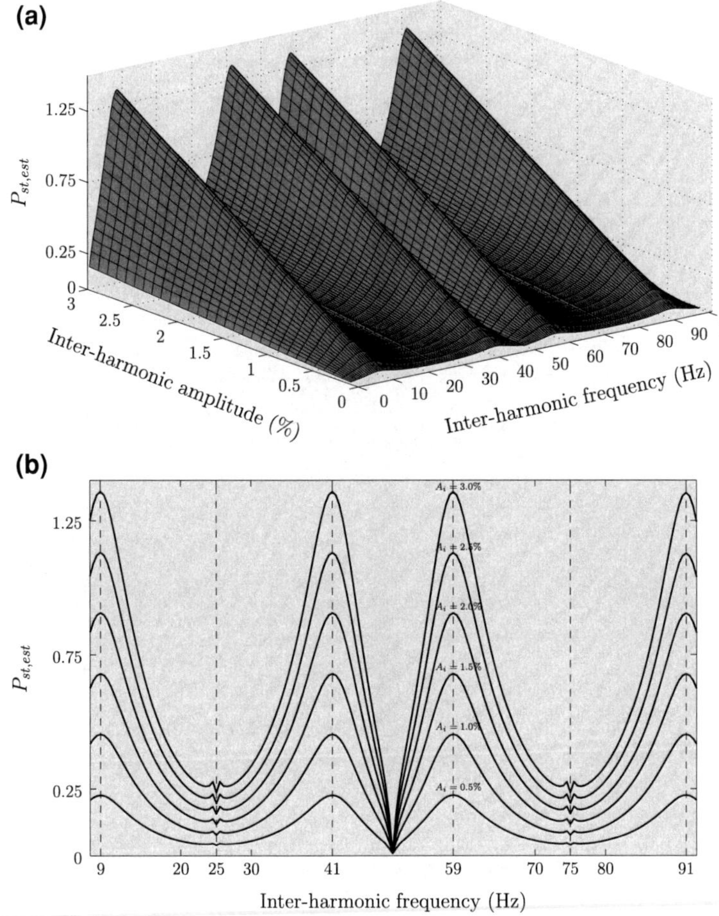

Fig. 8.10 P_{st} of the fictitious voltage in case of the ZCD technique when errors in the estimation of $u_0(t)$ occur. **a** 3D representation. **b** Contour lines from $A_i = 0.5\%$ to $A_i = 3.0\%$

shows how to design a discrete band-pass filter, assuming that $u_m(t)$ has been sampled at f_s samples per second.

8.4.2.1 Design of a Narrow Band-Pass Filter

The filter will be designed from a notch-filter conceived to suppress the fundamental component of $u_m(t)$. When the bandwidth for the continuous frequency, b_f, is small the transfer function of the notch filter can be analytically expressed as follows:

$$H_1(z) = \frac{z^2 - 2z\cos(\Omega_0) + 1}{z^2 - 2(1 - \frac{BW}{2})\cos(\Omega_0)z + (1 - BW)} \tag{8.17}$$

where $\Omega_0 = \frac{2\pi f_0}{f_s}$ is the discrete pulsation corresponding to $f_0 = 50\,\text{Hz}$, and $BW = \frac{2\pi b_f}{f_s}$ is the bandwidth expressed in discrete pulsation units. Figure 8.11 shows the module of $H_1(f)$ and the phase delay $\tau_1(f) = -\frac{\phi_1(f)}{f}$ for b_f 0.3 Hz. $\phi_1(f)$ represents the phase of $H_1(f)$. The frequency axis was scaled between 45 and 55 Hz in order to observe in detail the 50 Hz component.

The narrow band-pass filter can be obtained from the notch filter following the scheme of Fig. 8.12. From this scheme $H(z) = \frac{Y(z)}{X(z)} = 1 - H_1(z)$, resulting:

$$H(z) = BW \cdot \frac{\cos(\Omega_0)z - 1}{z^2 - 2(1 - \frac{BW}{2})\cos(\Omega_0)z + (1 - BW)} \tag{8.18}$$

Figure 8.13 shows the module of $H(f)$ and the phase delay $\tau(f) = -\frac{\phi(f)}{f}$ for a bandwidth $b_f = 0.3\,\text{Hz}$, where $\phi(f)$ represents the phase of $H(f)$. The frequency axis was also scaled between 45 and 55 Hz in order to observe in detail the pass band of the filter.

By using the same test signals as in previous tests, Figs. 8.14 and 8.15 show the estimated flicker severity, $P_{st,est}$ for the case of STFT and ZCD methods when the input signal is filtered by the narrow band-pass filter. The results are strongly better than in case of the STFT and ZCD methods without filtering $u_m(t)$. When $u_m(t)$ is not filtered the $P_{st,est}$ value in the worst case is above 1.25 for both techniques; by filtering $u_m(t)$ the estimated flicker values fall down below 0.03, much closer to the reference value 0.0086.

However, the main problem that $H(z)$ presents to obtain the fundamental component of $u_m(t)$ is the abrupt behavior of the phase delay around 50 Hz, that produces a delay of several samples in the output due to eventual small variations of the fundamental frequency around 50 Hz. This could cause appreciable errors in the $P_{st,est}$ of $u_{fic}(t)$. To confirm this hypothesis we changed the input test signal, including a small variation over the fundamental frequency, $f_0 = 50.05\,\text{Hz}$. In this case the intention was to model an analytical voltage at the terminals of the wind turbine, $u_m(t)$, near real conditions based on different studies into the measurement of perturbations produced by wind turbines [18, 20]. After analyzing those works a test voltage was configured according to the following analytical expression:

$$u_m(t) = \sqrt{\frac{2}{3}}U_n \cos(2\pi f_0 t) + \sqrt{\frac{2}{3}}U_n \sum_{i=1}^{N} a_i \cos(2\pi f_i t) \tag{8.19}$$

In this way, the test voltage was made up of a fundamental component of frequency f_0. Furthermore, N frequency components of amplitude a_i and frequency f_i were added including, 3rd, 5th, 7th, 9th and 11th harmonic components of f_0 with $a_i = 0.5\%$ of the fundamental's amplitude, and the interharmonic components from 40 to 60 Hz, with 1 Hz separation and with amplitudes $a_i = 0.1\%$ of the fundamental's amplitude.

Moreover, from the same bibliographic analysis used for $u_m(t)$, a test current was configured and derived from the following expression:

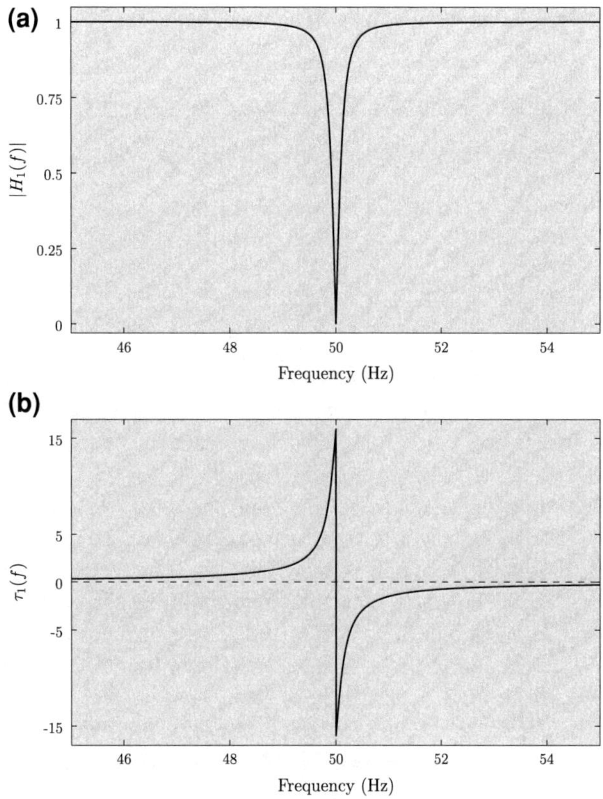

Fig. 8.11 Frequency responses of the notch filter. **a** Frequency response module of the notch filter. **b** Phase delay of the notch filter

Fig. 8.12 The narrow band-pass filter scheme

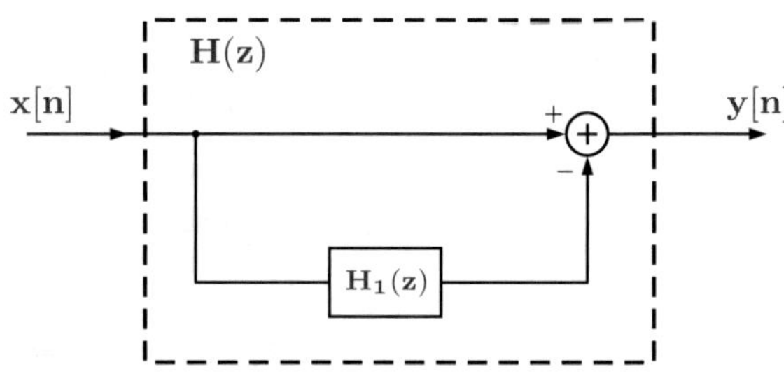

$$i_{\mathrm{m}}(t) = \sqrt{2}I_{\mathrm{n}}\cos\left(2\pi f_0 t + \alpha_{\mathrm{iu}}\right) + \sqrt{2}I_{\mathrm{n}}\sum_{\mathrm{k}=1}^{\mathrm{M}} b_{\mathrm{k}}\cos(2\pi f_{\mathrm{k}} t) \tag{8.20}$$

The configuration of the fundamental component was identical to the $u_{\mathrm{m}}(t)$ case, except for the inclusion of phase α_{iu}, which was calculated after considering

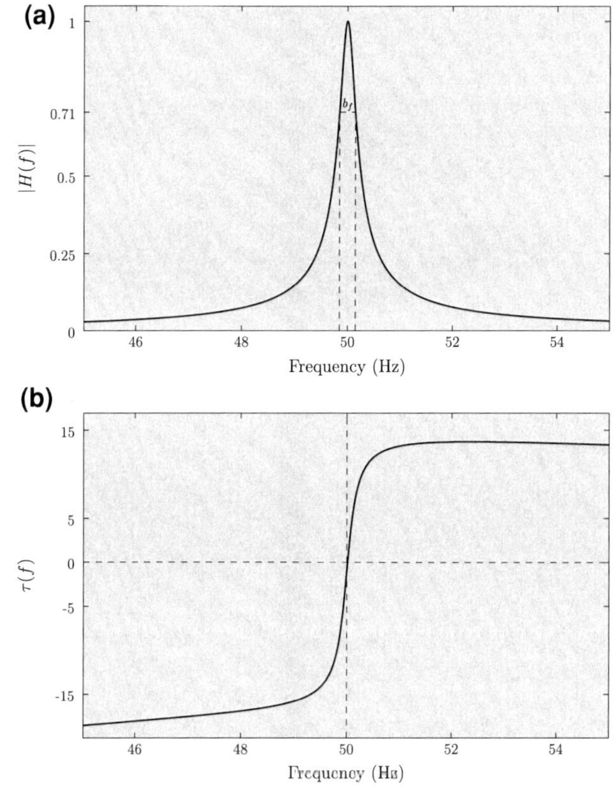

Fig. 8.13 Frequency responses of the band-pass filter. **a** Frequency response module of the band-pass filter. **b** Phase delay of the band-pass filter

a power factor of 0.95. In this case, M interharmonic components different from those of $u_m(t)$ were added including the 3rd, 5, 7, 9 and 11th harmonic components of f_0 with $b_k = 1.5\%$ of the fundamental's amplitude; the 20th, 21th, 22th, 23th, 24th and 25th harmonic components of f_0 with $b_k = 1.75\%$ of the fundamental's amplitude and, the interharmonic components from 1,025 to 1,225 Hz, with 50 Hz separation and with amplitude $b_k = 1.75\%$ of the fundamental's amplitude.

Using the new test signals (8.19) and (8.20) and the same wind turbine parameters (Table 8.2), the theoretical $u_{fic}(t)$ and the P_{st} associated were calculated, giving a value of $P_{st,ref} = 1.614$ for a fundamental frequency $f_0 = 50$ Hz and $P_{st,ref} = 1.611$ for a fundamental frequency $f_0 = 50.05$ Hz. This value was taken as the reference for the comparison of all the methods of estimation of $u_0(t)$ described previously. On the other hand a different $u_0(t)$ was estimated for each method, and the corresponding $u_{fic}(t)$ and their $P_{st,est}$ were calculated.

Table 8.3 gives details about the value of $P_{st,est}$ obtained for the $u_{fic}(t)$ calculated from the estimation of $u_0(t)$, as well as the error with respect to the

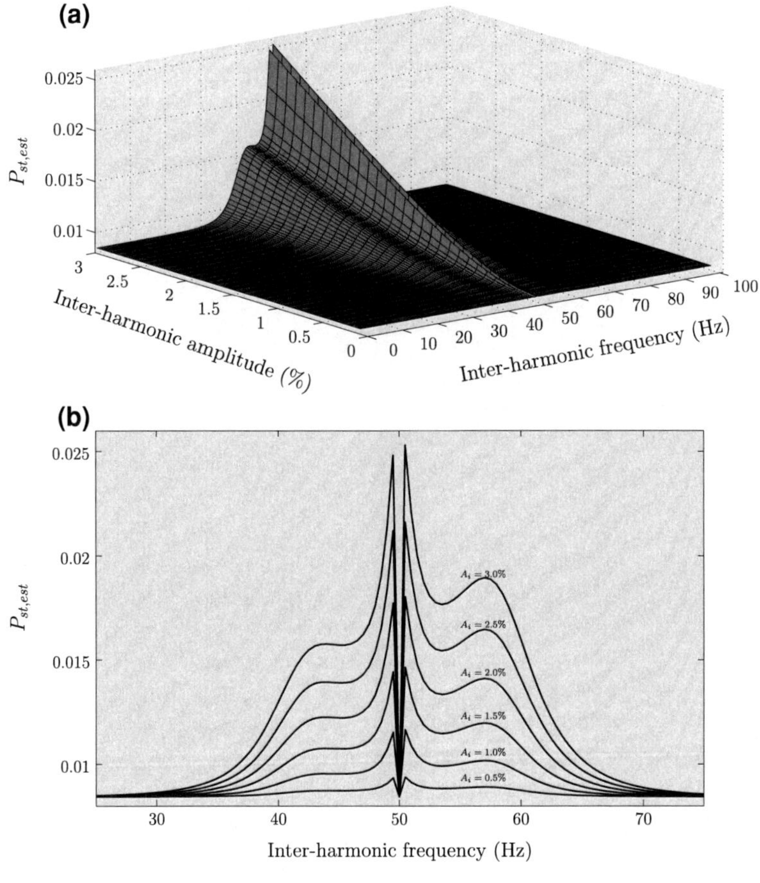

Fig. 8.14 P_{st} of the fictitious voltage in case of the STFT technique, with band-pass filtering of $u_{\mathrm{m}}(t)$, when errors in the estimation of $u_0(t)$ occur. **a** 3D representation. **b** Contour lines from $A_{\mathrm{i}} = 0.5\%$ to $A_{\mathrm{i}} = 3.0\%$

theoretical reference value, when the fundamental frequency is 50 Hz. In all cases, a sampling frequency of $f_{\mathrm{s}} = 3,200$ S/s was used.

On the other hand, Table 8.4 shows the results in case of a deviation of 0.1% in the fundamental frequency, $f_0 = 50.05$ Hz.

The results confirm the conclusions that were obtained from the application of both methods without filtering. These methods are quite consistent when there is no interference near the fundamental frequency. Hence, the errors in the estimation of the flicker produced by $u_{\mathrm{fic}}(t)$ are almost negligible when the fundamental frequency is exactly located at 50 Hz. It is also confirmed that when $u_{\mathrm{m}}(t)$ is filtered the results are closer to the reference values. However, when the fundamental frequency is not exactly located at 50 Hz, but slightly displaced to 50.05 Hz (0.1%), the influence of the phase delay of the narrow band filter produces a delay of several samples in the output. These variations produce errors in

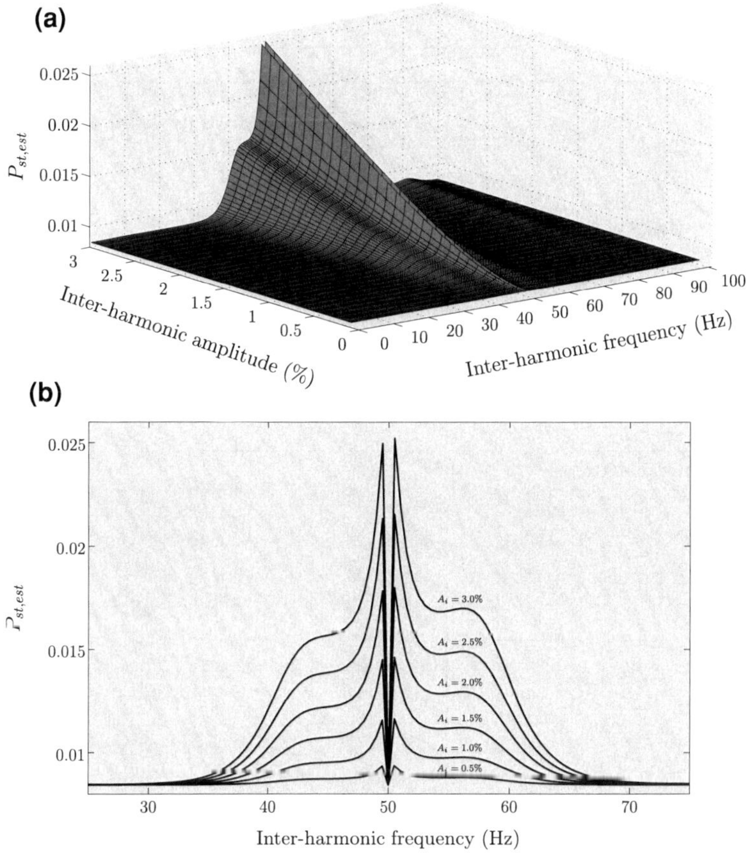

Fig. 8.15 P_{st} of the fictitious voltage in case of the ZCD technique, with band-pass filtering of $u_m(t)$, when errors in the estimation of $u_0(t)$ occur. **a** 3D representation. **b** Contour lines from $A_i = 0.5\%$ to $A_i = 3.0\%$

the estimation of the instantaneous phase of $u_0(t)$ and strongly affect the assessment of the flicker of $u_{fic}(t)$.

The problem of the nonlinear phase can be solved by applying a band-pass zero-phase IIR filter as it will be stated in the next section.

8.4.2.2 Anticausal Zero-Phase Filter Implementation

For IIR filters, such as $H(z)$, the phase distortion is usually highly nonlinear. To eliminate phase distortion, anticausal zero-phase filter implementation can be used. Consider the processing scheme in Fig. 8.16.

After filtering in the forward direction, the filtered sequence is reversed and run it back through the filter. The result has exactly zero-phase distortion. In fact, in

Table 8.3 Simulated results for analytical signals without deviation in the fundamental frequency

Method	$P_{st,est}$	Dev (%)[a]
STFT	1.620	0.32
ZCD	1.619	0.28
filtering + STFT	1.614	0.00
filtering + ZCD	1.614	0.00

[a]Percentage deviation from $P_{st,ref} = 1.614$

Table 8.4 Simulated results for analytical signals with a deviation in the fundamental frequency

Method	$P_{st,est}$	Dev (%)[a]
filtering + STFT	1.764	9.49
filtering + ZCD	1.766	9.59

[a]Percentage deviation from $P_{st,ref} = 1.611$

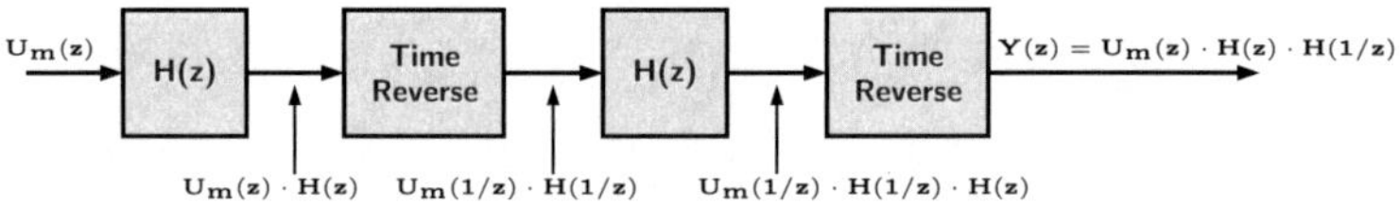

Fig. 8.16 The anticausal zero-phase filter scheme

the frequency domain, $Y(\Omega) = U_m(\Omega) \cdot |H(\Omega)|^2$. The magnitude is the square of the filter's magnitude response, and the filter order is double the order of $H(z)$.

This implementation can only be used in cases such as ours, where $u_m(t)$ is a finite duration signal known before being filtered. It is necessary to eliminate the transitory at both ends of the signal obtained.

To test the behavior of this technique we used the new test signals (8.19) and (8.20) and calculated the flicker value of $u_{fic}(t)$ obtained through the estimation of $u_0(t)$ by STFT and ZCD when $u_m(t)$ is previously filtered with an anticausal zero-phase filter (ZPF). These tests were made considering again a small deviation of 0.1% in the fundamental frequency. Table 8.5 shows the results of this experiment, confirming how this combination of methods provides a very precise estimation of $u_0(t)$ and leaves the errors in P_{st} practically nil.

In principle both methods, ZPF + STFT and ZPF + ZCD, provide similar results in the estimation of the flicker value. However, the computational requirements demanded by the implementation of the STFT (Discrete Fourier Transform of one cycle with a one-sample sliding window) leads us to suggest the ZPF + ZCD technique as the most advisable one in order to optimize the estimation of the flicker of $u_{fic}(t)$.

Table 8.5 Simulated results for analytical signals with a deviation in the fundamental frequency

Method	$P_{st,est}$	Dev (%)[a]
ZPF + STFT	1.610	-0.103
ZPF + ZCD	1.611	0.00

[a]Percentage deviation from $P_{st,ref} = 1.611$

8.5 Results Using Real Signals

To finalize the study, a comparison of the most relevant techniques brought to light in the previous analysis was carried out using signals from the real registers. To achieve this, several measurements were performed on two different wind turbines. The first corresponds to a machine with a double speed asynchronous generation system (four and six poles), fixed sail passage and a fixed generator velocity, nominal power of 600 kW and nominal voltage of 690 V. The second machine has a four-pole, synchronous generation system and electronic power control; variable sail passage and variable generator velocity, and provides a nominal power of 800 kW and a nominal voltage of 1,000 V.

A total of 45 three-phase registers of voltage, current and wind speed were taken in the first machine and 25 three-phase registers in the second one, all of them with a duration of 10 min. These registers were used for the comparison of the four most relevant methods for estimation of $u_0(t)$ through the calculation of the P_{st} of $u_{fic}(t)$. The techniques employed for the comparison of the real registers were ZCD, STFT, ZCD after narrow band filtering and ZCD after filtering using a zero-phase filter (ZPF + ZCD). For comparison purposes, the values of P_{st} obtained with the zero-crossing method after filtering using a zero-phase filter were used as reference.

Figure 8.17 shows the results for the first machine. This figure shows the percentage deviation in the P_{st} with respect to the reference value for each register corresponding to the techniques selected. The flicker values are represented as a function of the corresponding mean power of each 10 min register.

It can be observed that the average percentage of deviation is not excessive, less than 2%, when techniques without filtering are used. However, on filtering the fundamental component of $u_m(t)$ using a narrow band-pass filter, the average percentage of variation is greater than 5%, and even in numerous cases over 10%. This is the result of the variability of the fundamental component f_0 of $u_m(t)$ in the generation of the first machine. In these cases, the application of a nonlinear phase filtering causes errors in the estimation of the phase of $u_m(t)$, which affect the formation of $u_{fic}(t)$ and thus the calculation of the P_{st}.

Figure 8.18 shows the results for the second machine. This figure shows the values of P_{st} obtained for each register corresponding to the techniques selected, as a function of the corresponding mean power of each 10 min register.

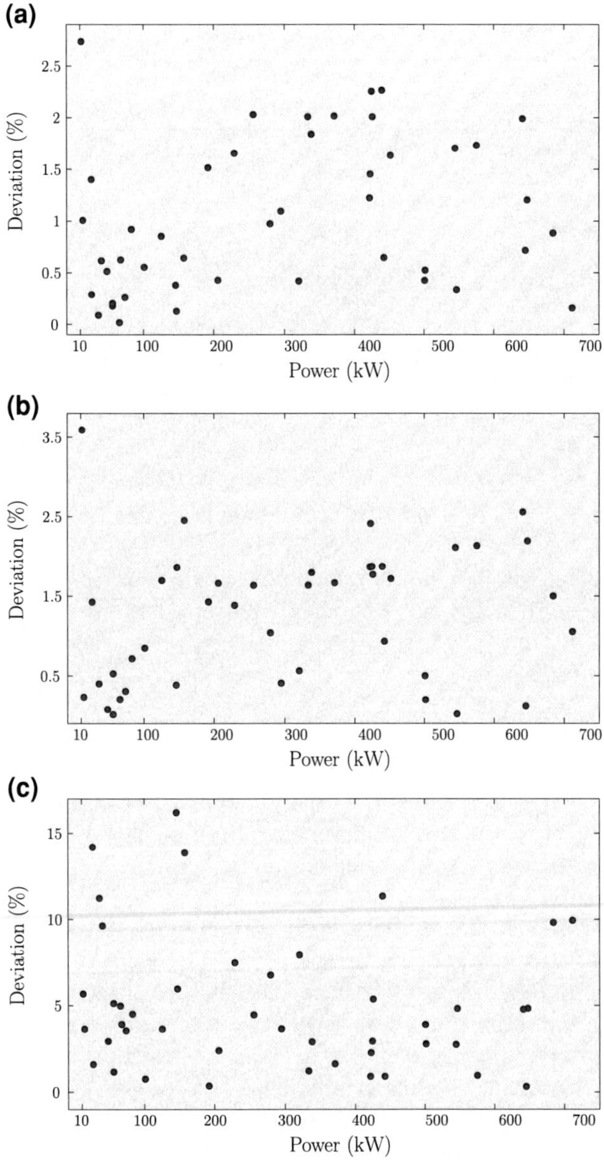

Fig. 8.17 Results for actual registers in first turbine. **a** ZCD method. **b** STFT method. **c** filtering + ZCD method

It can be observed in this case that the methods without filtering (STFT and ZCD) provide extremely large variations with respect to the reference method (ZPF + ZCD). This is caused by the significant interharmonic components near the fundamental frequency, present in $u_{\mathrm{m}}(t)$. Both methods that use the

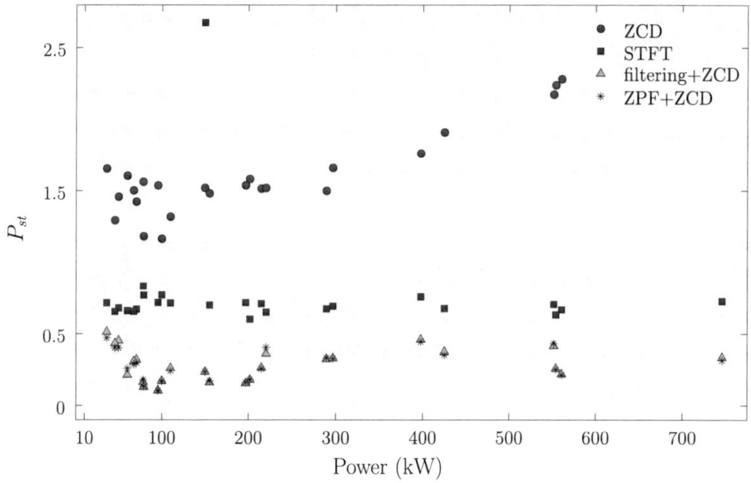

Fig. 8.18 Results for actual registers in the second turbine

narrow-band filter provide similar results because the fundamental component of $u_{\mathrm{m}}(t)$ in this machine fits very precisely with the nominal frequency.

8.6 Conclusions

Power injection from wind turbines affects substantially the power quality. The chapter deals with the measurement of the voltage flicker caused by the wind turbines connected to the network, specified by the IEC 61400-21 standard. The most sensible task of the flicker measurement procedure is the implementation of the fictitious grid. To analyze the sensitivity we suggested several signal processing techniques that produce different errors in the assessment of the flicker coefficients. Those techniques are mainly based on the estimation of the instantaneous phase of the fundamental component of the measured voltage, $u_{\mathrm{m}}(t)$, namely the STFT and the ZCD. Our study shows, using both analytical signals and real signals, that filtering the measured voltage, $u_{\mathrm{m}}(t)$, is essential for an accurate estimation of the phase of its fundamental component. Furthermore, a typical band pass filter could not be efficient enough to obtain good results in some real conditions. The filtering produces small phase delays in the estimation of the fictitious grid that could cause appreciable errors in the flicker values. To solve this problem, the Anticausal Zero-Phase Filter implementation can be used, eliminating the phase distortion.

Acknowledgements This work was supported by the Ministry of Science and Innovation (MICINN) of Spain through the project ENE2009-13978-C02-02. The authors would also like to thank SOTAVENTO GALICIA S.A. (Spain) for making the signals available free of charge for the purpose of this work.

References

1. Ackermann T (2006) Wind power in power systems. Multi-Science, Wiley, Chichester, vol 20. pp 447–449
2. Council G (2010) Global Wind 2009 Report. Online: http://www.gwe.. net
3. De Koster M, De Jaeger E, Vancoetsem W (1999) Light flicker Caused by Interharmonics. Proceedings on CIRED. [Online]. Available: http://www.cired-s2.org/Sessions/1999/Documents/Papers/2.2.pdf
4. Gunther E, Inc E, Knoxville T (2001) Interharmonics in power systems. Power engineering society summer meeting 2001. IEEE 2
5. IEC: 61400-21 (2008) Wind turbines part 21: measurement and assessment of power quality characteristics of grid connected wind turbines. Ed. 2.0
6. IEC: 61000-4-15 (2010) Electromagnetic compatibility (EMC) part 4: testing and measurement techniques - section 15: flickermeter functional and design specifications. Ed. 2.0
7. Jorgensen P, Tande J, Vikkelso A, Norgand P, Christensen J, Sorensen P, Kledal J, Sondergard L (1997) Power quality and grid connection of wind turbines. CIRED IEE. doi:10.1049/cp:19970476
8. Keppler T, Watson N, Arrillaga J, Chen S (2003) Theoretical assessment of light flicker caused by sub- and interharmonic frequencies. IEEE Trans Power Delivery 18(1):329–333
9. Larsson A (2002) Flicker emission of wind turbines caused by switching operations. Energy Convers, IEEE Trans 17(1):119–123
10. Larsson A (2002) Flicker emission of wind turbines during continuous operation. Energy Convers, IEEE Trans 17(1):114–118
11. Lee J, Devaney M (1994) Accurate measurement of line frequency in the presence of noiseusing time domain data. In: Proceedings of the 10th IEEE instrumentation and measurement technology Conference, pp 1016–1019
12. MEASNET: Power quality measurement procedure, Version 4
13. Mombauer W (1998) Flicker caused by interharmonics. EtzArchiv 12:391–396
14. Moreno-Munoz A (2007) Power quality: mitigation technologies in a distributed environment. Springer, Berlin
15. Piwko R, Camm E, Ellis A, Muljadi E, Zavadil R, Walling R, O'Malley M, Irwin G, Saylors S (2009) A whirl of activity. Power Energ Mag, IEEE 7(6):26–35
16. Ruiz J, Gutierrez J, Irusta U, Lazkano A (2009) A precise analysis of the IEC flickermeter when subject to rectangular voltage fluctuations. IEEE Trans Instrum Meas 58(11):3839–3846. doi:doi:10.1109/TIM.2009.2020837
17. Ruiz J, Gutierrez J, Lazkano A, Ruizde Gauna S (2010) A review of flicker severity assessment by the IEC flickermeter. IEEE Trans Instrum Meas 59(8):2037–2047
18. Sørensen P, Pedersen T, Gerdes G, Klosse R, Santier F, Robertson N, Davy W, Koulouvari M, Morfiadakis E, Larsson Å (2001) European wind turbine testing procedure developments. Task 2: Power quality. Risø National Laboratory
19. Tayjasanant T, Wang W, Li C, Xu W (2005) Interharmonic-flicker curves. IEEE Trans Power Delivery 20(2):1017–1024
20. Vilar C, Usaola J, Amaris H (2003) A frequency domain approach to wind turbines for flicker analysis. IEEE Trans Energy Convers 18(2):335–341
21. Xu W (2005) Deficiency of the IEC flicker meter for measuring interharmonic-caused voltage flickers. Power engineering society general meeting, 2005. IEEE pp 2285–2288

Chapter 9
Grey Predictors for Hourly Wind Speed and Power Forecasting

Tarek H. M. El-Fouly and Ehab F. El-Saadany

Abstract Wind energy resources, unlike dispatchable central station generation, produce power dependable on an external irregular source, the incident wind speed, which does not always blow when electricity is needed. This results in the variability, unpredictability, and uncertainty of wind resources. Therefore, the integration of wind facilities to utility electrical grid presents a major challenge to power system operators. Such integration has significant impact on the optimum power flow, transmission congestion, power quality issues, system stability, load dispatch, and economic analysis. Due to the irregular nature of wind power production, accurate prediction represents the major challenge to power system operators. This chapter investigates the usage of Grey predictor rolling models for hourly wind speed forecasting and wind power prediction.

9.1 Introduction

Wind speed depends on temperature, pressure differences, and the terrain. Wind forecasting is essential for the power utility to schedule the connection and disconnection periods of wind farms or the conventional generators with the utility network for an optimal operational cost and a low environmental impact. Also, wind speed forecasting is crucial to simulate the output power from wind farms and to solve the load flow problems of electrical power systems with embedded wind generation [1].

T. H. M. El-Fouly
CanmetENERGY, Natural Resources Canada, Varennes, QC J3X1S6, Canada
e-mail: telfouly@nrcan.gc.ca

E. F. El-Saadany (✉)
ECE Department, University of Waterloo, Waterloo, ON N2L3G1, Canada
e-mail: ehab@uwaterloo.ca

S. M. Muyeen (ed.), *Wind Energy Conversion Systems*,
Green Energy and Technology, DOI: 10.1007/978-1-4471-2201-2_9,

Wind power prediction is very important for several applications such as wind farm units' maintenance; energy reserves planning and scheduling; power system generators scheduling; optimal power flow between conventional units and wind farms; and electricity market bidding. Table 9.1 presents the various prediction horizons, the developed techniques for each prediction horizon, and their fields of application [2].

There are two main approaches for predicting wind power. The first is based on predicting the wind power directly from previously recorded historical wind production data while the second is based on forecasting wind speed at wind turbines locations and then using wind turbine power curves to predict wind power production. Several techniques and models have been developed and implemented for wind speed forecasting and wind power prediction. The following sections will briefly introduce some of these techniques.

9.1.1 Time Series-Based Techniques

Several time series models have been developed for wind speed forecasting, including Auto Regressive (AR) models, Auto Regressive Moving Average (ARMA) models, and Auto Regression Integrated Moving Average (ARIMA) models. Time series models based on Artificial Neural Networks (ANNs) have been used for wind speed forecasting. Among these developed ANN-based forecasting methods, the Elman Recurrent Network (ERN), the Adaptive Network-based Fuzzy Inference System (ANFIS), the Radial Basis Function network (RBF), and the Neural Logic Networks (NLN) methods [3–8].

These models require large sets of historical data for their parameter estimation and model training (up to weeks of recorded data) and have proven to be effective only for very short-term predictions (a few hours ahead), especially, one-step ahead prediction. Furthermore, it has been reported that ANN-based methods have some disadvantages such as; the absence of reliable theory to build the ANN structure and that the training process may be trapped in partial minimum and thus will not be able to reach the global optimization.

9.1.2 Spatial Correlation-Based Techniques

These techniques are based on using wind speed and data sets from neighbouring sites to the site under investigation for wind speed and wind power prediction. One of the methods is based on using a Fuzzy Expert System (FES) to forecast the wind speed and the electrical power at a wind farm site [9]. For this method, the wind speed and direction measurements from several wind stations, installed around and in the wind farm site via wireless modems to a central computer, are involved. This central computer runs the FES, which exploits any spatial

Table 9.1 Forecasting horizons for wind power prediction and their corresponding techniques and fields of application

Forecasting horizon	Techniques	Applications
0 (Nowcasting)	Time series/statistical approaches	Unit maintenance
Few seconds or minutes	Time series/statistical approaches	Control purposes
1–6 h	Time series/statistical approaches	Small power systems operation such as 1 h electricity market bidding and optimal power flow analysis
1–72 h	Time series/statistical approaches/physical models	Interconnected power systems operation such as generator scheduling and dispatch, unit commitments, and day ahead electricity market bidding, and reserve
1–7 days	Time series/statistical approaches/physical models	Maintenance and planning

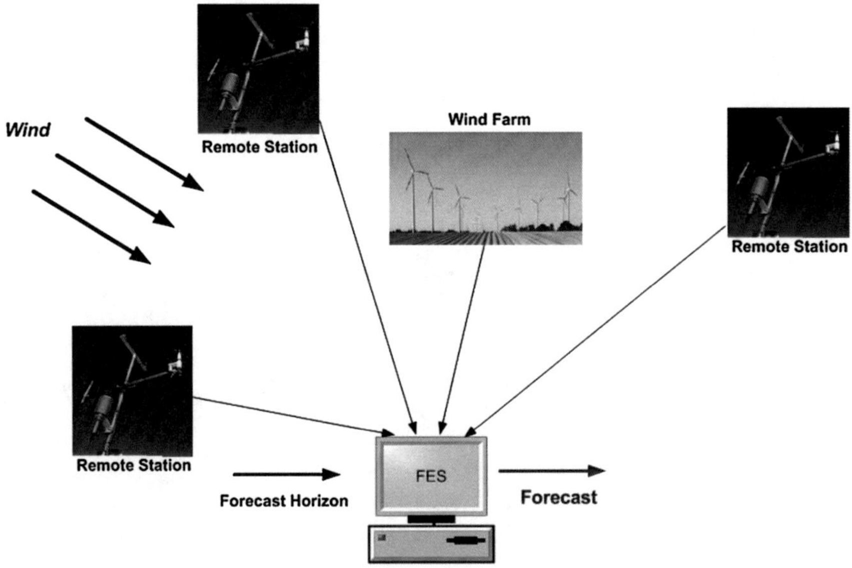

Fig. 9.1 FES-based wind forecasting system

correlation among the measuring stations' wind series. Figure 9.1 presents the wind forecasting technique, based on the FES.

Another spatial correlation-based method was developed using data of several sites to train the ANN model for forecasting wind speed and power for a few hours ahead [10]. The measurements are collected from sites 0.8 to 40 km apart. Advanced spatial correlation-based technique using local recurrent neural networks and advanced fuzzy models have also been developed for forecasting wind speed and wind power up to 36 h ahead [11–13].

These spatial correlation-based models require sets of data from more than one site in order to achieve reasonable prediction accuracy. Moreover, these techniques are direction dependent. Consequently, wind direction data is also necessary for the accurate application of such techniques in wind speed forecasting.

9.1.3 Physical Power Prediction Models

Several physical models have been developed that predict wind power production for days ahead [14–18]. These models are based on using Numerical Weather Prediction (NWP) models while taking into consideration several factors including local surface roughness and its changes, the effects of obstacles and orography, speed up or down, local wind speed scaling within wind farms, wind farm layout, and wind turbines power curves [15, 16].

As an example one of the developed physical models requires the use of the predictions of the wind speed from the High Resolution Limited Area Model (HIRLIAM) of the Danish Meteorological Institute, and is modified specifically for individual areas (sites of the wind farms) by using the geostrophic drag law to transform such predictions to the surface. Then the resulting surface wind speed is applied to a matrix, generated by sub-models of the $WA^{S}P$ (Wind Atlas Application and Analysis Program) for considering some local effects such as shelter from obstacles, local surface roughness and its changes, and the effects of orography, speed up or down. The output of the $WA^{S}P$ passes through a Model Output Statistics (MOS 1) that scales the local wind speed. Then, the PARK model is used to introduce the effect of the wind farm layout (wake effect) and the power curves of the wind turbines. The output of the PARK model is corrected by the MOS 2 model. Figure 9.2 is a flowchart of the model. This approach was used to predict the wind farm power production from 0 to 36 h ahead. The same approach has been modified and enhanced by using spatial smoothing effects to predict power production for as many as 48 h [17].

These models have been involved in wind farms' output power prediction, but are characterized by the following drawbacks [18]:

- Very complicated and very expensive.
- Not reliable when weather service forecasts are delayed.
- Generate large errors when there is a time shift between the forecast and the real data. This has been recorded to be very often the case.
- Not effective for very short-term prediction (few hours ahead).
- Dependent on NWPs which are often inaccurate.

9.2 Grey Predictor Rolling Models

Grey systems, that refer to any system with partially unknown information about its parameters, structure, and/or characteristics, were first introduced by Professor Julong Deng in 1982 [19–21]. They are characterized by their need for only few historical data points (as few as four historical data points), being highly adaptive to the dynamic behaviour of the data and their little computation effort and processing time [22].

Various Grey Predictor models have been developed and used in many prediction applications such as predicting objects position and targets tracking in which the grey system predicts the future trend of an object or target, based on few historical measurements [23], forecasting stock prices indices [24] and power system yearly peak loads [22], and predicting the changes in the inertia and damping coefficients of the mechanical parts of induction servo motors [25].

Two wind speed data sets are used in the analysis presented in this chapter. Each wind speed data set consists of 24 h (1 day) recorded over 15 min periods at the Madison weather station [26]. The average of the hourly wind speed time

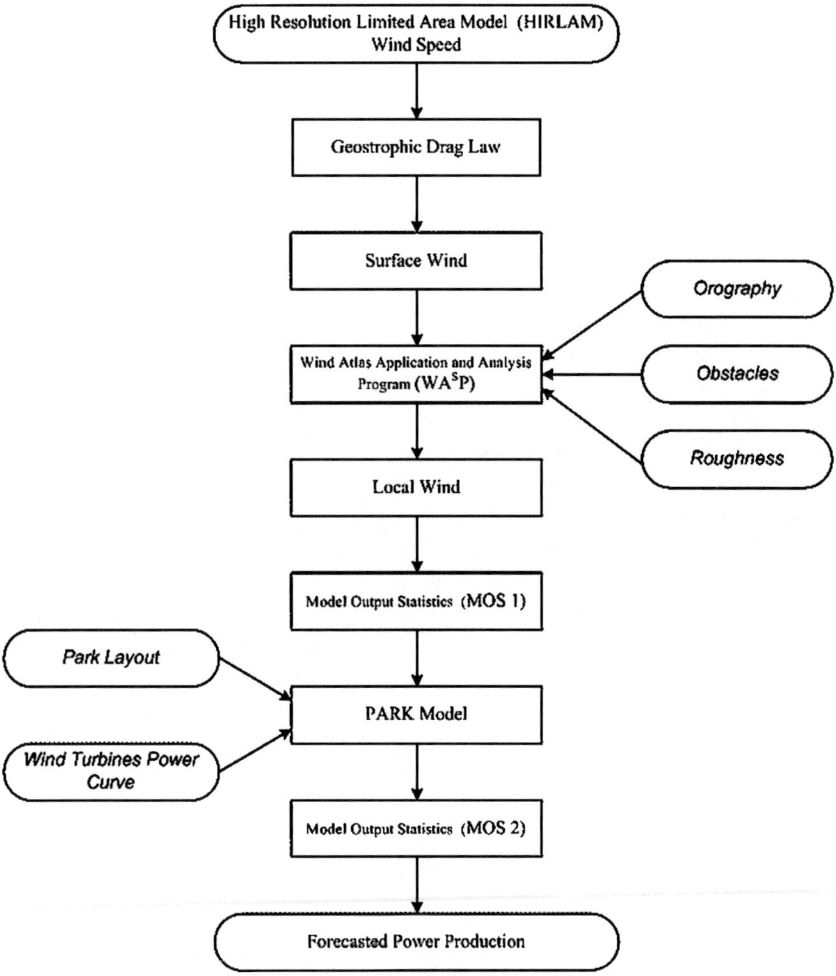

Fig. 9.2 Flowchart of the physical prediction model

series that are used in the analysis, presented in Fig. 9.3, were generated by averaging the four data sets recorded during each hour, as follows.

$$X_{\text{average}}(j) = \frac{1}{4}\sum_{i=1}^{4} X_{\text{recorded}}(i) \qquad \forall j = 1, 2, \ldots, 24\,, \tag{9.1}$$

where $X_{\text{average}}\ (j)$ is the hourly averaged data point during hour j, and $X_{\text{recorded}}\ (i)$ is the 15 min recorded data points during hour j.

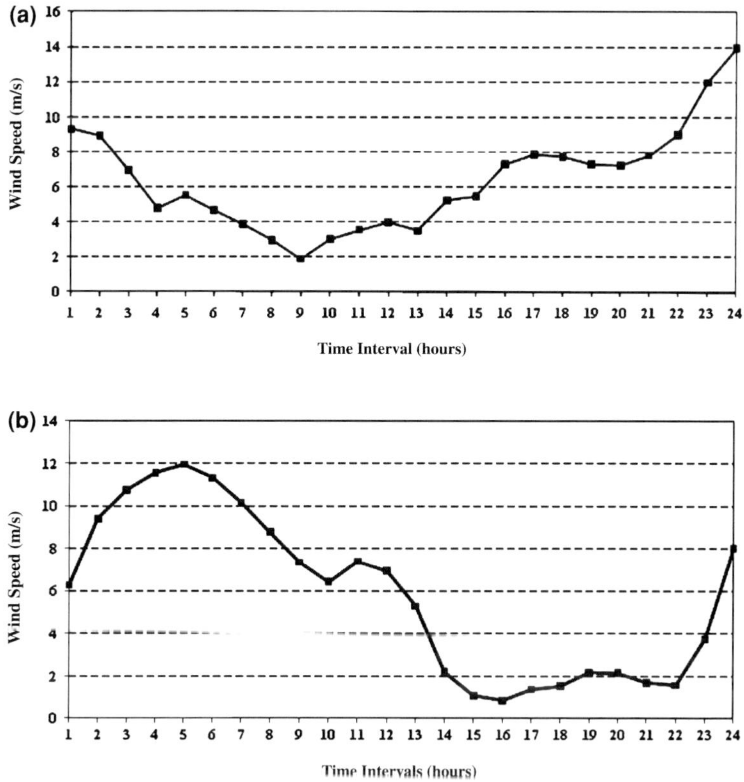

Fig. 9.3 24 h wind speed time series samples under investigation. **a** Sample 1. **b** Sample 2

9.2.1 Traditional Grey Rolling Model GM(1,1)

There are several models for the Grey predictors currently in use. This section investigates the usage of the most common model, the traditional GM(1,1) model [27, 28]. The procedure to develop this model and the related forecasting process are illustrated in Fig. 9.4. The mathematical formulations describing each stage are summarized as follows.

Stage 1: Accumulated Generating Operation The aim of this operation is to reduce the noise and randomness from the data set under analysis. This is carried out by transforming the original set of data $X^{(0)}$ into a new accumulated set $X^{(r)}$. This new data set is also characterized by a more smoothed regular pattern than the original set. Figure 9.5 illustrates the effect of this operation on the original set of data $X^{(0)}$. The equations used to generate the rth order Accumulated Generating Operation (AGO) series can be expressed as follows,

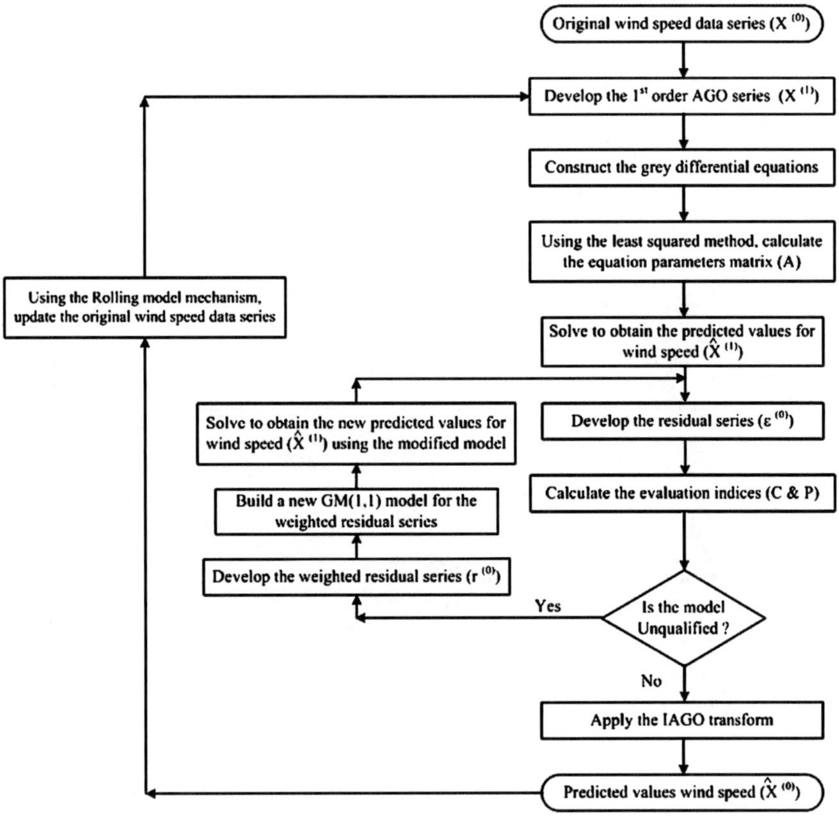

Fig. 9.4 Forecasting process using traditional GM(1,1) rolling model

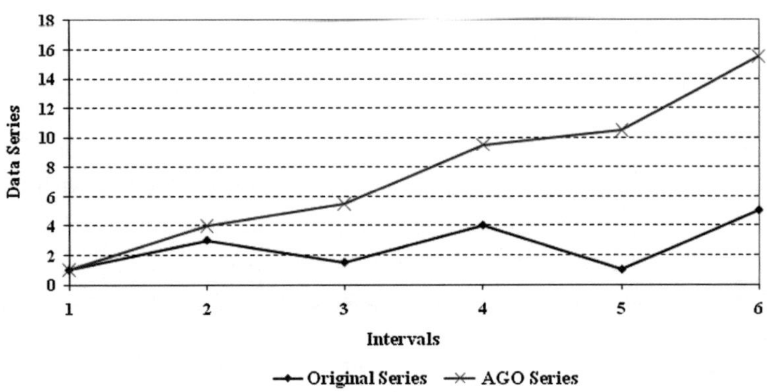

Fig. 9.5 First order AGO data series

$$X^{(r)}(k) = \sum_{i=1}^{k} X^{(r-1)}(i) \quad \forall\, k = 1, \ldots, n\,;\; r = 1,\ 2, \ldots \tag{9.2}$$

For the first order AGO series, Eq. 9.2 is formulated as,

$$X^{(1)}(k) = \sum_{i=1}^{k} X^{(0)}(i) \quad \forall\, k = 1, \ldots, n\,; \tag{9.3}$$

where $X^{(0)}$ represents the original data series, $X^{(1)}$ represents the first AGO series, $X^{(r)}$ represents the rth AGO series, n represents the sample data, and k and i represents the step for the AGO and the original series, respectively.

Stage 2: GM(1,1) Differential Equation This process is devoted for formulating the differential equation that relates the model's dependent variables with the independent ones. This equation is referred to as the Grey dynamic model. The general differential equation for a Grey model with m variable GM(1, m) is,

$$\frac{\mathrm{d}X_1^{(1)}}{\mathrm{d}t} + aX_1^{(1)} = b_1\, X_2^{(1)} + b_2\, X_3^{(1)} + \cdots + b_{m-1}\, X_m^{(1)} \tag{9.4}$$

where X_1 represents the independent variable, X_2, …, X_m represent the dependent variables, and m represents the number of variables. The coefficients a, b_1, b_2 …….b_{m-1} are determined using the least square method.

The GM(1,1) represents the model with one independent variable and no dependent variables. Therefore, Eq. 9.4 can be rewritten as follows,

$$\frac{\mathrm{d}X^{(1)}}{\mathrm{d}t} + aX^{(1)} = b \tag{9.5}$$

where X represents the independent variable of the GM(1,1), and a and b are the model coefficients determined using the least square method.

Stage 3: Equation Parameters Calculations The a and b parameters of the GM(1,1) are determined using the least squared method expressed as follows;

$$A = \begin{bmatrix} a \\ b \end{bmatrix} = \left[\beta^T.\beta\right]^{-1}.\beta^T Y\,, \tag{9.6}$$

where

$$\beta = \begin{bmatrix} -Z^{(1)}(2) & 1 \\ -Z^{(1)}(3) & 1 \\ \vdots & \vdots \\ -Z^{(1)}(n) & 1 \end{bmatrix}, \tag{9.7}$$

$$Y = \begin{bmatrix} X^{(0)}(2) \\ X^{(0)}(3) \\ \vdots \\ X^{(0)}(n) \end{bmatrix}, \tag{9.8}$$

and

$$Z^{(1)}(i) = \frac{X^{(1)}(i-1) + X^{(1)}(i)}{2}. \tag{9.9}$$

Stage 4: Forecasting Model This step is dedicated to calculate the forecasted or predicted values for the AGO series $\left(\hat{X}^{(1)}\right)$. The equation used in this process can be written as follows:

$$\hat{X}_1^{(1)}(i+1) = \left(X_1^{(0)}(1) - \frac{1}{a}\sum_{k=2}^{m} b_{k-1} X_k^{(1)}(i+1)\right)e^{-ai} + \frac{1}{a}\sum_{k=2}^{m} b_{k-1} X_k^{(1)}(i+1) \tag{9.10}$$

For the GM(1,1) model the forecasting model is expressed by,

$$\hat{X}^{(1)}(i+1) = \left(X^{(0)}(1) - \frac{b}{a}\right)e^{-ai} + \frac{b}{a} \tag{9.11}$$

where i represents the step, and $X^{(0)}$ represents the first data in the original time series.

Stage 5: Model Diagnostic Checking In [22] a model diagnostic checking process has been developed to evaluate and test the traditional Grey model's grade. This is carried out by calculating the residual (error) series $\varepsilon^{(0)}$ for the historical data, used in developing the model, as follows:

$$\varepsilon^{(0)}(i) = X^{(1)}(i) - \hat{X}^{(1)}(i) \tag{9.12}$$

The mean $(\bar{\varepsilon})$ and the standard deviation (S_1) of this series are calculated as follows:

$$\bar{\varepsilon} = \frac{1}{n}\sum_{i=1}^{n} \varepsilon^{(0)}(i) \tag{9.13}$$

and

$$S_1 = \sqrt{\frac{1}{n}\sum_{i=1}^{n}\left(\varepsilon^{(0)}(i) - \bar{\varepsilon}\right)^2}. \tag{9.14}$$

Table 9.2 Grey model grades

Grade	Evaluation indices	
	P	C
Good	> 0.95	< 0.35
Qualified	> 0.8	< 0.5
Just	> 0.7	< 0.45
Unqualified	≤ 0.7	≥ 0.65

The mean $(\bar{X})$ and the standard deviation (S_2) of the AGO series are calculated as follows:

$$\bar{X} = \frac{1}{n}\sum_{i=1}^{n} X^{(1)}(i) \tag{9.15}$$

and

$$S_2 = \sqrt{\frac{1}{n}\sum_{i=1}^{n} \left(X^{(1)}(i) - \bar{X}\right)^2} \ . \tag{9.16}$$

Two evaluation indices, C (the relative performance of a model with a scattering degree S_2 and a fitness degree S_1) and P (smaller errors probability index), are then calculated by

$$C = \frac{S_1}{S_2} \tag{9.17}$$

and

$$P = prob\left(\left|\varepsilon^0(i) - \bar{\varepsilon}\right| \leq 0.6745 S_2\right. \tag{9.18}$$

The model is then graded according to Table 9.2.

Stage 6: Modified Forecasting Model If the model is found to be unqualified (needs correction), a weighted residual series for a selected historical data points $(i = k', \ldots, n$ where $k' \geq 1)$ of the original residual series is developed using the following equation,

$$r^{(0)}(i') = w\ \varepsilon^{(0)}(i) \quad \forall\ i' = i - k' + 1 = 1, 2, \ldots, n - k' + 1 \tag{9.19}$$

where w is the assigned weight for the residual series, $w \in [0, 1]$, $r^{(0)}$ is the weighted residual series, and i' represents the weighted series step.

Another Grey model is then built to fit the generated weighted residual series over the selected data points following the same previously described stages (Stages 1–4). The modified prediction equation, after correction, for the GM(1, m) model can be expressed as follows:

$$\hat{X}_1^{(1)}(i+1) = \left(X_1^{(0)}(1) - \frac{1}{a}\sum_{k=2}^{m} b_{k-1}X_k^{(1)}(i+1)\right)e^{-ai} + \frac{1}{a}\sum_{k=2}^{m} b_{k-1}X_k^{(1)}(i+1)$$
$$+ s(-a')\left(r^{(0)}(1) - \frac{b'}{a'}\right)e^{-a'(i-k'+1)} \tag{9.20}$$

where a' and b' are the parameters for the weighted series $r^{(0)}$, s is a selector that is equal to 1 for $k' \geq i$ and equals 0 for $k' < i$. The last equation can be reformulated for the GM(1,1) as follows:

$$\hat{X}^{(1)}(i+1) = \left(X^{(0)}(1) - \frac{b}{a}\right)e^{-ai} + \frac{b}{a} + s(-a')\left(r^{(0)}(1) - \frac{b'}{a'}\right)e^{-a'(i-k'+1)} \tag{9.21}$$

The model after being corrected with the residual term is then re-passed by the same diagnostic check process, expressed in stage 5, to validate its grade, and so on.

Stage 7: Inverse Accumulated Generating Operation This operation is used to transform the forecasted AGO series of data back into the original series. It represents the inverse operation of the AGO. The mathematical formulation representing this operation can be expressed by

$$\alpha^{(0)}\left(X^{(r)}(k)\right) = X^{(r)}(k) - X^{(r)}(k-1) \tag{9.22}$$

$$\alpha^{(1)}\left(X^{(r)}(k)\right) = X^{(r-1)}(k) - X^{(r-1)}(k-1) \tag{9.23}$$

and so on, for the rth series

$$\alpha^{(r)}\left(X^{(r)}(k)\right) = X^{(0)}(k) - X^{(0)}(k-1) \quad k = 1, \ldots, n, \tag{9.24}$$

where α is the Inverse Accumulated Generating Operation (IAGO) operator.

For the GM(1,1) model, built to fit the first AGO series, the IAGO is expressed by

$$\hat{X}^{(0)}(1) = \hat{X}^{(1)} \quad (1) \tag{9.25}$$

$$\hat{X}^{(0)}(i+1) = \hat{X}^{(1)}(i+1) - \hat{X}^{(1)}(i) \quad \forall i = 1, 2, 3, \ldots \tag{9.26}$$

Stage 8: Rolling Modeling Mechanism The rolling modeling mechanism offers a tool for updating the used input data in developing the GM(1,1) model [29]. This mechanism ensures that the most recent historical data are used in the modeling process at each time interval. This technique is based on waiting until observing the actual data of the predicted interval. Then, updating the used historical data by eliminating the oldest historical data and adding the recent observation to predict the next future interval.

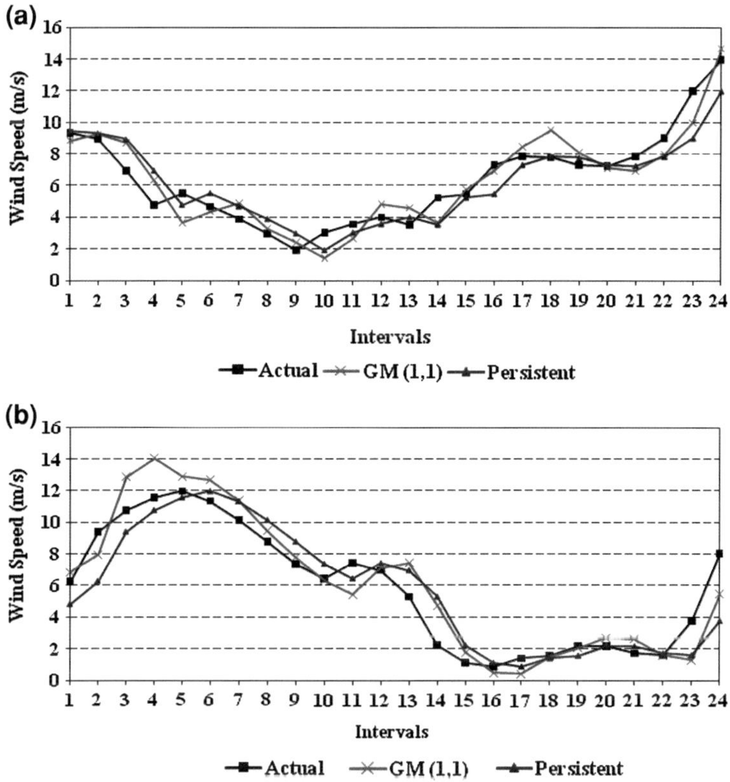

Fig. 9.6 Wind speed forecasting using traditional GM(1,1) rolling model and persistent model. **a** Sample 1 **b** Sample 2

The actual and forecasted wind speed time series for the two illustrative data samples are presented in Fig. 9.6. This Figure displays the results of the traditional GM(1,1) rolling and the persistent models (that assume that the future hour predicted value is equal to the current observation). The Figure demonstrates the effectiveness of the traditional GM(1,1) rolling model in tracking the actual time series better than that of the persistent model. The use of the traditional GM(1,1) rolling model results in the Mean Absolute Error (MAE) values of 0.94 and 1.12 m/s for Samples 1 and 2, respectively, which corresponds to 0.97 and 1.19 m/s, respectively, for the persistent model. Moreover, for the studied samples, the traditional GM(1,1) rolling model forecasts wind speed time series with an improvement, in the MAE, over the persistent model of 3.09 and 5.88% for Samples 1 and 2, respectively.

Figure 9.7 gives the C Indices for the developed models by using the traditional GM(1,1). These Figures reveal that the developed models' grades are good. This has also been confirmed by the P evaluation index that is consistently equal to unity over the studied 24 h data samples.

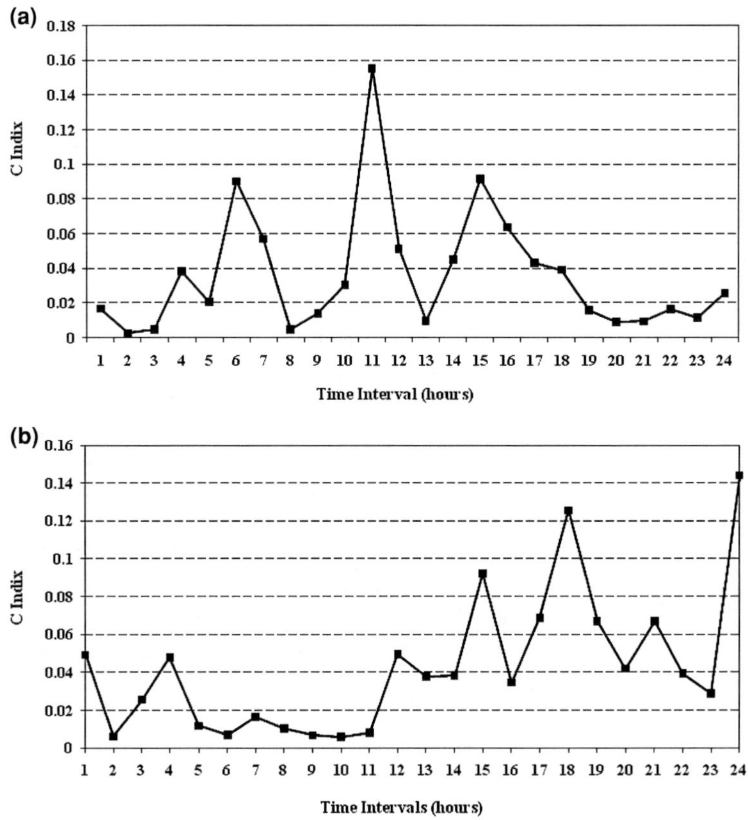

Fig. 9.7 C Indices for the developed models using the traditional GM(1,1). **a** Sample 1. **b** Sample 2

Despite the superiority of the newly developed model, the use of the traditional GM(1,1) in predicting continuously variable time series usually results in the occurrence of overshoots [29] that reduce the prediction accuracy. The following models were developed to reduce these overshoots and increase the overall prediction accuracy of the Grey model, hereafter, referred to as the adaptive alpha-based GM(1,1) model [29], the improved shifted model and the averaged Grey models [2].

9.2.2 Adaptive Alpha-Based GM(1,1) Model

The adaptive alpha-based GM(1,1) model was developed by following the same stages described in the previous section for the traditional GM(1,1) model while

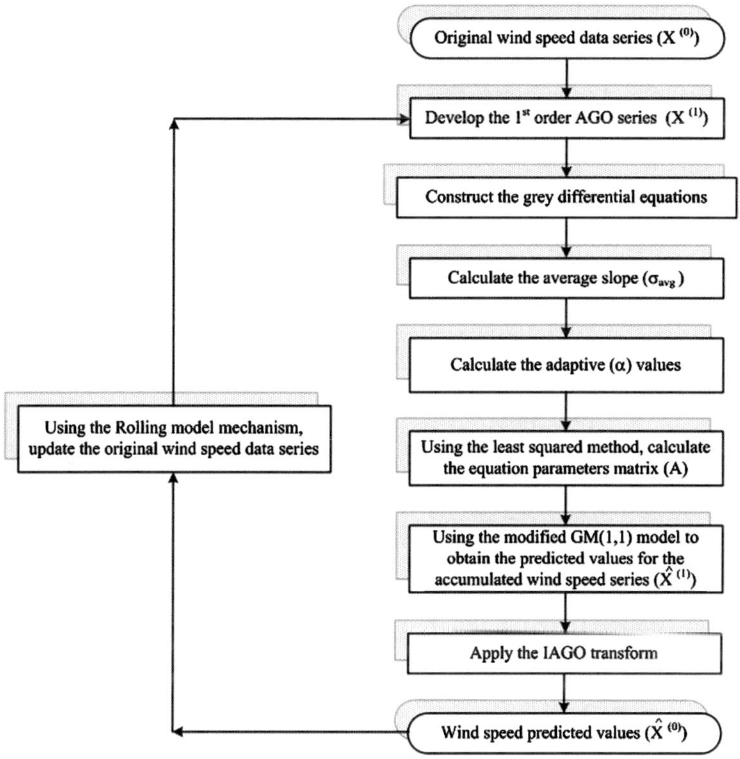

Fig. 9.8 Prediction process using adaptive alpha-based GM(1,1) rolling model

introducing a new formula for calculating the $Z^{(1)}(i)$ terms, as demonstrated in Fig. 9.8. In 1999, Chang et al. proposed an adaptive $\alpha(i)$ for the $Z^{(1)}(i)$ terms [30] so that the formula for calculating these terms is expressed as

$$Z^{(1)}(i) = [1 - \alpha(i)]X^{(1)}(i-1) + \alpha(i)X^{(1)}(i) \ , \tag{9.27}$$

where $\alpha(i)$ is a weighting factor within the range $0 \le \alpha(i) \le 1$. For the traditional GM(1,1) model, this factor is set to 0.5.

In this model, the weighting factor $\alpha(i)$ is determined by what is known as the "average system slope" technique [29]. The mathematical formulations describing each stage involved in developing this model are summarized as follows.

Stage 1: Accumulated Generating Operation Same as described in Sect. 9.2.1.

Stage 2: GM(1,1) Differential Equation Same as described in Sect. 9.2.1.

Stage 3: Average Slope Coefficient This stage is dedicated to calculate the average slope coefficient (σ_{avg}) as follows:

$$\sigma_{\text{avg}} = \sqrt[n-1]{\frac{X^{(0)}(n)}{X^{(0)}(1)}}, \tag{9.28}$$

where $X^{(0)}(n)$ and $X^{(0)}(1)$ represent the last and the first data points in the original data series, respectively, and n represents the total number of data points used in developing the GM(1,1) model.

Stage 4: Relative Position In this stage the relative positions (k_j) for the remaining data points (excluding the first and the last points that are forced to match the data values of the real system end points) are determined to force the data of the theoretical system model to equate the data of the real system model. This is mathematically formulated as follows (for four data points model):

$$X^{(0)}(1)\sigma_{\text{avg}}^{k_2} = X^{(0)}(2) \quad \Rightarrow \quad k_2 = \log\left(X^{(0)}(2)/X^{(0)}(1)\right)/\log\left(\sigma_{\text{avg}}\right) \tag{9.29}$$

$$X^{(0)}(1)\sigma_{\text{avg}}^{k_3} = X^{(0)}(3) \quad \Rightarrow \quad k_3 = \log\left(X^{(0)}(3)/X^{(0)}(1)\right)/\log\left(\sigma_{\text{avg}}\right) \tag{9.30}$$

and

$$X^{(0)}(1)\sigma_{\text{avg}}^{3} = X^{(0)}(4)\,. \tag{9.31}$$

Stage 5: Adaptive Alpha Set The adaptive alpha set is calculated using the following if rules:

(a)

$$\begin{array}{llll} IF & k_2 \leq 0 & then & \alpha(2) = 0 \\ \text{else} & IF & k_2 \in (0,1) & then \quad \alpha(2) = k_2 \\ \text{else} & IF & k_2 \geq 0 & then \quad \alpha(2) = 1 \end{array} \tag{9.32}$$

(b)

$$\begin{array}{llll} IF & k_3 \leq 1 & then & \alpha(3) = 0 \\ \text{else} & IF & k_3 \in (1,2) & then \quad \alpha(3) = k_3 - 1 \\ \text{else} & IF & k_3 \geq 2 & then \quad \alpha(3) = 1 \end{array} \tag{9.33}$$

(c)

$$\alpha(4) = 1\,. \tag{9.34}$$

Stage 6: Equation Parameters Calculations Same as described in Sect. 9.2.1 while introducing a new formula for calculating the $Z^{(1)}(i)$ terms.

Stage 7: Forecasting Model Same as described in Sect. 9.2.1.

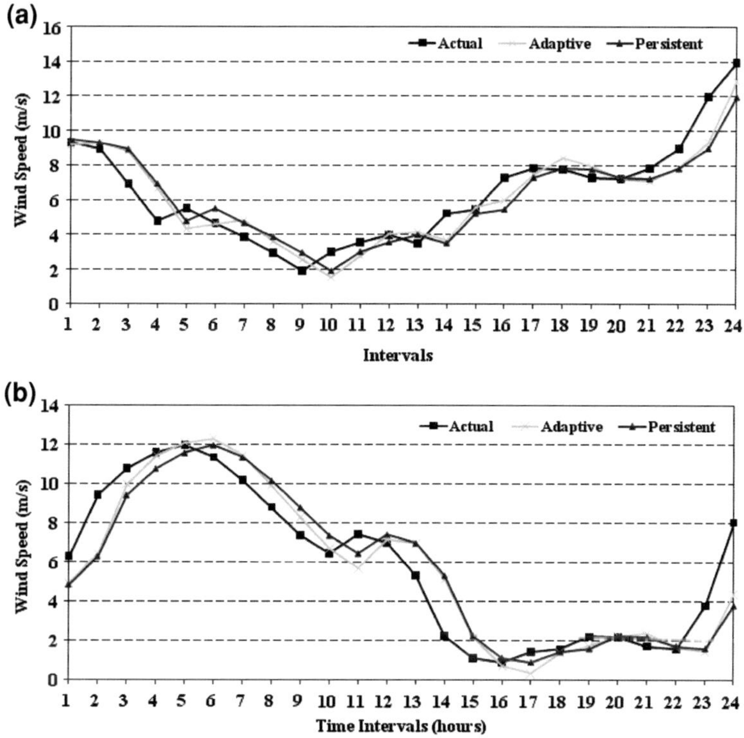

Fig. 9.9 Wind speed forecasting using the adaptive alpha-based GM(1,1) rolling model and persistent model. **a** Sample 1. **b** Sample 2

Stage 8: Inverse Accumulated Generating Operation Same as described in Sect. 9.2.1.

Stage 9: Rolling Modeling Mechanism Same as described in Sect. 9.2.1.

The corresponding actual and predicted wind speed time series, for the two illustrative data samples, and using the adaptive alpha based model, are presented in Figs. 9.9 and 9.10. These Figures reveal that the adaptive alpha-based model can reduce the overshoots in the predicted time series; however, for the intervals with no overshoots, the traditional model tracks the actual time series better. The adaptive alpha based model results in MAE values of 0.87 and 1.09 m/s for Samples 1 and 2, respectively, which correspond to an improvement, in the MAE, over the persistent model of 10.31 and 8.4%, respectively. The adaptive alpha-based model revealed more improvements in the MAE, over the persistent model, than the traditional model.

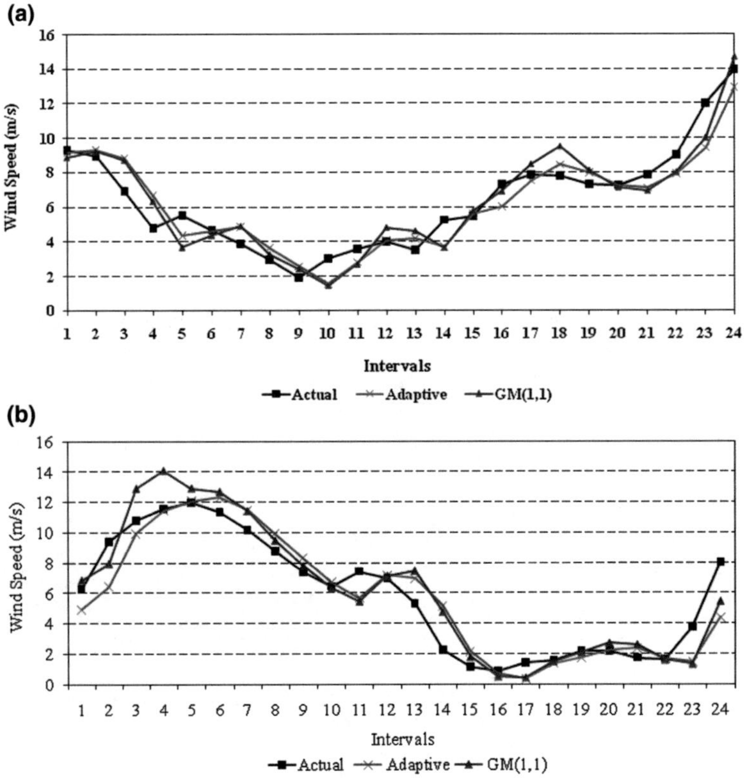

Fig. 9.10 Wind speed forecasting using the adaptive alpha-based GM(1,1) rolling model and traditional GM(1,1) model. **a** Sample 1. **b** Sample 2

9.2.3 Improved Shifted Grey Model

In 2007, EL-Fouly et al. [2] introduced the improved shifted Grey GM(1,1) model in an attempt to increase the prediction accuracy. This model is based on generating two shifted prediction models from the traditional GM(1,1) model, discussed in Sect. 9.2.1. Then, a hybrid model of these new shifted models is developed as the final prediction model using the Widro-Hoff delta rule. Figure 9.11 presents a flowchart illustrating the different stages for developing the improved shifted Grey model. The mathematical formulations, describing each stage, are summarized as follows.

Stage 1: Accumulated Generating Operation Same as described in Sect. 9.2.1.

Stage 2: GM(1,1) Differential Equation Same as described in Sect. 9.2.1.

Stage 3: Equation Parameters Calculations Same as described in Sect. 9.2.1.

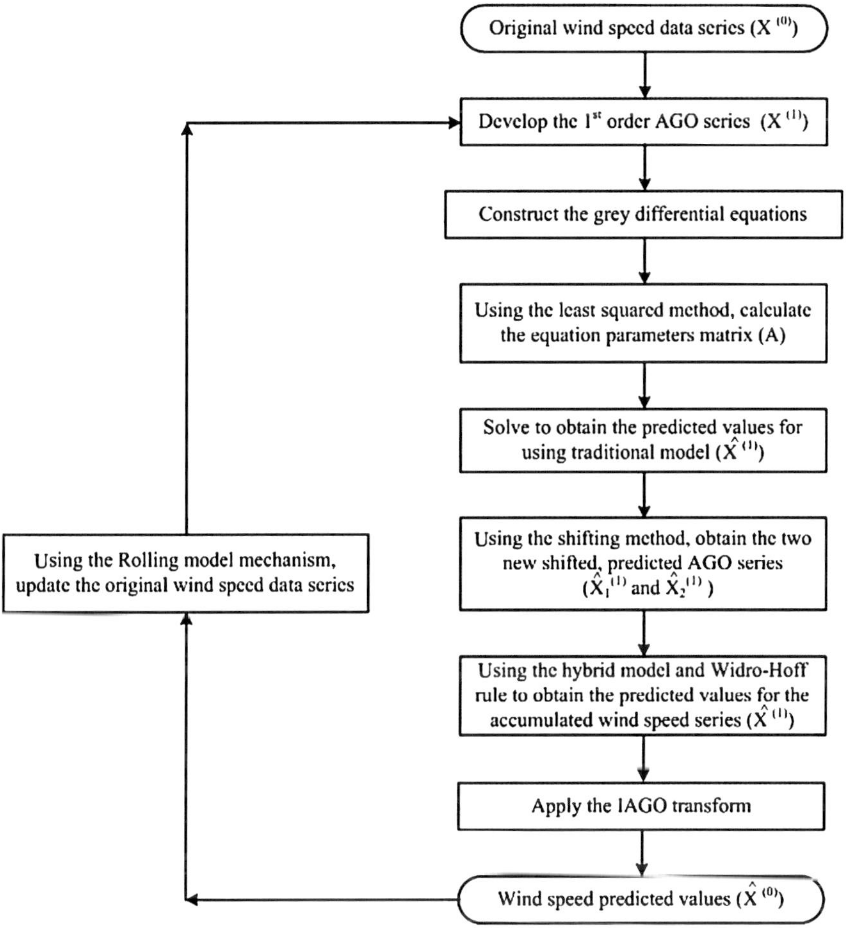

Fig. 9.11 Prediction process using improved shifted GM(1,1) rolling model

Stage 4: Forecasting Model Same as described in Sect. 9.2.1. This step ends by developing the traditional GM(1,1) forecasting model, to predicting the values for the AGO series $\left(\hat{X}_{\mathrm{GM}}^{(1)}\right)$, which is represented mathematically as

$$\hat{X}_{\mathrm{GM}}^{(1)}(i+1) = \left(X^{(0)}(1) - \frac{b}{a}\right)e^{-ai} + \frac{b}{a}. \tag{9.35}$$

Stage 5: Two Shifted Forecasting Models The goal of this stage is to generate an envelope-like time series that contains the actual AGO series. Two new shifted forecasting models $\left(\hat{X}_1^{(1)} \text{ and } \hat{X}_2^{(1)}\right)$ are generated as follows:

$$\hat{X}_1^{(1)}(i+1) = \hat{X}_{\mathrm{GM}}^{(1)}(i+1) + \mathit{shift} \tag{9.36}$$

and

$$\hat{X}_2^{(1)}(i+1) = \hat{X}_{\mathrm{GM}}^{(1)}(i+1) - \mathit{shift}\,, \tag{9.37}$$

where *shift* represents the value by which the traditional GM(1,1) model is shifted and is chosen to be equal to the last data point in the first AGO series; i.e., $X^{(1)}(n)$, where $n = 4$ (four data points are used to build the GM models) to ensure that the generated envelope contains the actual AGO series.

Stage 6: Hybrid Forecasting Model The two generated shifted models are then combined to create the hybrid improved Grey model $\left(\hat{X}_{\mathrm{imp}}^{(1)}\right)$ as follows:

$$\hat{X}_{\mathrm{imp}}^{(1)}(i+1) = w_1 \cdot \hat{X}_1^{(1)} + w_2 \cdot \hat{X}_2^{(1)}(i+1) \tag{9.38}$$

and

$$w_1 + w_2 = 1, \tag{9.39}$$

where w_1 and w_2 represent the weights for the shifted models. These weights are updated using the Least Mean Square (LMS) technique, known as the Widro-Hoff delta rule [31, 32], which is

$$w(i+1) = w(i) + \delta \cdot \frac{X^{(1)}(i) \cdot e(i)}{\left[X^{(1)}(i)\right]^T \cdot X^{(1)}(i)}\,, \tag{9.40}$$

where $w(i)$ and $w(i+1)$ represent the current and the updated weights vector, δ is the learning parameter, $X^{(1)}(i)$ and $\left[X^{(1)}(i)\right]^T$ represent the shifted AGO series vector and its transposed vector, and $e(i)$ is the error vector given by

$$e(i) = X^{(1)}(i) - \hat{X}_{\mathrm{imp}}^{(1)}(i) \tag{9.41}$$

Stage 7: Inverse Accumulated Generating Operation Same as described in Sect. 9.2.1.

Stage 8: Rolling Modeling Mechanism Same as described in Sect. 9.2.1.

Figures 9.12 and 9.13 display the corresponding actual and predicted wind speed time series, when the improved shifted Grey model is employed. This model achieves levels of the MAE of 0.95 and 1 m/s for Samples 1 and 2, respectively. This corresponds to an improvement over the persistent model of 2.06 and 15.97%, respectively. Figure 9.13 reveals that the improved model manages to reduce the overshoots; however, the tracking feature for the actual time series is still worse than that of the traditional Grey model. Moreover, for Sample 2 a drop in the percentage improvement of the MAE occurs, compared with that of the adaptive alpha-based model.

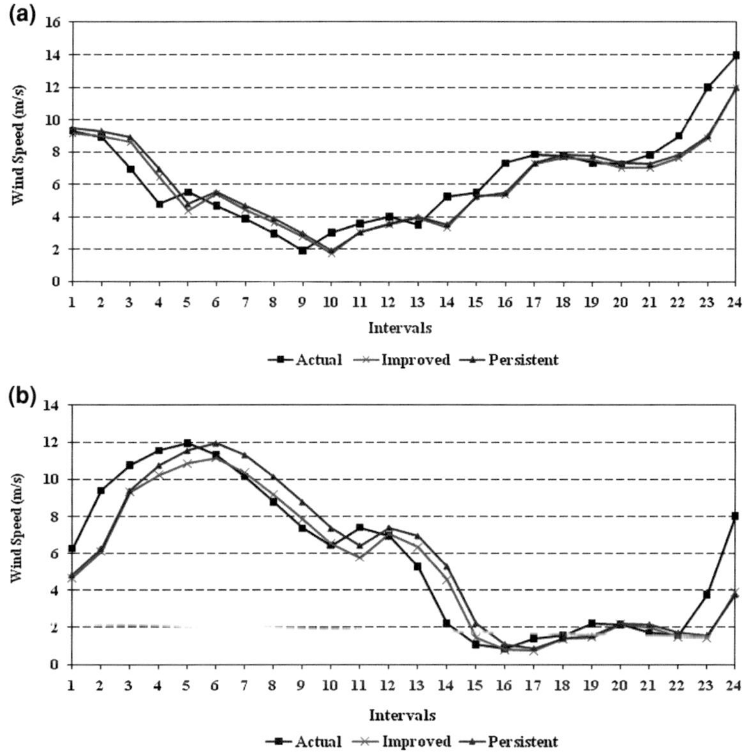

Fig. 9.12 Wind speed forecasting using the improved shifted GM(1,1) rolling model and the persistent model. **a** Sample 1. **b** Sample 2

9.2.4 Averaged Grey Model

To capture the good features of both the traditional GM(1,1) model (represented by good tracking and lower prediction error for the intervals without prediction overshoots) and the improved Grey model (represented by reducing the overshoots, and hence, reducing the prediction errors at the intervals where overshoots occur), EL-Fouly et al. [2] developed the averaged Grey forecasting model that is represented mathematically by

$$\hat{X}^{(1)}_{\text{avg}}(i+1) = \frac{\hat{X}^{(1)}_{\text{GM}}(i+1) + \hat{X}^{(1)}_{\text{imp}}(i+1)}{2}, \tag{9.42}$$

where $\hat{X}^{(1)}_{\text{avg}}(i+1)$ represents the predicted future point using the averaged Grey model.

Figures 9.14 and 9.15 present the corresponding actual and predicted wind speed time series, when the averaged Grey model is used. This model results in MAE values of 0.85 and 0.87 m/s for Samples 1 and 2, respectively, which

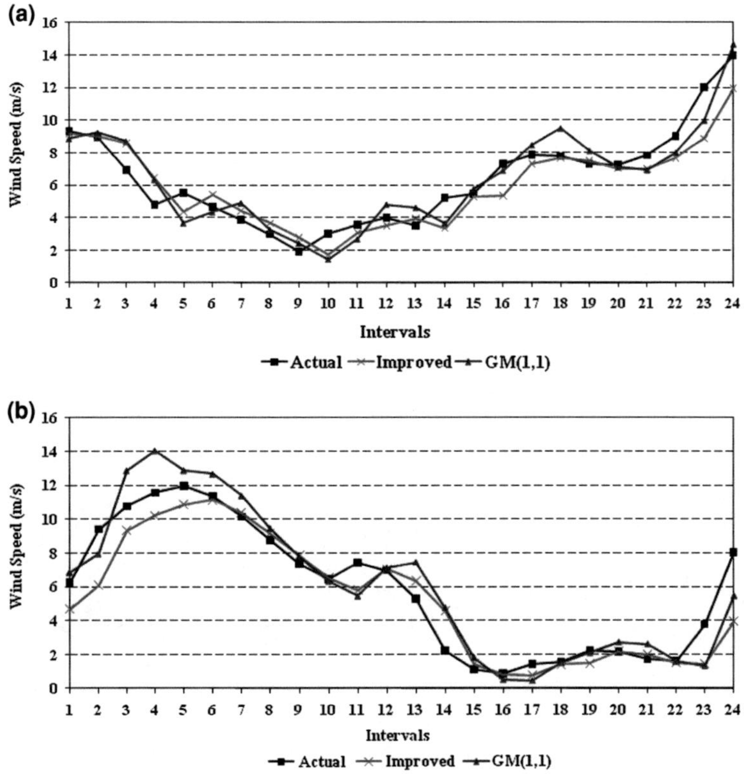

Fig. 9.13 Wind speed forecasting using the improved shifted GM(1,1) rolling model and the traditional GM(1,1) model. **a** Sample 1. **b** Sample 2

correspond to an improvement over the persistent model of 12.37 and 26.89%, respectively. This demonstrates the superiority of the averaged model in wind speed forecasting over the previously discussed Grey models in this section. Moreover, these Figures reveal that the tracking feature is enhanced when the averaged model is used and is overall superior, when compared with the persistent model and the other presented Grey models.

Table 9.3 provides a comparison of all the Grey models introduced in this section and the persistent model in terms of; the samples' MAE; and the samples' RMSE. Moreover, this table demonstrates the improvements of the averaged GM(1,1) model, compared with the persistent model. The table reveals that, for the two illustrative data samples, the highest percentage of improvement is achieved when the averaged GM(1,1) model is employed.

Figure 9.16 represents the scattering and the linear relationships between the predicted and the actual values for the wind speed of the four data sets, predicted by the averaged GM(1,1) model. This linear relationship is expressed as

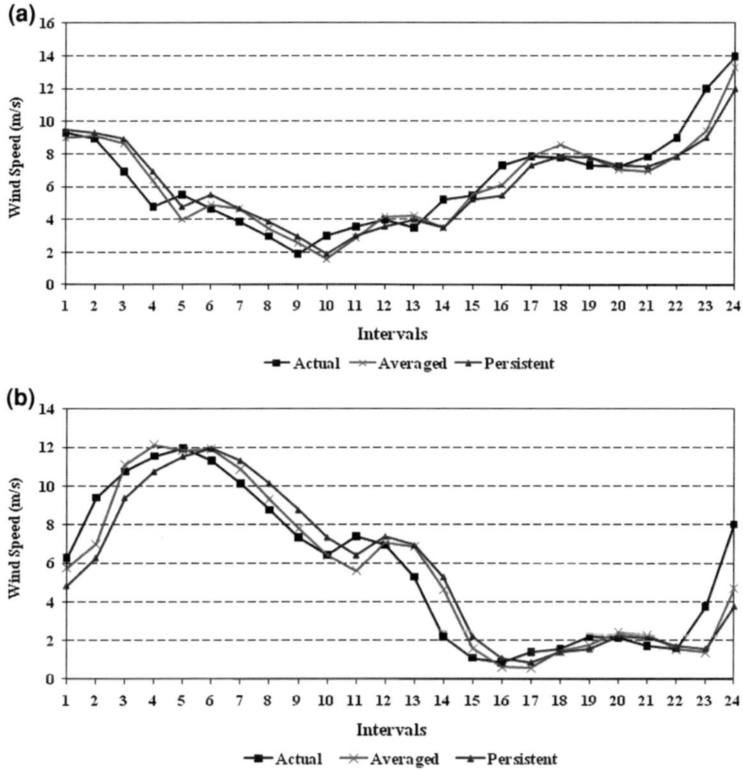

Fig. 9.14 Wind speed forecasting using the averaged GM(1,1) rolling model and the persistent model. **a** Sample 1. **b** Sample 2

$$\hat{y}_i = m x_i + c, \tag{9.43}$$

where $\hat{y}_i$ and x_i are the predicted and the actual wind speed or power values at time interval, i, m, and c are the linear relationship coefficients obtained by using the least square method, where m is the scaling factor (slope of the linear relationship), and c is the y-axis (prediction-axis) intercept of the linear relation. The best linear relationship between the actual and the predicted values is achieved when $c = 0$ and $m = 1$.

Figure 9.16 reveals a good scattering of the predicted values of wind speed. The corresponding linear relationships between the predicted and the actual values of the wind speed data are $0.618 + 0.876\ x_i$ and $0.618 + 0.876\ x_i$ for Samples 1 and 2, respectively. This indicates that the scaling factor is very close to unity, and that the y-axis intercept coefficient is very close to zero.

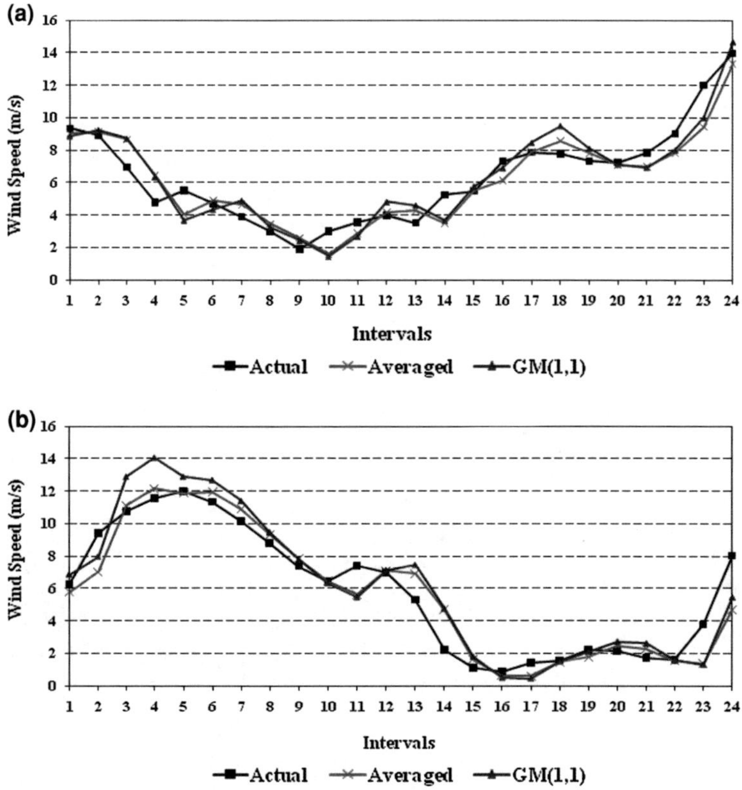

Fig. 9.15 Wind speed forecasting using the averaged GM(1,1) rolling model and the traditional GM(1,1) model. **a** Sample 1. **b** Sample 2

Table 9.3 MAE (m/s), RMSE (m/s), and improvements for wind speed prediction

	Parameter		Persistent	GM(1,1)	Adaptive	Improved	Averaged
Sample 1	*MAE*	Value	0.97	0.94	0.87	0.95	0.85
		Improvement (%)	–	*3.09*	*10.31*	*2.06*	*12.37*
	RMSE	Value	1.22	1.10	1.09	1.22	1.06
		Improvement (%)	–	*9.84*	*10.66*	*0.00*	*13.11*
Sample 2	*MAE*	Value	1.19	1.12	1.09	1.00	0.87
		Improvement (%)	–	*5.88*	*8.40*	*15.97*	*26.89*
	RMSE	Value	1.58	1.42	1.48	1.47	1.26
		Improvement (%)	–	*10.13*	*6.33*	*6.96*	*20.25*

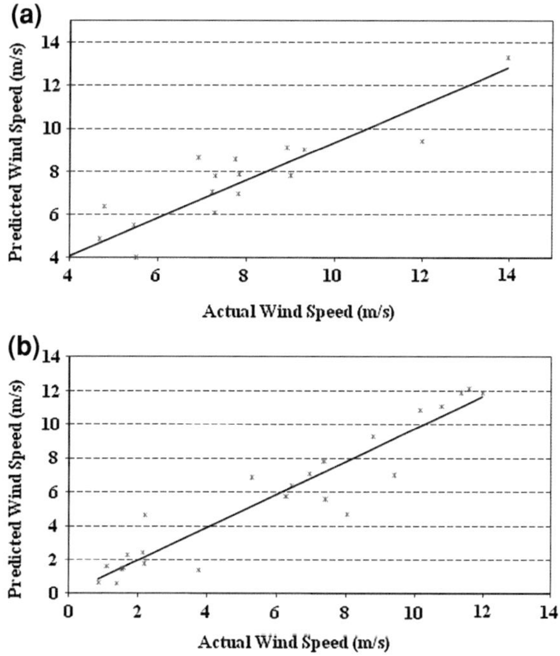

Fig. 9.16 Actual and predicted wind speed relationships using the averaged GM(1,1) rolling model. **a** Sample 1. **b** Sample 2

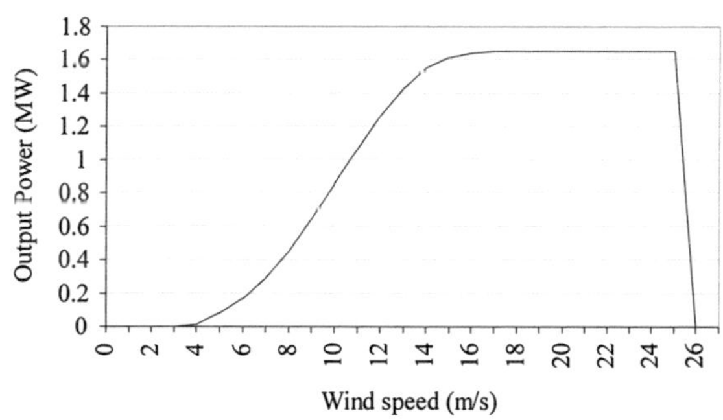

Fig. 9.17 Manufacturer wind power curve for the VESTAS V66-1.65 MW wind turbine

9.3 Hourly Wind Power Prediction

Forecasted wind speed data is used as an input to the wind turbine power curve. For illustrative purposes, the manufacturer power curve for the VESTAS V66-1.65 MW wind turbine, given in Fig. 9.17, is used in the analysis to predict the output wind power production. The wind turbine power curve under investigation is characterized by a cut in wind speed of 4 m/s, rated wind speed of 16 m/s, and a cut out wind speed of 25 m/s.

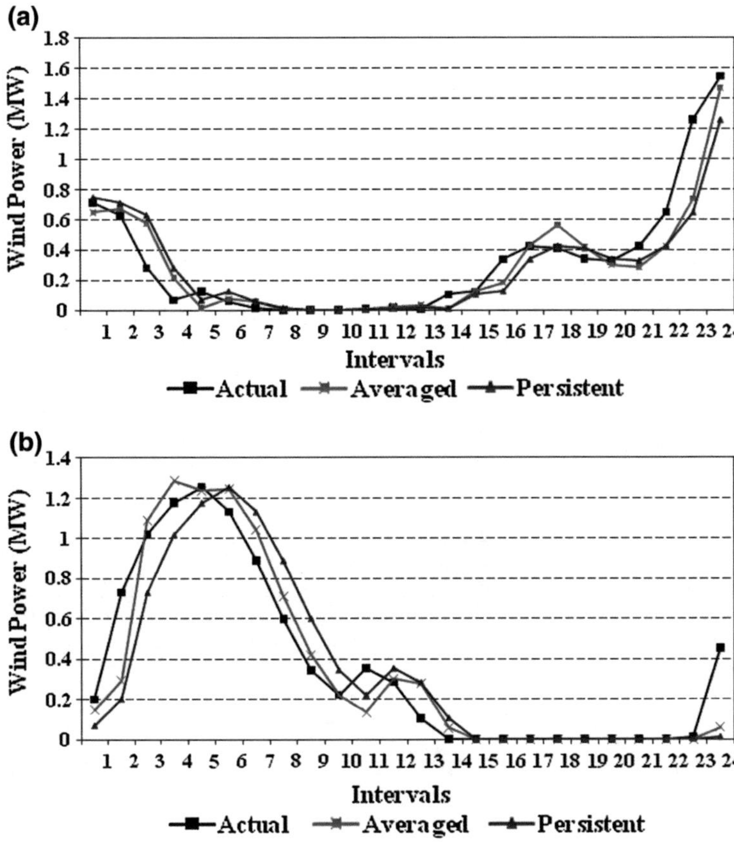

Fig. 9.18 Wind power prediction using the averaged GM(1,1) rolling model and the persistent model. **a** Sample 1. **b** Sample 2

In this section, the generated actual and predicted wind power time series, when the averaged GM(1,1) and the persistent models are employed, for the two samples are presented in Fig. 9.18. The Figure demonstrates the effectiveness of the averaged GM(1,1) rolling model for tracking the actual generated wind power time series.

Table 9.4 lists the MAE, RMSE, and average percentage error referred to the rated power (% error for the power prediction) for the generated results by the averaged GM(1,1) rolling model. This table also reveals a lower MAE, lower RMSE, and lower average percentage error values for the averaged GM(1,1) rolling model, compared with those of the persistent model. The table also reveals that the averaged model, for the illustrative data samples, predicts the wind power more accurately than the persistent model; in fact, up to 36.31% for the MAE, 25.83% for the RMSE, and 36.34% for the average percentage error.

Table 9.4 MAE (kW), RMSE (kW), average percentage error (% error), and improvements for wind power prediction

	Parameter	Persistent	Averaged	Improvement (%)
Sample 1	MAE	109.84	93.80	*14.60*
	RMSE	179.70	150.96	*15.99*
	error (%)	6.62	5.69	*14.05*
Sample 2	MAE	131.60	83.81	*36.31*
	RMSE	194.90	144.55	*25.83*
	error (%)	7.98	5.08	*36.34*

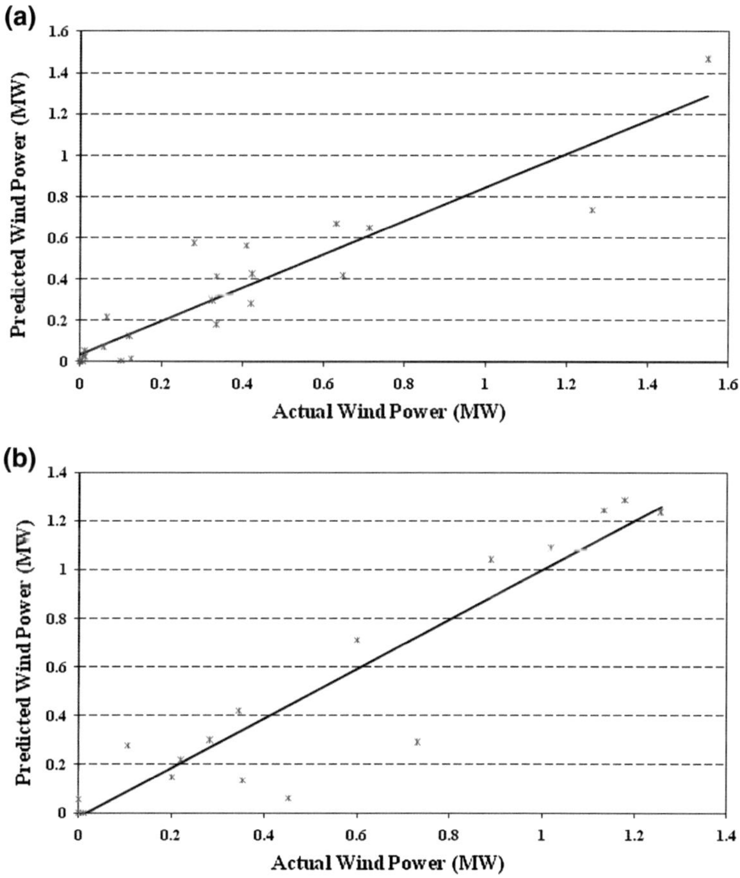

Fig. 9.19 Actual and predicted wind power relationships using the averaged GM(1,1) rolling model. **a** Sample 1. **b** Sample 2

Figure 9.19 presents the scattering and the linear relationships between the predicted and the actual values for the wind power prediction for the two data sets with the averaged GM(1,1) model. This Figure exhibits a good scattering of the predicted values of the wind power. The corresponding linear relationships between the predicted and the actual values of the wind power data are $0.037 + 0.809\, x_i$ and $-\, 0.016 + 1.015\, x_i$ for Samples 1 and 2, respectively. This indicates that the scaling factor and the y-axis intercept coefficients are very close to unity and zero, respectively.

9.4 Conclusions

Due to the irregular nature of wind power, production prediction represents one of the major challenges to power system operators. This chapter introduced one of the newly developed forecasting techniques for 1 h ahead average hourly wind speed and wind power using the Grey predictor models. Forecasting using the traditional GM(1,1) model revealed an improvement over the persistent model (the traditional reference model) in tracking the actual wind speed time series. However, this model is characterized by the occurrence of some overshoots in the predicted time series. Such overshoots can result in predicting wind speed time series worse than that of the persistent model. To overcome the problem of overshoots occurrence, three modified versions of the GM(1,1) model were introduced. The adaptive alpha-based grey model and the improved Grey model achieved higher levels of improvement over the persistent model than the traditional GM(1,1) model. However, those models lack the good tracking characteristic for the actual wind speed time series of the traditional model. The averaged Grey model attained the highest level of wind speed forecasting and wind power prediction accuracy, compared with the persistent model and the other presented Grey models, while demonstrating a very good tracking feature and a reduction in the overshoot occurrence.

References

1. EL-Fouly THM, EL-Saadany EF, Salama MMA (2004) A study of wind farms output power prediction techniques. In: Proceedings of the 36th North American power symposium, NAPS 2004, Idaho, USA, pp 249–254
2. EL-Fouly THM, EL-Saadany EF, Salama MMA (2007) Improved grey predictor rolling models for wind power prediction. IET Proc Gener Transm Distrib 1(6):928–937
3. Huang Z, Chalabi ZS (1995) Use of time-series analysis to model and forecast wind speed. J Wind Eng Ind Aerodyn 56(2–3):311–322
4. Kamal L, Jafri YZ (1997) Time series models to simulate and forecast hourly averaged wind speed in Quetta, Pakistan. Sol Energy 61(1):23–32
5. More A, Deo MC (2003) Forecasting wind with neural networks. Mar Struct 16(1):35–49
6. Sfetsos A (2000) A comparison of various forecasting techniques applied to mean hourly wind speed time series. Renewable Energy 21(1):23–35
7. Costa M, Pasero E (2001) Artificial neural systems for verglass forecast. In: Proceedings of the international joint conference on neural networks, IJCNN, '01, vol 1. pp 258–262

8. Miranda MS, Dunn RW (2006) One-hour-ahead wind speed prediction using bayesian methodology. In: Proceedings of the IEEE power engineering society general meeting
9. Damousis IG, Dokopoulos P (2001) A fuzzy expert system for the forecasting of wind speed and power generation in wind farms. In: 22nd IEEE power engineering society international conference on power industry computer applications, PICA 2001. Innovative computing for power—electric energy meets the market, pp 63–69
10. Alexiadis MC, Dokopoulos PS, Sahsamanoglou HS (1999) Wind speed and power forecasting based on spatial correlation models. IEEE Trans Energy Convers 14(3):836–842
11. Damoisis IG, Alexiadis MC, Theocharis JB, Dokopoulos PS (2004) A fuzzy model for wind speed prediction and power generation in wind parks using spatial correlation. IEEE Trans Energy Convers 19(2):352–361
12. Barbounis TG, Theocharis JB, Alexiadis MC, Dokopoulos PS (2006) Long term wind speed and power forecasting using local recurrent neural network models. IEEE Trans Energy Convers 21(1):273–284
13. Barbounis TG, Theocharis JB (2007) A locally recurrent fuzzy neural network with application to the wind speed prediction using spatial correlation. Neurocomputing 70: 1525–1542
14. Lange M, Focken U (2005) Physical approach to short-term wind power prediction. Springer, USA
15. Landberg L (1998) A mathematical look at a physical power prediction mode. Wind Energy 1(1):23–28
16. Landberg L (1999) Short-term prediction of the power production from wind farms. J Wind Eng Ind Aerodyn 80(1–2):207–220
17. Focken U, Lange M, Mönnich K, Waldl H, Beyer H, Luig A (2002) Short-term prediction of the aggregated power output of wind farms—a statistical analysis of the reduction of the prediction error by spatial smoothing effects. J Wind Eng Ind Aerodyn 90(3):231–246
18. Ackermann T (2005) Wind power in power system. Wiley, England
19. Deng JL (1982) Control problems of grey systems. Syst Control Lett 1(5):288–294
20. Wen K (2004) Grey systems: modeling and prediction. Yang's Scientific Press, AZ
21. Liu S, Lin Y (2005) Grey information: theory and practical applications. Springer Verlag London Ltd, London
22. Yang HT, Liang TC, Shih KR, Huang CL (1995) Power system yearly peak load forecasting: a grey system modeling approach. In: Proceedings of international conference on energy management and power delivery, EMPD, '95, vol 1. pp 261–266
23. Luo RC, Chen TM (1999) Target tracking by grey prediction theory and look-ahead fuzzy logic control. In: Proceedings of the 1999 IEEE international conference on robotics and automation, vol 2. pp 1176–1181
24. Chang BR, Tsai SF (2002) A grey-cumulative LMS hybrid predictor with neural network based weighting for forecasting non-periodic short-term time series. IEEE Int Conf Syst Man Cybern 6:5
25. Wai RJ, Duan RY, Chang LJ (2001) Grey feedback linearization speed control for induction servo motor drive. In: The 27th annual conference of the IEEE industrial electronics society, IECON, '01, vol 1. pp 580–585
26. Madison Weather Station. http://www.oardc.ohio-state.edu/centernet/stations/mahome.asp
27. EL-Fouly THM, EL-Saadany EF, Salama MMA (2006) Grey predictor for wind energy conversion systems output power prediction. IEEE Trans Power Syst (Letter) 21(3): 1450–1452
28. EL-Fouly THM, EL-Saadany EF, Salama MMA (2005) A novel technique for wind speed forecasting using grey predictor. In: Proceedings of the 10th Americas conference on wind engineering (10ACWE), Louisiana, USA
29. Yao AWL, Chi SC, Chen JH (2003) An improved grey-based approach for electricity demand forecasting. Electr Power Syst Res 67:217–224
30. Chang SC, Wu J, Lee CT (1999) A study on the characteristics of (k) of grey prediction. In: Proceedings of the 4th conference on grey theory and applications. pp 291–296

31. Marei MI, EL-Saadany EF, Salama MMA (2004) Envelope tracking techniques for flicker mitigation and voltage regulation. IEEE Trans Power Delivery 19(4):1854–1861
32. Dash PK, Swain DP, Routary A, Liew AC (1996) Harmonic estimation in a power system using adaptive perceptrons. IEE Proc Gener Trans Distrib 143(6):565–574

Chapter 10
Lightning Protection of Large Wind-Turbine Blades

F. Rachidi, M. Rubinstein and A. Smorgonskiy

Abstract We discuss in this chapter the salient issues related to lightning protection of large wind turbine blades. Lightning protection of modern wind turbines presents a number of new challenges due to the geometrical, electrical and mechanical particularities of turbines. Wind turbines are high structures and, like tall towers, they not only attract downward flashes but initiate upward flashes as well. The proportion between these types of flashes depends on many factors such as the structure height and the local terrain elevation. The rotation of the blades may also trigger lightning and result in considerable increase in the number of strikes to a wind turbine unit. Since wind turbines are tall structures, the lightning currents that are injected by return strokes into the turbines will be affected by reflections at the top, at the bottom, and at the junction of the blades with the static base of the turbine. This is of capital importance when calculating the protection of internal circuitry that may be affected by magnetically induced electromotive forces that depend directly on the characteristics of the current in the turbine. The presence of carbon reinforced plastics (CRP) in the blades introduces a new set of problems to be dealt with in the design of the turbines' lightning protection system. One problem is the mechanical stresses resulting from the energy dissipation in CRP laminates due to the circulation of eddy currents. The thus dissipated energy is evaluated and recommendations are given as to the number of down conductors and their orientation with respect to the CRP laminates so that the dissipated energy is minimized. It is also emphasized that the high static fields under thunderclouds might have an influence on the moving carbon fiber parts.

F. Rachidi (✉) · A. Smorgonskiy
Swiss Federal Institute of Technology (EPFL), Electromagnetic Compatibility
Laboratory EPFL-SCI-STI-FR, ELL-138, Station 11, CH-1015, Lausanne, Switzerland
e-mail: Farhad.Rachidi@epfl.ch

M. Rubinstein
University of Applied Sciences of Western Switzerland, Yverdon, Switzerland

S. M. Muyeen (ed.), *Wind Energy Conversion Systems*,
Green Energy and Technology, DOI: 10.1007/978-1-4471-2201-2_10,

Representative full scale blade tests are still complex since lightning currents from an impulse current generator are conditioned to the electrical characteristics of the element under test and return paths. It is therefore desirable to complement laboratory tests with theoretical and computer modeling for the estimation of fields, currents, and voltages within the blades.

10.1 Introduction

Wind energy is one of the fastest growing electric power generation technologies. It is well known that wind turbines are vulnerable to lightning, which can cause important damages to wind turbine components [1]. Currently available statistics reveal that for the wind turbines installed in Europe [2], the average number of faults per 100 turbine years is equal to six, whereas for the wind turbines installed in Japan [3], this value can reach even 36 due to the severe winter lightning parameters.

Lightning protection of traditional wind turbines has been addressed in detail in standards [4–6]. However, modern wind power generation units are characterized by ever taller turbines and wind turbine blades are now being produced with lengths of 60 m and beyond. As a result, it is expected that modern turbines will be more exposed to lightning strikes in the future [7].

In addition to the above, carbon fiber composite materials are being used extensively to reinforce the blades [6]. The presence of such composite materials will affect the efficiency of the lightning protection system (LPS) and should therefore be taken into account in its design phase.

Consequently, the design of lightning protection of modern wind turbines will be a challenging problem. In this chapter, we discuss the salient issues related to lightning protection of long wind turbine blades. The text is organized as follows: In Sects. 10.2, 10.3, 10.4, 10.5 and 10.6, we describe open questions, which need to be adequately addressed to achieve an efficient protection of modern wind turbines against lightning. In Sect. 10.2, we discuss theoretical and empirical approaches to evaluate lightning incidence to tall structures, with special emphasis on upward-initiated lightning flashes, which are predominant for tall structures. Section 10.3 deals with the effect of the rotation of the blades on the lightning behavior and protection of wind turbines. The interaction of lightning with tall structures, such as telecommunication towers, has recently attracted considerable attention of lightning researchers (e.g., [8–16]). It has been shown in particular that the lightning current along a tall structure exhibits multiple reflections occurring at the discontinuities and resulting in a nonuniform distribution of the current. The implication of this effect on lightning protection of wind turbines is discussed in Sect. 10.4. In Sect. 10.5, we discuss the potential impact of the presence of composite materials on the performance of wind turbine lightning protection systems. The effect of nearby (also called indirect) lightning is briefly discussed in Sect. 10.6. Finally, conclusions and recommendations are presented in Sect. 10.7.

10.2 Evaluation of Lightning Incidence to Wind Turbines

The design of an LPS should be based on the risk of lightning striking the structure in question. This risk is a function of the structure height, the local topography and the local level of lightning activity [6].

Elevated objects such as wind turbines experience both downward and upward flashes, the proportion being a function of object height [17]. The total annual lightning incidence N (in year^{-1}) is given by

$$N = N_u + N_d \tag{10.1}$$

where $N_u$and N_d are the annual number of upward flashes and downward flashes, respectively [17].

Based on observations of the lightning incidence to structures with heights ranging from 20 to 540 m situated on a flat surface in different regions of the world, Eriksson [18] derived the following equation for N

$$N = N_g \cdot 24 \cdot h_s^{2.05} \cdot 10^{-6} \tag{10.2}$$

where h_s is the height of the structure in meters and N_g is the ground flash density in km^{-2} year^{-1}.

IEC [6] recommends that wind turbines on a flat terrain be modeled as a tall mast with a height equal to the hub height plus one rotor radius, the equivalent attractive or collection area being defined as a circle with a radius of three times the turbine height

$$R_a = 3 \cdot h_s \tag{10.3}$$

In IEC [6], the overall number of lightning flashes to the wind turbine is calculated using the expression

$$N = N_g \cdot \pi \cdot (3 \cdot h_s)^2 \cdot 10^{-6} \tag{10.4}$$

It should be especially noticed that in the IEC standard [6], the term 'equivalent attractive radius' is used for calculating the overall lightning incidence and not only downward flashes as it is used in other methods [17] where the number of upward and downward flashes to the structure is calculated separately. Note that the empirical expression of Eriksson (10.2) is very similar to (10.4). To show this, we can rewrite (10.2) as

$$N \approx N_g \cdot \pi \cdot (3.2 \cdot h_s)^2 \cdot 10^{-6} \tag{10.5}$$

Wind turbine failures due to lightning depend strongly on the terrain where the wind parks are installed. As reported in [2], wind turbines installed in the low mountain areas in Germany have a higher risk of lightning damage (14 faults per 100 unit years) compared to wind turbines installed in the coastal areas (5.6 faults per 100 unit years).

The evaluation of lightning incidence to wind turbines situated in mountainous regions is much more difficult than on flat ground due to the fact that topological factors will play a major role in the enhancement of the electric field at the top of the wind turbine. Therefore, the structure's actual height above ground level should be replaced with the effective height h_{eff}, whose evaluation is a very complex task and calls for further research [19].

Four methods have been reported by Theethayi [19] for the evaluation of h_{eff} and applied to the Gaisberg tower which is 100 m tall and is situated on a mountain of about 750 m height. The resultant h_{eff} varies from 300 to 1000 m. Another example is Berger's 70 m tall tower [20] situated on a prominent summit of the Monte San Salvatore, which rises 640 m above Lake Lugano. Eriksson [18] evaluated its effective height to be 350 m. A summary of the different techniques for the estimation of h_{eff} can be found in [21].

For complex terrains, IEC [6] recommends to add the height of the wind turbine position to the actual height of the turbine for the evaluation of h_{eff}. This recommendation is reasonable for the particular example shown in the standard, where the wind turbine is situated on a hill with a height comparable to that of the blade length. However, for wind parks situated on mountains, this recommendation could give an overestimated value of lightning incidence and, therefore, will complicate and increase the cost of required lightning protection measures.

10.3 Impact of Rotating Tower Blades

The fact that the blades in wind turbines rotate introduces new effects and complexity to the design of efficient lightning protection systems for them as discussed in what follows.

Lightning can be initiated by fast, upward moving objects connected to the ground (e.g., [22]). Small rockets trailing grounded wires have been used for many years to trigger lightning for research purposes (e.g., in France, at St. Privat d'Allier, in the United States at the Kennedy Space Center, at Camp Blanding's ICLRT in Florida and at Fort MacLellan in Alabama, and in Japan at Okushishiku [23]). A number of similarities between rocket triggered lightning and tall wind turbines with their rotating blades suggest that wind turbines may trigger their own lightning. Indeed, just like the small rockets used in classical triggered lightning, the tips of the rotating blades in wind turbines are connected to the ground by way of a down conductor which, in wind turbines, is used for lightning protection purposes. In rocket triggered lightning, the rockets reach an altitude of 200 m or so before setting off a lightning. The tips of the rotating blades are beginning to reach similar altitudes, as the height of the nacelle is higher than 100 m and the blades themselves are as long as 60 m or even longer, as mentioned earlier in this chapter. For classical rocket triggered lightning, the average speed of the rockets can be estimated by dividing the 200 m distance the rockets reach before the initiation of the lightning by the time it takes from the launch of the rocket to the initiation. This time is of the order of 2 s, as can be inferred from filmed sequences shot at the ICLRT in Camp Blanding, Florida, resulting in an average speed estimate of 100 m/s. The maximum speed of the tip of wind turbines is 80 m/s.

A number of differences between rocket triggered lightning and wind turbines must however be pointed out. The bottom 100 m or more of the ground conductor in wind turbines is fixed. In addition, the moment at which the tip of a particular blade in a wind turbine attains its highest vertical speed corresponds to the blade being horizontal and, therefore, at the height of the hub which, at least for current turbines, is much lower than 200 m. When the tip of the blade is at its highest point, the speed of the tip is purely horizontal. Moreover, the down conductor is designed to withstand lightning currents and it therefore may not vaporize when the continuous current from the initial phase of the lightning passes through it. As a consequence, the initial part of the current, called by Wang et al. [24, 25] the initial current variation ICV, will only seldom exhibit a current drop due to the melting of the wire. The question arises as to the way one could test the hypothesis that rotating blades on wind turbines trigger lightning. One possible test method is to compare the number of strikes to rotating blades with the number of strikes to still turbines in the same wind turbine park. The strike frequency for the operating blades should be statistically higher than that for inactive ones. Another possible method is to correlate the instantaneous position and sense of movement of blades struck by lightning to determine whether blade tips that are moving up are struck more often than descending blades.

Figure 10.1 presents the basic geometry for studies of lightning strikes to wind turbine systems particularized for the case where one of the blades points straight up at the time of a strike.

With the tip of the blades moving at speeds of up to 80 m/s and with inter-stroke intervals lasting from a few milliseconds to many tens of milliseconds or more, the angular position of a blade for two different strokes in the same flash may vary as much as tens of degrees, creating the conditions for the effects to be discussed in the following paragraph and in Sect. 10.4.

It is possible that different return strokes within the same flash flow down different blades as illustrated in the following scenario: Suppose one return stroke strikes the tip of a blade near the blade position shown in Fig. 10.1. If the inter-stroke interval is long, the continuing current (if present) may extinguish and the ionized channel that led to the original blade may shift as the rotating tip moves away from it. A subsequent leader may, in that case, attach to a different blade as illustrated in Fig. 10.2.

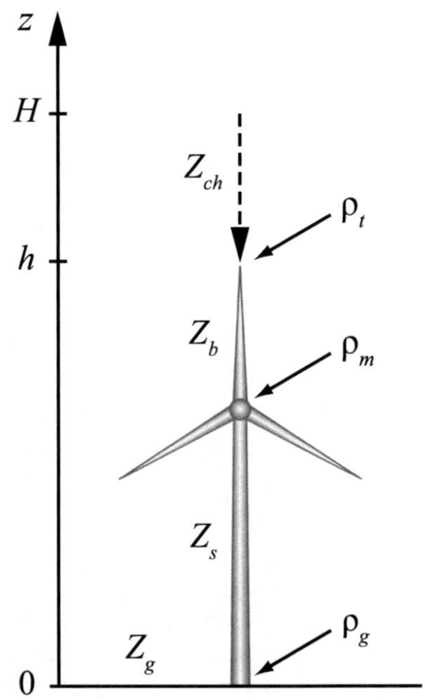

Fig. 10.1 Geometry for studies in lightning to wind turbine systems. (Adapted from [7] ©2010 IEEE)

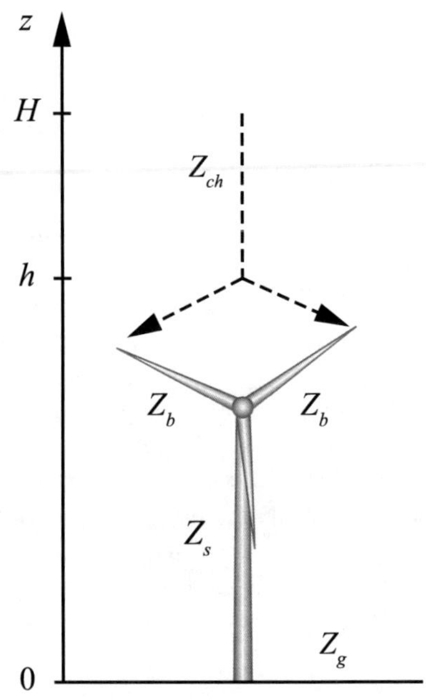

Fig. 10.2 Geometry for studies in lightning to wind turbine systems—rotation of the blades for subsequent lightning returns strokes. (Adapted from [7] ©2010 IEEE)

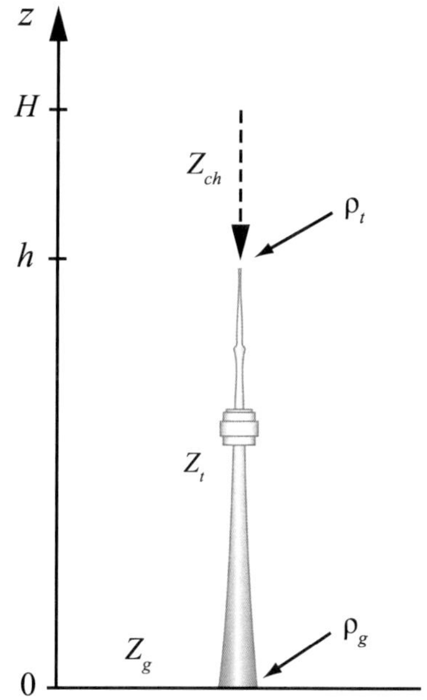

Fig. 10.3 Typical geometry for studies in lightning to elevated strike objects. (Adapted from [7] ©2010 IEEE)

10.4 Lightning Current Transient Behavior Within the Turbine

When lightning strikes an elevated tower or wind turbine, the transient phenomena in the strike object introduce changes in the original lightning current waveform. The observed current parameters within the structure of the strike object are therefore 'disturbed' by the presence of current reflections and the degree of disturbance depends on the physical and electrical characteristics of the strike object.

The influence of the strike object on the current waveform has been recently investigated by a number of research groups around the world and several of the so-called engineering return-stroke models have been extended to take into account the presence of the elevated strike object (see e.g., [26]). In some of these studies, the strike object was modeled as an ideal, uniform transmission line. Figure 10.3 presents a typical geometry employed for studies in lightning to elevated strike objects.

We can use the models developed for lightning return strokes to static elevated strike objects such as towers to discuss the way in which they should be adapted to the case of wind turbines.

In the return stroke models that take into account the strike structure, it is often assumed that the propagation speed of current pulses along the strike object is equal to the speed of light c and that the current reflection coefficients at its extremities (ρ_t at the top and ρ_g at the bottom) are constant. Further, the existence of upward-connecting leaders and any reflections at the return stroke wavefront are neglected.

The bottom reflection coefficient for the current in the tower can be expressed in terms of the characteristic impedance of the tower, Z_t, and the grounding system impedance, Z_g, as follows

$$\rho_g = \frac{Z_t - Z_g}{Z_t + Z_g} \tag{10.6}$$

Similarly, the top reflection coefficient for the current in the tower can be expressed in terms of the characteristic impedance Z_t and the equivalent impedance of the lightning return stroke channel Z_{ch},

$$\rho_t = \frac{Z_t - Z_{ch}}{Z_t + Z_{ch}} \tag{10.7}$$

For a lightning return stroke initiated at the top of the strike object, the current along it and along the lightning channel for a given height z were derived by Rachidi et al. [27] and it is given by

For $0 \leq z \leq h$:

$$\begin{aligned} i(z,t) = (1-\rho_t)\sum_{n=0}^{\infty}\Bigg[\rho_t^n\rho_g^n i_0\left(h, t - \frac{h-z}{c} - \frac{2nh}{c}\right) \\ \cdot u\left(t - \frac{h-z}{c} - \frac{2nh}{c}\right) + \rho_t^n\rho_g^{n+1} i_0\left(h, t - \frac{h+z}{c} - \frac{2nh}{c}\right) \\ \cdot u\left(t - \frac{h+z}{c} - \frac{2nh}{c}\right)\Bigg] \end{aligned} \tag{10.8}$$

and for $h < z < H_{tot}$:

$$\begin{aligned} i(z,t) = \Bigg[P(z-h)\, i_o\left(h, t - \frac{z-h}{v^*}\right) - \rho_t i_o\left(h, t - \frac{z-h}{c}\right) \\ + (1-\rho_t)(1+\rho_t)\sum_{n=0}^{\infty}\rho_g^{n+1}\rho_t^n i_o\left(h, t - \frac{h+z}{c} - \frac{2nh}{c}\right) \\ \cdot u\left(t - \frac{h+z}{c} - \frac{2nh}{c}\right)\Bigg]\, u\left(t - \frac{z-h}{v}\right) \end{aligned} \tag{10.9}$$

where $i_o(h, t)$ is the so-called 'undisturbed' current, defined as the current that would be measured at the top of the strike object (lightning attachment point) if both reflection coefficients ρ_t and ρ_g were equal to zero, z is the height along the strike object for Eq. (10.8) and along the channel for Eq. (10.9), c is the speed of light, v is the return stroke speed, H_{tot} is the total height, obtained by adding the lengths of the lightning channel and of the elevated strike object, $P(z')$ is a model-dependent attenuation function, v^* is the current-wave speed in the lightning channel, and $u(t)$ is the Heaviside unit step function.

Table 10.1 $P(z')$ and v^* for different return-stroke models (Adapted from [34])

Model	$P(z')$	v^*
BG	1	∞
TCS	1	$-c$
TL	1	v
MTLL	$1-z'/H_{tot}$	v
MTLE	$\mathrm{Exp}(-z'/\lambda)$	v

Expressions for $P(z')$ and v^* for some of the most commonly used return- stroke models are summarized in Table 10.1, in which λ is the attenuation height for the MTLE model.

In Table 10.1, the different return-stroke models are defined as Bruce-Golde (BG) model [28], Traveling Current Source (TCS) model [29], Transmission Line (TL) model [30], MTLL model [31], and MTLE model [32, 33].

For lightning to wind turbines, the previous model needs to be adapted to take into account the discontinuity between the body and the blades of the wind turbine system at the hub. The value of the discontinuity is a function of a number of electrical and geometrical parameters of the blades, of the hub, of the nacelle and of the base. The instantaneous angle of the struck blade with respect to the base may strongly influence the reflection and transmission coefficients at the discontinuity. This last parameter introduces a level of complexity that is specific to wind turbines since, as discussed in Sect. 10.3, the angle (and, therefore, the reflection coefficient) can vary appreciably from one stroke to another within the same lightning flash.

10.5 Impact of Carbon Reinforced Plastic

In [6], Carbon Reinforced Plastic materials are considered as electrical conductors and it is recommended to bond CRP to other conducting components for lightning protection purposes. However, this recommendation raises two questions which need to be addressed: (1) are CRP components able to conduct lightning current without being damaged? and, (2) how should the bonding between CRP and LPS be made?

Another issue related to the use of CRP in wind turbine blades is their response to the static electric field below a thundercloud. It is indeed well known [17] that the electric field E_g at ground level below thunderclouds can reach values of about -5 to -15 kV/m. It is likely that the CRP material in the blade experiences fields which vary (with the rotation of the blade) from E_g (when the blade tip is near the ground) to a value than can reach a few times E_g, due to the field enhancement effect, when the blade tip is at its highest position. Further testing is needed to evaluate the behavior of CRP materials in the presence of such electric fields. Another important issue is the mechanical stresses resulting from the energy

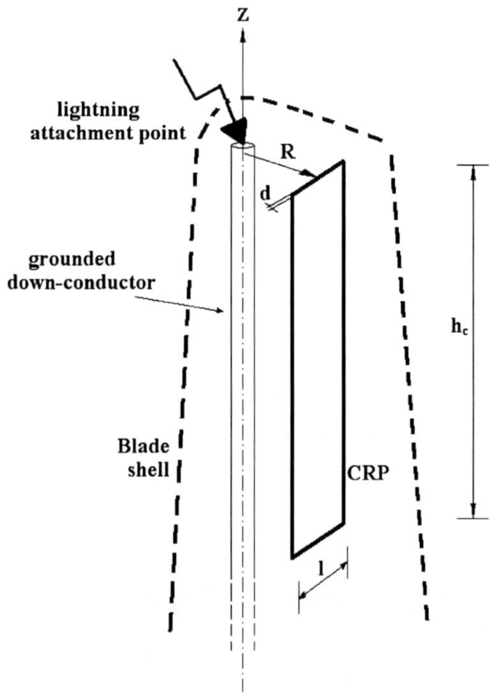

Fig. 10.4 Geometry for the evaluation of eddy currents in a CRP laminate. (Adapted from [7] ©2010 IEEE)

dissipation in CRP laminates due to the circulation of eddy currents, which are induced by magnetic field changes produced by lightning currents flowing in other conductive parts within the blade. The circulation of eddy currents results in an energy dissipation as heat, which can generate mechanical stresses. To evaluate the eddy current losses in CRP laminates, we consider the geometry presented in Fig. 10.4.

The CRP laminate is defined by a volume of thickness d, width l and length h, parallel to, and at a distance R from the lightning down conductor. The average loss due to eddy currents in a laminate is given by [35, 36]:

$$P(f) = \frac{k(f)df}{2} B_{\text{mav}}(\omega)^2 \frac{\sinh(k(f)d) - \sin(k(f)d)}{\cosh(k(f)d) - \cos(k(f)d)} \tag{10.10}$$

where

$$B_{\text{mav}}(f) = \mu \frac{I(f)}{2\pi R} \frac{\tanh\left(\sqrt{j2\pi f \mu\sigma}\frac{d}{2}\right)}{\sqrt{j2\pi f \mu\sigma}\frac{d}{2}} \tag{10.11}$$

$$k(f) = \sqrt{\frac{2\pi f \mu\sigma}{2}} \tag{10.12}$$

μ is the medium permeability and σ is the conductivity of the CRP.

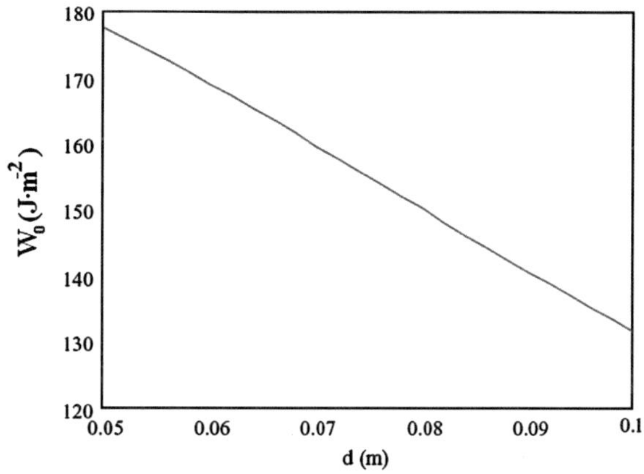

Fig. 10.5 Specific dissipated energy W_o in a CRP laminate of width d. (Adapted from [7] ©2010 IEEE)

The dissipated energy per unit volume is given by

$$W' = 2 \int_0^{f_{max}} |P(f)| \, df \tag{10.13}$$

and the total dissipated energy in the laminate is, therefore

$$W = W' \cdot l \cdot d \cdot h \tag{10.14}$$

Figure 10.5 presents the specific energy $W_o = W/(l.h)$ in J/m^2 as a function of width d and considering the following parameters used in [7]: $R = 0.2$ m, $\sigma = 7.246 \times 10^4$ S m^{-1} and $\mu_r = 1$. The return stroke current corresponds to a typical first return stroke and has an amplitude of 30 kA.

It can be seen that the resulting energy could reach significant values. One way of reducing the dissipated energy is the use of two down conductors instead of one, as illustrated in Fig. 10.6b, which results in a considerable reduction of magnetic field inside the blade.

One could think of two ways to orient carbon fiber laminates within the blades: either parallel or perpendicular to the plane formed by the conductors (see Fig. 10.7). Increasing the resistance of the circuit associated to the eddy currents results in a decrease in the dissipated energy. This corresponds to the perpendicular orientation illustrated in Fig. 10.7b.

10.6 Effect of Indirect Lightning on a Wind Turbine

Since lightning flashes to the ground in the vicinity of a wind turbine or to other turbines belonging to the same wind farm are more frequent than direct flashes, they are believed to constitute a higher risk than direct strikes [37]. Scheibe et al. [38] state that 30% of the damages to wind turbines are caused by direct lightning and the remaining 70% is due to indirect lightning.

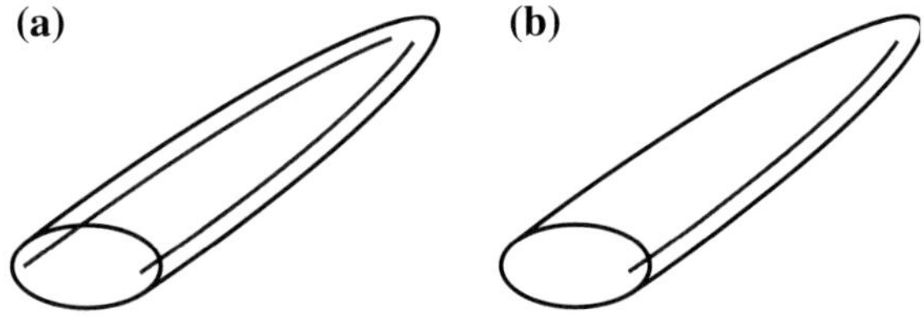

Fig. 10.6 **a** One down conductor. **b** Two down conductors. The magnetic field inside the blade is reduced for configuration **b**. (Adapted from [7] ©2010 IEEE)

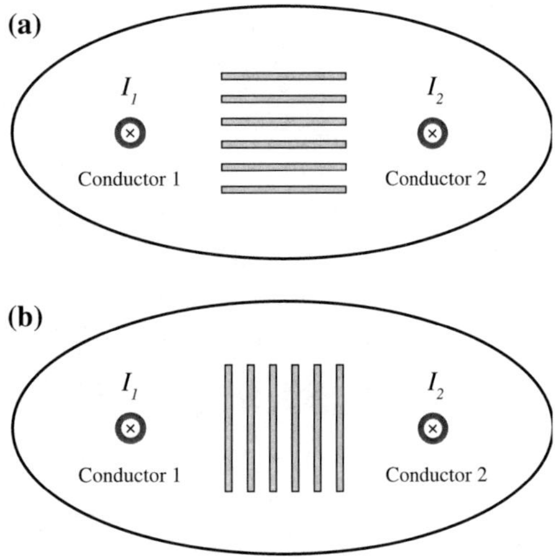

Fig. 10.7 Location of the fiber laminates inside the blade: **a** Parallel to the plane formed by the conductors. **b** Perpendicular to the plane formed by the conductors. (Adapted from [7] ©2010 IEEE)

A lightning discharge generates electric fields in its immediate vicinity of several hundreds of kV/m. These strong fields could cause severe stress to dielectric materials of conductive parts to be insulated. Electric and magnetic fields near a wind turbine will also cause a number of electromagnetic compatibility (EMC) problems to the various electronic control systems of a wind turbine. In order to avoid wind-turbine outages or more severe damages due to control malfunctions, control devices and their layout should be designed to enable them to work properly under these electromagnetic perturbations. For EMC design of susceptible circuitry, IEC standards (e.g., [4–6]) would be applied, and also several aeronautic standards and guides could be useful (e.g., [39,40]).

10.7 Conclusions and Recommendations

Lightning protection of wind turbines presents a number of new challenges due to the geometrical, electrical, and mechanical particularities of the turbines. This is especially true for modern units since they are becoming taller and because carbon fiber composite materials are being used to reinforce them. The main conclusions of the chapter are summarized hereunder:

1. The risk assessment for the purpose of lightning protection system design is based on empirical formulas for the estimation of the number of flashes to a tall object on a flat terrain. Therefore, the risk analysis for the modern wind turbines located in mountainous regions is a difficult task since it involves the evaluation of the "effective height", which is not clearly defined, and thus may result in both over- and underestimation of the actual number of strikes to a tall wind turbine.
2. The rotation of the blades may have a considerable influence on the number of strikes to the blades of large wind turbines as these may be triggering their own lightning. The impact of this mechanism on the overall lightning incidence to a tall turbine can be evaluated by comparing the number of strikes to rotating blades with the number of strikes to still turbines in the same wind turbine park or by studying the probability of strikes to blade tips as they move compared to the frequency of strikes to blades when the tip is moving down.
3. Since wind turbines are tall structures, the lightning currents that are injected by return strokes into the turbines are affected by reflections at the top, at the bottom, and at the junction of the blades with the static base of the turbine. This is of capital importance when calculating the protection of internal circuitry that may be affected by magnetically induced electromotive forces that depend directly on the characteristics of the current in the turbine.
4. The presence of carbon reinforced plastics in the blades introduces a new set of problems to be dealt with in the design of the turbines' lightning protection system. In particular, the way in which these laminates are bonded to other conducting components in the blade has to be carefully considered. Questions about the influence of the high static fields under thunderclouds on the moving carbon fiber parts need to be addressed by lightning protection researchers and engineers.
5. The presence of eddy currents in CRP laminates causes important energy dissipation, which might result in mechanical stresses. In order to reduce such effects, it is desirable to have two lightning down conductors (instead of one), allowing a reduction of the magnetic field inside the blade. Moreover, a perpendicular orientation of laminates with respect to the plane formed by the down conductors allows additional reduction in energy dissipation.

Acknowledgment This chapter is partially based on the Ref. [7] ©2010 IEEE. Thanks are due to J. Montanya, J.L. Bermudez, R. Rodríguez Sola, G. Solà, and N. Korovkin.

References

1. Cotton I, McNiff B, Soerenson T et al (2000) Lightning protection for wind turbines. In: Proceedings 25th international conference on lightning protection (ICLP 2000), Rhodes
2. Faulstich S, Hahn B, Lyding P (2010) Electrical subassemblies of wind turbines—a substantial risk for the availability. In: Proceedings european wind energy conference and exhibition (EWEC), Warsaw

3. Wada A, Yokoyama S (2004) Lightning damages of wind turbine blades in winter in Japan—lightning observation on the Nikaho-Kogen wind farm. In: Proceedings 27th international conference on lightning protection (ICLP 2004), Avignon
4. IEC 61024-1 (1993) Protection of structures against lightning. Part 1: general principles. International Electrotechnical Commission, Geneva, Switzerland
5. IEC 61312-1 (1995) Protection against lightning electromagnetic impulse. Part 1: general principles. International Electrotechnical Commission, Geneva, Switzerland
6. IEC 61400-24 (2010) Wind turbine generator systems. Part 24: lightning protection. International Electrotechnical Commission, Geneva, Switzerland
7. Rachidi F, Rubinstein M, Montanya J et al (2008) A review of current issues in lightning protection of new generation wind turbine blades. IEEE Trans Ind Electron 55:2489–2496. doi:10.1109/TIE.2007.896443
8. Motoyama H, Janischewskyj W, Hussein AM et al (1996) Electromagnetic field radiation model for lightning strokes to tall structures. IEEE Trans Power Deliv 11:1624–1632. doi:10.1109/61.517526
9. Baba Y, Ishii M (2001) Numerical electromagnetic field analysis of lightning current in tall structures. IEEE Trans Power Deliv 16:324–328. doi:10.1109/61.915502
10. Rachidi F, Janischewskyj W, Hussein AM et al (2001) Current and electromagnetic field associated with lightning return strokes to tall towers. IEEE Trans Electromagn Compat 43:356–367. doi:10.1109/15.942607
11. Bermudez JL, Rachidi F, Rubinstein M et al (2005) Far-field–current relationship based on the TL model for lightning return strokes to elevated strike objects. IEEE Trans Electromagn Compat 47:146–159. doi:10.1109/TEMC.2004.842102
12. Baba Y, Rakov VA (2005) Lightning electromagnetic environment in the presence of a tall grounded strike object. J Geophys Res 110:D09108. doi:10.1029/2004JD005505
13. Pavanello D, Rachidi F, Rubinstein M et al (2007) On return stroke currents and remote electromagnetic fields associated with lightning strikes to tall structures: 1. computational models. J Geophys Res 112:D13101. doi:10.1029/2006JD007958
14. Pavanello D, Rachidi F, Janischewskyj W et al (2007) On return stroke currents and remote electromagnetic fields associated with lightning strikes to tall structures: 2. experiment and model validation. J Geophys Res 112:D13122. doi:10.1029/2006JD007959
15. Baba Y, Rakov VA (2008) Influence of strike object grounding on close lightning electric fields. J Geophys Res 113:D12109. doi:10.1029/2008JD009811
16. Mosaddeghi A, Shoory A, Rachidi F et al (2010) Lightning electromagnetic fields at very close distances associated with lightning strikes to the gaisberg tower. J Geophys Res 115:D17101. doi:10.1029/2009JD013754
17. Rakov VA, Uman MA (2003) Lightning: physics and effects. Cambridge University Press, Cambridge
18. Eriksson AJ (1978) Lightning and tall structures. Trans South Afr Inst Electr Eng 69:859–870
19. Theethayi N, Diendorfer G, Thottappillil R (2004) On determining the effective height of Gaisberg tower. In: Proceedings european electromagnetics (EUROEM 2004), Magdeburg
20. Berger K (1972) Mesungen und resultate der blitzforschung auf dem Monte San Salvatore bei Lugano, der jahre 1963–1971. Bull SEV 63:1403–1422
21. Zhou H, Theethayi N, Diendorfer G et al (2010) On estimation of the effective height of towers on mountaintops in lightning incidence studies. J Electrost 68:415–418. doi:10.1016/j.elstat.2010.05.014
22. Fisher RJ, Schnetzer GH, Thottappillil R et al (1993) Parameters of triggered-lightning flashes in Florida and Alabama. J Geophys Res 98:22887–22902. doi:10.1029/93JD02293
23. Rakov VA (1999) Lightning discharges triggered using rocket-and-wire techniques. Recent Res Dev Geophys 2:141–171
24. Wang D, Rakov VA, Uman MA et al (1999) Characterization of the initial stage of negative rocket-triggered lightning. J Geophys Res 104:4213–4222. doi:10.1029/1998JD200087
25. Wang D, Rakov VA, Uman MA (1999) Attachment process in rocket-triggered lightning strokes. J Geophys Res 104:2143–2150. doi:10.1029/1998JD200070

26. Rachidi F (2011) Electromagnetic environment in the vicinity of a tall tower struck by lightning. In: Proceedings 3rd international symposium on winter lightning (ISWL), Tokyo, Japan
27. Rachidi F, Rakov VA, Nucci CA, Bermudez JL (2002) Effect of vertically extended strike object on the distribution of current along the lightning channel. J Geophy Res 107:4699. doi:10.1029/2002JD002119
28. Bruce CER, Golde RH (1941) The lightning discharge. J Inst Electr Eng, Part II 88:487–505. doi:10.1049/ji-2.1941.0065
29. Heidler F (1985) Traveling current source model for LEMP calculation. In: Proceedings 6th symposium and technical exhibition on electromagnetic compatibility, Zurich
30. Uman MA, McLain DK (1969) Magnetic field of lightning return stroke. J Geophys Res 74:6899–6910. doi:10.1029/JC074i028p06899
31. Rakov VA, Dulzon AA (1987) Calculated electromagnetic fields of lightning return strokes. Tekhnicheskaya Elektrodinamika 9(1):87–89
32. Nucci CA et al (1988) On lightning return stroke models for LEMP calculations. In: Proceedings 19th international conference on lightning protection (ICLP), Graz
33. Rachidi F, Nucci CA (1990) On the Master, Uman, Lin, Standler and the modified transmission line lightning return stroke current models. J Geophys Res 95(20):389–394. doi:10.1029/JD095iD12p20389
34. Rakov VA, Uman MA (1998) Review and evaluation of lightning return stroke models including some aspects of their application. IEEE Trans Electromagn Compat 40:403–426. doi:10.1109/15.736202
35. Jufer M (2004) Electromecanique. Presses Polytechniques Universitaires Romandes, Lausanne
36. Slemon GR, Straughen A (1980) Electric machines. Addison Wesley, Reading, MA
37. Eriksson AJ, Stringfellow MF, Meal DV (1982) Lightning—induced overvoltages on overhead distribution lines. IEEE Trans Power Appar Syst PAS-101:960–968. doi:10.1109/TPAS.1982.317162
38. Scheibe K, Schimanski J, Wetter M (2004) Lightning protection for wind turbines and its test with a high performance lightning current generator. In: Proceedings 27th international conference on lightning protection (ICLP), Avignon
39. ARP5412 (2005) Aircraft lightning environment and related test waveforms ARP, 2005
40. Fisher FA, Perala RA, Plumer JA (1985) Lightning protection of aircraft. Lightning technologies, inc., 10 Downing Parkway, Pittsfield, MA 012011999

Chapter 11
Lightning Surge Analysis of a Wind Farm

Yoh Yasuda

Abstract The rapid expansion of wind power generation has brought problems involving lightning strikes to the fore. Many such incidents have damaged not only the wind turbine that was actually struck but also other turbines that were not, a phenomenon that is yet to be fully explained. In this chapter, the author presents a case study using a wind farm model with multiple wind turbines connected to a power system. The aim is to clarify the influence of the earthing system on surge propagation from a wind turbine that has been struck to others which have not, during a winter lightning strike. When one of the wind turbines in a wind farm is struck by lightning, the phenomenon of surge invasion to the collection system is categorised as "back-flow surge". It has been reported that this back-flow surge sometimes burns out surge protection devices (SPDs) or breaks low-voltage circuits even far from the point where the lightning struck. In practice, many such incidents that have occurred not only involved the wind turbine that was actually struck but also other affected wind turbines that had not been struck. This chapter will analyse incidents of burnout to SPDs resulting from winter lightning at wind farms using ARENE and PSCAD/EMTDC. Calculations were performed to clarify the mechanism of how the back-flow surge propagates to other turbines from the directly struck wind turbine. The calculations also clarified that burnout incidents could easily occur even in a turbine that had not been struck by the lightning. It also became evident that burnout incidents can be reduced when interconnecting earthing wires are installed between wind turbines.

Y. Yasuda (✉)
Department of Electrical Engineering and Computer Science,
Kansai University, 3-3-35 Yamate-cho, Suita, Osaka 564-8680, Japan
e-mail: yasuda@mem.iee.or.jp

S. M. Muyeen (ed.), *Wind Energy Conversion Systems*,
Green Energy and Technology, DOI: 10.1007/978-1-4471-2201-2_11,

11.1 Introduction

11.1.1 Incidents Due to Lightning in Wind Turbines

As wind power generation undergoes rapid, worldwide growth, lightning incidents involving wind turbines have come to be regarded as a serious problem. Because of their distinctive shape and the fact that they are very tall, open-air structures, wind power generators are often struck by lightning. It has been suggested that such facilities tend to be vulnerable to lightning damage because of their configuration. It is thought that it will be necessary to employ protective measures of a different kind from those used with conventional electrical equipment.

In addition to serious damage to blades, breakdown of low-voltage and control circuits have frequently occurred in many wind farms throughout the world. According to a report from NEDO (New Energy and Development technology Organization, Japan) [1], the most frequent failures, more than 50%, in wind turbine equipment are those occurring in low-voltage, control, communication circuits. Indeed, many dielectric breakdowns of low-voltage circuits and burnout incidents of surge protection devices (SPDs) in wind turbine are reported. Such frequent problems in the low-voltage circuits may cause a deterioration of the utilisation rate and consequently cause increases in the cost of power generation.

11.1.2 Winter Lightning and "Back-Flow Surge"

Although some reports describe these incidents and suggest methods for lightning protection system (LPS), there appear to have been few investigations into insulation schemes, earthing designs and transient analysis for the latest generation of apparatuses. A vigorous debate on LPS for wind turbines is currently underway, especially in Japan, which has a unique and somewhat relentless environment that includes the notorious "winter lightning" in the coastal area of the Sea of Japan [2–5]. While blade protection has been relatively well discussed, the behaviour of wind farms experiencing surge invasion during a lightning strike has not been clarified. Thus much work remains to be done in this area.

Winter lightning is a special lightning that typically has upward flashes initiated by an upward leader from an earthed structure and has a duration longer than several ten milliseconds. Therefore, winter lightning sometimes has heavy electric charge more than 300 coulombs, which is the maximum value specified in IEC 61400-24:2010 [6]. In fact, a lot of incidents of wind turbine blades have been reported due to the winter lightning in the coastal area of the Sea of Japan during the last two decades [7–9].

NEDO has investigated the damages of wind turbines from April 2002 to March 2006 and obtained the interesting result that the number of damages in control and electric facilities is shared in 42% in the summer season and 58% in the winter season. One of NEDO's reports [1] shows that lightning in winter season is more dangerous for wind turbines than in summer.

The phenomenon of surge invasion from a wind turbine that is struck by winter lightning to the collection system in a wind farm is quite similar to the case of "back-flow surge", where the surge is that flowing from a customer's structure such as a communication tower to the distribution line [10]. High resistivity soil often makes SPDs at tower earthing systems operate in reverse and allow back-flow of surge current to the grid. It is reported that this back-flow surge can sometimes burn SPDs or break down low-voltage circuits even on an electric pole far from the point where the lightning struck.

Several breakdown and burnout incidents in low-voltage circuits and SPDs at wind farms are thought to be the result of the above-mentioned "back-flow surge". In practice, many of the incidents not only involved the wind turbine that had actually been struck by lightning but also other wind turbines that had not been struck. The reason why turbines that were not struck were nevertheless damaged has not been fully explained.

11.1.3 Necessity of Reinforcement of a Collection System in Wind Farm

The author has, therefore, concentrated on investigating surge analysis on a wind farm since 2000 [11–13]. In this chapter, the author presents a case study of surge analysis using a wind farm model with multiple wind turbines connected to a power system. The aim of the present analysis is to clarify the influence of the electrical and geometrical configuration of a wind farm when lightning surge is propagated through a collection system from a wind turbine that has been struck to other turbines that have not.

11.2 Wind Farm Model for Lightning Surge Analysis

11.2.1 Wind Farm Model

Figure 11.1 shows a wind farm model made up of ten wind power generators, identical in performance and condition. In this model, it is assumed that;

- A high-voltage transformer (6.6 kV/66 kV) interconnects the power grid system and an array of ten wind power generators of the 1 MW class at 0.6 km intervals.
- Although a wind power generator consists of a gearbox, a synchronous or an induction generator, rectifier, 3-phase inverter and so on, for this simulation a stable synchronous generator is presumed for simplicity.

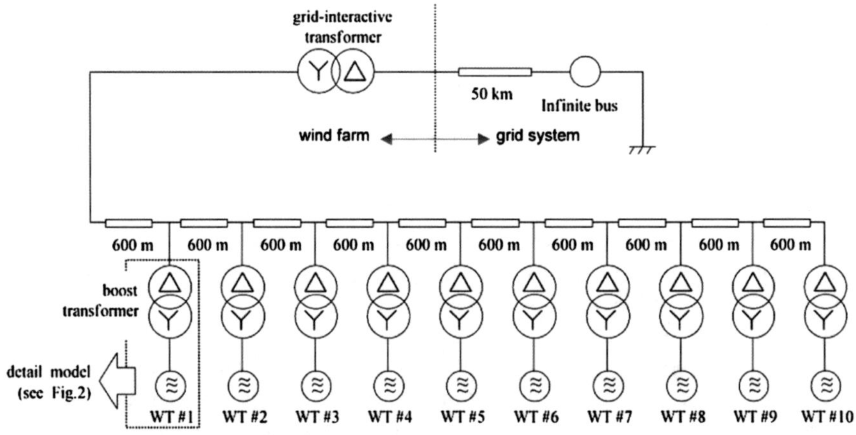

Fig. 11.1 Wind farm model with ten wind turbines

Fig. 11.2 Wind turbine model for surge analysis

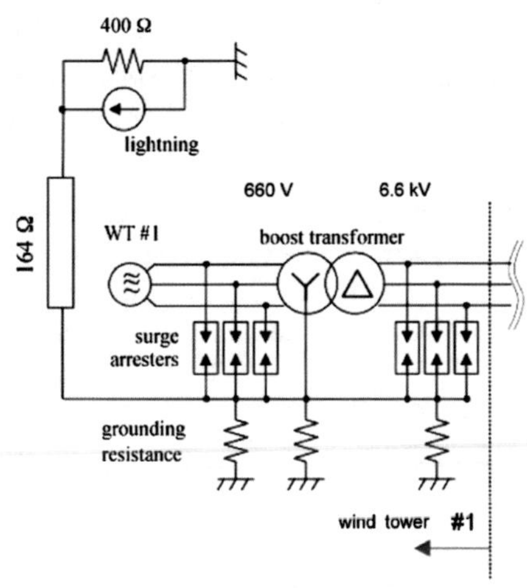

- Boost transformers for the generators (660 V/6.6 kV) are installed inside the wind turbine towers, or in their vicinity. SPDs are attached to the primary and secondary terminals and connected to a common earth, as shown in the detailed diagram in Fig. 11.2.
- The earth resistance of each earthing point is simulated as 20 or 50 Ω. Thus, the total value of the interconnected earthing system of the wind turbines becomes 6.67 or 16.6 Ω.
- Other details and constants used in the model are shown in Table 11.1.

In addition, since what we wished to observe in this simulation was the burnout of SPDs at wind turbines that had not themselves been struck, we assumed that

Table 11.1 Analysis conditions

Wind turbine (synchronous generator) model		
Rating power (MVA)	1.00	
Direct-axis reactance X_d, X_d', X_d'' (p.u)	2.00, 0.25, 0.20	
Quadrature-axis reactance X_q, X_q', X_q'' (p.u)	1.90, 0.50, 0.20	
Time constants T_{do}', T_{do}'', T_{qo}', T_{qo}'' (s)	6.0, 0.03, 0.50, 0.06	
Transformer model	Boost	Grid-interactive
Connection method	Y (neutral earthing)/Δ	
Rating power (MVA)	1.0	10.0
Rating voltage (V)	600/6,600	6,600/66,000
Frequency (Hz)	60	60
Percentage impedance (%)	15.7	1.57
Mutual leakage inductance (mH)	18.2	18.2
Collection line model		
Positive-/zero- phase resistance (Ω/km)*	0.00105/0.0210	
Positive-/zero- phase inductance (mH/km)*	0.83556/2.50067	
Positive-/zero- phase capacitance (nF/km) *	12.9445/6.4723	
Frequency-dependent characteristics	Considered	

* : values at 60 Hz

blade burnout or explosive destruction and dielectric breakdown at the turbine that had actually been struck was prevented by certain measures.

11.2.2 Model for Winter Lightning

A standard summer lightning event is generally assumed to have a crest peak of 30 kA, a crest width of 2 μs and a wave tail of 70 μs. By contrast, since winter lightning has varying crest widths and crest peaks, a standardised model has not been established. Therefore, in this report, the modelling of winter lightning is based on the model described in Ref. [10]. The parameters for crest peak and duration of wave tail are determined from a cumulative frequency distribution of the lightning current wave shape.

The assumptions concerning lightning in the present analysis are as follows;

- Summer lightning:
  - Crest peak: 30 kA
  - Crest width: 2 μs
  - Duration of wave tail: 70 μs
  - Polarity: positive (for comparison with winter lightning)
- Winter lightning:
  - Crest peak: 51 kA (the 16% value of the cumulative frequency distribution of lightning)

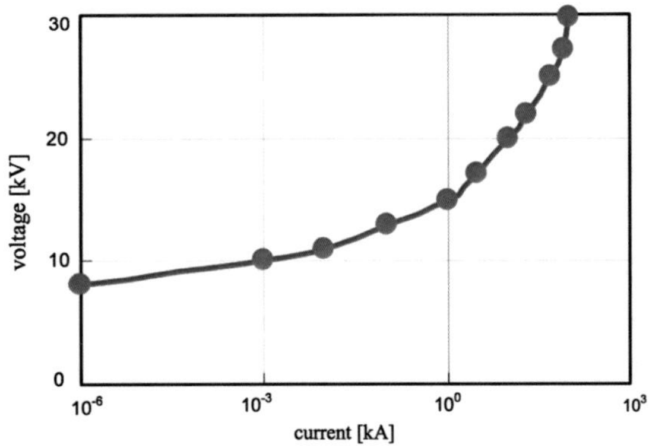

Fig. 11.3 SPD model for surge analysis

- Crest width: 2 μs
- Duration of wave tail: 631 μs (the 16% value of the cumulative frequency distribution of lightning)
- Polarity: positive
- Phase angle at lightning strike: $\theta = 0°$.

11.2.3 Model for Surge Protection Device

To provide protection from surge invasion, it was assumed that SPDs were installed in both the primary (low-voltage side; wind turbine side) and secondary (high-voltage side; grid side) terminals of the boost transformer near to each wind turbine. The nominal discharge current of the SPD was assumed to be 2.5 kA and its characteristic curve, starting at $V_{1\mathrm{mA}} = 8$ kV, is shown in Fig. 11.3.

The burnout of an SPD depends on whether the heat produced by the current flowing through the SPDs exceeds the thermal limit of the SPD. To calculate the total heat absorbed by the SPD E_{ar} (kJ) in the present analysis, it is necessary to specify the simultaneous power $p_{\mathrm{ar}}(t)$ (W) derived from the SPD's terminal voltage $v_{\mathrm{ar}}(t)$ (V) and the current flowing through the SPD $i_{\mathrm{ar}}(t)$ (A). Then, the total electric energy W_{ar} (Wh) can be calculated by integrating $p_{\mathrm{ar}}(t)$ from 0 to the time T(s) when $i_{\mathrm{ar}}(t)$ converges to 0 kA. The total thermal energy absorbed in the SPD E_{ar} (kJ) is given by unit conversion from W_{ar} (Wh). This sequence is described by the following equations:

$$p_{\mathrm{ar}}(t) = v_{\mathrm{ar}}(t) \times i_{\mathrm{ar}}(t) \tag{11.1}$$

$$W_{\mathrm{ar}} = \int_0^T p_{\mathrm{ar}}(t)\mathrm{d}t/3,600 \tag{11.2}$$

$$E_{\mathrm{ar}}\,(\mathrm{kJ}) = 3.6 \times W_{\mathrm{ar}}\,(\mathrm{Wh}). \tag{11.3}$$

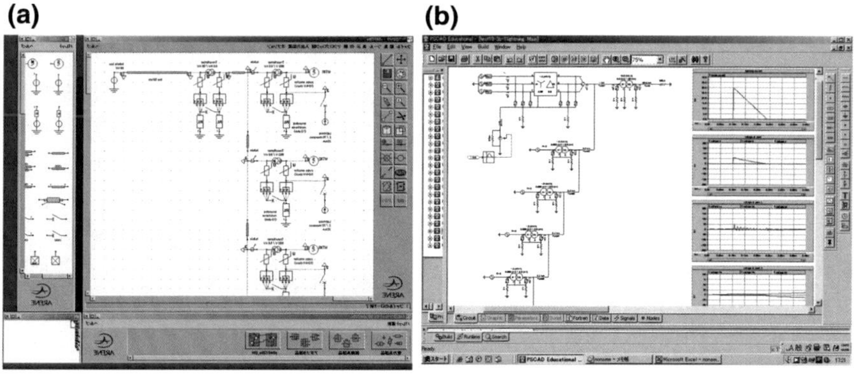

Fig. 11.4 User interface of ARENE and PSCAD/EMTDC. **a** ARENE. **b** PSCAD/EMTDC

11.2.4 Model Description by ARENE and PSCAD/EMTDC

In the present analysis, two different digital transient simulators were employed; ARENE (non real-time PC version) produced by EDF and PSCAD/EMTDC by Manitoba HVDC Research Centre. Both the simulators have similar algorism of Schneider–Bergeron method as well as the conventional EMTP. Figure 11.4 illustrates each GUI (Graphical User Interface) description of the present wind farm model by ARENE and PSCAD/EMTDC, respectively. Of course, it is possible to simulate the above-proposed model by the conventional EMTP and ATP.

11.3 Lightning Surge Analysis I: Comparison Between ARENE and EMTDC

11.3.1 Basic Comparison of the Surge Waveforms

The results of the two simulators are shown in Fig. 11.5. The figure displays the analysis results of surge voltage at the boost transformer's secondary terminal on each turbine's when a lightning strikes wind turbine (WT) #1, which is the nearest turbine to the grid. The graphs in column (1) show the results calculated by ARENE and those in column (2) correspond to that by PSCAD/EMTDC. Both results indicate that the further the turbine from the lightning-struck turbine is, the larger the magnitude of surge that the turbine suffers. Comparing the waveforms calculated by both ARENE and EMTDC, it is recognised that the surge altitudes and wave shape by both simulators are relatively close to each other in spite of slight differences in surge durations and the complexity of reflection waveforms.

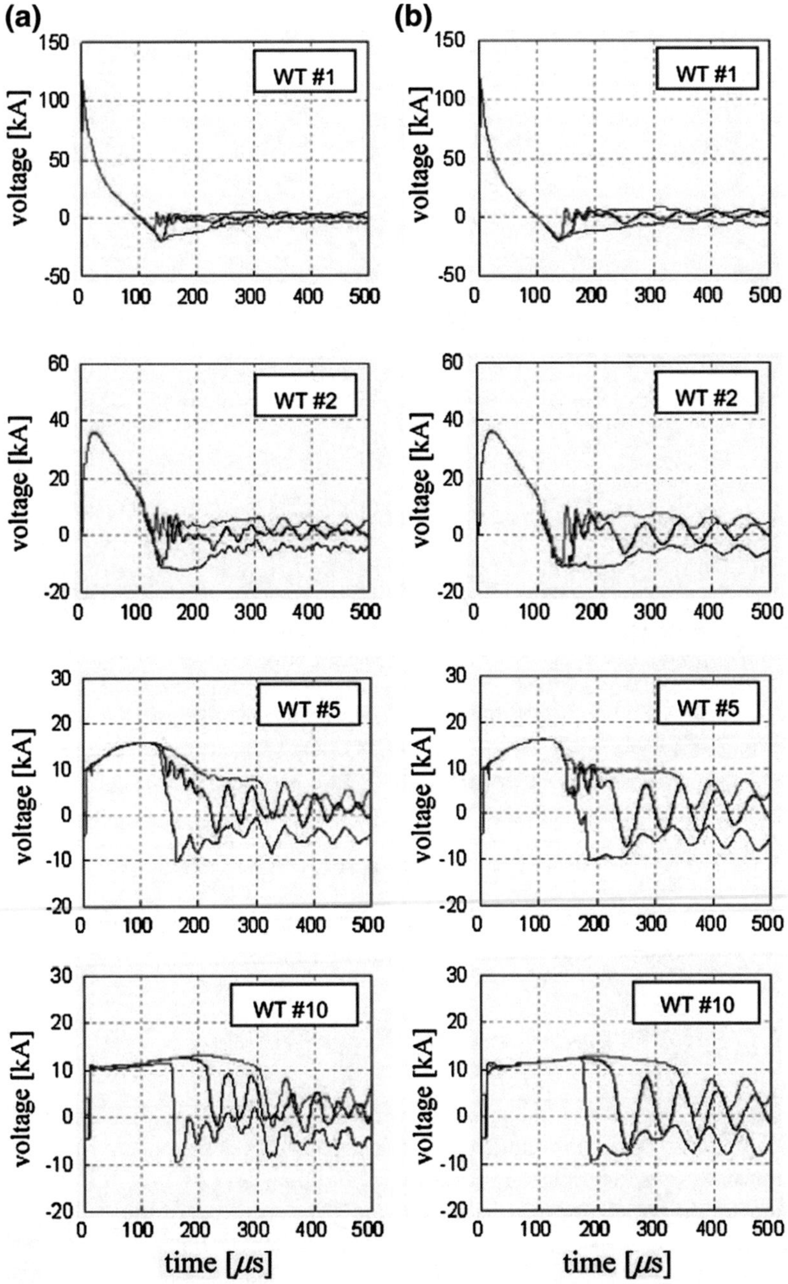

Fig. 11.5 Comparison results of voltage waveforms (lightning strikes at WT#1, R_{g0}= 6.77). **a** by ARENE. **b** by EMTDC

11.3.2 Comparison of the Surge Propagations in the Wind Farm

Summarising the results of the surge altitudes reaching at each wind turbine, the graph can be drawn as Fig. 11.6. The figures illustrate the graphs of impulse height of surges reaching at each wind turbine. Figure 11.6 gives clearly the coincidence from the viewpoint of aspect of surge propagation calculated by two different simulators, ARENE and EMTDC. It shows the accuracy of the present model and the calculation for the lightning analysis on the wind farm.

11.3.3 Tendency of the Surge Propagations in the Wind Farm

Figure 11.6 also shows the interesting tendency of the surge propagation in the wind farm. It is clear that the surge from WT#1, which was struck by lightning, propagated to not only the adjacent turbine but also to quite distant turbines far from the lightning-struck one. Moreover, it is recognised that the further the wind turbine from the lightning-struck one, the longer the surge duration reached to the turbine. It suggests that a turbine that was not struck by lightning directly even at the farthest point from the lightning-struck one has the possibility to suffer from discharge incident due to the "back-flow surge" propagating along the collection line within the wind farm.

The mechanism of the back-flow surge propagation and SPD's incidents will be discussed in detail in the following chapter.

11.4 Lightning Surge Analysis II: SPD's Incidents Due to Back-Flow Surge

11.4.1 Analysis of the Surge Propagations in the Wind Farm

In this chapter, a comparison is made between the energy consumption of SPDs during a summer lightning strike and that during a winter lightning one. It is supposed that the lightning strikes either wind turbine No.1 (WT#1), which is the nearest turbine to the grid, or wind turbine No.10 (WT#10), the furthest turbine from the grid.

Figures 11.7 and 11.8 are the results of analyses done by ARENE in the cases of summer and winter lightning strikes, respectively, to a wind turbine with an earth resistance of 6.67 Ω. In both figures, graphs (a) and (b) indicate the results in the cases of lightning strikes to WT#1 and WT#10, respectively.

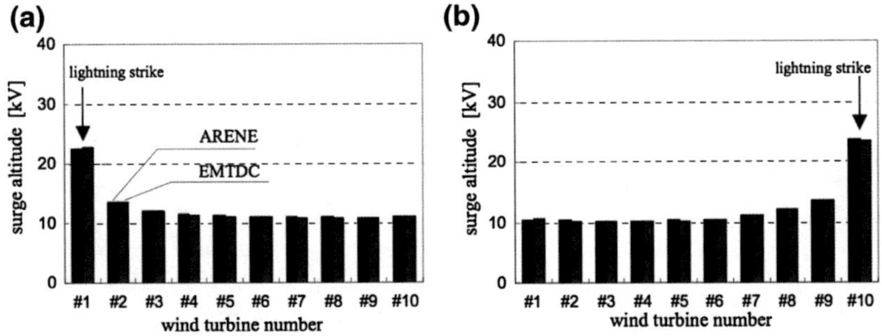

Fig. 11.6 Comparison results of surge propagations ($R_{g0} = 1.68\ \Omega$). **a** Case 1: Lightning on WT#1. **b** Case 2: Lightning on WT#10

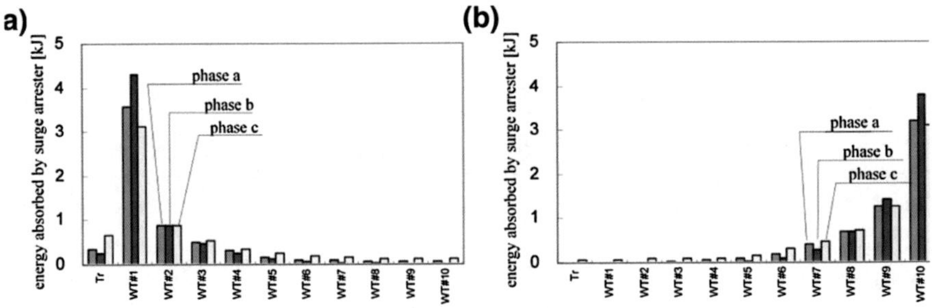

Fig. 11.7 Energy absorbed by SPDs in each wind turbine (Summer lightning $R_{g0} = 6.77\ \Omega$). **a** Case 1: Lightning on WT#1. **b** Case 2: Lightning on WT#10

Second, the effect on the different phases should be discussed. As can be seen in both Figs. 11.7 and 11.8, the SPD in phase *b* at the wind turbine that was actually struck consumed the largest energy. This is because the potential difference of phase *b* is the largest at the moment of the lightning strike. In contrast, at the other turbines, the SPD in phase *c* tends to absorb the most energy. The further the turbine is from the turbine that was struck, the more evident this tendency becomes. The reason why the energy absorbed in the SPD of phase *c* tends to be the largest is thought to be the fact that the energy imbalance among three phases depends upon the timing of lightning strike, i.e. phase angle of each phase conductor, at the struck turbine. When the line voltage in phase *c* at the moment of the lightning strike is the highest among the three phases at the struck turbine, the voltage between both terminals of the SPD tends to become the highest among the three phases. The further away the turbine is, the stronger the tendency becomes.

Further, to examine in more detail the results for winter lightning, another analysis was performed. Figure 11.9 is the result of the case where the earth resistance of the wind turbines was set at 16.7 Ω. In Case 1, the energy consumption of the SPDs surpassed their thermal limitation only in the WT#1, which had been directly struck, as shown in Fig. 11.9a.

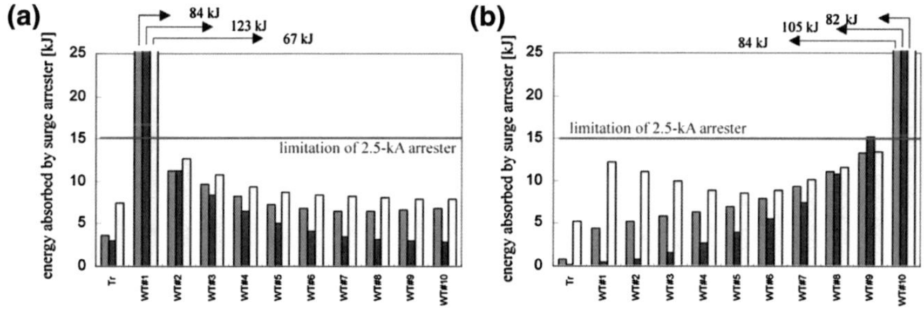

Fig. 11.8 Energy absorbed by SPDs in each wind turbine (Winter lightning $R_{g0} = 6.77\ \Omega$). **a** Case 1: Lightning on WT#1. **b** Case 2: Lightning on WT#10

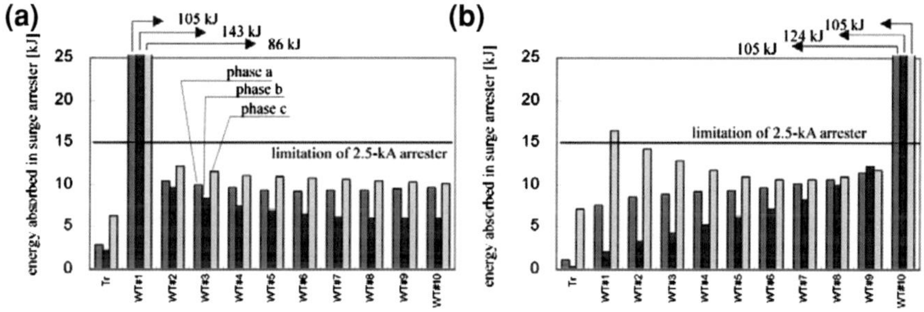

Fig. 11.9 Energy absorbed by SPDs in each wind turbine (Winter lightning $R_{g0} = 16.6\ \Omega$). **a** Case 1: Lightning on WT#1. **b** Case 2: Lightning on WT#10

On the other hand, in the case of a lightning strike to WT#10, the energy consumption in phase *c* of the SPDs is higher than the limit not only at WT#10 itself, but also at WT#1, the turbine furthest away from the lightning strike. As can be seen in Fig. 11.9b, it is not always the case that the further a wind turbine is from the turbine that was struck, the lower the energy absorbed by the SPD. From Fig. 11.9b, it is evident that, as far as the energy absorbed by phase *c* is concerned, from WT#6 onwards the further away the turbine is, the higher the energy absorbed. This tendency was also observed using other values of earth resistance. The longer the tail of the lightning strike, the clearer the tendency of this phenomenon becomes.

11.4.2 Analysis of Surge Waveforms

To clarify the reason for the reversal phenomenon, waveforms of voltage, current and simultaneous power at each wind turbine need to be drawn. Figure 11.10 is a set of graphs of the waveforms at WT#5 and WT#10 in the case of a lightning strike to WT#1 (Case 1). Correspondingly, Fig. 11.10 is also the set of waveforms

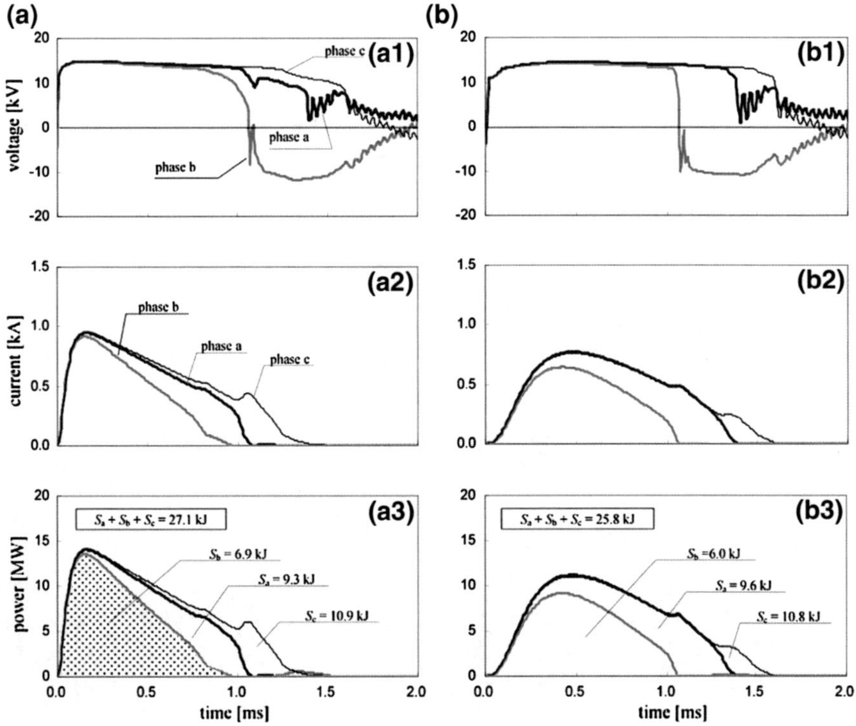

Fig. 11.10 Waveforms of voltage, current and simultaneous consumption power at SPDs in each wind turbine (Winter lightning $R_{g0} = 16.6\ \Omega$, Case 1). **a** WT#5. **b** WT#10

at WT#1 and WT#5 when WT#10 has been directly struck (Case 2). Comparing the two graphs, it is evident that the summation of SPD's absorbed energy of all phases at any wind turbine is almost the same value of about 26 kJ. What is interesting is that the simultaneous power curve of phase c at WT#1 in Case 2 (Fig. 11.10a3) exceptionally forms into a large arc and makes a large area. The absorbed energy calculated from the integrated area is up to 15 kJ which is a thermal limitation of the present 2.5 kV class SPD. In this case, the current inrushing to WT#1 did not flow through the SPD in phase b but in phase c. The reason is considered to be the difference in the potential at the moment of the lightning strike, where the initial phase θ of phase a in the present calculation is set at 0°, i.e. $V_a = 0$ V.

Thus, from the above discussion, it becomes clear that during strong winter lightning strikes with a long tail there is a possibility of SPD burnout incidents not only at the turbine that the lightning strikes and those nearby, but also at turbines far away from the one that was actually struck.

First, comparisons between the case of summer lightning and winter lightning are discussed. Comparing Figs. 11.7 and 11.8, it is clear that the energy absorbed by each SPD is 10–30 times larger in the case of winter lightning than in the case

Table 11.2 Cumulative distribution of lightning current parameters

	50% value	16% value
Peak (kA)	24	51
Wave tail (μs)	89	631

of summer lightning. Since the wave tail of winter lighting is 631 μs, which is about nine times longer than that of summer lightning, even more than the 1.7-times higher crest peak, it is likely that the energy absorbed becomes significantly larger.

11.4.3 Analysis of Burnout Ratio of SPDs

In this section, the burnout ratio of the SPDs installed between the high-voltage terminal of the boost transformer of each wind turbine and the low-voltage terminal of the grid-interactive transformer during a winter lightning strike to either WT#1 or WT#10 is analysed.

The definition of the burnout ratio of an SPD is essentially that obtained from Ref. [10]. The burnout ratio is defined as "the probability that a SPD suffers from burnout during a lightning stroke". First, "a limit curve of SPD burnout" should be drawn, which is the curve connecting the minimum limit crest peak for each lightning event that causes SPD burnout against the corresponding wave tail. The curve is obtained by repetitive calculation using digital transient simulators by ARNE. Thus, the burnout ratio P_f can be given by the following equation:

$$P_f = \int_{0}^{+\infty} \left\{ \int_{y(Tt)}^{+\infty} f(I_p) dI_p \right\} g(T_i) dT_i \tag{11.4}$$

where, $f(I_p)$ is a probability density function of the crest peak of lightning I_p, $g(T_t)$ is a probability density function of the wave tail of lightning T_t, $y(T_t)$ is a function curve of the burnout ratio of the SPD I_p (A) against a certain wave tail T_t (μs). In addition, the common logarithm of both probability density functions is assumed to be normally distributed obeying the parameters in Table 11.2. As indicated in Fig. 11.11, the probability density is distributed according to a surface of normal distribution function whose centre is 89 μs of wave tail and 24 kA of crest peak. Hence, it is evident that the lower the position on the burnout limit curve, the higher the burnout ratio of the SPD.

In the present analysis, since the SPDs installed at the high-voltage terminals of the boost transformer of each wind turbine are of the 2.5-kA class, the threshold energy for burnout is estimated to be 15 kJ. To simplify the modelling and the present calculation, the SPDs after burnout are assumed not too short and continue to conduct current according to the characteristic I–V curve shown in Fig. 11.3.

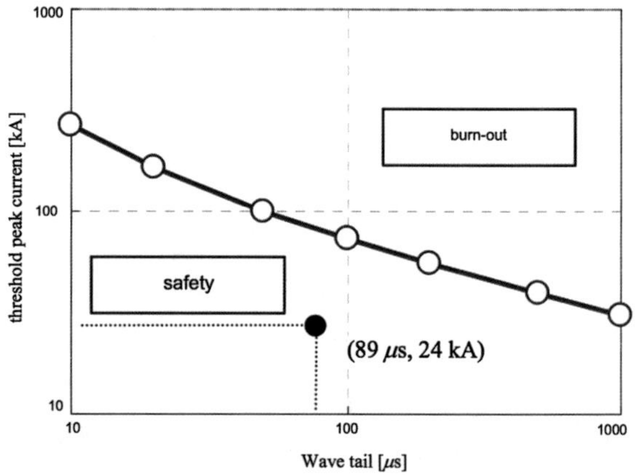

Fig. 11.11 Limitation curve for SPD burnout

Table 11.3 Minimum peak for burnout against each wave tail

	Earth resistance (Ω)	Peak (kA)						
		Wave tail						
		10 μs	20 μs	50 μs	100 μs	200 μs	500 μs	1,000 μs
WT #1	50	65	47	31	24	18	14	13
	30	85	61	41	30	23	18	15
	20	110	77	51	37	28	22	19
	10	184	120	76	56	42	32	27
	5	322	200	121	89	66	50	43
WT #10	50	1,635	797	317	162	85	39	21
	30	2,691	1,308	513	255	128	52	28
	20	3,881	1,887	737	363	177	68	35
	10	6,939	3,375	1,319	645	311	113	54
	5	12,300	5,990	2,349	1,154	556	203	95

Table 11.3 shows sample results for the calculation of "minimum crest peak for burnout against wave tail" as defined by Eq. 11.4. After calculation for each wind turbine, a burnout limitation curve for each turbine can be plotted, as shown in Fig. 11.12. In the present analysis, phase θ at the moment of lightning stroke is set as $\pi/2$, which is the strictest case whether burnout will occur or not at phase a. Also, crest width is assumed to be 1 μs as the strictest condition.

Figure 11.12 shows two graphs of burnout ratio against earth resistance in the case of winter lightning strikes to WT#1 and #10, respectively. From both graphs, it is confirmed that, when earth resistance is 2 Ω, the burnout ratio is less than 2% even in the turbine that was struck. It can also be seen from Fig. 11.12 that the ratio in the adjacent WT#2 becomes higher than 10% in case of 8 Ω when lightning strikes WT#1; moreover, the ratio in the adjacent WT#9 reaches 10% in the case of even 5 Ω when lightning strikes WT#10. This suggests that there certainly is a possibility of SPD burnout even in case of relatively low resistance. It is remarkable that the possibility of burnout at the furthest away turbine, WT#1,

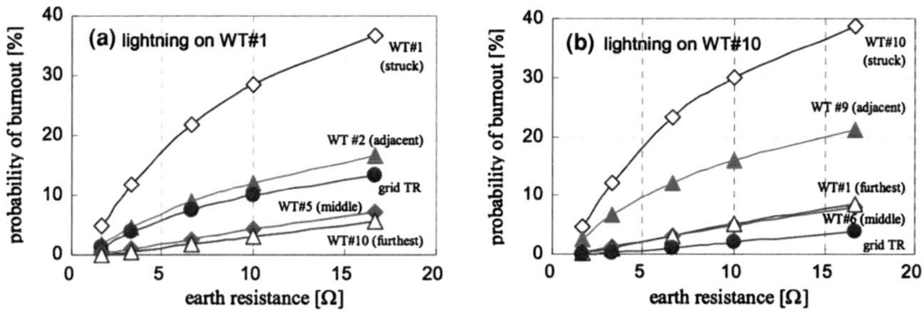

Fig. 11.12 Possibility of SPD burnout in each wind turbine. **a** Case 1: Lightning on WT#1. **b** Case 2: Lightning on WT #10

becomes higher than that at the centre (WT#6) in the case of a lightning strike to WT#10 at the end of the wind farm. This tendency is also confirmed by the results shown in the previous section.

11.4.4 Conclusive Discussions

This chapter described an analysis of incidents of burnout to SPDs resulting from winter lightning at wind farms. Calculations were performed to clarify the mechanism of how "back-flow surge" propagates to other turbines from the wind turbine that has been directly struck by winter lightning.

The calculations demonstrated, with various lightning point and earth resistance values, from which it becomes clear that burnout incidents could quite easily occur even in a turbine far from the lightning-struck turbine. Consequently, it also becomes evident that the possibility of burnout incidents can become high in the case of high resistance and/or long-tail winter lightning.

11.5 Lightning Surge Analysis III: Effect of Overhead Earthing Wire(s)

The aim of the analysis in this chapter is to clarify the influence of earthing wire(s) of the collection line in a wind farm. Reference [10] also noted that earthing wire(s) can reduce burnouts of SPDs in the case of a communication tower. This paper tries to clarify whether there is a similar effect from the installation of earthing wire(s) for wind farm LPS.

The phenomenon of surge invasion to the collection system from a wind turbine that is struck by lightning is quite similar to the "back-flow surge" reported in Ref. [10]. In that report, the surge flowed from a customer's structure such as a communication tower into the collection line in a wind farm. High resistivity soil

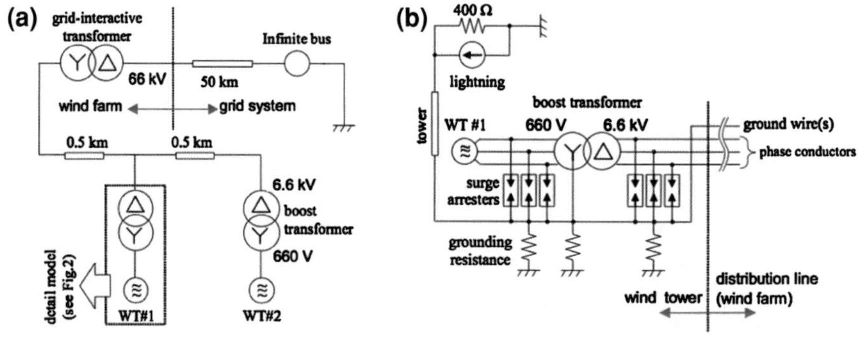

Fig. 11.13 Wind farm and Wind turbine model. **a** Wind farm model with two turbines. **b** Wind turbine model

often creates SPDs for tower earthing systems to operate in reverse and allow reflux of the surge current to the collection line. It is reported that this back-flow surge can sometimes burnout SPDs or break down low-voltage circuits even on an electric pole far from the point where the lightning struck.

Several breakdown and burnout incidents in low-voltage circuits and SPDs at wind farms are thought to be the result of the above back-flow surge. In practice, many of the incidents that have occurred not only involved the actual lightning-struck wind turbine but also other wind turbines that had not been struck. The reason why turbines that were not struck were nevertheless damaged has not been fully explained.

11.5.1 Model of a Collection Line in a Wind Farm

Comparing the previous two chapters, the simpler model for a wind farm with two turbines is employed as illustrated in Fig. 11.13. The main different point of a model of a collection system is the existence of earthing wire(s) as shown in Fig. 11.13b. The collection line in the present wind farm model is assumed as an overhead line with three phase conductors installed 10 m over the ground. Configuration details are shown in Fig. 11.14 and a set of parameters is shown in Table 11.4.

The main aim of the present analysis is to confirm the effect of earthing wire(s). Various conditions, therefore, are simulated: (1) “Case GW0”: no earthing wires are installed above the overhead line. (2) “Case GW1”: one earthing wire is tensioned 1 m above the three phase conductors. Both terminals of the earthing wire are connected to the common earthing system of the wind turbines and the grid-interactive transformer. (3) “Case GW2”: two earthing wires horizontally separated at 0.7 m length above the conductors are installed. In PSCAD/EMTDC,

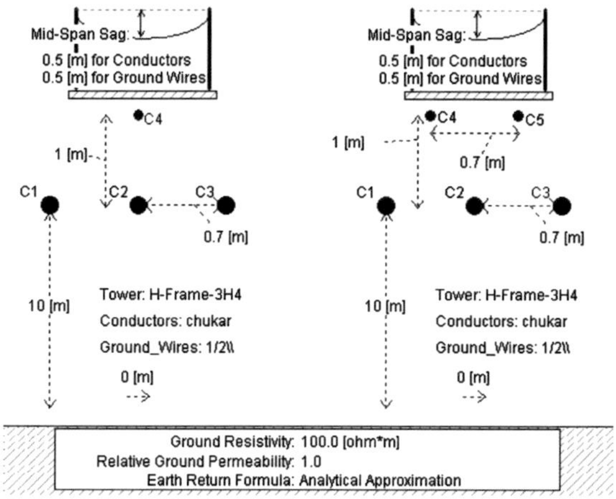

Fig. 11.14 Collection line model in wind farm by PSCAD description (*Left*: 1-earthing-wire model, *Right*: 2-earthing-wires model)

the calculation model of an overhead collection line obeys the Bergeron Method, which is similar to the widely used EMTP/ATP calculation.

In the following section, comparisons will be made between the energy consumption of the SPDs among the three cases with various number of earthing wires. It is assumed that the lightning strikes wind turbine No.1 (WT#1), which is the nearest turbine to the grid.

11.5.2 Observation of Waveforms around SPDs

Figure 11.15 sets out the results of EMTDC calculations in the cases of winter lightning strikes. Column (A) denotes the various waveforms measured around the SPD (phase *a*) installed at the high-voltage terminal of the boost transformer of WT#1. In the WT#1 SPD, since the lightning surge invades to the common earthing system and operates the SPD in reverse, from earth to the line, the polarity of each waveform was inverted. Also, Columns (B) and (C) correspond to the waveforms around the SPD at WT#2 and the grid-interactive transformer, respectively. The phenomena of "back-flow surge" in the wind farm can easily be recognised.

On the other hand, the graphs in Row (1) show voltage waveforms between the terminals of the respective SPDs. Row (2) is for current waveforms flowing through the SPD, and Row (3) is for simultaneous power according to Eq. 11.1, i.q. Every graph in Fig. 11.15 has three curves due to the various conditions, i.e. Case WG0, Case WG1 and Case WG2.

Comparing Case WG0 and Case WG1, it can be clearly seen that the effect of the earthing wire to reduce the surge reaching the next turbine and the grid transformer is quite significant. From the three graphs in Row (1), it is clear that the surge duration at every point in the wind farm is reduced by half. The surge

Table 11.4 Analysis conditions

Wind turbine (synchronous generator) model			
Rating power (MVA)		1.0	
Rating voltage (kV)		0.66	
Impedance (R-L-C model)	Resistance (Ω)	0.002	
	Inductance (mH)	0.231	
	Capacitance (μF)	0.001	
Transformer model		Boost	Grid-interactive
Connection method		Y/Δ	Y/Δ
Rating power (MVA)		1.0	10.0
Rating voltage (kV)		0.66/6.6	6.6/66
Frequency (Hz)		60	60
No load losses (p.u.)		0.0	0.0
Copper losses (p.u.)		0.005	0.005
Positive sequence leakage inductance (p.u.)		0.15	0.15
Saturation		no	no
Aircore reactance (p.u.)		0.2	0.2
Magnetising current (%)		1.0	1.0
Collection line model in wind farm			
Phase conductors	Height of all conductors (m)	10	
	Configuration of conductors	Horizontal	
	Spacing between phases (m)	0.7	
	Conductor radius (mm^2)	20.3	
	Sag for all conductors (m)	0.5	
	Number of sub-conductors in a bundle	1	
Earthing wire	Earthing wire radius (mm^2)	5.5	
	Number of earthing wire(s)	0, 1, 2	
	Height of earthing wire(s) (m)	11	
	Spacing between earthing wires (m) (in case of two wires)	0.7	
	Sag for all earthing wires (m)	0.5	
Resistivity of earth (Ωm)		100	

current passing through the SPD shown in Row (2) is also cut down by almost half or two-thirds. Consequently, the simultaneous power produced in the SPD becomes much lower, as shown in Row (3).

11.5.3 Evaluation of the Possibility of the SPD's Burning out

In Row (3) of Fig. 11.15, the integral area surrounding the simultaneous power curve becomes equal to the thermal energy produced in the SPD. Summarizing the above integration, the bar graphs shown in Fig. 11.16 are drawn to evaluate the possibility of a burnout incident at the SPDs. Column (1) in Fig. 11.16 illustrates the results of the integration area of simultaneous power curves, i.e. the energy

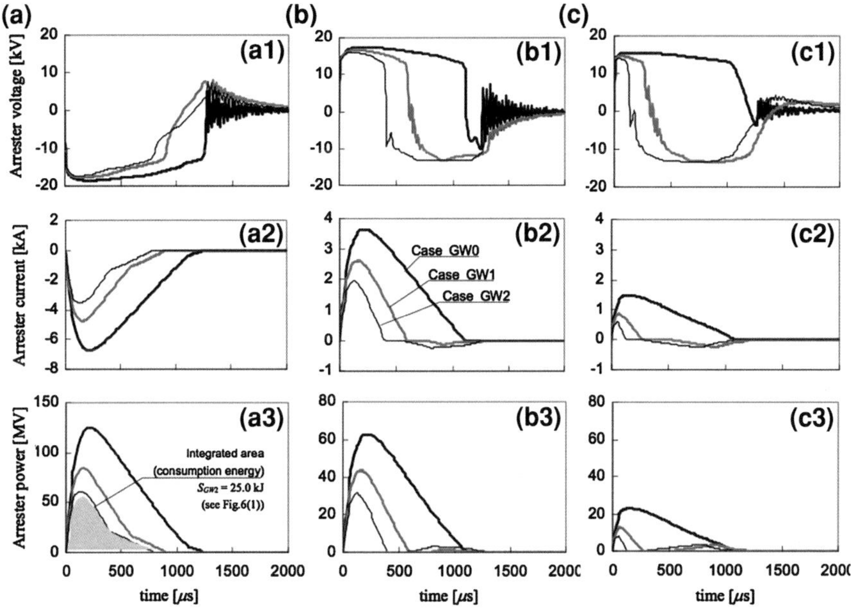

Fig. 11.15 Calculated waveforms at SPDs at various points among the wind farm in case of winter lightning (2/631 rs, 51 kA). (Column (A) wind turbine No.1 (WT#1), Column (B): wind turbine No.2 (WT#2), Column (C): grid-interactive transformer (Tr), Row (1): voltage between terminals of SPD, Row (2): current through SPD, Row (3): simultaneous power consumed at SPD)

consumption in each SPD. The graph of WT#1, which is directly struck by lightning, displays a tendency to produce huge thermal energy in the SPDs. This suggests that there is a definite possibility of burnout incidents under the conditions found with huge winter lightning strikes. Since the total earth resistance of the wind tower is assumed as 3.33 Ω in the present case, it becomes clear that a lower resistance or a higher rate for the SPD is needed to avoid burnouts.

The most important result presented in the present report is shown in Row (B) in Fig. 11.16. From this graph of the energy consumption in the SPD of WT#2, the successful effect of installing earthing wire(s) is evident. If an earthing wire was not installed (Case GW0), a huge quantity of energy could surge in, even to the adjoining turbine that was not directly struck by the lightning. By contrast, in the case of earthing wire employment (Case GW1), the surge energy invading to WT#2 is cut down to less than half. Moreover, the results for Case GW2 shows that the surge energy can be suppressed by much less than 15 kJ, which is the thermal limitation of a 2.5 kA class SPD.

A comparison between the different numbers of earthing wires also gives an interesting result. From the results in Fig. 11.16, a multiple earthing wire strategy provides a further margin of safety against lightning surge. A similar tendency to that in the present calculation for the back-flow surge is noted in the case of a

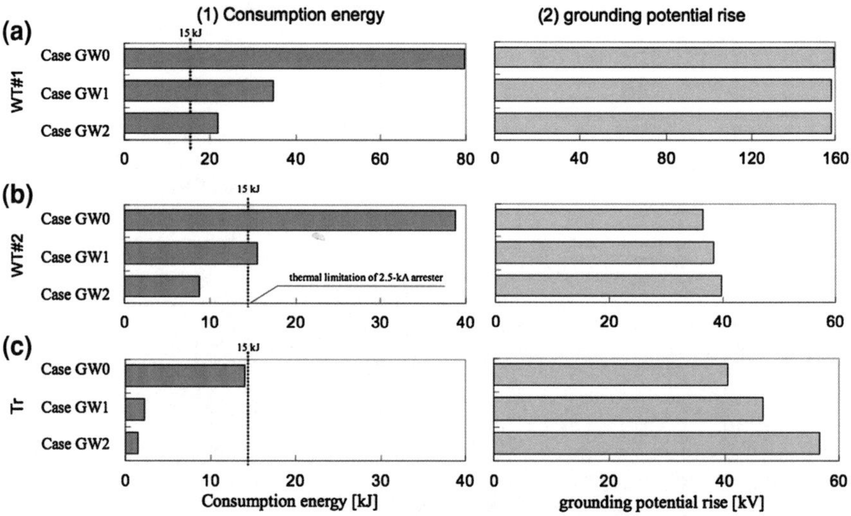

Fig. 11.16 Energy consumption at SPDs and maximum earth potential rise (EPR) at points around the wind farm in case of summer lightning (2/631 μs, 51 kA). (Column 1: total consumption energy at SPD, Column 2: maximum EPR, Row A: wind turbine No.1 (WT#1), Row B: wind turbine No.2 (WT#2), Row C: grid-interactive transformer (Tr).)

communication tower in Ref. [10], it becomes evident that a back-flow surge in a wind farm can be reduced by the installation of earthing wire(s).

11.5.4 Evaluation of Potential Rise of Earthing System

Finally, we need to also mention that a negative influence from an earthing wire(s) installation. Column (2) of Fig. 11.16 shows a surprising result. The graph of an EPR in WT#2 demonstrates an upward trend according to increases in the number of earthing wires. The same tendency can be recognised in the result in the grid-interactive transformer (Tr). Even worse is that the additional installation of earthing wires does not contribute very much to a reduction of the EPR in WT#1's earthing system.

Another result of calculation in case of summer lightning (2/70 μs, 30 kA) is shown in Fig. 11.17, where there is barely any possibility of an SPD's burnout because the total energy of back-flow surge due to summer lightning is much smaller than that of winter lightning. From Fig. 11.17, it is also evident that the EPR at the equipments that are not struck by lightning tends to increase because of the installation of earthing wire(s). Comparing with the case of winter lightning, the altitude of the EPR due to summer lightning is relatively small. However, there still remains the negative impact given by the earthing wire(s).

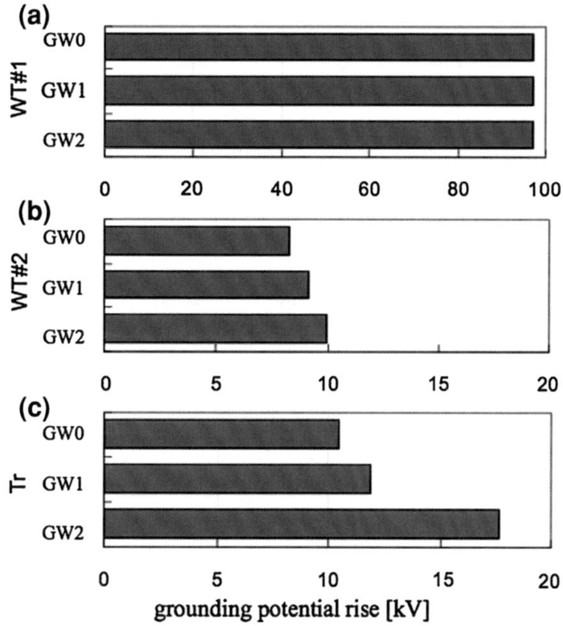

Fig. 11.17 Maximum EPR at points around the wind farm in case of summer lightning. (2/70 μs, 30 kA)

The reason for this negative impact seems to be because an earthing wire of 0.4 km has relatively strong inductive impedance against a high frequency domain of more than 1 MHz, such as found in a lightning surge. However, since the current flowing through the interconnecting earthing wire tends to be large, the EPR of the next turbine or a grid transformer displays an increasing tendency. This may give rise to a possibility of breakdowns of low-voltage circuits inside the wind turbine. As indeed already suggested in several reports [14–16], the present result agrees with these reports and confirms the problem. While it can be concluded that a earthing wire strategy is very effective against winter lightning, it may cause unexpected effects in the case of summer lightning.

11.5.5 Conclusive Discussions

This chapter presented an analysis concerning incidents of burnouts of SPDs resulting from winter lightning at wind farms using PSCAD/EMTDC. Calculations were performed to clarify the mechanism of how back-flow surge propagates from the wind turbine directly struck by lightning to other turbines.

The calculations, with various conditions, e.g. the number of interconnecting earthing wires, demonstrated that burnout incidents can be reduced by installing multiple earthing wires to the collection line in a wind farm. However from the viewpoint of the EPR, the earthing wire does not help to reduce the EPR of wind turbines and the grid- interactive transformer.

11.6 Conclusions

In this chapter lightning surge analyses on a wind farm was performed using ARENE and PSCAD/EMTDC. The analyses are simulated using a simple model with wind turbines, boost transformers, SPDs (surge protection devices) and a gird-interactive transformer connected with a collection system. From the results of several analyses the following conclusions were drawn:

1. A "back-flow surge" can propagate to other turbines from the wind turbine that has been directly struck by winter lightning via the earthing system and the collection line.
2. Burnout incidents of SPDs could quite easily occur even in a turbine far from the lightning-struck turbine.
3. Installing multiple earthing wires to the collection line in a wind farm can reduce the burnout incidents due to winter lightning.
4. The earthing wire does not help, especially for summer lightning, to reduce the EPR (earth potential rise) of wind turbines and the grid- interactive transformer.

Consequently, the results of the present calculations suggest that an accurate earthing design and an LPS (lightning protection system) strategy must be implemented for wind turbines situated in wind farms. If a wind farm is to be constructed in an area affected by heavy winter lightning, multiple earthing wires and higher rated SPDs should be installed to avoid burnouts of the SPDs and other equipment. If the wind farm also potentially suffers from summer lightning, the installation of earthing wire(s) is not recommended because the interconnection of earthing wires does not have a good effect on reducing the EPR. In both cases, trials to reduce earth resistance should be selectively done for the particular turbine that would tend to suffer from lightning because of the prevailing wind direction or geographical condition.

References

1. NEDO (2007) Activity report on study of lightning protection measures for wind turbine generator. New Energy and Development technology Organization, Japan
2. Miyake K, Suzuki T, Takashima M, Takuma M, Tada T (1990) Winter lightning on Japan Sea coast-lightning striking frequency to tall structures. IEEE Trans Power Deliv 5(3): 1370–1376
3. Miyake K, Suzuki T, Shinjou K (1992) Characteristics of winter lightning current on Japan Sea Coast. IEEE Trans Power Deliv 7(3):1450–1457
4. Motoyama H, Shinjo K, Matsumoto Y, Itamoto N (1998) Observation and analysis of multiphase back flashover on the okushishiku test transmission line caused by winter lightning. IEEE Trans Power Deliv 13(4):1391–1398
5. Yokoyama S, Miyake K, Suzuki T, Kanao S (1990) Winter lightning on japan sea coast–development of measuring system on progressing feature of lightning discharge. IEEE Trans Power Deliv 5:1418–1425

6. IEC (2010) Wind turbine generation system–24: lightning protection. IEC61400-24. International Electro-technical Commission, Geneva
7. Natsuno D, Yokoyama S, Shindo T, Ishii M, Shiraishi H, (2010) Guideline for lightning protection of wind turbines in Japan. Proceedings of 30th international conference on lightning protection (ICLP2010), No.SSA-1259
8. Shindo T, Suda T (2009) Lightning risk of wind turbine generator system. IEEJ Trans Power Energy 129(10):1219–1224
9. Wada A, Yokoyama S, Numata T, Hirose T (2004) Lightning observation on the nikaho-kogen wind farm. Proceedings of international workshop on high voltage engineering (IWHV'04), vol 1. pp 51–55 (The papers of joint technical meeting on electrical discharges, switching and protecting engineering and high voltage engineering, IEE Japan, ED- 04-118, SP-04-29, HV-04-59)
10. Nakada K, Wakai T, Taniguchi H, Kawabata T, Yokoyama S, Yokota T, Asakawa A (1999) Distribution arrester failures caused by lightning current flowing from customer's structure into distribution lines. IEEE Trans Power Deliv 14(4):1527–1532
11. Yasuda Y, Funabashi T (2004) Lightning analysis on wind farm: sensitivity analysis on earthing. Proceedings of 27th international conference on lightning protection (ICPL 2004), pp 1041–1046
12. Yasuda Y, Funabashi T (2007) Analysis on back-flow surge in wind farms. Proceedings of the international conference on power systems transients (IPST 07)
13. Yasuda Y, Uno N, Kobayashi H, Funabashi T (2008) Surge analysis on wind farm when winter lightning strikes. IEEE Trans Energy Convers 23(1):257–262
14. Cotton I (1998) Windfarm earthing. Proc 11th Int Symp High Volt Eng (ISH99) 2:288–291
15. Hermoso B (2006) Wind farm earthing installations: rated and lightning frequencies behaviour. Proceedings of international conference on grounding earthing (GROUND' 2006), pp 411–414
16. Sekioka S, Funabashi T (2010) A Study on effective length for practical design of grounding system in a wind turbine. Proceedings of 30th international conference on lightning protection (ICLP2010), No.5B-1085

Chapter 12
Electric Grid Connection and System Operational Aspect of Wind Power Generation

Bharat Singh Rajpurohit, Sri Niwas Singh and Lingfeng Wang

Abstract Grid integration of wind power is one of the prime concerns as wind power penetration level is increasing continuously. New grid codes are being set up to specify the relevant requirements for efficient, stable, and secure operation of power system and these specifications have to be met in order to integrate wind power into the electric grid. This chapter discusses several issues of advanced grid codes relating to the wind turbines integration into power system. New grid codes for wind power integration of different countries are presented and compared. In India, share of wind power as percentage of installed generation capacity has exceeded 10% in many states and grid connection standards for wind power are in the process of establishment. Grid code requirements for wind power for Indian power sector have been suggested. System operational aspects of wind power generation with Indian experience have discussed.

B. S. Rajpurohit (✉)
School of Computing & Electrical Engineering, Indian Institute of Technology Mandi, Mandi 175001, India
e-mail: brajpurohit@gmail.com

S. N. Singh
Department of Electrical Engineering, Indian Institute of Technology Kanpur, Kanpur 208016, India
e-mail: snsingh@iitk.ac.in

L. Wang
Department of Electrical Engineering & Computer Science College of Engineering, The University of Toledo, Toledo, OH 43606, USA
e-mail: lingfeng.wang@utoledo.edu

S. M. Muyeen (ed.), *Wind Energy Conversion Systems*,
Green Energy and Technology, DOI: 10.1007/978-1-4471-2201-2_12,

12.1 Introduction

Grid integration is one of the major concerns in recent years as wind power generation level keeps on increasing continuously in the power system. Therefore, wind farms will have a significant influence on the operation and control of emerging power system which is expanding in terms of interconnections, network capacity, load demand, and addition of new devices for improving the performance of the system. New grid codes are being set up in several countries by specifying the relevant requirements for efficient, stable, and secure operation of power system and these specifications have to be met in order to integrate wind power into the modern electric grid. Several states of India have high penetration of wind power and specific grid connection requirements (GCR) for wind power are yet to be established. This work proposes GCR for wind power integration in India and discusses several technical and operational issues arising due to high penetration of wind power generation in Indian power systems [1].

During the last decade, the installed capacity of wind power generation has been increased tremendously all over the world. Wind power, which is the fastest-growing source of electric power generation, is proved to be a potential source for generation of electricity with minimal environmental impact. At the end of 2010, the wind installed capacity stands at over 194.4 GW, worldwide, which is more than 35.8 GW from the capacity in 2006. Global growth of installed capacity of wind power is shown in Fig. 12.1 [2].

Wind projects, today, are large enough to have a significant impact on transmission network security, operation, and planning. Rapid installation, increased turbine size, and large-scale wind farms development worldwide demand an integration of wind power projects in the existing power system. Increased penetration of wind power into the electricity grid gives rise to new challenges for the entire system and, in particular, to the transmission system operators in maintaining reliability and stability of electricity supply. Due to the intermittent nature of wind power generation, maintaining frequency and voltage within the prescribed operating limits become difficult task. The number of medium and large wind farms (more than 50 MW) connected to the high voltage transmission system is likely to increase dramatically, especially with offshore wind farms.

In the past, GCR for wind turbines or wind farms was not necessary due to low level of wind power penetration. IEEE Standard 1001 '*IEEE Guide for Interfacing Dispersed Storage and Generation Facilities with Electric Utility Systems*' was the only guideline for the connection of generation facilities to the distribution networks. The standard included the basic issues of power quality, equipment protection, and safety. The standard expired and, therefore, in 1998, the IEEE Working Group SCC21 P1547 started to work on a general recommendation for the interconnection of distributed generation, i.e., *IEEE Standard for Interconnecting Distributed Resources with Electric Power Systems* [3].

Conventional power plants employ synchronous machines, which assist in maintaining transient stability, voltage control, reactive power support, frequency

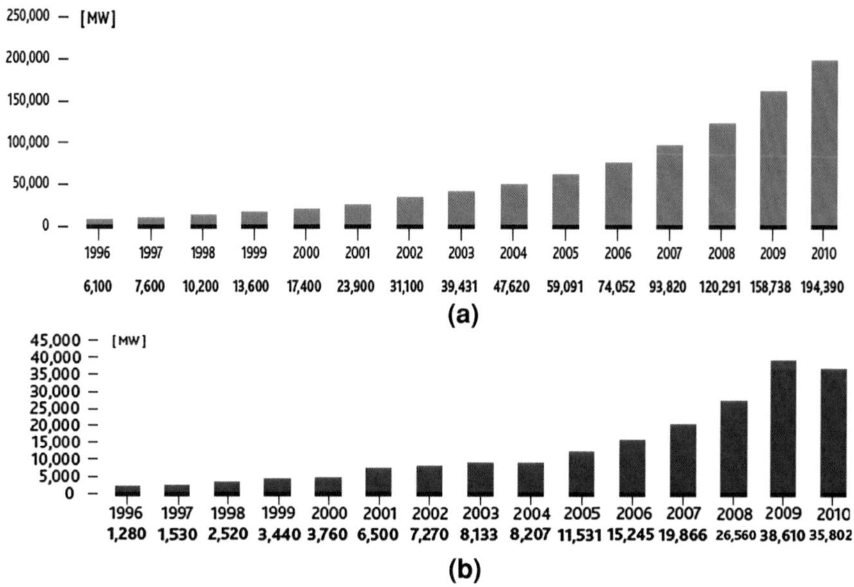

Fig. 12.1 **a** Global cumulative installed capacity 1996–2010. **b** Global annual installed capacity 1996–2010 [2]

control, and fault ride-through capabilities defined by the transmission system operators (TSOs). The counterpart to synchronous machines in wind farms are mainly fixed speed asynchronous generators, doubly fed induction generators (DFIGs) and synchronous generators with back to back converters. The first generation of commercial grid connected wind turbines in the 1980s was dominated by the fixed speed concept mainly using asynchronous induction generators, which were supplemented with a capacitor bank for reactive power compensation. Through the 1990s, different types of variable speed concepts became popular in the market. According to the trends, the DFIG concept was the most successful variable speed concept with more than 45% market share in 2002. Technical characteristics of DFIG can be made very close to those of synchronous machines by employing power electronic converters and proper control mechanism, thus being able to enhance the performance to meet the connection requirements defined by TSO.

The interconnection rules were continuously reformulated because of the increasing wind power penetration and the rapid development of wind turbine technology, as shown in Fig. 12.2. Not only was the increased size of the wind turbines but also was the increased size of the wind farms, which resulted in interconnection requests at the transmission level. Hence, interconnection rules for wind farms to be connected to the transmission level are required. The main focus in the electricity grid codes has been on the fault ride-through issue, where the TSO requires wind power generators to stay connected to the grid during and after a fault in the transmission system. Another important requirement to the wind

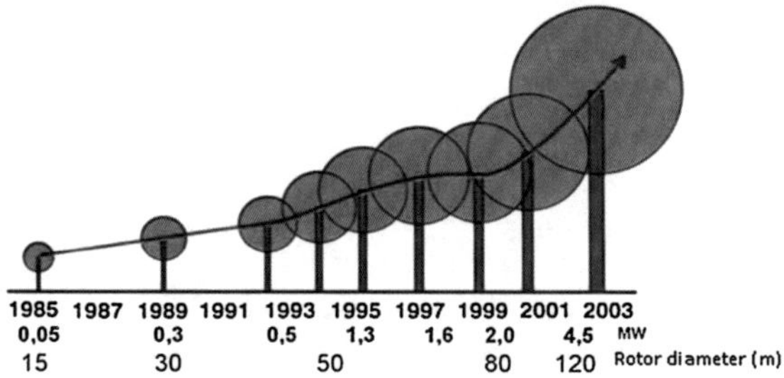

Fig. 12.2 Technology development of single wind turbine from 1985 to 2003

power installations is on active and reactive power (voltage) control capability, to make the wind power installations able to support the control of grid frequency and grid voltage. New grid codes are being set up to specify the relevant requirements for efficient, stable, and secure operation of power system and these specifications have to be met in order to integrate wind power generation into the electric grid.

In this chapter, several issues of advanced grid codes relating to the wind turbines integration into power system have been presented. GCR of several countries (India, Denmark, Germany, USA, Ireland, UK, Sweden, and Scotland), which are proactively meeting the challenge of considerable wind power penetration, are analyzed. Some countries have more than one power system operators. Table 12.1 presents the GCR issued by different TSO in the world. In India, with increasing penetration of wind energy, share of wind power as percentage of installed generation capacity has exceeded 10% in many states and grid connection standards for wind power are in the process of establishment. Grid code requirements for wind power for Indian power sector have been suggested. System operational aspects of wind power generation with Indian experience have discussed.

12.2 Grid Connection Requirements

This section will provide a brief description of various controls which are required for reliable and secure operation of power system. The normal practices for wind power control adopted in various countries are also discussed. Owing to the complexity of the regulations, only a few specific interconnection procedures are explained.

12.2.1 Active Power Control

This is a requirement for generating units to be able to deliver power and remain connected to the network even if the system frequency deviates from specified one. Active (real) power control is used to control the system frequency by changing

Table 12.1 Grid codes from different TSO

Country	TSO	Author	Title	Issue year
China	All	CEPRI	Technical rules for connecting wind farm to power system	2005
Canada	Alberta electric system operator	Alberta electric system operator,	Wind power facility, technical requirements	2004
Denmark	Eltra/Elkraft	Eltra/Elkraft	Regulation TF 3.2.5, wind turbines connected to grids with voltages above 100 kV	2004
Finland (Nordic countries)	FINGRID OYJ	FINGRID OYJ	Connection code for connection of wind power plants to finnish power system	2009
Germany	E.ON.	E.ON.	Grid code high and extra high voltage	2006
Ireland	EirGrid	EirGrid	EirGrid grid code: WFPS1- Wind farm power station grid code provisions (ver. 3)	2007
Poland	PSE	PSE	Traffic and maintenance manual transmission	2006
Scotland	Scottish hydro electric	Scottish hydro electric	Guidance note for the connection of wind farm	2002
Sweden	Svenska kraftnät (SvK)	Svenska kraftnät (SvK)	Swedish power transmission utility regulations and general advice on the reliability design of generating	2005
U.K.	NGET	NGET	Grid code	2008
U.S.A.		FERC	FERC order No. 661-A, interconnection for wind energy	2005

the power injected into the grid. Using real power control, overloading of lines can also be minimized, power quality standards can be maintained, and large voltage steps and inrush currents during start-up and shutdown of wind turbines can be avoided. Modern technologies allow some control over the active power to improve frequency of the network; however, this is not always possible as wind generators are normally operated at their maximum power point.

Active power control is also important for transient and voltage stability enhancement. If the power can be reduced efficiently as soon as a fault occurs, the over-speeding of turbines can be prevented. Another concern from the viewpoint of the power system operator is the rate at which power is ramped up after clearing a fault. The requirement for ramp up rates is made to avoid power surges and ramp down rates to avoid generation unavailability [4]. Real power control is one of the required controls in all considered GCRs. It is noticed that the Scottish guidance note and *ESBNG* (Electricity supply board national grid) in Ireland set requirements on maximum active power change during start-up, shutdown, and wind speed change in order to avoid impacts on system frequency.

Eltra & Elkraft of Denmark and SvK of Sweden require active power change in order to ensure sufficiently fast down regulation in case of necessity (e.g., over frequencies). *E.ON.* of Germany regulations define both maximum permissible active power change and minimum required active power reduction capability. *Eltra* and *Eltra & Elkraft* require active power control such that 1 min average of active power should not exceed the set-point by more than 5% of maximum power of the wind farm. *E.ON.* requires active power reduction of minimum 10% of registered capacity per minute. In the case of grid voltage loss, power has to be ramped up with a gradient of not more than 10% of rated power per minute. This ramp can be realized in steps (reconnection of single WT) if the step size does not exceed 10% of rated power per minute. *ESBNG* of Ireland requires that, in any 15-minute period, the active power output change is limited as follows: 5% of registered capacity per minute for WFs $<$ 100 MW, 4% for WFs $<$ 200 MW, and 2% for WFs $>$ 200 MW.

For Indian grid scenario, the following active power control mechanism is suggested. The wind power generating units are normally operated to maximum power using maximum power point tracking algorithm and remain connected to the network even if the system frequency deviates from specified one. Active (real) power control is used to control the system frequency by changing the power injected into the grid. The active power production from the wind farm must be controllable to maintain the security and stability of the electric grid. The following control functions must be available [5].

- An adjustable upper limit to the active power production from the wind farm shall be available whenever the wind farm is in operation. The upper limit control of active power production, measured as a 15 min average value, does not exceed a specified level and the limit shall be adjustable by remote signals. It must be possible to set the limit to any value with an accuracy of ±5%, in the range from 20 to 100% of the wind farm rated power.
- Ramping control of active power production must be possible to limit the ramping speed of active power production from the wind turbine in upwards direction (increased production due to increased wind speed or due to changed maximum power output limit) to 10% of rated power per minute. There is no requirement to down ramping due to fast wind speed decays, but it must be possible to limit the down ramping speed to 10% of rated power per minute, when the maximum power output limit is reduced by a control action.
- Fast down regulation should be possible to regulate the active power from the wind turbine down from 100 to 20% of rated power in less than 5 s. This functionality is required for system protection schemes. Some system protection

schemes implemented for stability purposes require the active power to be restored within short time after the down regulation. For that reason, disconnection of a number of wind turbines cannot be used to fulfill this requirement.

- Automatic control of the wind turbine active power production as a function of the system frequency must be possible. The control function must be proportional to the frequency deviations with a dead-band. The detailed settings can be provided by the State utilities (SU).
- During under-frequency (it shows the deficit in the generation), wind power can increase the power output without affecting the network congestion.

In India, the system frequency has controlled by the State Load Dispatch Centers (SLDC) in coordination with Regional Load Dispatch Centers (RLDC) at about 50 Hz, within the range of 49.2–50.3 Hz band [5]. Wind farms must be capable of operating continuously for 49.2–50.3 Hz frequency band and allowed to be disconnected during over frequency as per the wind turbine specifications. In addition, the wind turbines can reduce power at frequency of above 50.3 Hz as detailed settings provided by the SU.

12.2.2 Frequency Control

This is a requirement for generating units to be able to increase or decrease power output with falling or rising frequency. In the power system, the frequency is an indicator of the imbalance between production and consumption. For the normal power system operation, the frequency should be close to its nominal value. In the case of an imbalance between supply and demand, the primary and secondary controls are used to reduce the imbalance of power.

In a power system, conventional generating units are normally equipped with the governor control which works as primary load frequency control. The time span for this control is 1–30 s. In order to restore the frequency to its nominal value and release of used primary reserves, the secondary control is employed with a time span of 10–15 min. The secondary control, thus, results in a slower increase or decrease of generation. Some regulation requires wind farms to be able to participate in secondary frequency control. During the over-frequencies, this can be achieved by shutting down some of the turbines in the wind farm or by reducing the power output using pitch control. Since wind cannot be controlled, power production at normal frequency would be intentionally kept lower than possible in order for the wind farm to be able to provide secondary control at under-frequencies [6].

Following major disturbances, power system may experience large excursions in the voltages and frequency. Beyond specific frequency limits, system stability cannot be guaranteed and generators as well as consumers may be at the risk of equipment damage. In this case, disconnection from the grid seems to be the best strategy. Figure 12.3 shows the overview of frequency control requirements in different counties [3].

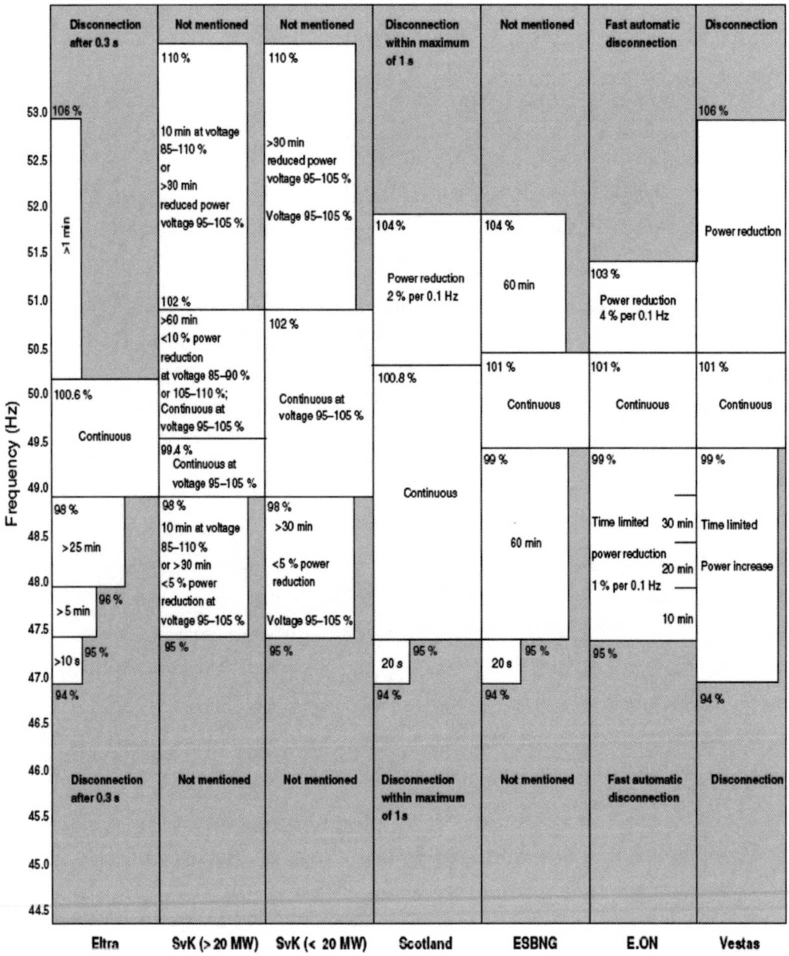

Fig. 12.3 Overview of frequency control requirements [3]

According to the *E.ON.*, wind turbines have to stay connected in the grid within the frequency range of 47.5–51.5 Hz. Outside this range, disconnection without any time delay is necessary. Figure 12.4 shows the active power reduction allowed (Fig. 12.4a) and the available voltage (operation) range (Fig. 12.4b) depending on the frequency. Also the minimum time span for guaranteed operation under the various scenarios is shown in these figures. In addition, the wind turbines have to reduce power at frequency of about 50.2 Hz as shown in Fig. 12.5 [7, 22].

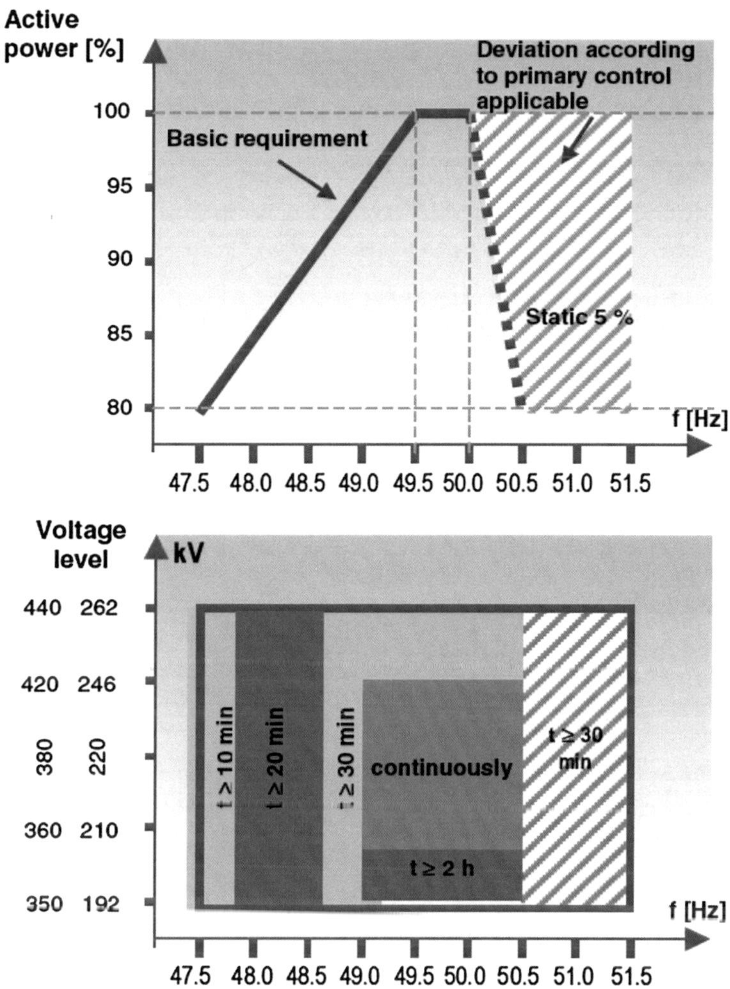

Fig. 12.4 a, b Operating requirements depending on network frequency [7]

12.2.3 Voltage Control

The voltage control requirement is used for generating units to supply lagging/leading reactive power at the grid connection point. Wind turbine should be capable of supplying a proportion of the system's reactive capacity, including the dynamic capability and should contribute to maintain the reactive power balance. Requirements of the grid codes for reactive power capability demand that the power factor be maintained in the specified range.

Fig. 12.5 Frequency characteristic of wind power generation [9]

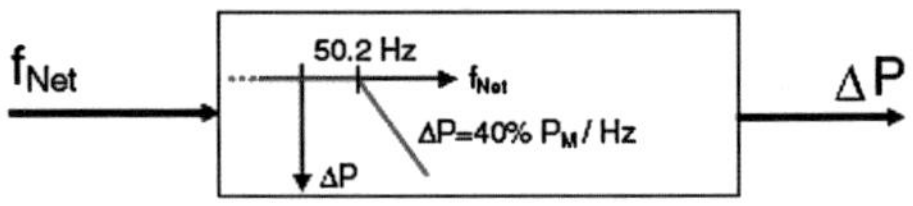

$$\Delta P = 20\,P_M \frac{50.2\,\text{Hz} - f_{Netz}}{50\,\text{Hz}} \quad \text{at } 50{,}2\ \text{Hz} \leq f_{Net} \leq 51{,}5\ \text{Hz}$$

P_M Available power

ΔP Power reduction

f_{Net} Frequency

Between 47.5 Hz $\leq f_{Net} \leq$ 50.2 Hz no limitation

At $f_{Net} \leq$ 47.5 Hz and $f_{Net} \geq$ 51.5 Hz separation from grid

Fig. 12.6 Comparison of power factor ranges as required by the different GCR [4]

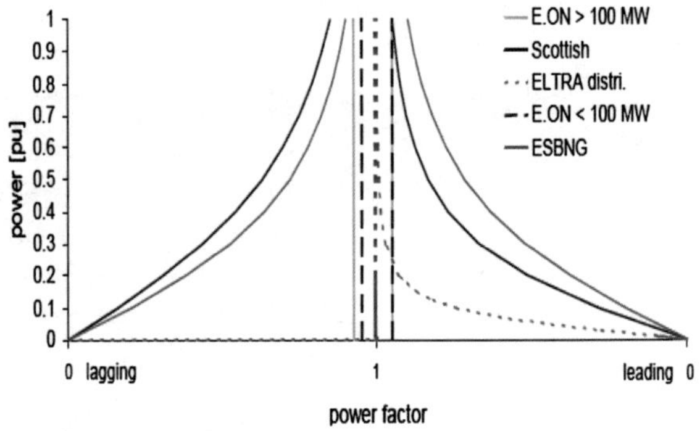

In Fig. 12.6, the reactive power requirements are compared in terms of power factor [4]. Note that ‘lagging’ refers to the production of reactive power, and ‘leading’ to the absorption of reactive power. In Fig. 12.6, only the operating limits are considered without considering the voltage conditions. According to the German grid code, the wind turbines must provide, as a mandatory requirement, voltage support during voltage dips. The corresponding voltage control characteristics are summarized in Fig. 12.7 [7]. According to this requirement, the wind turbines have to supply at least 1.0 pu. reactive current when the voltage falls below 50%. A dead band of 10% is introduced to avoid undesirable control actions. However, for the wind farms connected to the high voltage grid, the continuous voltage control without dead band is also under consideration.

Voltage regulators and the control of reactive power at the generators and at the connection points are used in order to keep the voltage within the required limits and avoid voltage stability problems. Wind turbines have to contribute to voltage regulation in the system to either maintain voltage at the point of connection of a wind turbine or to compensate reactive power [3].

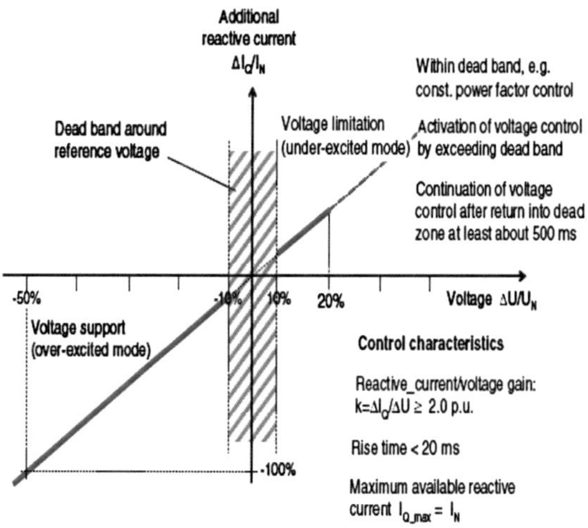

Fig. 12.7 Voltage control requirements [9]

Older wind turbines based on conventional induction generators require a reactive power support from the power systems. These often degrade the system voltage performance rather than supporting it. Several grid codes require the need for wind plants to support power system voltage by having the capability to control their reactive power within the 0.95 leading to 0.95 lagging range. Many modern wind generators provide this dynamic capability directly from the power electronics devices that control the real power operation of the machine. These can provide excellent voltage control for the power system. The majority of modern wind turbines use doubly-fed induction generator (DFIG). The stator of these machines is connected directly to the grid, whereas the three phase rotor windings are supplied through a voltage source converter. By varying the voltage magnitude and frequency, the active and reactive power generated can be controlled and thus, the optimal rotor speed can be adjusted for any wind speed. Because of the limited speed variability required for wind turbines, the converters have to be designed for 20–30% of the total generator power only [1].

In the Danish grid code for wind farms connected to the transmission system, it is required that a wind farm owner supplies a PQ diagram showing the regulation capability for reactive power of the installation at the connection point. This is understood as the steady-state capability, because there are other requirements to the dynamic behavior during the faults. The reactive power capability of a wind farm depends a lot on the capability of the wind turbines, although the impact of the grid should also be considered in a PQ diagram for the whole wind farm. The reactive power capability of the wind turbines depends on its type or technology for grid connection. For the synchronous generator, the boundaries defined by the field current limitation, the armature current limitation and the mechanical power limitation can easily be calculated from the ratings and the reactance. A similar description of a DFIG is presented in literature, as shown in Fig. 12.8 [8].

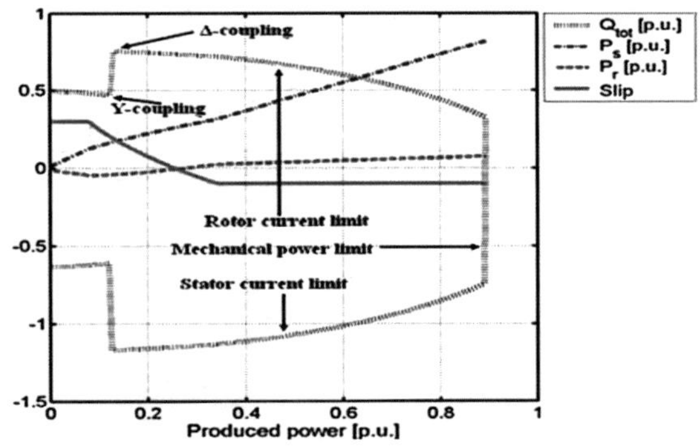

Fig. 12.8 Capability curve of DFIG [8]

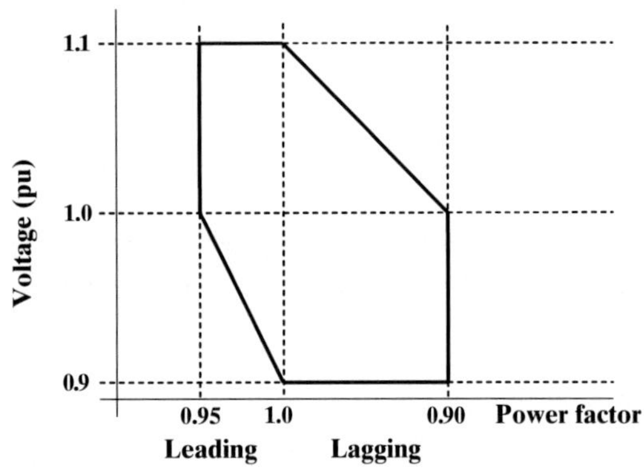

Fig. 12.9 Operating range of power with voltage [5]

For Indian grid conditions, wind farms shall be capable of operating at rated output for power factor varying between 0.9 lagging (over-excited) and 0.95 leading (under-excited). Figure 12.9 shows the operating range of wind farms at different voltage levels. Thc above performance shall also be achieved with voltage variation of ±10% of nominal, frequency variation of +1.6%, and −0.06% and combined voltage and frequency variation of ±10%. Wind farms are required to have sufficient reactive power compensation to be neutral in reactive power at any operating point. In India, the SLDC (and users), ensure that the grid voltage remains within the operating limits as specified in Indian Electricity Grid Code (IEGC) Sect. 5.2, as shown in Table 12.2, and hence it is required from the wind turbine to remain connected and deliver power for the specified voltage ranges and put efforts to maintain it.

Also, wind farms shall make available the up-to-date capability curves indicating restrictions to the SLDC/RLDC, to allow accurate system studies and effective operation of the state transmission system. The reactive power output of the wind farm must be controllable in one of the two following control modes according to SU specifications.

Table 12.2 Grid voltage operating limits [5]

Nominal system voltage (kV)	Grid voltage		
	Variation limits (%)	Maximum (kV)	Minimum (kV)
400	+5/−10	420	360
220	±10	245	200
132	±10	145	120
33	±10	36	30

- The wind farm shall be able to control the reactive exchange with the system at all active power production levels. The control shall operate automatically and on a continuous basis.
- The wind farm must be able to automatically control its reactive power output as a function of the voltage at the connection point for the purpose of controlling the voltage.

The detailed settings of the reactive power control system will be provided by the respective SU. The wind farm must have adequate reactive power capacity to be able to operate with zero reactive exchange with the network measured at the connection point, when the voltage and the frequency are within normal operation limits.

12.2.3.1 Wind Farm Protection and Fault Ride-Through

This is the requirement for generating units to revert to normal operation when fault on power system is cleared. Fault ride-through (FRT) requirement is imposed on a wind power generator so that it remains stable and connected to the network during the network faults. Disconnection from grid may worsen a critical grid situation and can threaten the security standards when wind penetration is high.

In Germany, the wind generating plants are expected to acquit themselves during a low-voltage disturbance as summarized in a voltage versus time curve shown in Fig. 12.10 [9]. Wind turbines are required to stay on the grid within areas 1 and 2. If a wind turbine faces overloads, stability or other kinds of technical problems in area 2, it can be disconnected itself from the grid provided a resynchronization can take place after 2 s. Moreover, it must be able to increase the active power output following the resynchronization by gradients of at least 10% of the nominal power per second. The comparison of the different under-voltage operating ranges only the outermost operating limits and their corresponding trip times shown in Fig. 12.11 [4].

Eltra, *Eltra & Elkraft*, *Scottish grid code* and *E.ON* specify not only voltage operating limits, but also mentions specifically ride-through of transient faults to sustain generation. When a voltage dip occurs during normal operation of the WT, the current rises in order to export the same amount of power as before the voltage dip. This implies that the whole WF must be designed for currents bigger than the

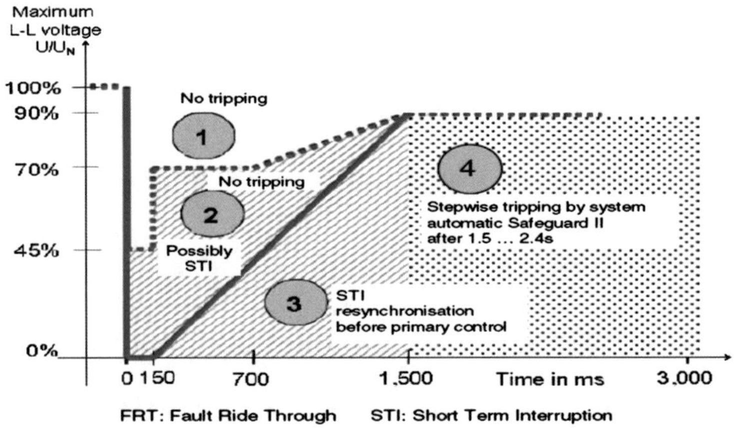

Fig. 12.10 Definition of FRT requirements [7]

Fig. 12.11 Requirement for voltage operating range [4]

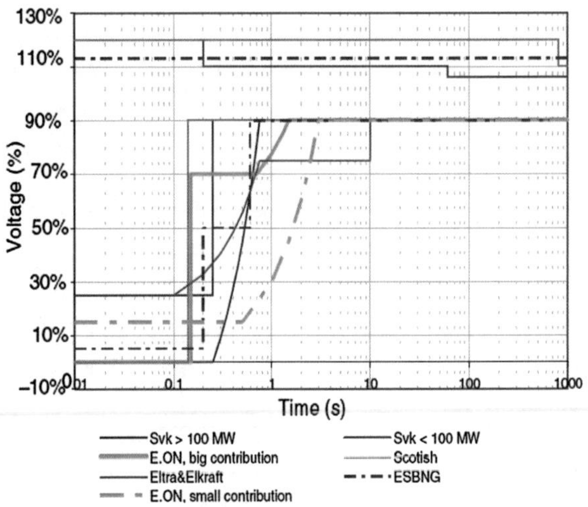

rated current. Similar discussion is applied for high voltage ride through (HVRT). Situations with over-voltages may arise due to load shedding or unbalanced faults. The resulting over-voltages may have different magnitudes and durations, depending on the disturbance scenario. Therefore, the international grid code requirements concerning HVRT slightly differ. In Australia, grid codes stipulate the wind turbines to withstand even an overvoltage of 1.3 pu. for 60 ms (Fig. 12.12) [10, 11].

For Indian grid conditions, disconnection from grid may worsen the situation and can threaten the security standards at high wind penetration. The wind farm must be able to operate satisfactorily during and after the disturbances in the

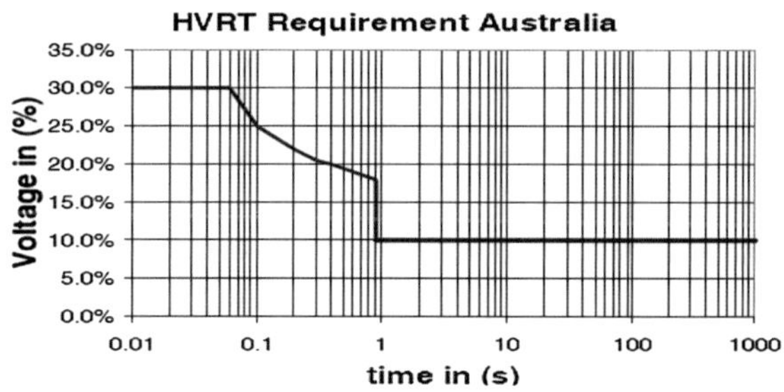

Fig. 12.12 HVRT requirement in the Australian grid code [10, 11]

distribution/transmission network. This requirement applies under the following conditions:

- The wind farm and the wind turbines in the wind farm must be able to stay connected to the system and to maintain operation during and after clearing faults in the distribution/transmission system.
- The wind farm may be disconnected temporarily from the system, if the voltage at the connection point during or after a system disturbance falls below the certain levels.

The fault, where the voltage at the connection point may be zero, duration is 100 ms for 400 kV and 160 ms for 220 and 132 kV. Wind turbines are required to be equipped with under-frequency and over-frequency protection, under-voltage and over-voltage protection, differential protection of the generator transformer, and backup protection (including generator over-current protection, voltage-controlled generator over-current protection, or generator distance protection). Prevalent practice shall be followed according to Regulations 2007 [15].

12.2.4 Communication Requirements

Wind farms must be controllable from remote locations by telecommunication system. Control functions and operational measurements must be made available to the Load Dispatch Centers (LDCs). The TSO specifies the required measurements and other necessary information to be transmitted from the wind farm. Reliable and efficient communication systems shall be provided to facilitate necessary communication and data exchange, and supervision/control of the grid by the dispatch centres, under normal and abnormal conditions.

12.2.5 Supervisory Control and Data Acquisition (SCADA)

SCADA is recommended for the remote control of wind power and telemetry of the parameters important for scheduling and forecasting is obtained. In most regulations, the wind farm owner is required to provide the signals necessary for the operation of the power system. The requirements on communication are quite similar in all considered documents. Information's required generally from wind farms are voltage, current, frequency, active power, reactive power, operating status, wind speed, wind direction, regulation capability, ambient temperature and pressure, frequency control status and external control possibilities.

12.2.6 Other Requirements

12.2.6.1 Metering

Recording instruments such as data acquisition system/disturbance recorder/event logger/fault locator (including time synchronization equipment) shall be installed at each wind farm for recording of dynamic performance of the system. Agencies shall provide all the requisite recording instruments as specified in the connection agreement according to the agreed time schedule. These requirements are similar for conventional power sources and mentioned in detail in Central electricity authority (CEA) (Installation and operation of meters, Regulation 2006) [18], Indian electricity grid code (IEGC) [14], and respective state electricity grid codes [16].

12.2.6.2 Start and Stop

It is recommended, that the wind farms be designed so that the wind turbine does not stop simultaneously within the wind farm due to high wind speeds.

12.2.6.3 Modeling and Validation

Prior to the installation of a wind turbine or a wind farm, a specific test programme must be agreed with the SU in the area regarding the capability of the wind turbine or wind farm to meet the requirements in this connection code. As a part of the test programme, a simulation model of the wind turbine or wind farm must be provided to the SU in a given format and the model shall show the characteristics of the wind turbine or wind farm in both static simulations (load flow) and dynamic simulations (time simulations). The model shall be used in feasibility studies prior to the installation of the wind turbine or wind farm and the commissioning tests for the wind turbine or the wind farm shall include a verification of the model. These requirements are similar to the conventional power sources and mentioned in detail in IEGC and respective state electricity grid codes.

12.3 System Operational Aspects of Wind Power Generation: Indian Experience

Wind power program in India is highly successful and occupies the fifth position in the world having wind power installed capacity of 13 GW as on March 2011. However, the percentage of wind power generation is very small due to heavy generation from other sources of energy. Indian government policy and regulatory framework both at the state and central levels are encouraging power generation from new and renewable energy sources. This sector has been growing at over 35% in the last four years. The wind power program in India was initiated towards the end of the sixth five-year plan (1983–1984) and aims at catalyzing commercialization of wind power generation in the country [13]. In India, with increasing penetration of wind energy, share of wind power as percentage of installed generation capacity has exceeded 10% in many states, however, the common specific grid connection standards for wind power are in the process of establishment. Technical guidelines and requirements for wind power generation are varying with one state to other states and not good enough for the large wind power integration into the grid.

In order to promote wind power and to maintain common grid discipline Indian agencies have issued several notifications and regulations. In this work, Indian experience of system operational aspects of wind power generation are summarized and some technical and operational issues of high penetration of wind power are addressed.

12.3.1 Remuneration for Wind Power

The national tariff policy, which was notified by the Ministry of Power in January 2006, in continuation with the Electricity Act 2003 and the National Electricity Policy 2005, emphasizes the importance of setting renewable energy targets and preferential feed-in tariffs for renewable energy procurement by the respective State electricity regulatory commission (SERC). Several SERC, in turn, provided details on concessional feed-in tariffs (mostly decided by *cost-based approach*), wheeling (transporting electric power over transmission lines), banking of energy for future use, third party sale and power evacuation facilities, as shown in Table 12.3 with other fiscal incentive, as shown in Table 12.4. Recently, Central electricity regulatory commission (CERC) issues regulations of terms and conditions for tariff determination from renewable energy sources, Regulations 2009 [19] which is single part tariff consisting of the following fixed cost components:

- Return on equity;
- Interest on loan capital;

Table 12.3 Preferential tariffs/policy introduced announced by the SERC's for wind

Sources	Wind power		
Items states	Wheeling charges	Banking	Buy-back (INR/kWh)
Andhra Pradesh	2% of energy	12 months	3.37
Gujarat	4% of energy	–	3.37 fixed for 20 years.
Haryana	2% of energy	Allowed	4.08 + Escalation 1.5%
Karnataka	2% of energy	2%/month for 12 month	3.40 fixed for 10 years
Kerala	5% of energy	9 months (Jun–Feb)	3.14 fixed for 20 years
Maharashtra	2% of energy +5%trans.loss	12 months	3.50 + Escalation of 0.15 for 13 years from documentation of the project
Madhya Pradesh	2% of energy	Not allowed	3.97 (with decrease of 0.7 up to 4th year) then fixed at 3.30 from 5th year onwards uniformly for 20 years
Rajasthan	10% of energy	3 months	3.59 for Jaisalmer, Jodhpur etc. and 3.67 for other districts
Tamil Nadu	5% of energy	5%, 12 months	2.90 (levelised)
West-Bengal	INR 0.3/kWh	6 months	To be decided on case to case with a cap of 4

Table 12.4 Implemented fiscal incentives for wind power generation

Item	Description
Accelerated depreciation	80% depreciation in the first year can be claimed for the following equipment: *for Wind:* Extra 20% after march 2005 for new plant & machinery
Tax holiday	Ten years tax holidays.
Customs duty	Concessional customs and excise duty exemption for machinery and components for initial setting up of projects.
Sales tax	Exemption is available in certain states

- Depreciation;
- Interest on working capital;
- Operation and maintenance expenses.

It also considers the incentive or subsidy offered by the Central/State government to the renewable energy power plants while determining the tariff under these regulations. Wind power is fed into the grid when and where available on priority basis and these shall be treated as *must run* power plants and shall not be subjected to 'merit order dispatch' principles in order to maximize generation and to gainfully utilize wind power already installed. Recently government of India announced generation based incentives (GBI) schemes for new wind farms. Under this scheme, a GBI will be provided to wind electricity producers INR 0.50 per unit of electricity fed into the grid for a period not less than 4 years and a maximum period of 10 years in parallel with fiscal incentives such as accelerated depreciation on a mutually exclusive manner, with a cap of INR 6.2 million per MW.

Recently, CERC issued a notification, terms and conditions for recognition and issuance of renewable energy certificate for renewable energy generation, regulations, 2010 [20]. The concept of REC seeks to address the mismatch between availability of renewable energy sources and the requirement of obligated entities to meet their renewable purchase obligations. The REC mechanism mainly aims at promoting investment in the renewable energy projects and to provide an alternative mode to the RE generators for recovery of their costs.

There are two categories of certificates. First, solar certificates issued to eligible entities for generation of electricity based on solar as renewable energy source, and non-solar certificates issued to eligible entities for generation of electricity based on renewable energy sources other than solar. The solar certificate shall be sold to the obligated entities to enable them to meet their renewable purchase obligation for solar, and non-solar certificate shall be sold to the obligated entities to enable them to meet their obligation for purchase from renewable energy sources other than solar.

But the condition laid down is that these *renewable energy sources do not have any power purchase agreement for the capacity related to such generation to sell electricity at a preferential tariff.*

Reactive power compensation should ideally be provided locally, by generating reactive power as close to the reactive power consumption as possible. The regional entities except generating stations are therefore expected to provide local VAr compensation/generation such that they do not draw VArs from the EHV grid, particularly under low-voltage condition. To discourage VAr drawals by regional entities except generating stations, VAr exchanges with ISTS shall be priced as follows:

- The regional entitiy except generating stations pays for VAr drawal when voltage at the metering point is below 97%;
- The regional entitiy except generating stations gets paid for VAr return when voltage is below 97%;
- The regional entitiy except generating stations gets paid for VAr drawal when voltage is above 103%;
- The regional entity except generating stations pays for VAr return when voltage is above 103%;
- Provided that there shall be no charge/payment for VAr drawal/return by a regional entitiy except generating stations on its own line emanating directly from an ISGS.

The charge for VArh shall be at the rate of 25 cent INR/kVArh w.e.f. 01.04.2010, and this will be applicable between the regional entity, except generating stations, and the regional pool account for VAr interchanges. This rate shall be escalated at 0.25 cent INR/kVArh per year thereafter, unless otherwise revised by the Commission.

12.3.2 Operational Issues

With increasing penetration of wind power, it is equally important to address concerns of grid operations. In case, information about likely wind power generation forecast is available then, it will facilitate grid operation. Internationally, such information about wind energy generation forecast is available through sophisticated software and extensive data analysis and simulation techniques. In Indian system, it should be made mandatory for all non-firm renewable energy generating sources (RES), especially wind power, shall furnish the tentative day-ahead hourly generation forecast (MWh) for the energy availability at *inter connection point* to the concerned RLDC/SLDC to facilitate better grid co-ordination and management. Also with the provisions to update and correct forecast it by hour-ahead forecast to minimize the real time scheduling error. Further, it has been clarified that above forecasts shall be used for calculating deviation from such scheduled forecasts and must be subjected to unscheduled interchange (UI) mechanism outlined under CERC UI Regulations 2009 [17], but with suitably selected price cap on wind power generation decided in conjunction with fixed price paid for wind power. The argument for putting such ceiling for wind power generators is that cost of energy generation through wind is lower than the ceiling price specified by the CERC and it was considered that there is sufficient incentive for the wind power generator even with a low ceiling price and more important to prevent any gaming. Wind farm owners are in-charge of balancing its own production balance by market-based means or by developing technical capabilities. Unscheduled interchange mechanism is a best mechanism, exercised in India, and can make wind power (or other non-firm renewable energy sources) semi-competitively dispatchable. In this proposed manner, wind farm owners continually get fixed return on wind power accurately dispatched and get paid/charges for UI power. Wind farm owners can optimally schedule their generation slightly lower than actually forecasted wind power to avoid any charges. Sufficient return on wind power will ensure promotion to wind power in longer term and UI mechanism will ensure the competitiveness and technological innovation. As there is huge demand-supply gap prevails in India, frequency remains mostly in lower side of range specified for UI mechanism and hence remunerate much more, for UI injection of power, compare to fixed price received by wind power in next future. Recently CERC introduces complementary commercial mechanisms (discussed in Sect. 1.3.4) to schedule wind and solar power into the Indian grid.

12.3.3 Grid Connectivity and Evacuation Arrangements

Grid connectivity has posed a major challenge in harnessing the renewable energy as most of the renewable energy sources, particularly wind and small hydro sites are in remote areas wherein transmission and distribution network is sparse. As per the provisions of Electricity Act 2003, it is the responsibility of concerned licensee or respective State utility (SU) to provide grid connectivity to the generating stations. However, due to various reasons, there have been difficulties for developing the infrastructure for evacuation of energy generated from renewable energy sources. Further, Electricity Act 2003 under Section 86(1) (e) specifically empowers State electricity regulatory commission (SERC) to take suitable measures for ensuring the grid connectivity to the renewable energy projects or wind farms. However, in most of the cases, responsibility of licensee and wind farm developer in developing the evacuation infrastructure varies across the states.

In most of the states, *inter connection point* stretches up to nearest grid substation and associated cost for development of such evacuation infrastructure is required to be borne by the wind farm developer. *Inter connection point* can be defined as an interface point to the transmission or distribution network, to which the wind turbine or wind farm is to be connected. For wind energy projects, *inter connection point* is to be located and specified by the respective SU. General connectivity conditions elaborated in Regulations 2007 must be held valid for wind farms.

Therefore, it is preferred that evacuation infrastructure from generator terminal up to grid *inter connection point* shall be developed by the wind farm developer and beyond *inter connection point* the concerned licensee shall develop the network. The concerned licensee or SU shall be responsible for providing grid connectivity to the wind farms from the *inter connection point*, on payment of wheeling or transmission charges as the case may be, in accordance with the regulations of the respective SERC.

12.3.4 Complementary Commercial Mechanisms [21]

Wind and solar energy being of variable nature, needs to be predicted with reasonable accuracy for proper scheduling and dispatching of power from these sources in the interconnected system. Hence wind and solar generation forecasting is necessary for increased penetration. Wind power generation forecasting can be done on an individual developer basis or joint basis for an aggregated generation capacity of 10 MW and above connected at a connection point of 33 kV and above. If done jointly, the wind/solar forecasting facility shall be built and operated by wind/solar developers in the area and sharing of the cost shall be mutually agreed. The wind/solar energy forecasting system shall forecast power based on wind flow/solar insolation data at the following time intervals:

- Day ahead forecast: Wind/solar power forecast with an interval of 15 min for the next 24 h for the aggregate wind/solar farms.
- The schedule of wind and solar power generating stations supplying inter-state power under long-term access and medium-term and short-term open access may be revised by giving advance notice to RLDC. Such revisions by wind and solar power generating stations shall be effective from sixth time-block, the first being the time-block in which notice was given. There may be maximum of 8 revisions for each 3 h time slot starting from 00:00 h during the day.

The wind and solar generators shall be responsible for forecasting their generation up to an accuracy of 70%. Therefore, if the actual generation is beyond ±30% of the schedule, wind or solar generator would have to bear the UI charges. For actual generation within ±30% of the schedule, no UI would be payable/receivable by generator, The state which purchases power from the wind generators, shall bear the UI charges for this variation, i.e., within ±30%. Examples for commercial settlement for wind generation are given as: (cases 12.1, 12.2, 12.3, 12.4, 12.5, 12.6)

- For all of the above scenarios, the UI rate shall be the normal UI rate applicable at that particular time block except in case 6 i.e., for over injection beyond 50%. In this case the wind generator shall get UI for injection beyond 50% of the schedule at the UI rate applicable for frequency interval below 50.02 and not below 50.00 Hz, in accordance with the CERC unscheduled interchange charges and related matters Regulations 2009, as amended from time to time.
- The balance in the renewable regulatory fund (RRF) on account of UI by the wind generators in all states, whether surplus or deficit, shall be shared by all the states of India in the ratio of their peak demands in the previous month. The RRF shall be operated by the NLDC on a national basis.
- For all above cases, it is assumed that purchasers drawl is as per schedule.

12.3.5 *Special Dispensation for Scheduling of Wind and Solar Generation*

Scheduling of wind and solar power generation plants would have to be done where the sum of generation capacity of such plants connected at the connection point to the transmission or distribution system is greater than 10 MW and connection point

Case 12.1 Actual as per generation schedule

Schedule (MW)	Actual generation (MW)	Implication on purchaser	UI	
			Implication on host state	Implication on wind generator
100	100	Purchaser to pay wind generator for 100 MW at contracted rate.	No implication on host state.	No implication on wind generator.

Case 12.2 Under injection by the wind generator within 30% variation

Schedule (MW)	Actual generation (MW)	Implication on purchaser	UI	
			Implication on host state	Implication on wind generator
100	70	Payment to be made by purchaser for 70 MW (as per actual) at contracted rate and for 30 MW to Renewable Regulatory Fund (RRF) at UI rate of his region.	For 30 MW UI liability on the host state, as a result of under generation by the wind generator embedded in the state system, the same shall be received by the host state from RRF.	No implication on wind generator.

Case 12.3 Under injection by the wind generator beyond 30% variation

Schedule (MW)	Actual generation (MW)	Implication on purchaser	UI	
			Implication on host state	Implication on wind generator
100	60	To pay for 70 MW to wind generator (since, in this range, the wind generator comes under UI mechanism) at contracted rate. 30 MW by purchaser at UI rate in his region, to RRF	Out of 40 MW liability of UI on host state on account of under generation by wind generator, UI for 30 MW shall be received by the host state from RRF and UI of 10 MW would be received from the UI pool.	UI rate for 10 MW payable by wind generator to UI Pool

Case 12.4 Over injection by the wind generator within 30% variation

Schedule (MW)	Actual generation (MW)	Implication on purchaser	UI	
			Implication on host state	Implication on wind generator
100	130	To pay for 130 MW to wind generator at contracted rate. purchaser shall receive payment for 30 MW from RRF at UI rate of his region.	For 30 MW, UI benefit for the host state on account of overgeneration by wind generator to be passed on to the RRF.	No implication on wind generator.

Case 12.5 Over injection by the wind generator from 130 to 150% generation

Schedule (MW)	Actual generation (MW)	Implication on purchaser	UI	
			Implication on host state	Implication on wind generator
100	140	To pay for 130 MW at contracted rate. Purchaser shall receive payment for 30 MW from RRF at UI rate of his region.	For 40 MW UI benefit for the host state on account of over generation by wind generator, UI for 30 MW to be passed on to the RRF and UI for 10 MW to be passed to UI pool.	UI for 10 MW to be received from UI pool.

Case 12.6 Over injection by the wind generator beyond 150% (as compared to schedule)

Schedule (MW)	Actual generation (MW)	Implication on purchaser	UI	
			Implication on host state	Implication on wind generator
100	160	To pay for 130 MW at contracted rate. Purchaser shall receive payment for 30 MW from RRF at UI rate of his region.	For 60 MW benefit for the host state from UI Pool on account of higher generation by wind, UI for 30 MW to be passed on to RRF and UI for 30 MW to be passed on to UI pool.	UI for 20 MW to be received by wind generator from UI pool at the UI rate applicable at that particular time and for 10 MW UI to be received by wind generator from UI pool at the UI rate applicable for frequency interval below 50.02 and not below 50.00 Hz.

is 33 kV and above, where PPA has not yet been signed. For capacity and voltage level below this, as well as far old wind farms, it could be mutually decided between the wind generator and the transmission or distribution utility, as the case may be. The schedule by wind and solar power generating stations supplying inter-state power under long-term access and medium-term and short-term open access may be revised by giving advance notice to RLDC. Such revisions by wind and solar power generating stations shall be effective from sixth time-block, the first being the time-block in which notice was given. There may be maximum of 8 revisions for each 3 h time slot starting from 00:00 h during the day.

Concerned RLDC and SLDC shall maintain the record of schedule from renewable power generating stations based on type of renewable energy sources i.e., wind or solar from the point of view of grid security. While scheduling generating stations in a region, system operator shall aim at utilizing available wind and solar energy fully.

12.4 Discussion

From the information presented above, it can be observed that the interconnection regulations vary considerably from country to country. It is often difficult to find a general technical justification for the exiting technical regulations that are currently in use worldwide due to the different wind power penetration levels in different countries and operational methodology of power systems. For instance, countries with a weak power system, such as Scotland or Ireland, have considered the impact of wind power on network stability issues, which means that they require fault ride-through capabilities for wind turbines already at a lower wind power penetration level compared with countries those have very robust systems. It is interesting to know that inclusion of FRT regulations for DFIG increase overall cost by 5%.

European wind energy association (EWEA) recommends that European grid connection (or other nations) regulations are to be developed in a more consistent and harmonized manner [12–23]. Harmonized technical requirements will bring the maximum efficiency for all parties and should be employed wherever possible and appropriate. While this applies for all generation technologies, there is a particular urgency in the case of wind power. As wind penetration is forecasted to increase significantly in the short to medium term, it is essential that grid code harmonization process is to be done immediately. It will help the manufacturers to internationalize their products/services, the developers to reduce the cost and the TSO's to share experience, mutually, in operating power systems.

It is also important that national GCR should aim at an overall economically efficient solution i.e., the costly technical requirements such as 'fault ride-through' capability for wind turbines should be included only if they are technically required for reliable and stable power system operation. Hence, it can be summarized that GCR should be harmonized at least in the areas that have little impact on the overall costs of wind turbines. In other areas, GCR should take into account the specific power system robustness, the penetration level and/or the generation technology. Moreover, interconnection standards of different countries may also vary in future.

Several states of India have high penetration of wind power and it is expected that this share will increase with time. Present scenario can be observed that the interconnection regulations vary considerably from state to state. Harmonized technical GCR are required and it will bring the maximum efficiency for all parties and should be employed wherever it is possible and appropriate. While this applies for all generation technologies, there is a particular urgency in the case of wind power. As wind penetration is forecasted to increase significantly in the short-to-medium term, it is essential that grid code harmonization process is to be done immediately. It will help the manufacturers to internationalize their products/services, the developers to reduce the cost and the system operators to share experience, mutually, in operating power systems.

It can be stated that GCR should be harmonized at least in the areas those have little impact on the overall costs of wind turbines. In other areas, GCR should take into account the specific power system robustness, the penetration level and/or the generation technology. The technical harmonization in the following areas is recommended [4]

- Wind power farm behavior in normal network conditions;
- Behaviour during and after network disturbances;
- Frequency response/active power control;
- Voltage control/reactive power;
- Verification and testing;
- Site-related aspects;

Harmonization in GCR will help in achieving following goals:

- For setting of proper regulations for the connection of wind power technology to the electricity grid;
- For facilitating the internationalization of manufacturers and developers; and
- For developing new standards, codes and verification procedures, interaction between GCR issuing working groups.

12.5 Conclusions

The use of renewable energy resources is growing in electric power generation and the emphasis is given to the cost effective utilization of these energy resources. Wind power is one of the most popular renewable energy sources as it provides a number of economic and technical advantages. With increased penetration of wind power generation in the electric power system, the efficient, stable, economical, and secure operation of power systems is becoming a major concern.

Wind farms have a significant influence on the operation of power systems. Grid codes are being set up to specify the relevant requirements for efficient and secure operation of power system for all network users and these specifications have to be met in order to integrate wind turbines into the grid. In this chapter, existing GCR of several countries, which are proactively meeting the challenge of considerable wind power penetration, are analyzed.

GCR's discussed, in this chapter, are active power control, frequency control, voltage control, wind farm protection (fault ride-through control) etc. These interconnection regulations for wind turbines or wind farms tend to add the following requirements:

- to maintain operation of the turbine during a fault in the grid, known as 'fault ride-through' capability;
- to operate the wind turbine in the predefined frequency range;
- to control the active power during frequency variations (active power control);

- to limit the power increase to a certain rate (power ramp rate control);
- to supply or consume reactive power depending on power system requirements (reactive power control);
- to support voltage control by adjusting the reactive power, based on grid measurements (voltage control).

These interconnection requirements can increase the total cost of a wind turbine or wind farm. Hence, interconnection regulations should be enforced for secure and economic operations of power system.

References

1. Rajpurohit BS (2009) Reactive power capability and performance analysis of grid connected unified DFIG for wind power application. Ph.D. Thesis, IIT Kanpur, India
2. Global wind energy Council. Online Available: http://www.gwec.net
3. Ackermann T (2005) Wind power in power system. Wiley, England, pp 115–142
4. Jauch C, Matevosyan J, Ackermann T, Bolik SM (2005) International comparison of requirements for connection of wind turbines to power systems. Wind Energy 8(3):295–306
5. Singh B, Singh SN (2011) Development of grid connection requirements for wind power generators in India. Renew Sustain Eng Rev 15(3):1669–1674
6. Singh B, Singh SN (2009) Wind power interconnection into power system: a review on grid code requirements. The Electricity J 22(5):54–63
7. Winter EW, Dittrich A (2006) Advanced grid requirements for the integration of wind turbines into the German transmission system. IEEE PES General Meeting, Montreal, Canada
8. Lund T, Sorensen P, Eek J (2007) Reactive power capability of a wind turbine with doubly-fed induction generator. Wind Energy 10(4):379–394
9. E.on Netz (2006) Grid code, high and extra high voltage, April 2006. Online Available: http://www.eon-netz.com/
10. AEMC (2007) national electricity rules Version 13. Online Available: http://www.aemc.gov.au
11. Feltes C, Englehardt S, Kretschmann J, Fortmann J, Koch F, Erlich I (2008) High voltage ride through of DFIG based plants. IEEE PES General Meeting, Pittsburgh, USA
12. Ministry of new and renewable energy. Online Available: http://www.mnes.nic.in
13. Singh B, Singh SN, Østergaard J (2009) Renewable energy generation in India: present scenario and future prospects. IEEE PES General Meeting, Calgary, Canada
14. Central electricity regulatory commission (2006) Indian electricity grid code (IEGC) 2006
15. Central electricity authority (2007) Technical standards for connectivity to the grid, Regulations 2007
16. Rajasthan electricity regulatory commission (2008) Rajasthan electricity grid code (REGC) 2008
17. Central electricity regulatory commission (2009) Unscheduled Interchange charges and relate matters, Regulations
18. Central electricity authority(2006) Installation and operation of meters, Regulation 2006
19. Central electricity regulatory commission (2009) Terms and Conditions for Tariff determination from Renewable Energy Sources Regulations
20. Central electricity regulatory commission (2010) Terms and conditions for recognition and issuance of renewable energy certificate for renewable energy generation, Regulations
21. Central electricity regulatory commission (2010) Indian Electricity Grid Code 2010
22. Erlich I, Bachmann U (2005) Grid code requirements concerning connection and operation of wind turbines in Germany. IEEE PES General Meeting, San Francisco, USA
23. EWEA working group on grid code requirements-position paper. European Grid Code requirements for Wind Power Generation. Online Available: http://www.ewea.org

Chapter 13
Application of Pumped Storage to Increase Renewable Energy Penetration in Autonomous Island Systems

Stefanos V. Papaefthymiou, Stavros A. Papathanassiou and Eleni G. Karamanou

Abstract Pumped storage is generally viewed as the most promising technology to increase renewable energy source (RES) penetration levels in power systems and particularly in small autonomous island grids, where technical limitations are imposed by the conventional generating units. In this chapter, an operating policy is proposed for hybrid wind-hydro power stations (HPS) in island grids, in order to increase wind penetration levels, while at the same time minimizing the impact on the conventional generation system and ensuring the viability of the HPS investment. The proposed operating strategy is applied to different autonomous island systems using a dedicated logistic model, in order to evaluate the effect on the overall operation and economics of the island systems and to assess the feasibility of HPS investments. In addition, the real study case of the HPS in Ikaria island, Greece, which is currently in the construction stage and will be one of the first wind-hydro-pumped-storage hybrid stations in the world, is examined and the expected benefits from its operation according to the proposed policy are presented. The material presented in this chapter is based on publications Papaefthimiou et al. (IET Renew Power Gener 3:293–307, 2009, IEEE Trans Sustainable Eng 1:163–172, 2010) available in IET-Renewable Power Generation and IEEE Transaction on Sustainable Energy.

S. V. Papaefthymiou · S. A. Papathanassiou (✉) · E. G. Karamanou
Greece School of Electrical and Computer Engineering, Electric Power Division 9, National Technical University of Athens (NTUA), Iroon Polytechniou st, 15780 Zografou, Athens, Greece
e-mail: st@power.ece.ntua.gr

S. M. Muyeen (ed.), *Wind Energy Conversion Systems*, Green Energy and Technology, DOI: 10.1007/978-1-4471-2201-2_13,

13.1 Introduction

In the Aegean Sea, there exist today more than 50 islands, which are not interconnected with the mainland power system. These island systems isolated or interconnected to each other in groups, present a peak load demand from a few hundred kW up to approximately 700 MW in the case of Crete and are currently powered by oil-fired autonomous stations. High generation costs, dependence on oil products and environmental considerations have been a powerful driver for the increasing exploitation of the renewable energy potential of the islands during the last decades [1–3], wind energy being the most significant. Nevertheless, in spite of the high wind potential existing in the Aegean Sea, wind penetration levels in autonomous island grids are currently limited, as a result of the output power limitations imposed on the installed wind farms [4, 5]. These limitations are related to the technical constraints introduced by the conventional generating units (typically diesel generators) and specifically to their minimum loading limit and dynamic response considerations. Based on the current operating practices, a realistic upper limit for RES energy penetration is estimated around 15–20% of the annual load demand.

The effect of high wind penetration on the operation and control of power systems has been the subject of numerous publications, e.g. [6–8], where the technical, operational and economic implications have been identified, mainly resulting from the intermittency and limited predictability of wind power generation. The introduction of energy storage is considered as the most effective means to remove obstacles in order to significantly increase wind penetration levels in electric power systems, particularly in the case of isolated island grids [9–29]. For power system sizes of a few MW and beyond, conventional battery storage, flywheels and other similar means become unappealing, pumped storage being the most mature and economically viable alternative. Such systems require in concept a pump/turbine station and two water reservoirs at sufficient altitude separation (typically a few hundred meters), as it is schematically illustrated in Fig. 13.1. The concept underlying the introduction of pumped storage is that wind energy which would otherwise be discarded, due to the penetration limits imposed (e.g. during periods of low load and high wind), can be stored by pumping water to the upper reservoir. This energy is subsequently recovered via the hydro turbines, in a controllable manner, permitting thus the substitution of thermal energy and capacity. Notably, the combination of wind and hydro generation, even in the absence of storage facilities, may contribute to the reduction of wind power variability and uncertainty. This renders possible the provision of firm capacity by the combined wind-hydro generating plant and entails significant economic benefits in the electricity market environment [30, 31].

Small isolated systems combining renewable and conventional generation and usually energy storage, as well, are referred to as hybrid systems. Here the term Hybrid Power Station (HPS) is used to describe a combined station, comprising a wind farm and pumped storage facilities, which is owned and operated by a single

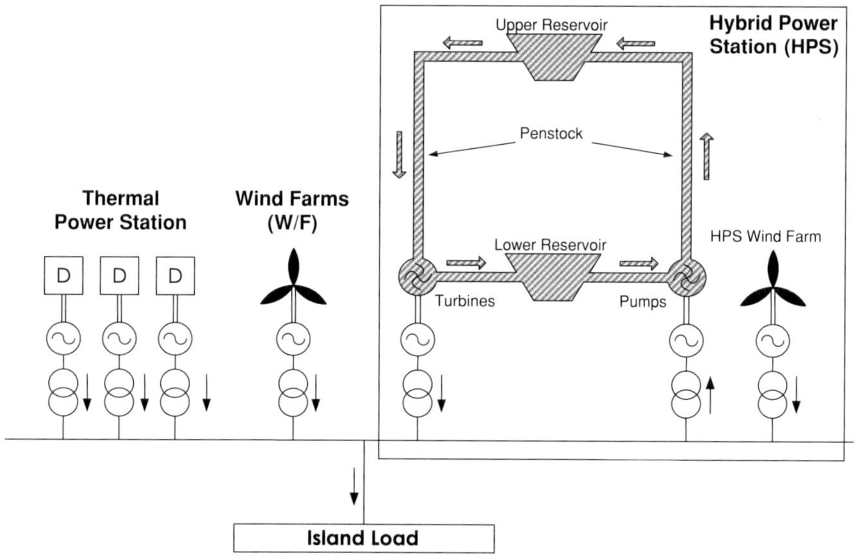

Fig. 13.1 Schematic of a hybrid power station (HPS) in an isolated island system ([32], © [2009] IET)

entity, the HPS producer. This reflects the current legal framework in Greece [33, 34], which sets the landscape for such investments in the existing electricity market environment. Notably, this definition implies that the storage facilities are perceived by the law-maker as the tool for introducing significant additional wind capacity in saturated island systems, rather than as a means for increasing the energy yield of wind farms already existing on the islands.

Over the past years, application of pumped-storage to facilitate wind power integration has been analysed from several perspectives. The impact of wind power penetration in electricity grids from an economic perspective is analysed in [9], where hydraulic storage is presented as the optimum solution to constrain the rising system costs due to high wind penetration levels, by alleviating the need for additional peak-load generation. In [24], the operation of a hybrid wind/hydro power system is examined for the Greek island of Ikaria, concluding that the cost of electricity is considerably reduced when conventional thermal units are replaced by the hybrid system. In [25], the strategy for maximizing the profit of a combined wind-hydro pumped storage plant is investigated, taking into account the Portuguese energy remuneration rules. In [26], the sizing of a hybrid system and the applicable operating policy is investigated for the Canary islands, to optimize exploitation of the available hydraulic and wind potential, with no adverse effect on the reliability of the power system. Application of pumped-storage has been analysed for several autonomous island grids in Greece, adopting a variety of operating policies, which however lack a concrete correlation to a specific regulatory framework [17–24]. In [29], an operating strategy is presented, based on the current legislation in Greece, according to which the plant is required to provide firm power during certain periods of the day, while it may also generate over the remaining hours, if the stored energy suffices. Sizing of the hybrid system components, given a specific operating policy, is another perspective investigated in the literature. In [27, 28], a numerical methodology is presented for the optimum sizing of the various components of an HPS, adopting a simple operating policy, where the pumps consume excess wind energy and the hydro-turbines provide firm power over an agreed time interval every day.

In this chapter, an operating policy for hybrid wind-hydro power stations in island grids is proposed, which ensures satisfactory HPS economics and increase of RES penetration levels, without disturbing the operation of the conventional generation system and the existing wind farms (WFs) of the island. It is directly correlated to the existing legal framework in Greece [33, 34], which is briefly outlined in Sect. 13.3, and has provided the basis for the formulation of a specific regulatory framework on hybrid power stations [35, 36]. The rationale and the details of the operating strategy for the overall generation system of an island including HPS, as well as the internal management of the HPS for maximization of its expected benefit, are described in Sect. 13.4. The proposed policy is applied to three islands of different sizes and characteristics, using a dedicated logistic model with wind and load time series as inputs, to simulate the operation of the system over long time periods (typically one year). The simulation results, presented in Sect. 13.5, demonstrate the impact from the introduction of HPS on the operation of conventional generation and existing wind farms, as well as on the overall system economics. They also provide the basis for assessing the feasibility of HPS investments under the current regulatory framework, as it is presented in Sect. 13.6. In Sect. 13.7, the Ikaria HPS, the first HPS project under construction in Greece and probably worldwide, is described and the expected benefits from its operation according to the proposed policy are presented and analyzed. A brief discussion on regulation and control issues is presented in Sect. 13.8 and the main conclusions from the analysis presented in this chapter are summarized in Sect. 13.9.

13.2 Short Description of the System

A conceptual representation of the simulated autonomous island power systems is presented in Fig. 13.1. The size of the autonomous grids in Greek islands varies from the sub-MW level up to several hundred MWs. In Fig. 13.1, the system is supplied from a conventional thermal power station, which typically comprises several internal combustion engines. Larger island systems, such as those of Crete and Rhodes, include also steam turbines, combined-cycle plants and gas turbines for peaking duty. Heavy oil is the primary fuel, while light diesel is used as well. Several wind farms are already installed on the islands and operated by independent power producers. Currently, installed wind capacities vary up to approximately 30% of the annual peak load, while wind energy penetration levels reach 15% of the annual load demand.

The HPS includes one (or possibly more) wind farms and pumped storage facilities, which are not necessarily installed at the same or adjacent locations. The turbine and pump station typically comprises several individual turbine and pump units. The components of the HPS are directly connected to the grid and include independent energy metering facilities. The lower and upper water reservoirs are located at sufficient altitude separation, to ensure a hydraulic head of several hundred meters (typically more than 300 m), and their capacity depends on the available head and the energy storage specifications of the station.[1] An HPS with double penstock and separate turbine and pumping facilities, as shown in Fig. 13.1, has the capability of simultaneously generating and absorbing power, a characteristic important for its performance, as will be explained in the following sections. On the other hand, the HPS might comprise a single penstock arrangement and reversible pump-turbine units, reducing thus the investment cost but lacking the advantage of simultaneous pump and turbine operation.

The main operating modes of an HPS are the following:

- *Turbine Mode*: Energy stored at the upper water reservoir is recovered via the hydro turbines. HPS pumps and wind turbines are idle.
- *Pumping Mode*: Wind energy produced by the HPS wind farm is stored via pumping action (essentially a *Wind-Pumping Mode*). Hydro turbines are not operating.
- *Grid Pumping Mode*: A variation of the Pumping Mode, where the energy stored is produced by conventional generation, rather than by the HPS wind farm.

The following additional operating modes are also possible:

- *Wind Generation Mode*: The HPS wind farm injects power directly to the grid (i.e. without simultaneous pumping, as in the Pumping Mode). This is permissible only when the wind power absorption margin of the system exceeds the capacity of all wind farms external to the HPS. This mode is of minor significance, as it occurs infrequently in saturated island grids.
- *Turbine-Pump Mode*: A combination of the Turbine and Pumping Modes, where energy is simultaneously drawn by the pumps and produced by the turbines (water is circulated within the HPS storage system). Its significance lies with the fact that, when the station operates in Turbine Mode, available wind energy from the HPS wind turbines can be stored via pumping action. This mode is feasible with double penstock arrangements and constitutes an advantage of this

[1] The storage capacity is often stated as the equivalent operating time (number of hours) at the rated turbine output. In the case of relatively large HPS, rated more than 10 MW, physical limitations and cost considerations reduce the storage capacity to less than one day. Values around 20 h are encountered in many design studies available in practice. In any case, this will depend very much on the topography of the location and the possible utilization of water reservoirs, already existing for irrigation purposes.

configuration, since the Turbine Mode is active for several hours per day (typically 4–8 h).
- *Hydro-Wind Mode*: When the hydro turbines are in operation, providing firm power, part of their output is substituted by wind power generated by the HPS wind farm. The hydro turbines provide the primary reserve to compensate variations of wind power, a concept already explored in the literature [30, 31]. In this way, wind power is provided directly to the load, avoiding the energy losses in the pumped storage facilities. Single-penstock plants can also make use of the available wind power when the hydro turbines are operating (since the Turbine-Pump Mode is not feasible in this case).

13.3 Outline of the Regulatory Framework

A brief description is provided in this section of the regulatory framework recently introduced in Greece for HPS operating in isolated island grids [33, 34]. As already mentioned, pumped storage is viewed as a means to install additional wind capacity in saturated island grids (rather than to increase the efficiency of existing installations, i.e. reduce their energy curtailments). Hence, an HPS comprises storage facilities and one (or more) new RES stations, all connected to the island network. The fundamental operating principle is that the energy generated by the HPS wind farm is stored, rather than directly fed to the load, since the latter would reduce the penetration margin available to other existing or licensed wind power stations. The energy thus stored at the upper water reservoir can be subsequently recovered via turbine action, in a fully dispatchable manner, preferably during the peak load hours of each day, substituting expensive peak units.

Since hydro turbines are fully dispatchable generation, they can substitute not only energy but also installed capacity of conventional (thermal) units. To make this possible, the provision of "guaranteed power" (i.e. firm capacity) needs to be ensured in case of prolonged low-wind periods, in which case the reserves of the HPS storage system are exhausted. For this purpose, a limited amount of "grid-pumping" is allowed, using conventional energy from the grid, which essentially constitutes a peak shaving mode of operation, transferring energy from the peak to the valley of the daily load curve. Obviously, grid pumping is reasonable to be scheduled at low load intervals, such as the late night hours.

Remuneration of an HPS is based both on energy and capacity payments. Energy is metered independently for the wind farm, the hydro turbines and the pumps. Pricing is related to the avoided system costs by the operation of the HPS:

- Wind energy is sold at the standard wind energy tariff for islands (currently 99.44 €/MWh).
- Hydro turbine energy pricing is based on the mean annual variable operating cost of the peak units of the system (defined as the units operating for less than 30% of the time over the year).

- Pump energy pricing is based on the mean annual variable operating cost of the base units of the system (defined as the units operating for more than 70% of the time over the year).

Variable operating costs include fuel, operation and maintenance (O&M) expenses and the cost for CO_2 emissions rights. All tariffs are fixed over the year and are the same for all HPS operating on an island. Notably, "net metering" is allowed for wind farm and pump energy, in which case the HPS is billed for the net energy produced or consumed, after deducting the energy drawn by the pumps from that generated by the HPS wind farm. Obviously, net-metering would be chosen by the HPS producer when the cost for pumping (i.e. the variable cost of the base thermal units) exceeds the wind energy tariff, which is the case in most islands today.

Since an HPS can provide firm capacity to the island system, capacity payments are also foreseen based on the declared "guaranteed power" of the HPS, which is typically the rated capacity of its hydro turbines. Remuneration for capacity credit corresponds to the annualized capacity cost of an equivalent conventional power station (including the annual capital amortization expenses and its fixed operating cost).

13.4 Operating Policy for Autonomous Island Systems with HPS

13.4.1 Discussion of Alternative Policies

The regulatory framework outlined in the previous section is quite general and does not stipulate any specific operating policy for the HPS. Space limitation reasons do not permit an extended presentation of alternative policies modelled and simulated [35]. Instead, a general discussion is first made on the principles and available options, leading to the proposed operating policy, presented in the following section.

A simple operating policy for HPS, already evaluated in relevant studies (e.g. [26–28]), is the "peak-shaving" mode, which consists in scheduling operation of the hydro turbines at a fixed power (usually the rated) during the peak load hours, on a daily basis. The energy required for this purpose is stored by pumping wind power, but also via a large extent of grid pumping during the low-load hours. This is a fundamentally simple concept, as it creates a standard and dependable income from the daily disposal of hydro energy, while grid pumping may also be profitable when the gap between turbine and pump energy tariffs is large (typically in excess of 1.5:1, as often happens in island systems).

This policy presents serious drawbacks, particularly when dealing with large HPS. First of all, excessive grid pumping results in a waste of conventional energy and fuel. For this reason it should be allowed only in periods of high system load (e.g. in the summertime tourist season), when the provision of guaranteed capacity from the HPS is indeed necessary, rather than on an everyday basis. This reduces dramatically the extent of grid pumping, from around 30% to less than 5% of the total energy absorbed by the pumps on an annual basis.

A second drawback of this simple policy is that dispatch of the hydro turbines at a fixed output power reduces the penetration margin for wind farms external to the HPS. This happens because the hydro turbines substitute conventional thermal units, which provide flexibility in their loading and the required spinning reserve for the integration of wind power to the system. This drawback is quite essential, because it conflicts with the contractual provisions of existing wind power stations. To effectively resolve this issue, dispatched hydro turbines should operate in exactly the same manner as the substituted thermal units, i.e. they should provide primary reserve and balancing power for the integration of wind power, rather than operate at a fixed output power. Further, they must provide ancillary services, such as frequency and voltage regulation, under the same regime as the conventional units.

The operating policy to be adopted should also exhibit flexibility in the dispatch and operation of all units and fit in the framework of a daily market operation, where applicable. Two families of operating policies can be envisaged:

- Policies in which the HPS producer submits hourly power bids for the hydro turbines and load declarations for the pumps, when grid pumping is foreseen, based on the internal optimization of the HPS operation, as assumed in [29]. Dispatch of the turbines is performed according to the schedule submitted by the producer, since the hydro turbines have precedence over the thermal units, on the grounds that they utilize clean wind energy previously stored.
- Policies in which the HPS producer submits a daily energy offer, as well as a daily load declaration for grid pumping, whenever applicable. The Island System Operator (ISO) has to respect the submitted daily energy offer, but the hourly dispatch schedule is determined so as to optimize the operation of the overall system. Similarly apply for the pump energy, when a load declaration has been submitted.

Evaluation of the two families of operating policies has shown that the hourly power bid strategy is more profitable for the HPS, particularly in the case of relatively small water reservoirs, because it permits the HPS to fully optimize its intra-day operation, even when employing a moderately accurate wind power forecasting. With this strategy, the HPS turbines are practically dispatched at full capacity as soon as the water reservoir is full, freeing up storage capacity for subsequent pumping of wind energy. Thus the amount of wind energy discarded is minimized.

However, when dealing with large HPS, this strategy has a very adverse effect on the conventional generation, which has to adapt to the operating schedule of the hydro turbines. For illustration purposes, a characteristic example is shown in Fig. 13.2a, from the simulation of the operation of the power system of Crete with a 250 MW HPS, during one week in the year 2012. Apparently, the operation of the thermal units is entirely disrupted, due to the high frequency of

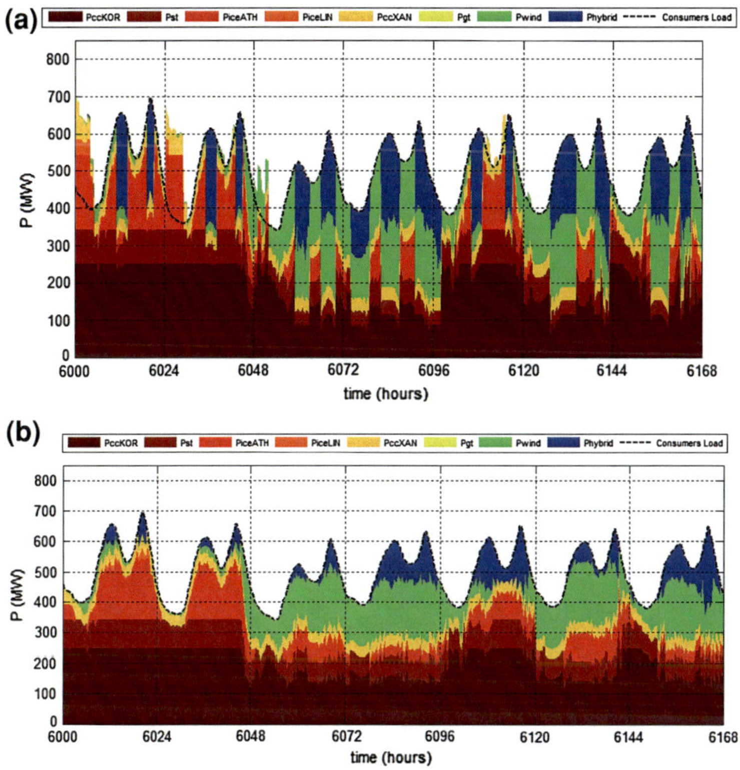

Fig. 13.2 Simulation of the power system of Crete in year 2012 (one week of operation), assuming a 250 MW HPS and two operating policies. **a** The HPS submits hourly power bids. **b** The HPS submits a daily energy offer. Brown/red/yellow hues used for thermal units, green for wind farms external to the HPS and blue for the HPS hydro turbines, ([32], © [2009] IET)

the start–stop operations and the high output power ramp rates. Therefore, adoption of this policy drastically limits the size of the HPS that can be integrated in the island systems.

The daily energy offer policy, on the other hand, achieves a far superior quality of operation for the generation system of the island, as is evident from Fig. 13.2b, where the HPS effectively shaves the peaks of the load curve, permitting a more efficient operation of the conventional units. The only drawback of this policy is that the required reservoir capacity of the HPS is relatively increased (indicatively by 25–30% [35]), for achieving the same results as the hourly power bid strategy. The reason is that the hydro turbines are dispatched by the Island System Operator only at peak load hours, i.e. at noon and in the evening. Hence, all wind energy generated during the night and morning hours has to be stored, until the turbines are dispatched. Nevertheless, this drawback is fully offset by the great flexibility offered by this policy, which improves essentially the operation of the conventional generation system and permits the integration of large HPS to the island grid.

13.4.2 Proposed Operating Policy for the Island System

Based on the discussion of the previous section, a specific policy can now be formulated for an island system with HPS, based on a daily generation dispatch schedule, determined by the ISO at the end of each day, for the next 24 h interval. More specifically, the operating policy consists of the following steps:

Step 1: Guaranteed power and energy request from the ISO

The ISO demands guaranteed power and energy, P_G and E_G, from the HPS for the next dispatch day (24 h). This occurs only at high load periods (rather than on a daily basis), to cover deficits in conventional generation capacity. The guaranteed power requested cannot exceed the declared "guaranteed power" of the HPS.

Step 2: HPS energy offer

The HPS submits an energy offer (E_O) for the next 24 h, based on the hydraulic energy available for electricity generation. Possibly, wind power forecasts for the HPS WF may also be included in the energy offer, applying suitable safety factors. The daily energy offer cannot be less than the guaranteed energy, E_G, demanded by the ISO.

Step 3: HPS load declaration

If the guaranteed energy, E_G, requested by the ISO exceeds the potential energy production, E_H, of the HPS using the water already available in the upper reservoir, then grid pumping is allowed for the HPS to secure the provision of the requested energy, E_G. In this case, the HPS will submit a load declaration[2] for the energy, E_D, to be drawn from the grid during the next 24 h. To avoid excessive grid pumping, E_D is limited to the amount of energy required to complement E_H up to E_G. Hence:

$$E_D = \max\left\{0, \frac{E_G - E_H}{n_{P-H}}\right\} \tag{13.1}$$

where n_{P-H} is the overall round-trip efficiency of the pumped storage facilities. A load declaration cannot be submitted if the ISO has not requested the provision of guaranteed energy ($E_G = 0 \Rightarrow E_D = 0$).

[2] The term "load declaration" signifies the declared intention of the HPS producer to dispatch pumps in order to absorb energy from the grid (rather than its own wind production) during the next day. This additional pumping load will increase the total demand of the system.

Step 4: Dispatch of HPS turbines and pumps by the ISO

The ISO determines the dispatch of the HPS hydro turbines over the next 24 h, in order to absorb the energy offered by the HPS. Scheduling is performed by the ISO, typically at peak load hours, to optimize the operation of the overall system, rather than at specific intervals selected by the HPS producer. Similarly, the ISO schedules the dispatch of the pump units, according to the load declaration submitted by the HPS, in the valley of the load curve (night hours).

Step 5: Dispatch of APS conventional units by the ISO

The ISO proceeds to the dispatch of conventional units, to meet the forecasted load demand. If $P_{L,i}$ is the demand for hour i, $P_{PV,i}$ the forecasted PV generation,[3] $P_{E,i}$ the scheduled generation of the HPS turbines and $P_{D,i}$ the pump load, then the required conventional capacity $P_{Cn,i}$ is given by:

$$P_{Cn,i} \geq (1 + \varepsilon) \cdot [(P_{L,i} - P_{PV,i}) + P_{D,i} - P_{E,i}] \tag{13.2}$$

where ε reflects the spinning reserve policy followed by the power station operators (e.g. $\varepsilon = 10\%$).

Step 6: Output power constraints for WFs external to the HPS

Given the load demand and the dispatch schedule for conventional and hydro generators, output power limitations can then be calculated for wind farms external to the HPS. For this purpose, the standard practice is applied [4], to ensure that the minimum loading restriction of thermal units is observed and an "instantaneous" wind power penetration upper limit is not exceeded:

$$P_{W,i} \leq P^{T}_{Wmax,i} = (P_{L,i} - P_{PV,i}) + P_{D,i} - \sum_{j \in A_{C,i}} c_{Tj} P_{Cnj} - c_{HT,e} \sum_{j \in A_{H,i}} P_{Hnj} \tag{13.3}$$

$$P_{W,i} \leq P^{D}_{W\,max,i} = \sum_{j \in A_{C,i}} c_{Dj} P_{Cnj} + c_{HD} \sum_{j \in A_{H,i}} P_{Hnj} \tag{13.4}$$

$$P_{W\,max,i} = \min\left\{ P^{T}_{W\,max,i}, P^{D}_{W\,max,i} \right\} \tag{13.5}$$

where (index i denotes in all cases the hour i of the next day):

$P^{T}_{W\,max,i}$ wind power restriction due to the minimum loading constraint of conventional units

$P^{D}_{W\,max,i}$ wind power restriction due to dynamic response constraints of conventional units

$P_{Wmax,i}$ overall wind power penetration limit for hour i

$P_{W,i}$ output of all WFs external to the HPS

---

[3] Power limitations are not imposed to PV stations. PVs are practically embedded in the total load of the island and therefore forecasting of the net demand P_L-P_{PV} may also be performed, [37, 38].

P_{Cnj} rated power of conventional unit j
P_{Hnj} rated power of hydro turbine j
c_{Tj} minimum loading of conventional unit j (% of rated)
c_{Dj} dynamic reserve of conventional unit j (% of rated)
$c_{HT,e}$ equivalent minimum loading of hydro turbines (% of rated), selected to match the technical minimum of the substituted conventional units
c_{HD} dynamic reserve of hydro turbines (% of rated)
$A_{C,i}$,$A_{H,i}$ set of conventional units and hydro turbines dispatched for hour i

Equation 13.3 signifies that hydro turbines operate in the same mode as the substituted conventional units, i.e. providing reserve for wind power absorption. Hence, hydro turbines may be dispatched for their rated power, but they may operate at reduced output to accommodate wind power from WFs external to the HPS. Equation 13.4 ensures that wind power injected to the system will not exceed the dynamic reserve capabilities of the regulating units (conventional and hydro units[4]).

The overall constraint, $P_{Wmax,i}$, of (13.5) is then allocated to the various WFs in proportion to their installed capacity. If the penetration limit exceeds the installed WF capacity, the residual (excess) margin is made available to the HPS WF, in order to inject power directly to the load. WF set-point calculations may be performed during operation of the system, i.e. in real time, rather than in the previous day.

13.4.3 HPS Internal Management Decisions

After determination by ISO of the 24 h generation and pump load (if grid pumping is needed) schedule for HPS, the producer will proceed to the dispatch of turbine and pump units according to this schedule and will also decide the exploitation mode of the available wind power in order to maximize the expected benefits, as explained hereafter.

Regarding the exploitation of the available wind power from the HPS WF, in principle there exist three possibilities:

Mode 1 Available wind power is stored via pumping
Mode 2 Direct supply of wind power to the load when a set-point (penetration margin) is allocated to the HPS WF

[4] The dynamic response of hydro turbines is a serious concern, both for the integration of wind power external to the HPS, as well as for the hydro-wind mode of operation discussed in Sect. 13.2. Generally, the speed of response of hydro turbines and the level of fast primary reserves they can afford depends on the hydraulic configuration of the hydroelectric plant and it may be faster or slower than the response of conventional units. In present analysis, it has been assumed that hydro turbines respond fast enough to provide a similar level of primary reserve as the conventional thermal units they substitute.

Mode 3 Direct supply of wind power to the load, substituting scheduled hydro generation. In this mode, hydro turbines and the HPS WF operate as a combined "virtual" power plant, generating the scheduled HPS output. This mode is subject to additional control requirements and it also presupposes a rapid dynamic response of the hydro turbines to compensate fast variations of wind power

Selection of wind power exploitation mode depends on the applicable energy tariffs. If the tariff for hydroelectric energy is higher than the fixed wind energy tariff, divided by the round-trip efficiency (n_{P-H}) of the storage facilities, which is often the case in island systems, directing all available wind power to the storage facilities is preferred. Otherwise, supplying wind power directly to the load (Mode 3) would be more profitable. The examined island systems in the subsequent sections are saturated and practically all wind penetration margin is dispensed to WFs external to the HPS, hence Mode 2 operation is rather infrequent. Mode 3 is also not favored, being selected only when the available wind power would otherwise be discarded (e.g. due to insufficient pumping capacity or the water reservoir being full or during hydro-turbine operation for single-penstock stations).

A basic idea underlying the proposed operating policy is that, during operation, the HPS will be neutral to the overall power equilibrium of the system, i.e. the net HPS generation or load will track the scheduled values.[5] To maintain this equilibrium in Mode 1 operation, wind and pump powers need to be balanced. For this purpose, short-term wind power forecasting is employed (up to 1 h ahead, using the persistence method), to schedule the dispatch of pump units. Wind power variations are tracked by the pumping station, using the regulation margin provided by variable speed pumps (~40–50% of their rated capacity). With fixed-speed pumps, this is possible to a certain extent by switching in and out the appropriate number of pump units. Large wind power increases, exceeding the capacity of operating pumps, are dealt with by curtailing wind power. Large unforeseen reductions are dealt with by disconnecting pump units (should the regulation margin provided by the variable speed pumps in operation not suffice).

In case of Mode 3 operation, hydro turbines should compensate for wind power variations, so that the net power delivered to the system follows the scheduled generation level. For this purpose, hydro turbines need to participate in the primary frequency regulation or their output power should be controlled using direct measurements of the HPS WF output power. When hydro turbines reach their technical minimum, wind power will be curtailed.

---

[5] Mode 2 operation is the only exception, since wind power from the HPS WF is directly injected into the system, in excess of the HPS generation schedule, exploiting the capability of the system to absorb additional wind power.

13.5 Evaluation of the Proposed Policy

13.5.1 Modelling of the System

To evaluate the proposed operating strategy, a time-series based simulation model is used, which reflects the power equilibrium in the system, ignoring all dynamic phenomena. Hence, long time simulations (typically annual) are possible. The simulation model structure is outlined in the flowchart of Fig. 13.13 and it reproduces as faithfully as possible the operating strategy presented in the previous Section.

Load demand is represented by time series, expressed as hourly average values and interpolated to the simulation time-step (typically 20 min). Conventional generating units are described by their basic characteristics (maximum and minimum output power, specific fuel consumption as a function of their output power, start–stop and operating time restrictions, CO_2 emissions etc.).

Wind farms are represented by the power curves of individual wind turbines. Inputs to the model are the wind speed time series, referred to the hub height of the turbines, which are translated into output power time series after deducting the various losses (here considered equal to 10%).

For PV stations, the annual time series of meteorological data (insolation, temperature etc.) are used to derive time series of PV output power (hourly values), using the model of [39].

Pump and hydro turbine units can be simulated by simplified models, characterized by their average efficiencies. The capacity of the water reservoirs is then expressed in energy units (MWh). This approach is adopted for the modelling of HPS in this section.

On the other hand, in Sect. 13.7, dealing with the real study case of the Ikaria HPS, more detailed hydraulic models are employed, which take into account the efficiency curves of hydro turbines and pumps as a function of their operating point and the variations of the net hydraulic head (see Appendix 1, Tables 13.1, 13.2). Given the electrical power, the water flow rates of a turbine or pump unit are calculated from:

$$P_H = \rho \cdot g \cdot Q_H \cdot (H_{0,H} - K_H \cdot Q_H^2) \cdot n_H \cdot n_G \cdot n_{Htr} \tag{13.6}$$

$$P_P = \frac{\rho \cdot g \cdot Q_P \cdot (H_{0,P} + K_P \cdot Q_P^2)}{n_P \cdot n_M \cdot n_{Ptr}} \tag{13.7}$$

where:

P_H, P_P electric power of hydro turbines and pumps
Q_H, Q_P discharge rates of hydro turbines and pumps
$H_{0,H}$, $H_{0,P}$ hydraulic heads of hydro turbines and pumps (dependent on altitude separation and reservoir water levels)
K_H, K_P friction coefficients of turbine penstock and pump discharge pipes
n_H, n_P hydro turbine and pump unit efficiencies
n_G, n_M electrical generator and pump motor drive efficiencies

Table 13.1 Basic characteristics of the study case island systems, ([32], © [2009] IET)

	Large island	Medium island	Small island
Island system data			
Max load	820 MW	78 MW	11.9 MW
Min load	200 MW	14.4 MW	1.6 MW
Load factor	53.3%	51.0%	42.7%
Installed wind capacity	220 MW	20 MW	2.45 MW
Mean annual wind speed	8.5 m/s	8.5 m/s	11 m/s
HPS data			
Turbines	4 × 25 MW	4 × 5 MW	2 × 1 MW
Pumps	5 × 25 MW	9 × 2.5 MW	3 × 0.75 MW
Wind farm	130 MW	25 MW	2.4 MW
Reservoir(*)	18 h	18 h	18 h
Energy losses	HPS: ~ 30%, WF: 10%		
Penstock	Single/Double		
	Pricing		
Turbine energy	137 €/MWh	166 €/MWh	183 €/MWh
Pump energy	81 €/MWh	100 €/MWh	146 €/MWh
Wind energy	87.4 €/MWh	87.4 €/MWh	87.4 €/MWh
Capacity credit	103 €/kW year	186 €/kW year	186 €/kW year

*Capacity given in equivalent turbine operating time at rated power

Table 13.2 Maximum capacity credit and HPS component sizing restrictions for each island, ([32], © [2009] IET)

		Large island	Medium island	Small island
Max load	MW	820	78	11.9
Min load	MW	200	14.4	1.0
Load factor	%	53.3	51.0	42.7
Min conventional capacity ($P_{conv,min}$)	MW	652.3	59.9	8.4
Max capacity credit ($P_{CC,max}$)	MW	167.7	18.2	3.5
Required pump rating ($P_{Pump,min}$)	MW	223.2	25.8	3.2
Min reservoir capacity	h	6.5	6.3	4.4

n_{Htr}, n_{Ptr} hydro turbine and pump transformer efficiencies
p water density (1,000 kg/m^3) and
g gravity acceleration (9.81 m/s^2)

In the case of constant speed pumps, from their technical specifications the relation is known between input power and resulting water discharge rate, given the total head, including pipe losses (Appendix 1, Table 13.6). In the simulation, the water levels of the reservoirs are known (and hence the head $H_{0,P}$), but the resulting flow rate cannot be directly calculated, because the head losses are needed (term $K_P \cdot Q_P^2$). For this purpose, an iterative solution is implemented in the model, employing (13.7) and the pump characteristics.

13.5.2 Case Studies

The proposed operating strategy is applied to three different Greek islands, representative of the small, medium and large island cases. The basic characteristics of the three island systems and HPS data used in the simulations are presented in Table 13.1. The HPS energy and capacity tariffs calculated for the three island systems [35], are also included in the table.[6] The large island has an extended generation system, including many different types of units (steam turbines, combined cycle plants, internal combustion engines and gas turbines, using heavy oil and diesel fuel oil). The medium-size island includes heavy oil-fired diesel engines, as well as gas turbines for peaking service. The small island is powered by internal combustion engines, using diesel fuel oil. In all cases, HPS with single and double penstock arrangements have been simulated. The configuration of the HPS in each case has been selected by a sizing analysis [35], such as presented in [27–29].

Indicative simulation results are illustrated in Fig. 13.3 for the medium-size island. In diagram 13.3a, the output of the conventional units, the HPS and the wind farms is shown, summing up to the total load demand. The effective smoothing of the load curve is apparent (basically peak shaving, but also valley filling, during the third night, where grid pumping takes place). In the other three diagrams, the operation of the HPS is presented in more detail. In diagram 13.3b, the output of the hydro turbines is shown, operating when scheduled by the ISO. Hydro-Wind mode of operation does not occur, as a result of the applicable tariffs, as discussed in Sect. 13.4.3. The operation of the HPS pumps is presented in diagram 13.3c. While their energy is mainly provided by the HPS wind turbines, an occasion for grid pumping occurs at the beginning of the 3rd day, when wind generation is not enough to ensure availability of guaranteed power and energy, requested by the ISO. Simultaneous hydro generation and pumping (Turbine-Pump mode) is possible for the double penstock HPS simulated here. The fourth diagram illustrates the variation of the HPS reservoir capacity over the simulated interval.

The main purpose for introducing HPS in island grids is to increase wind energy penetration and therefore reduce the use of conventional fuel. This is illustrated in Fig. 13.4, where the energy annually produced by the conventional units is shown for the three island systems, with and without an HPS. The achieved reduction depends on the size of the HPS. The significant reduction after installation of the HPS indicates a corresponding increase in RES penetration, which is substantial

---

[6] Calculation of energy and capacity prices was performed via simulation of each island system without an HPS over a year. Actual or realistic data was employed for each conventional generating unit, including specific fuel consumptions, GHG emission coefficients, operation and maintenance variable costs, costs for CO_2 emission rights etc. From the simulation, the peak and base-load generating units are identified (defined as mentioned in Sect. 13.3) and the respective mean annual variable operating costs are calculated to determine the tariffs for turbine and pump energy.

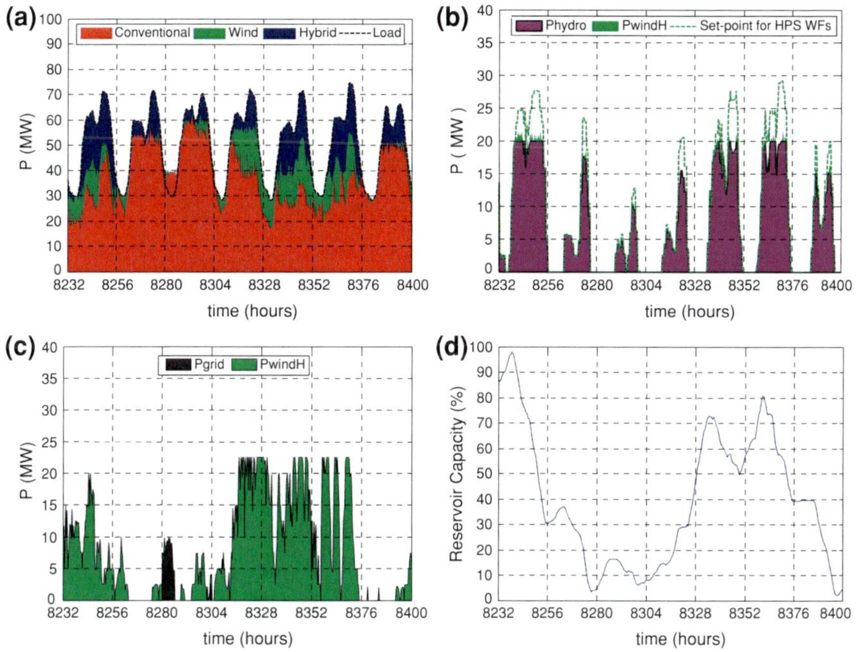

Fig. 13.3 Medium-size island system with an HPS. Operation during a typical week. **a** Output power of conventional units, hydro turbines and wind farms. **b** Output of HPS hydro turbines and wind farm. **c** HPS pump load and origin of pump energy. **d** Level of the HPS water reservoir, ([32], © [2009] IET)

comparable to the wind energy penetration achieved by all other wind farms on the island. The double penstock arrangement is much more efficient, since wind energy available when the hydro turbines are in operation can be better exploited (Turbine-Pump Mode).

The effect from the operation of an HPS on the overall generation cost of the system, although not dramatic, is still noticeable, as the pricing of the HPS energy does not reflect exactly the avoided conventional generation cost. This is shown in Fig. 13.5, where the annual operating cost includes fuel, O&M and CO_2 emission costs of thermal units, energy payments to the existing wind farms and the HPS energy and capacity remuneration. Ignoring the HPS capacity payments (Fig. 13.5a), the increase in the overall cost of each system is rather small. Inclusion of the HPS capacity payments (as in Fig. 13.5b) creates a more significant effect on system cost. Nevertheless, this may be misleading, because the firm capacity provided by the HPS permits deferral of conventional generation investment, whose cost is comparable to the annual capacity payments to the HPS.

Figure 13.6 confirms that the integration of an HPS, which operates according to the principles described, does not affect considerably the efficiency of wind

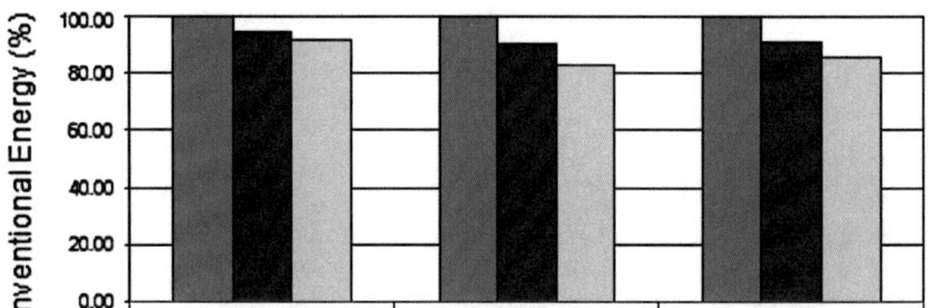

	Large (100 MW HPS) (E100%=3217 GWh)	Medium (20 MW HPS) (E100%=96 GWh)	Small (2 MW HPS) (E100%=37 GWh)
No HPS	100.00	100.00	100.00
Single Penstock	94.38	90.35	91.02
Double Penstock	91.37	82.73	85.16

Fig. 13.4 Conventional energy savings due to the operation of the HPS. Annual energy production of conventional units for the three study case islands

power stations external to the HPS. Any changes in the capacity factor are related to differences in the technical characteristics of HPS hydro turbines and the substituted conventional units.

In Fig. 13.7 the efficiency of the HPS wind farm is examined, to ensure that the proposed policy results in the expected utilization of the installed wind capacity. It is observed that the wind farm of the double-penstock HPS is as efficient as the other wind farms on the island (Fig. 13.6). However, for the single-penstock configuration a large amount of wind energy is discarded when the hydro turbines are in operation, resulting in a capacity factor reduced by approximately 1/3. Notably, the energy eventually delivered to the load is further reduced, by the amount of losses in the pumped storage station (approximately 30%).

Regarding the extent of grid pumping, i.e. the conventional energy stored and then returned to the system, this is negligible in all cases (less than 2% of the total energy stored by pumping), especially in the case of double-penstock arrangements, where the utilization of available wind energy is optimized.

As a rule of thumb, the simulation results indicate that, for double penstock arrangements, the HPS wind farm will operate with approximately the same capacity factor as the wind farms external to the HPS.[7] Since the generated wind energy will be circulated in the pumped storage facilities, the net energy yield of the HPS is approximately 2/3 of the energy generated by the HPS wind farm. The net income of the HPS is then equal to this energy times the tariff applicable for the hydro turbine output energy (assuming that net metering is adopted for the wind farm and pump energy), plus the income due to guaranteed capacity payments.

[7] For single-penstock arrangements the wind farm capacity factor can be reduced by approximately 1/3.

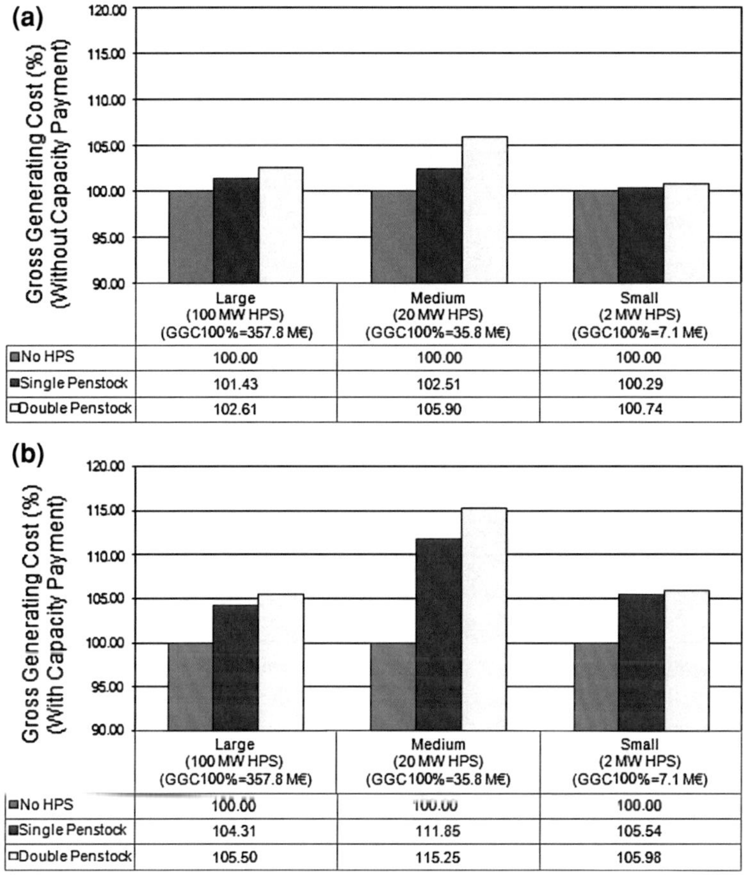

(a)

	Large (100 MW HPS) (GGC100%=357.8 M€)	Medium (20 MW HPS) (GGC100%=35.8 M€)	Small (2 MW HPS) (GGC100%=7.1 M€)
No HPS	100.00	100.00	100.00
Single Penstock	101.43	102.51	100.29
Double Penstock	102.61	105.90	100.74

(b)

	Large (100 MW HPS) (GGC100%=357.8 M€)	Medium (20 MW HPS) (GGC100%=35.8 M€)	Small (2 MW HPS) (GGC100%=7.1 M€)
No HPS	100.00	100.00	100.00
Single Penstock	104.31	111.85	105.54
Double Penstock	105.50	115.25	105.98

Fig. 13.5 Annual operating cost of the three island systems (variable cost of conventional units, including CO_2 emission rights and payments to wind farms and to the HPS). **a** Only energy payments to the HPS. **b** Energy and capacity payments to the HPS

13.6 HPS Investment Evaluation

13.6.1 Capacity Credit Calculation

The capacity payments to an HPS are based on its "guaranteed power", i.e. the firm capacity provided to the system via the hydro turbines, which permits deferring the installation of equivalent conventional generation. Guaranteed peak capacity is creditable only when the associated peak energy is also guaranteed, regardless of the prevailing wind conditions. For this reason grid pumping is allowed, effectively shifting load energy from the peak to the valley of the load curve, reducing the day-time peak, but increasing at the same time the demand during the night hours. This is illustrated in Fig. 13.8a, for a typical daily load

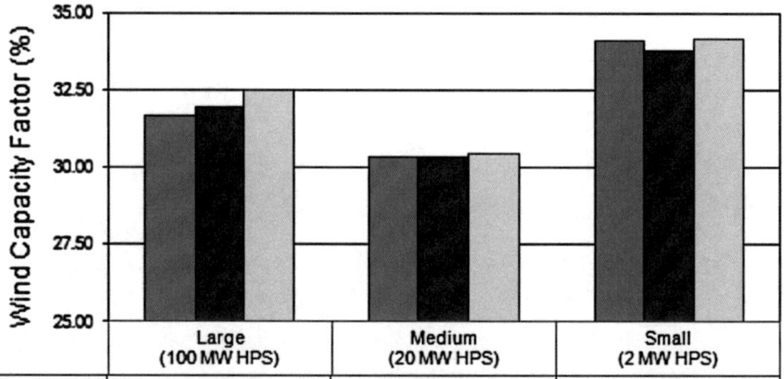

	Large (100 MW HPS)	Medium (20 MW HPS)	Small (2 MW HPS)
No HPS	31.67	30.29	34.08
Single Penstock	31.95	30.31	33.75
Double Penstock	32.47	30.45	34.15

Fig. 13.6 HPS effect on the capacity factor of existing wind farms on the three island cases

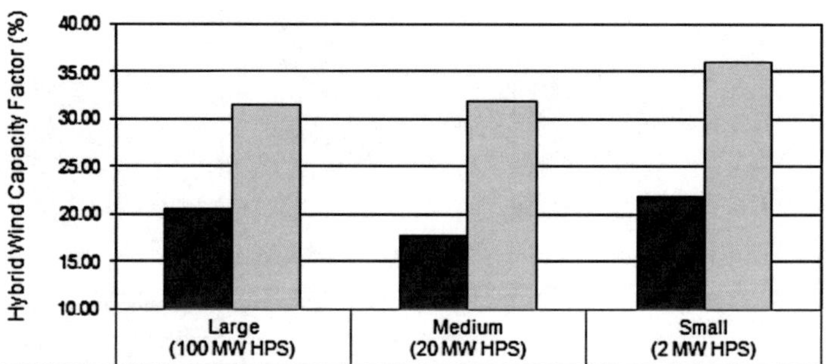

	Large (100 MW HPS)	Medium (20 MW HPS)	Small (2 MW HPS)
Single Penstock	20.57	17.73	21.83
Double Penstock	31.55	31.90	35.94

Fig. 13.7 Capacity factor of the HPS wind farm in each of the three island cases

curve. The presence of the HPS transforms the load curve, reducing the required conventional generation capacity by $P_{conv,0} - P_{conv}$, i.e. by an amount equal to the maximum power P_H delivered by the hydro turbines (typically their rated capacity). Therefore, the capacity credit attributable to the HPS should be equal to the rated capacity of its hydro turbines.

However, this rationale is not applicable when the size of the HPS becomes large compared to the size of the system. In Fig. 13.8b the transformation of the daily load curve is shown for increasing HPS sizes. For the largest HPS (curve 3) it is observed that the curve of the resulting net load now presents a new peak during night-time. Hence, although the day-time peak is indeed reduced by an amount equal to the rated capacity of the HPS turbines, P_{H3}, the capacity credit of this HPS

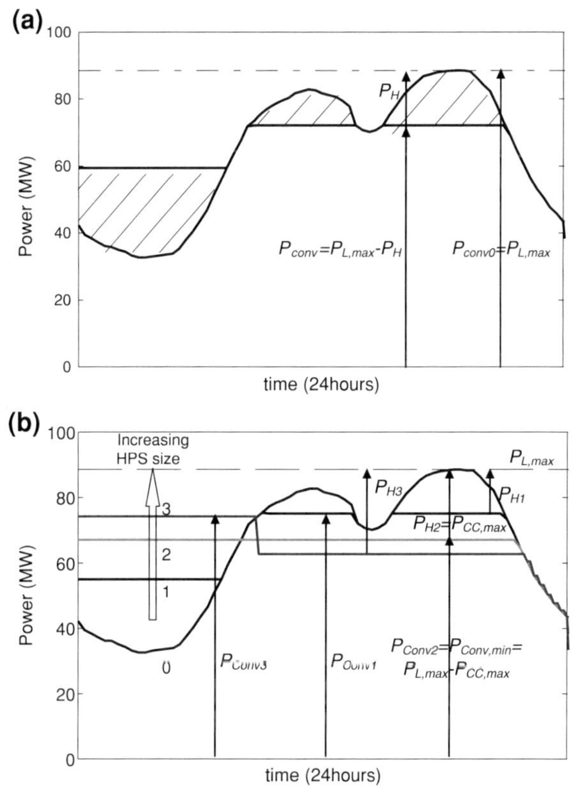

Fig. 13.8 **a** Daily load curve before and after introduction of the HPS: peak-shaving and equivalent valley-filling under no wind conditions (delivered peak energy stored via grid pumping). **b** Transformation of the daily load curve for increasing HPS capacities (0: no HPS, 3: max HPS size) under no wind conditions, ([32], © [2009] IET)

is lower than P_{H3}, because additional thermal capacity is necessary to meet the new night-time peak.

From Fig. 13.8b it is observed that the minimum conventional generation requirements, $P_{conv,min}$, ideally occur when the load curve becomes flat (curve 2 in the diagram). This determines the maximum capacity credit attributable to a HPS, regardless of its size. Notably, in these calculations the no wind (worse case) scenario is adopted, i.e. it is assumed that all energy supplied in the peak hours is pumped during the previous night hours. Energy losses in the pumped storage facilities are taken into account.

Based on this reasoning, it is clear that large HPS do not necessarily receive capacity payments for their total installed turbine capacity. Rather, there exists a capacity credit limit which is characteristic for the island (being determined by the shape of the load curve) and needs to be calculated on an annual basis. For this purpose, the daily calculation outlined in Fig. 13.8b needs to be repeated for all the 365 days of a year. Then, the minimum conventional generation requirements on an annual basis, $P^{A}_{con,min}$, are found from the respective minimum daily requirements, $P^{j}_{conv,min}$:

Table 13.3 Basic assumptions for the economic evaluation, ([32], © [2009] IET)

		Large island	Medium island	Small island
Capital investment	€/kW	3,000	4,000	5,000
Annual income	M€/MW	0.48	0.59	0.68
Net annual energy yield	MWh/MW	2,553	2,518	2,690
O&M cost (% investment)	%		2%	
Discount rate	%		7.4%	
Economic life	Years		20	
Tax rate	%		25%	
Equity	%		30%	
Subsidy	%		25%	
Loan	%		45%	
Loan interest rate	%		6%	
Loan duration	Years		15	

$$P^{\mathrm{A}}_{\mathrm{conv,min}} = \max_{\mathrm{j}} P^{\mathrm{j}}_{\mathrm{conv,min}} \tag{13.8}$$

In the absence of an HPS, the conventional generation requirements of the system are conceptually determined by the peak annual load demand, $P_{\mathrm{L,peak}}$. Hence the maximum capacity credit, $P_{\mathrm{CC,max}}$, attributable to any HPS on this island is

$$P_{\mathrm{CC,max}} = P_{\mathrm{L,max}} - P^{\mathrm{A}}_{\mathrm{conv,min}} \tag{13.9}$$

It is interesting to note that the line of thought followed for determining $P_{\mathrm{CC,max}}$ also generates restrictions on the sizing of the other components of the HPS, if the latter is to be eligible for full compensation. First of all, the energy provided during the peak hours needs to be previously stored in the upper water reservoir, which must have the required capacity. Then, pumping of this energy during the low-load night hours, implies that the HPS pumps have the required capacity for this purpose, as shown in Fig. 13.8b (their electrical rating should exceed the difference between the flat load curve 2 and the minimum load demand).

This investigation has been performed for the study case island systems and the results are summarized in Table 13.2. Apart from the maximum capacity credit for each island, the associated sizing requirements are also included for the pumps and the water reservoir of an HPS, in order for it to be eligible for maximum capacity payment.

13.6.2 Economic Evaluation of HPS Investments

The economic feasibility of the HPS investments in the three islands is investigated in this section. Data and assumptions of the analysis are presented in

Table 13.3. For the evaluation, the basic financial indices of Appendix 2 are calculated, namely the net present value (NPV) of the investment, its internal rate of return (IRR), the discounted payback period (DPP) and the HPS generation cost (c).

The annual HPS income in each case is calculated from the simulation results of Sect. 13.5 and the applicable pricing for energy and capacity on each island (Table 13.1). The restrictions of the previous section regarding the HPS capacity credit have been taken into account (in the medium-size island, the HPS receives only partial remuneration for the installed capacity of its hydro turbines). The resulting annual income of the HPS is presented in Fig. 13.9, per installed MW of the hydro turbines. It is observed that the double-penstock configurations yield a significantly higher income, which compensates for their increased initial investment cost.

Based on these results, the economic evaluation indices are then calculated and presented in Fig. 13.10 for double-penstock HPS configurations. The HPS investments are highly attractive in all cases. In the small island case, the payback period is comparatively longer, although the annual income is higher, because the initial investment is much higher. For the same reason the HPS generation cost is higher in this case. For single-penstock arrangements the economics are less favorable, particularly in the small island case, where the calculated IRR is below 10%.

A sensitivity analysis has been conducted for the IRR and the NPV of the investments in the case of the medium-size island, with respect to variations of the initial capital investment, the annual income, the O&M expenses and the investment subsidy rate. The results presented in Fig. 13.11 indicate that the feasibility of the project is influenced considerably by variations in the capital investment and the annual income, which constitute the most decisive factors. Subsidy rates are translated into corresponding reductions of the capital investment and therefore also affect the economics of the projects, while the O&M expenses have a negligible impact, being relatively low. In any case, however, the HPS investments remain highly profitable, exhibiting IRR values in excess of 10%.

13.7 A Real-World Study Case: The HPS of Ikaria Island

In spite of the great potential promised by pumped-storage, no such installation has been put in operation so far and the reported investigations mostly deal with hypothetical case studies (with the exception of the El-Hierro station in the Canary Islands [40]). Therefore, both field experience and detailed feasibility studies of mature projects are still missing, to provide realistic data or reliable and concrete study results.

Such a project is currently under way in Greece, for the construction of the first hybrid wind-hydro-pumped storage power station in the isolated island grid of Ikaria. The station is already in the construction stage, by PPC Renewables S.A.,

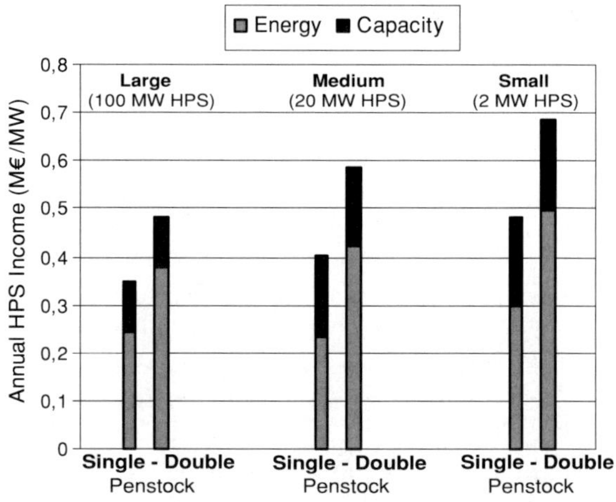

Fig. 13.9 Annual energy and capacity payments to the HPS, per installed MW of the hydro turbines, for the three island cases, ([32], © [2009] IET)

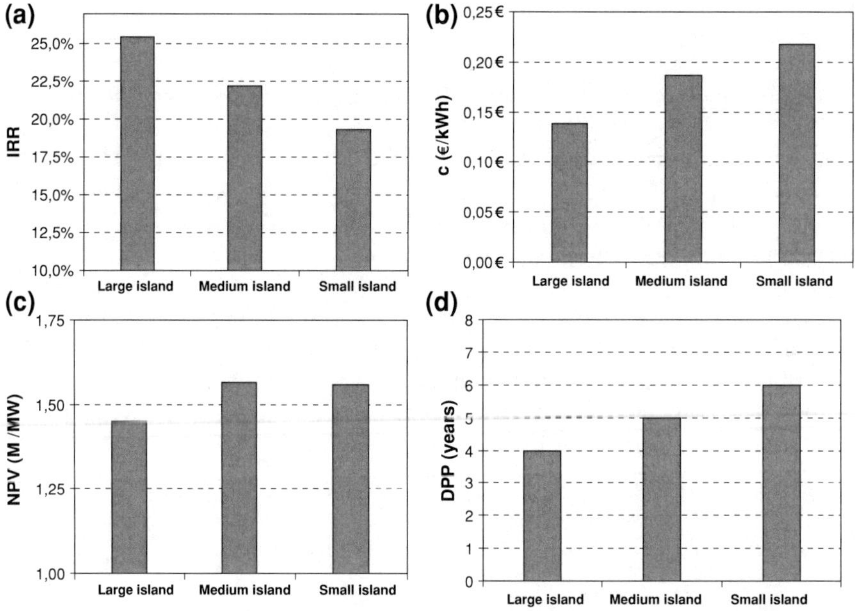

Fig. 13.10 Main financial indices for the evaluation of the HPS investment. **a** Internal rate of return. **b** HPS generating cost. **c** Net present value (per installed MW). **d** Discounted payback period, ([32], © [2009] IET)

subsidiary of the Greek utility PPC S.A. Licensing of the station according to the specific regulatory framework for Hybrid Power Stations (HPS) in Greece [33, 34, 36], necessitated extended feasibility studies, including the determination of a specific operating policy for the station and the island system and the evaluation of

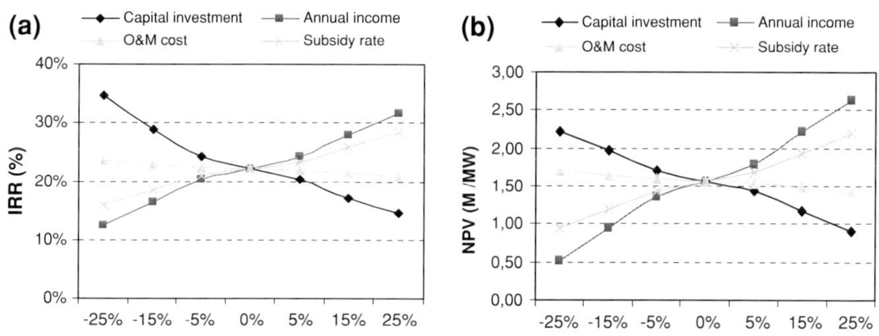

Fig. 13.11 Sensitivity analysis for **a** the IRR and, **b** the NPV of the HPS investment on the medium-size island, with respect to the capital investment, the expected annual income, the O&M expenses and the subsidy rate. X-axis values denote variation with respect to the base case values of Table 13.3 ([32], © [2009] IET)

the expected benefits to the energy equilibrium of the island [41, 42]. Results from these studies are presented in this section, based on the actual station characteristics and island power system data.

13.7.1 Description of the Ikaria HPS and Power System

The generation system of Ikaria, including the HPS, is presented in Fig. 13.12. The conventional autonomous power station (APS) at Agios Kirikos comprises heavy fuel oil (mazut) diesel engines, with a total installed capacity of 15.85 MW. Existing or licensed wind farms on the island amount to 1,835 kW. A foreseen photovoltaic (PV) station capacity of 1,040 kW is also taken into account. The peak load demand of the island for the future year when the HPS is expected to enter into service, is estimated at 9 MW, with a load factor of 42%. All stations are connected to the Medium Voltage (MV) distribution network of the island.

The Ikaria HPS, shown in Fig. 13.12, comprises three water reservoirs at sufficient altitude separation (elevations above sea level are noted in Fig. 13.12):

- An existing reservoir at Pezi (900,000 m^3), currently used for irrigation and water supply. Excess water from this reservoir will be exploited for energy generation purposes.
- Two new reservoirs at Proespera and Kato Proespera (approx. 80,000 m^3 each), for pumped storage purposes. The two pumped storage reservoirs are hydraulically connected using double penstock, to make turbine and pumping operation independent of each other.

The HPS includes two small hydroelectric plants (SHP), both equipped with Pelton turbines, one at Proespera (1 × 1.05 MW), to exploit excess water from Pezi dam and another at Kato Proespera (2 × 1.55 MW), which exploits excess water from Pezi and also participates in pumped storage operation. A pumping

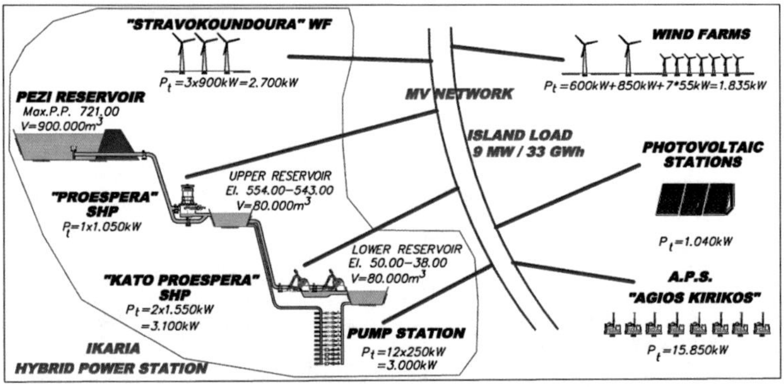

Fig. 13.12 HPS and other power stations on Ikaria island (future situation)

station is located at Kato Proespera, comprising 8 × 250 kW constant speed and 4 × 250 kW variable speed pumps (rated electric motor capacities). A 3 × 900 kW WF at Stravokoundoura is also part of the HPS. Technical data is given in the Appendix 1.

A particularity of the Ikaria HPS, which differentiates it from the usual HPS configuration described in Sect. 13.2, is that it combines two forms of renewable energy (wind via pumped storage and hydroelectric, via exploitation of excess water from the uppermost reservoir), using the same hydraulic infrastructure. This calls for a suitable internal operating policy of the HPS which will permit efficient exploitation of both resources.

The operation of the SHPs is subject to restrictions imposed by the environmental terms of the station. While the pumped-storage mode of operation, cycling water between the two lower reservoirs, is permitted throughout the year, hydroelectric operation is only possible from October 1 to April 30 (in the so-called "winter period"). During this period, a specified minimum water level must be ensured at the Pezi reservoir, while all excess water can be used for energy generation (hydro operation). In the "summer period" (May 1–September 30) the SHP at Proespera does not operate.

13.7.2 Internal Operating Policy of Ikaria HPS

Following the procedure described in Sect. 13.4.2, the ISO determines a generation schedule (also a pump load schedule, if grid pumping is needed) for the HPS hydro turbines. The HPS producer will then decide the allocation of the scheduled generation to the turbines of the two SHPs, as well as the exploitation mode of the available wind power, in order to maximize the expected benefit. A distinction is made between summer (May–September) and winter (October–April) periods, since the Proespera SHP operates only during the latter period (see Fig. 13.13).

When the operation of Proespera SHP is not possible (during summer season, as well as in the days of the winter period when no excess water is available in the Pezi dam), the internal operating policy of the HPS is exactly as described in Sect. 13.4.3 for the typical HPS configurations.

On the other hand, when excess water is available in the Pezi dam, the need arises to allocate the scheduled HPS generation to the two SHPs. Prioritizing operation of the Proespera SHP, improves the exploitation of excess water for hydro generation. At the same time, however, the water level of the Proespera reservoir increases, reducing the potential for pumped-storage operation, which is more profitable. Hence, the objective is to maintain the capacity of the Proespera reservoir high enough to maximize exploitation of excess water from Pezi, without compromising the pumped-storage potential, i.e. rejecting wind power available for pumping.

To achieve this objective, a target water level is set for the Proespera reservoir, equal to 85% of its capacity, and the dispatching and loading priorities of the two SHPs are alternated to maintain this level (the Kato Proespera SHP is prioritized to release water from the middle reservoir, while the Proespera SHP is prioritized to increase its level). A 5% hysteresis margin is used to avoid continuous priority alterations. This practice leaves a margin of about 12,000 m^3 (15% of the total reservoir capacity) for pumped storage purposes, which proves to be sufficient for the characteristics (discharge rates) of the installed pumps.

Concerning the exploitation of the available wind power during winter season, priority is given to pumping (Mode 1) because direct injection of wind power to the load via Mode 3 operation leads to frequent de-loading of the Proespera SHP and hence to unsatisfactory exploitation of the available water in the Pezi dam.

13.7.3 Results Achieved by the Operation of the HPS

In order to evaluate the effects of the HPS on the energy equilibrium of the island, the power system of Ikaria including the HPS was modelled according to Sect. 13.5.1 and simulated for the future year, when the HPS is expected to enter into service. The simulation model structure is outlined in the flowchart of Fig. 13.13.

Island load demand presents a peak of 9 MW and a load factor of 42%. Hydrological data, based on long-term measurements, includes time series of water inflow to the upper reservoir (Pezi). Three typical annual time series, characterized by high, medium and low inflows, are employed in the simulations. Regarding wind data, a measured high wind speed time series (mean annual speed 9.75 m/s at 55 m) is used for all WFs on the island. For the HPS WF at Stravokoundoura, a second, lower speed series is also used (mean annual speed 7.95 m/s at 55 m), to provide the envelope of the expected wind potential at Stravokoundoura. All technical data of the Ikaria HPS and APS used in the simulations are provided in Appendix 1.

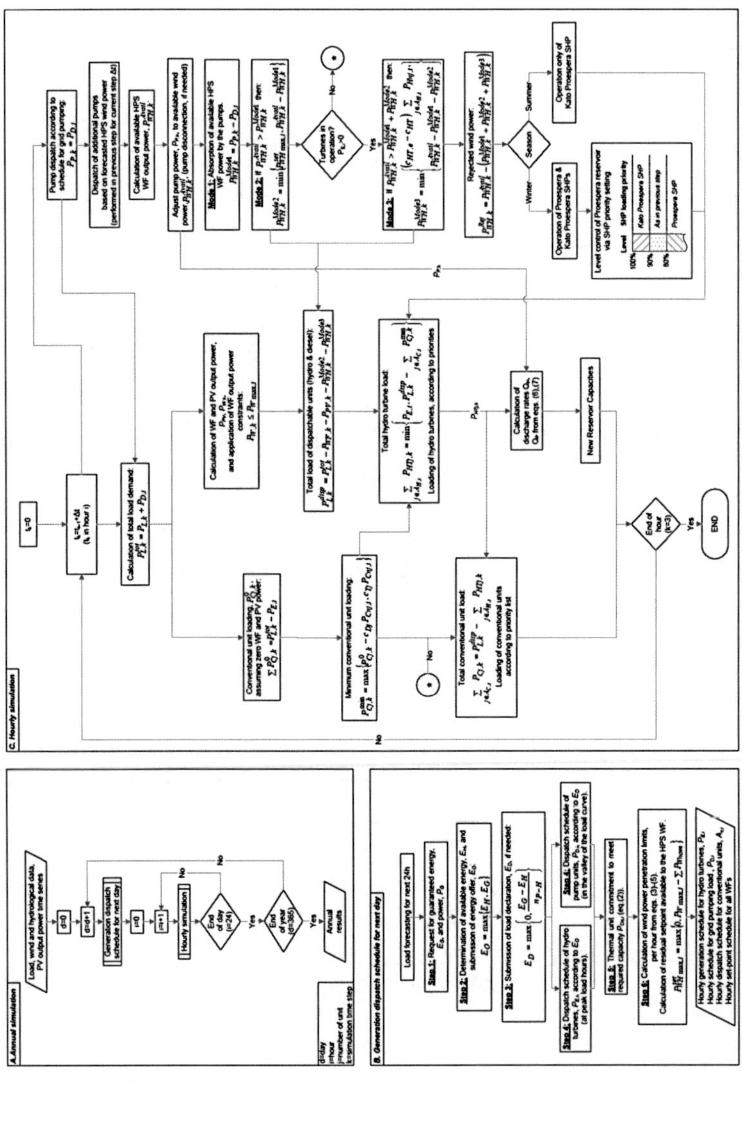

Fig. 13.13 Flowchart of the simulation algorithm ([42], © [2010] IEEE)

In Figs. 13.14 and 13.15, the operation of the Ikaria system is presented for two indicative weeks, considering low wind and average hydrological conditions. Figures 13.14a and 13.15a show the contribution of each station to the load demand (APS, HPS, other WFs and PV stations). During low-load periods in winter (Fig. 13.14), the load is primarily served by the HPS turbines, since water levels in the Pezi reservoir are high and operation of both SHPs is permitted. Conventional units operate only in the high load hours of each day, typically at their technical minimum. During the high-load summer season (Fig. 13.15), hydroelectric operation using water from Pezi is not permitted. Hence, energy offers of the HPS are based only on pumped wind energy and the HPS turbines are dispatched only at the peak load hours of each day, shaving the peaks of the load curve. It is important that, in spite of the relatively large HPS size, the dispatch and operation of the hydro turbines has no adverse effect on the conventional units (actually, it effectively smoothes the load curve of the APS).

Diagrams 13.14b and 13.15b present the hydro turbine output power (blue area) and the portion of the HPS WF power that is directly delivered to the load (green area), in operating Modes 2 and 3. The scheduled HPS generation is provided mainly by the hydro turbines. The variations of pump input power are shown in Figs. 13.14c and 13.15c, along with the water level in the Proespera reservoir. In the low-load week (Fig. 13.14c), pump energy is exclusively provided by the HPS wind turbines (green area). On the other hand, during the high-load week (Fig. 13.15c), grid pumping (black area) does occur at the beginning of the sixth day. This happens because the ISO demands guaranteed supply from the HPS, while the water level in the Proespera reservoir is not enough to ensure availability of the requested energy. In diagram 13.15a, grid pumping results in valley-filling of the load curve.

The reservoir water level regulation policy followed in the winter is evident in Fig. 13.14c. The reservoir level normally fluctuates between 80 and 90%, increasing further during high wind intervals, such as at the end of the week, when the full pumped storage potential is exploited. During the summer period, Fig. 13.15c, water levels remain lower and are fully used when guaranteed supply is requested.

In the following, the benefits from the operation of the HPS are quantified on an annual basis, for various operating scenarios.

The resulting increase in RES penetration is illustrated in Fig. 13.16, which presents the contribution of all generating stations of the island to the annual load demand. Results for the base case scenario of operation without the HPS are shown by the left group of bars. Operation with the HPS, for the three hydrological and the two wind regime conditions, are presented by the other six groups.

In the absence of the HPS, the wind and PV stations operating on the island result in an annual RES energy penetration around 17%, which is close to the maximum achievable limit without storage, following the current operating practice of island systems. The capacity factor (CF) of the installed wind farms is approximately 25%, in spite of the favourable wind potential, due to the imposed output power curtailments (up to 40% of their potential annual energy yield).

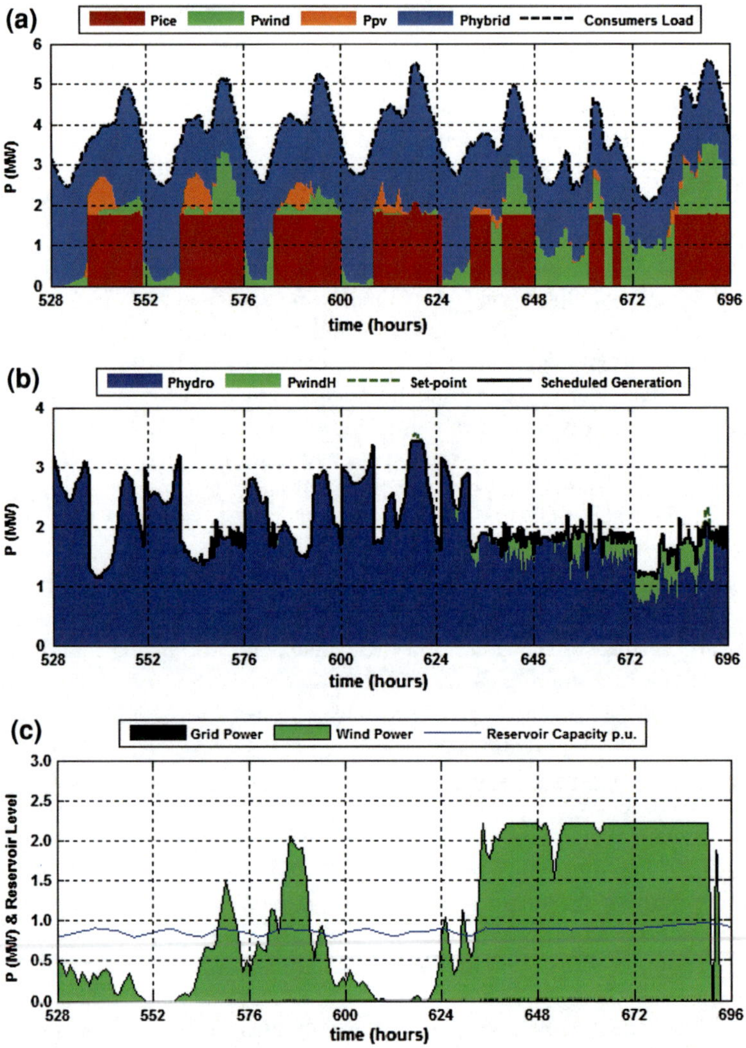

Fig. 13.14 Operation of the system for a typical low-load week (in January). **a** Output power of conventional units (pice), HPS (phybrid), WFs external to the HPS (pwind) and PV stations (ppv). **b** Contribution of hydro turbines and WF to the HPS output. **c** HPS pump power origin (wind/grid pumping) and proespera reservoir level (1 = max reservoir capacity), ([42], © [2010] IEEE)

Increasing the installed wind capacity to achieve higher penetrations would decrease further the wind farm CF, denying the viability of the investments. The HPS drastically boosts the total RES penetration by 18–36%, depending on the assumed hydrological and wind conditions, reaching overall penetration levels even in excess of 50%. It is important that this increase comes with no adverse effect on the operation of existing or licensed WFs on the island, whose CF

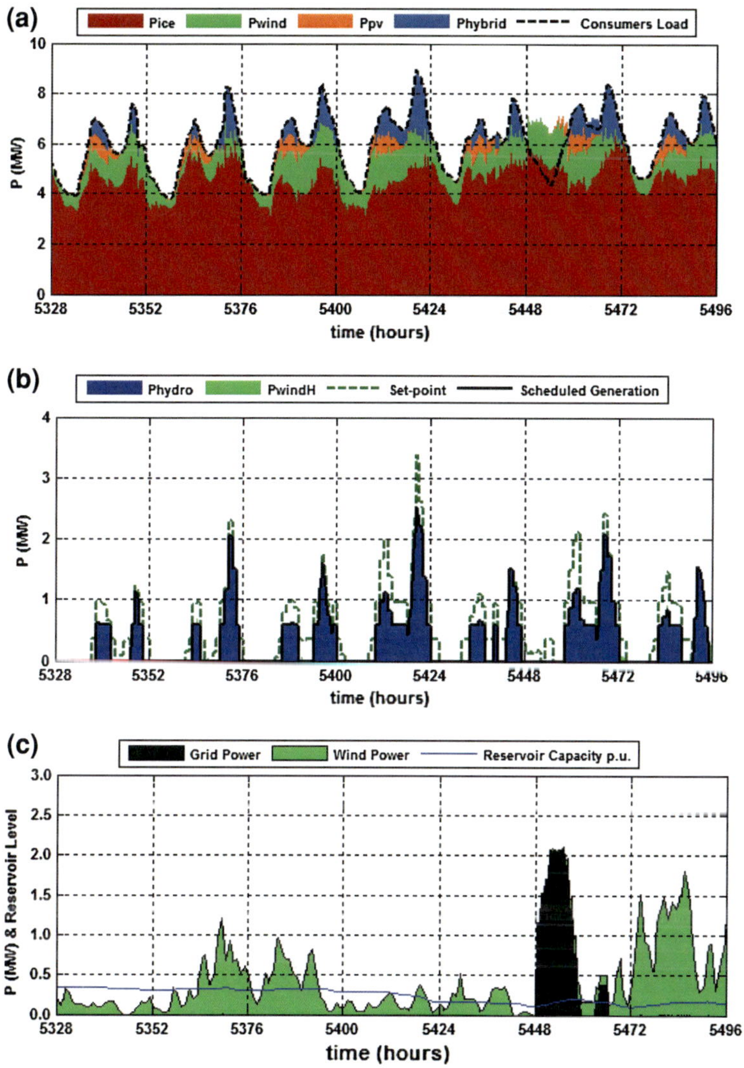

Fig. 13.15 Operation of the system for a typical high-load week (in August). **a** Output power of conventional units (pice), HPS (phybrid), WFs external to the HPS (pwind) and PV stations (ppv). **b** Contribution of hydro turbines and WF to the HPS output. **c** HPS pump power origin (wind/grid pumping) and Proespera reservoir level (1 = max reservoir capacity), ([42], © [2010] IEEE)

remains practically unaffected in the simulated scenarios. Obviously, the same applies for PV stations, whose output is not subject to curtailments.

As commented in Sect. 13.4.2, grid pumping is permitted to ensure guaranteed supply of power and energy from the HPS, but should be minimized to reduce waste of conventional fuel. The proposed operating policy indeed results in

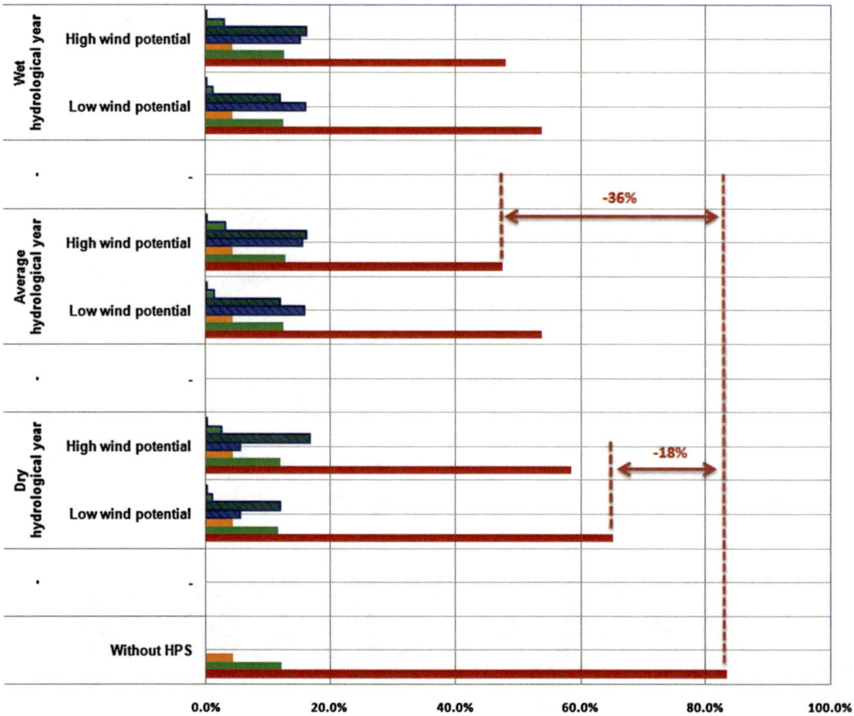

	Without HPS	-	Dry hydrological year Low wind potential	Dry hydrological year High wind potential	-	Average hydrological year Low wind potential	Average hydrological year High wind potential	-	Wet hydrological year Low wind potential	Wet hydrological year High wind potential
HPS Wind - Mode 3	0.0%		0.0%	0.2%		0.0%	0.2%		0.0%	0.2%
HPS Wind - Mode 2	0.0%		1.1%	2.6%		1.3%	3.2%		1.3%	3.1%
HPS Turbines - Wind	0.0%		12.1%	16.7%		12.1%	16.2%		12.1%	16.3%
HPS Turbines - Hydro	0.0%		5.6%	5.6%		16.0%	15.6%		16.0%	15.2%
PV	4.4%		4.4%	4.4%		4.4%	4.4%		4.4%	4.4%
Wind	12.2%		11.6%	12.0%		12.5%	12.9%		12.5%	12.7%
Conventional	83.3%		65.2%	58.5%		53.7%	47.4%		53.7%	48.0%

Fig. 13.16 Contribution of conventional units, HPS and RES stations (WFs, PVs) to the annual load demand without and with the HPS, under different hydrological and wind regime conditions

negligible grid-pumping, representing 0.5–2.5% of the total energy absorbed by the pumping station over the year. The higher percentage occurs in the low wind regime, when availability of stored wind energy is generally reduced.

Besides the energy benefits, the Ikaria HPS will also provide firm capacity to the island system, relaxing the need for installation of new thermal capacity in the future. The guaranteed capacity of the HPS is 2.655 MW (rather than the rated of 2 × 1.55 MW), which is the maximum output power of the Kato Proespera SHP at lowest hydraulic head conditions (minimum water level at the Proespera reservoir).

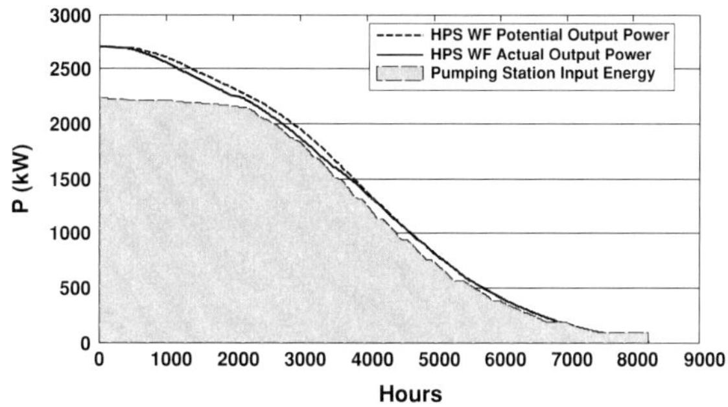

Fig. 13.17 Annual duration curves of the pumping station input power and the HPS WF potential and actual output power

In wind-pumped-storage mode of operation, the Ikaria HPS essentially constitutes a wind power station, equipped with storage facilities in a virtual "series connection" to the wind farm, since practically all wind energy produced is pumped and then delivered to the load. An efficient operating policy should ensure the best possible utilization of the installed wind capacity. Indeed, the achieved CF values are approximately 29 and 43% for the low and high wind time series, whereas the discarded wind energy on an annual basis is only 1.5 and 3.5% of that available. Obviously, the energy eventually delivered to the load is reduced, due to the inevitable losses in the pumped storage facilities (average full cycle efficiency less than 60% for the particular station)

In Fig. 13.17, the annual duration curves of the pumping station input power and the HPS WF potential and actual output power are shown. Available wind power is discarded mainly when it exceeds the actual maximum pump power (approx. 2.2 MW). The largest part of the excess wind power (area in Fig. 13.17 between potential WF output and pump power) is delivered to the load, via operation in Mode 2 or 3, whenever this is possible. Hence the actual losses are drastically reduced (small area in Fig. 13.17 between potential and actual WF output).

Wind energy losses also occur due to storage reservoir capacity limitations. However, the sizing of reservoirs and the adopted water level regulation policy result in very low losses of this type. This is verified by the duration curves of the Proespera reservoir water level, shown in Fig. 13.18 for all simulated scenarios. The capacity of the reservoir is exhausted for a very small fraction of the time (less than 2%) and even then, a certain potential for wind power exploitation still exists, via operation in Mode 3.

The efficiency of the HPS as a small hydroelectric station is evaluated by analysing water overflows[8] from the Pezi reservoir, whose annual duration curves are presented in Fig. 13.19 for the average hydrological year, with and without the HPS. The area between the solid and dashed lines in Fig. 13.19 represents the

[8] Water overflows occur when inflows to the reservoir exceed outflows through the penstock and the maximum water level is reached. Then, excess water overflows via the dam spillway, without being exploited for energy generation (or any other purpose, such as irrigation).

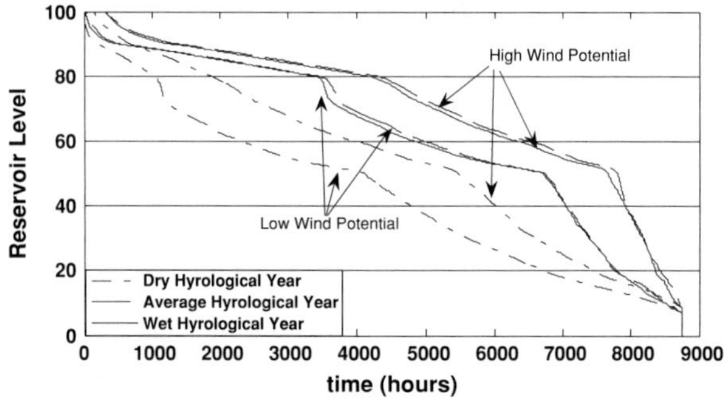

Fig. 13.18 Annual duration curves of the Proespera reservoir water level for the simulated scenarios

Fig. 13.19 Hydro potential exploitation. Annual duration curves of water overflows from the Pezi dam (average hydrological year), with and without the HPS, ([42], © [2010] IEEE)

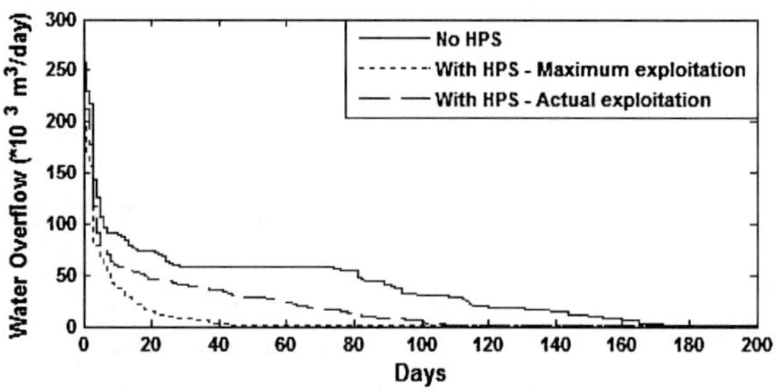

amount of overflows exploited by the HPS (approx. 48%). The dotted line corresponds to the hypothetical case of unconstrained hydro turbine operation, whenever there is excess water in Pezi. This sets an upper limit for water overflow exploitation, for the given dimensioning of the hydro plants (82%), which is not achievable in practice due to power system restrictions and the need to accommodate the pumped storage mode of operation, using the same hydraulic infrastructure (reservoirs, turbines, penstock).

13.8 Discussion on Technical Issues

The implementation of the proposed operating strategy necessitates the presence of an advanced Energy Control Centre (ECC), which will automate decision making and management of the HPS and APS units. A SCADA system for monitoring and control of all generation and storage facilities will be an essential part of the ECC. To permit fast and reliable communication among the individual components, fibre-optics will be installed along the interconnecting lines of the HPS components, to ensure real-time communication capabilities, which are essential e.g. for maintaining the power balance between pumping and wind generation.

Operation of the island system under the strategy described in this chapter imposes requirements on the control systems of the HPS individual units, in order to ensure the satisfactory dynamic behaviour of the system. Although dynamic response and regulation issues are under investigation and are not dealt with in this chapter, a brief commentary is possible. Since the HPS hydro turbines are going to substitute conventional generators, they must be equipped with speed governors and voltage regulators, which should provide similar regulation and response capabilities to the diesel units they replace. This is particularly important for periods when the island system is supported only by the HPS turbines. Such requirements need to be considered in the hydraulic design and selection of mechanical equipment for such a plant. Active power regulation capability is needed for the pumps (variable speed units), to effectively track the available wind power. Pumps might further be required to provide primary frequency response and dynamic reserve for balancing wind power variations, as the island system will operate at very high wind power penetration levels. For the same reason it is essential that the WF of the HPS exhibits full grid code compatibility [43]. Besides the Low-Voltage-Ride-Through (LVRT) capability, which is essential, the WTs should demonstrate fast output power control, accept max power and ramp-rate limits and provide fast frequency response in order to contribute to the primary frequency control of the system.

13.9 Conclusions

In this chapter, the prospect of implementing wind-hydro hybrid power stations (HPS) in autonomous island grids, in order to achieve high RES penetration levels, has been discussed and a specific operating policy has been proposed and evaluated. Three specific island systems of different sizes and characteristics were used as case studies and the benefits provided by the installation of HPS, along with their effect on the operation and economics of the island systems, have been investigated. Then, the HPS currently under construction on the island of Ikaria, Greece, was presented and the expected effects from its operation on the energy equilibrium of the island power system have been quantified.

The simulation results indicate that the HPS can increase substantially wind energy penetration in saturated autonomous power systems, while they can also provide firm capacity, substituting expensive peak units. The proposed operating policy permits the integration of large HPS in isolated island grids, minimizing their impact on conventional generation, as well as on the existing wind farms. Further, under the current pricing framework for energy and capacity, HPS constitute attractive investments, without penalizing excessively the operating cost of the overall island systems.

Acknowledgment The work associated with the determination of a suitable operating policy for autonomous island systems with HPS, as presented in the first six sections of this chapter, has been financially supported by the Regulatory Authority for Energy (RAE) of Greece. The feasibility studies for the Ikaria HPS, results of which are presented in Sect. 13.7, have been funded by PPC Renewables S.A.The first six sections of this chapter are based on material originally presented in [32] and published here under the permission granted by IET. The material of Sect. 13.7 was originally presented in [42] and it is reproduced here with the permission of IEEE.

Appendix 1: Technical Data of the Ikaria HPS and APS

All data presented in this Appendix come from the technical specifications of the HPS project under construction in Ikaria island, Greece, by PPC Renewables S.A.

A. *Water reservoirs*

The exploitable capacities of the three water reservoirs are noted in Fig. 13.13a. Specific relations between measured water level and volume of water in each reservoir are also available, based on the reservoir geometric characteristics.

B. *Proespera SHP*

It comprises one vertical shaft, four-nozzle Pelton turbine. Guaranteed efficiencies are given in Table 13.3 for various loading levels and net heads (including head losses between water intake and turbine distributor, equal to $30.82 \cdot Q^2$). It is located at 555.20 m above sea level (a.s.l.) and its maximum and minimum water discharge rates are 0.7 and 0.07 m^3/s respectively. The turbine is directly coupled to a synchronous generator, rated 1,300 kVA at 0.80 power factor, with 95.83% average weighted efficiency. A 6.3/20 kV, 1,300 kVA unit transformer is used, with an average weighted efficiency of 98.70%.

C. *Kato Proespera SHP*

It comprises two horizontal shaft, two-nozzle Pelton turbines. Guaranteed efficiencies are given in Table 13.4. The same apply as for Table 13.3 (head losses are $79.18 \cdot Q^2$). The turbines stand at 50.70 m a.s.l. and their maximum and minimum water discharge rates are 0.35 and 0.035 m^3/s. Turbine generators are rated 1,900 kVA at 0.80 power factor with an average weighted efficiency of 94.30%. Unit transformers 6.3/20 kV, 1,900 kVA are used with an average weighted efficiency of 98.78%.

D. *Pump station*

It comprises 8 constant speed and 4 variable speed pumps. Table 13.6 refers to constant speed pumps and gives efficiencies, discharge rates and input powers for different net heads (taking into account head losses, equal to $185 \cdot Q^2$). Variable speed pumps have the same characteristics and a minimum acceptable loading 50% of their rated capacity. Each pump is driven by a squirrel cage induction motor, rated 250 kW with average weighted efficiency 96%. Pump motors are

Table 13.4 Efficiencies of hydro turbine in the Proespera SHP, ([42], © [2010] IEEE)

Turbine power (% of max turbine output for a given head)									Max turbine output (kW)	Net head (m)
10	20	40	50	60	70	80	90	100		
89.8	89.5	89.9	89.9	89.9	89.8	90.1	90.0	89.8	1,031	167.2
90.1	89.8	90.2	90.2	90.2	90.1	90.4	90.3	90.1	928	150.0
89.6	89.3	89.7	89.7	89.8	89.6	89.9	89.8	89.6	841	136.7

Table 13.5 Efficiencies of hydro turbines in the Kato Proespera SHP, ([42], © [2010] IEEE)

Turbine power (% of max turbine output for a given head)									Max turbine output (kW)	Net head (m)
10	20	40	50	60	70	80	90	100		
82.0	87.0	90.1	90.1	89.6	90.2	90.5	90.6	90.4	1,563	503.4
82.0	87.0	90.1	90.1	89.7	90.3	90.5	90.6	90.4	1,521	490.0
81.8	87.0	90.1	90.0	89.7	90.3	90.5	90.6	90.4	1,490	480.0
81.5	86.8	89.8	89.7	89.6	90.2	90.4	90.5	90.2	1,406	454.0

Table 13.6 Basic characteristics of constant speed pumps, ([42], © [2010] IEEE)

Net head (m)	Efficiency (%)	Discharge rate (m^3/h)	Input power (kW)
525.40	70.94	85	171.54
509.00	73.37	95	179.58
494.30	74.64	102	184.06

connected to the grid via two transformers 0.4/20 kV, 2,000 kVA each, with 98.96% average weighted efficiency.

E. *HPS wind farm*

It consists of 3 × 900 kW variable speed WTs (Enercon E44). Power curve is available at [44].

F. *Full-cycle efficiency of the HPS*

The full-cycle efficiency of the HPS depends on the loading level of the generating units and pumps, as well as on water levels of the reservoirs. The minimum overall efficiency, needed for determining the load declaration upper limit of (1.1), is calculated equal to 50.3%. The mean annual efficiency is approximately 56%, as calculated using the simulation results.

G. *APS generator data*

The composition of the Ikaria APS in the future reference year and the basic characteristics of the conventional units are presented in Table 13.7. All units consume heavy fuel oil (mazut), with a CO_2 emission coefficient of 3.109 kg CO_2/kg

Table 13.7 Basic characteristics of APS conventional units, ([42], © [2010] IEEE)

Dispatch order	Type	Max power (kW)	Min power (kW)	Fuel consumption (gr/kWh)		
				50%	75%	100%
1A	CEGIELSKI SULZER	3,500	1,750	235.7	226.5	224.3
1B	CEGIELSKI SULZER	3,500	1,750	235.7	226.5	224.3
2	CEGIELSKI SULZER	2,900	1,550	235.7	226.5	224.3
3	CEGIELSKI SULZER	2,200	1,100	245.4	225.8	220.9
4A	FIAT	750	500	255.0	239.0	237.5
4B	FIAT	750	500	255.0	239.0	237.5
4C	FIAT	750	500	255.0	239.0	237.5
4D	FIAT	750	500	255.0	239.0	237.5
5	SACM	750	600	267.5	252.6	246.0

fuel. It is noted that the installed thermal capacity reflects the current planning of the ISO and does not constitute the necessary or optimal APS synthesis. In the simulations performed, only the first three diesel units in Table 13.7 are dispatched, for the given load demand of the island.

Appendix 2: Basic Financial Indices

The basic financial indices used for the evaluation of the HPS investment are briefly presented in the following. More details can be found in any financial mathematics textbook (e.g. [45]).

The net present value (*NPV*) of an investment is calculated from the following equation:

$$NPV = -K_0 + \sum_{t=1}^{N} \frac{F_t}{(1+i)^t} \tag{13.10}$$

where K_0 is the equity
F_t is the net annual income for year t (after taxes and loan amortization)
N is the economic lifetime of the investment and
i is the annual discount rate

The Internal Rate of Return (IRR) is the discount rate that makes the present worth of the investment over the entire evaluation period equal to zero. The IRR is the solution of (13.11) for the discount rate i, given the investment lifetime N:

$$-K_0 + \sum_{t=1}^{N} \frac{F_t}{(1+i)^t} = 0 \tag{13.11}$$

The Discounted Payback Period (DPP) is the time required for the present worth of the investment to become zero. It is the solution of (13.11) with respect to the lifetime N, given a specific discount rate i. If F_t is constant over the whole period, then (13.11) is transformed into (13.12), from which the analytical solution of (13.13) is derived for the DPP.

$$\sum_{t=1}^{N} \frac{1}{(1+i)^t} = \frac{K_0}{F} \tag{13.12}$$

$$DPP = \frac{-\ln(1 - i\frac{K_0}{F})}{\ln(1+i)} \tag{13.13}$$

The generation cost c of the HPS is calculated from (13.14):

$$c = \frac{K_{0,tot} \cdot R + OMC}{E_{net}} \tag{13.14}$$

where $K_{0,\mathrm{tot}}$ is the total investment cost
OMC is the annual operation and maintenance cost
E_{net} is the net annual energy yield and
R is the annuity factor, given by:

$$R = \frac{i}{1-(1+i)^{-N}} \tag{13.15}$$

References

1. Kaldellis JK, Zafirakis D (2007) Present situation and future prospects of electricity generation in Aegean Archipelago islands. Elsevier Eng Policy 35(9):4623–4639
2. Tsioliaridou E, Bakos GC, Stadler M (2006) A new energy planning methodology for the penetration of renewable energy technologies in electricity sector application for the island of Crete. Elsevier Eng Policy 34(18):3757–3764
3. Kavadias KA, Zafirakis D, Kondili E, Kaldellis JK (2007) The contribution of renewables on reducing the electricity generation cost in autonomous island networks. Proceedings of International conference on clean electrical power, Capri, Italy, pp 777–783
4. Papathanassiou S, Boulaxis N (2006) Power limitations and energy yield calculation for win farms operating in island systems. Elsevier Renewable Energy 31(4):457–479
5. Kaldellis JK, Kavadias KA, Filios AE, Garofallakis S (2004) Income loss due to wind energy rejected by the crete island electrical network–the present situation. Appl Eng 79(2):127–144

6. Maddaloni JD, Rowe AM, Cornelis van Kooten G (2008) Network constrained wind integration on Vancouver Island. Elsevier Eng Policy 36(2):591–602
7. Lund H (2005) Large scale integration of wind power into different energy systems. Eng 30(13):2402–2412
8. Wassier D, Garcia RS (2005) Instantaneous wind energy penetration in isolated electricity grids: concepts and review. Renewable Energy 30(8):1299–1308
9. Benitez LE, Benitez PC, Cornelis van Kooten G (2008) The economics of wind power with energy storage. Elsevier Eng Econ 30(4):1973–1989
10. Kaldellis JK, Zafirakis D, Kavvadias K (2009) Techno-economic comparison for energy storage systems for island autonomous electrical networks. Renew Sustain Energy Rev 13(2):378–392
11. Hennessy T, Kuntz M (2005) The multiple benefits of integrating electricity storage with windenergy. Power Eng Soc Meeting IEEE 2:1952–1954
12. Tsikalakis A, Tassiou I, Hatziargyriou N (2004) Impact of energy storage in the secure and economic operation of small islands. MedPower04, Larnaca, Cyprus
13. Ummels BC, Pelgrum E, Kling WL (2008) Integration of large-scale wind power and use of energy storage in the netherlands' electricity supply. IET Renew Power Gener 2(1):34–46
14. Brown PD, Peas Lopes JA, Matos MA (2008) Optimization of pumped storage capacity in an isolated power system with large renewable penetration. IEEE Trans Power Syst 23(2):523–531
15. Lu N, Chow JH, Desrochers AA (2004) Pumped-storage hydro-turbine bidding strategies in a competitive electricity market. IEEE Trans Power Syst 19(2):834–841
16. Garcia-Gonzalez J, De la Muela RMR, Santos LM, Gonzalez AM (2008) Stochastic joint optimization of wind generation and pumped-storage units in an electricity market. IEEE Trans Power Syst 23(2):460–468
17. Hatziargyriou N, Stefanakis J, Alexandridis S et al (2001) Safe increase of wind energy penetration in island systems by pumped storage units—the crete case. Proceedings of RENES 2001, Athens, Greece
18. Protopapas K, Papathanasiou S (2006) Application of pumped storage to increase wind penetration in isolated island grids. Proceedings of EWEC 2006, Athens
19. Caralis G, Zervos A (2007) Analysis of the combined use of wind and pumped storage systems in autonomous Greek islands. IET Renew Power Gener 1(1):49–60
20. Boulaxis NG, Papadopoulos MP (2003) Assessment of the contribution of hybrid systems in renewable energy penetration in islands. Proceedings of ISAP 2003 conference, Lemnos, Greece
21. Theodoropoulos P, Zervos A, Betzios G (2001) Hybrid systems using pump-storage implement tation in Ikaria island. Proceedings of OPET island international conference. Chania, Greece
22. Mantas Z, Theodoropoulos P, Betzios G, Zervos A (2003) Hybrid system using pump-storage for maximum wind energy penetration in Serifos island. Bull Hellenic Assoc Mech Electr Eng 6:
23. Kaldellis J, Kavvadias K, Vlachou D (2002) Electricity load management of APS using wind-hydro solution. Proceedings of MedPower 2002, Athens, Greece
24. Bakos GC (2002) Feasibility study of a hybrid wind/hydro power-system for low-cost electricity production. Appl Eng 72(3–4):599–608
25. Castronuovo ED, Pecas Lopes JA (2004) Optimal operation and hydro storage sizing of a wind hydro power plant. Int J Electr Power Energy Syst 26(10):771–778
26. Bueno C, Carta JA (2006) Wind powered pumped hydro storage systems, a means of increase ing the penetration of renewable energy in the Canary Islands. Renew Sustain Eng Rev 10(4):312–340
27. Anagnostopoulos J, Papantonis D (2007) Pumping station design for a pumped-storage wind-hydro power plant. Eng Convers Manag 48(11):3009–3017

28. Anagnostopoulos J, Papantonis D (2008) Simulation and size optimization of a pumped-storage power plant for the recovery of wind-farms rejected energy. Renewable Energy 33(7):1685–1694
29. Anagnostopoulos J, Papantonis D (2008) Pumped storage wind/hydro plants in non-interconnected grids. Proceedings of hydro 2008 conference, Ljubliana, Slovenia
30. Angarita J, Usaola J (2007) Combining hydro-generation and wind energy: biddings and operation on electricity spot markets. Electr Power Syst Res 77(5–6):393–400
31. Electricity generation from renewable energy sources and high efficiency co-generation of elec tricity and heat, Greece, Law 3468/2006, Official Gazette A′, 129/27.6.2006
32. Report prepared for PPC renewables S.A., NTUA Oct 2009 Technical and economic feasibility study for the HPS of Ikaria
33. Regulation about authorizations for electricity generation from renewable energy sources and high efficiency co-generation of electricity and heat, Greece, Ministerial Decree Δ6/Φ1/*οικ*.5707, Official Gazette B′, 438/3.4.2007
34. Report prepared for the regulatory authority for energy NTUA (2008) Technical and economic issues for the integration of hybrid stations in non-interconnected Isl and grids
35. Principles for the operation, management and pricing of HPS in non-interconnected island grids, Public consultation document, Greek regulatory authority for energy (www.rae.gr), July 2008
36. Papaefthimiou S, Karamanou E, Papathanassiou S, Papadopoulos M (2009) Operating policies for wind-pumped storage hybrid power stations in island grids. IET Renew Power Gener 3(3):293–307
37. Karamanou E (2006) PV penetration potential in autonomous island grids with installed wind power. Diploma thesis, NTUA
38. King D, Boyson W, Kratochvil J (2004) Photovoltaic array performance model. Sandia national laboratories, Report SAND2004-353
39. http://www.insula-elhierro.com
40. Tsili M, Papathanassiou S (2009) A review of grid code technical requirements for wind farms. IET Renewable Power Generation 3(3):308–332
41. Papaefthymiou SV, Karamanou EG, Papathanassiou SA, Papadopoulos MP (2010) A Wind-Hydro-Pumped Storage Station Leading to High RES Penetration in the Autonomous Island System of Ikaria. IEEE Trans Sustainable Eng 1(3):163–172
42. Karamanou E, Papathanassiou S, Boulaxis N (2008) Operating policies for autonomous island grids with PV penetration. Proceedings of 4th European PV-hybrid and mini-grid conference, Athens
43. http://www.enercon.de
44. DeGarmo EP, Sullivan WG, Bontadelli JA (1988) Engineering economy, 8th edn. Macmillan, New York
45. Jaramillo OA, Borja MA, Huacuz JM (2004) Using hydropower to complement wind energy: a hybrid system to provide firm power. Renewable Energy 29(11):1887–1909

Chapter 14
Grid Frequency Mitigation Using SMES of Optimum Power and Energy Storage Capacity

M. R. I. Sheikh and J. Tamura

Abstract Considering the viewpoint of cost-effectiveness, a computational method to determine the SMES power rating needed to minimize the grid frequency fluctuation is analyzed in this chapter. Moreover, the required minimum energy storage capacity of SMES unit is determined. Finally, simulation results using pulse width modulation (PWM) based voltage source converter (VSC) and two-quadrant DC–DC chopper-controlled SMES system are presented. It is seen that the proposed SMES system with required minimum energy storage capacity can significantly decrease the voltage and output power fluctuations of wind farm, which consequently mitigate the grid frequency fluctuation.

14.1 Introduction

14.1.1 Renewable Energy

We are now at a crucial cross board of our global energy scenario. Energy has been the life hood of the continual progress of human civilization. Since the industrial revolution of the two centuries ago, global energy consumption has increased by leaps and bounds to improve our living standards, particularly in the industrialized nations of the world.

M. R. I. Sheikh (✉)
EEE Department, Rajshahi University of Engineering & Technology, Rajshahi 6204, Bangladesh
e-mail: ris_ruet@yahoo.com

J. Tamura
EEE Department, Kitami Institute of Technology, 165 Koen-cho, Kitami 090-8507, Japan

S. M. Muyeen (ed.), *Wind Energy Conversion Systems*,
Green Energy and Technology, DOI: 10.1007/978-1-4471-2201-2_14,

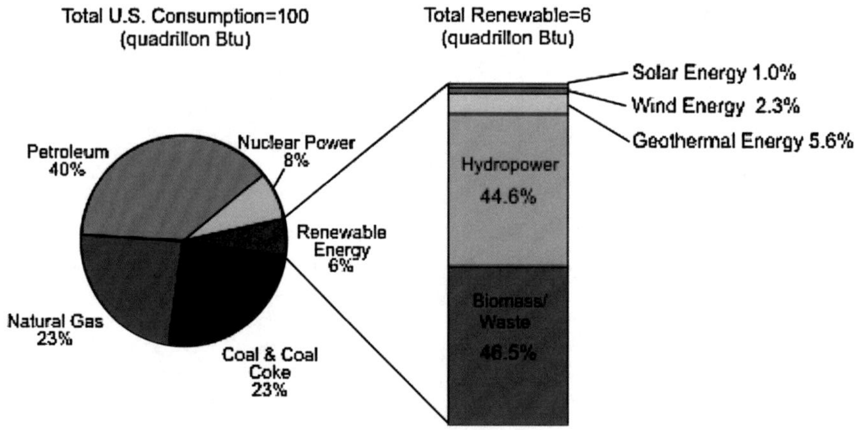

Fig. 14.1 Energy resources used in the United States [1]

According to the current energy resources used in United States as shown in Fig. 14.1, about 86% of the total energy is generated from fossil fuels, 8% is generated in nuclear plants, and remaining 6% comes from renewable sources (mainly hydro, biomass and wind power) [1]. Unfortunately, the world has limited amounts of fossil fuel and nuclear power resources. According to current estimates, natural uranium for nuclear power will last only about 50 years; oil will last no more than 100 years; gas, 150 years; and coal, 200 years [2]. Will the wheels of our civilization come to a screeching halt after the twenty-third century when fossil and nuclear fuels become totally exhausted?

Besides, our overdependence on fossil and nuclear fuels is causing environmental pollution and safety problems, which are now becoming dominant issues in our society. Rising pollution levels and worrying changing climate, arising in great part from energy-producing process, demand the reduction of ever-increasing environmentally damaging emissions. This impact of environmental pollution on global warming and resulting climate changes can have serious disastrous consequences in the long run [2].

At this juncture, we should be turning more and more to environmentally clean and safe renewable energy sources. Generating electricity, particularly by making use of renewable resources, allows the attainment of notable reductions of environmental pollution. Thereby, in addition to hydro-power used all over the world, the immense potentials of solar and wind energies assume great importance. Their promise is, however, subject to time-dependent process of nature. The systems needed to exploit them are still in their infancy. To establish themselves in a marketplace of high technical standards, a corresponding period for the development of these environmentally friendly technologies is particularly necessary [3].

The world has enormous resources of wind energy. This worldwide potential of wind power means that its contribution to electricity production can be of significant proportions. In some countries, the potential for wind energy

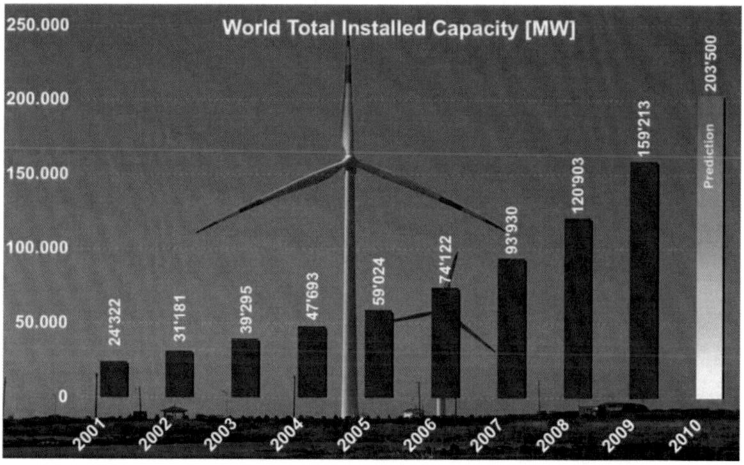

Fig. 14.2 World total wind power installed capacity [5]

production exceeds by far the local consumption of electricity. It has been estimated that tapping barely 10% of the wind energy available could supply all of the electricity needed in the world [3]. Good prospects and economically attractive expectations for the use of wind power are, however, linked to the incorporation of this weather-dependent power source into existing distribution networks.

14.1.2 The Scenarios for the Future on Wind Energy in the World

The worldwide market for wind energy has been growing faster than any other form of durable energy. Its installed power capacity in the world grew from only 4,800 MW in 1995 to 59,000 MW at the end of 2005 [4] that is, an increase of more than 1,200% in ten years. In the mean time, there are three scenarios worked out about how we can further expand wind energy and what benefits it will bring. Figure 14.2 shows the predicted world total installed capacity that is around 203,500 MW in 2010 [5]. Thus the total wind capacity will exceed 200,000 MW within the year 2010. Based on the accelerated development and further improved policies, world wind energy association (WWEA) is predicting that a global capacity of 1,900,000 MW will be possible by the year 2020 [5].

These predictions are impressive. In the expanding scenario we will be able to deliver 34% of our electricity from the wind energy by 2050 [4]. The cost price of wind energy can drop to 3 cents/kWh, the amount of jobs in the wind energy industries will result in a growth of 2.1 million, and the CO_2 emissions will decrease by 3,100 million tons. Wind energy is perhaps the most advanced among the "new" renewable energy technologies, but there is still much work to be done.

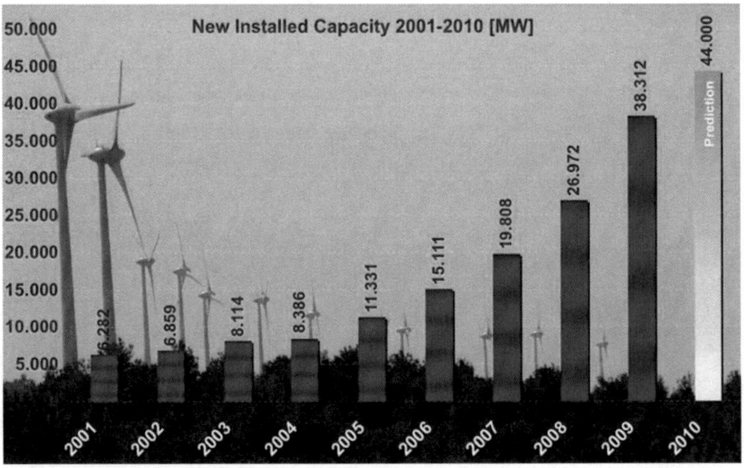

Fig. 14.3 World total new installed capacity [5]

This identifies the key tasks that must be undertaken in order to achieve a vision of over 2,000 GW of wind energy capacity by 2050 [6]. Figure 14.3 shows the world total new installed capacity up to 2010.

Therefore, a large number of wind turbine generators are going to be connected to the power system in the near future, percentage of wind farm output to the total power system capacity is expected to be fairly large. Wind farm composed of induction generators is considered in this work as it has some superior characteristics such as brushless and rugged construction, low cost, maintenance, and operational simplicity. But wind power is unsteady because wind speed is influenced by natural as well as meteorological situations. As the output power from wind farm fluctuating due to wind speed variations becomes large, fluctuations of the network frequency and voltage also become large. Though speed-governor system and pitch control system [7] can smooth the grid frequency and the wind farm output fluctuations up to a certain percentage, however, they are not sufficient to maintain network frequency to the desired level when the total wind power penetration into the grid is high. In this case, FACTS/ESS, i.e., FACTS with energy storage system (ESS), have recently emerged as more promising devices for power system applications [8].

Though every system has some advantages and at the same time some disadvantages, comparing among the ESS, superconducting magnetic energy storage systems (SMES) have received much attention among the researchers. The SMES is well known to be a system where energy is stored within a magnet that is capable of quickly releasing megawatt amounts of power. Since the successful commissioning test of the Bonneville Power Administration (BPA) 30 MJ unit [9], SMES systems have received much attention in power system applications. Thus SMES applications have been considered as new options to solve a variety of transmission, generation, and distribution system problems such as improvement of voltage and angular stability, increasing power transfer capability of existing grids, damping subsynchronous oscillations, damping inter-area oscillations, load leveling, etc. [10–12]. The SMES system is combined with the voltage-source IGBT converter which is capable of effectively controlling and instantaneously injecting both active and reactive powers into the power system. This ability of injecting/absorbing real or reactive power substantially enhances the controllability and provides operation flexibility to a power system and is therefore a prospective option in building FACTS. Therefore, SMES seems to be a viable and alternate solution to resolve the frequency fluctuation caused by wind farm.

Therefore, in this chapter, a relationship between SMES power rating and the smoothing ability is analyzed considering multiple wind generator-based wind farm model. Because the output power of real wind farm is, in general, much smoother than that of a single wind generator [13], and hence the required SMES power rating can be smaller than 55%, which is the result in [14]. It is expected that large SMES capacity gives better smoothing performance. However, large capacity will definitely increase the system overall cost. Therefore, the optimum size determination of SMES is one of the key points from the viewpoint of cost-effectiveness. So, in this chapter, an evaluation method of SMES power rating is presented in light of wind farm real power fluctuation. Moreover, the minimum energy storage capacity of SMES unit to mitigate the frequency fluctuation is determined. Finally, performances of the proposed SMES with required power rating and minimum energy storage capacity to mitigate the frequency fluctuation are evaluated by using PSCAD/EMTDC [15].

14.2 Overview of SMES

A superconducting magnetic energy storage system is a DC current device for storing and instantaneously discharging large quantities of power. The DC current flowing through a superconducting wire in a large magnet creates the magnetic field. The large superconducting coil is contained in a cryostat or Dewar consisting of a vacuum vessel and a liquid vessel that cools the coil. A cryogenic system and the power conversion/conditioning system with control and protection functions [16] are also used to keep the temperature well below the critical temperature of the superconductor. During SMES operation, the magnet coils have to remain in the superconducting status. A refrigerator in the cryogenic system maintains the required temperature for proper superconducting operation. A bypass switch is used to reduce energy losses when the coil is on standby. And it also serves other purposes such as bypassing DC coil current if utility tie is lost, removing converter from service, or protecting the coil if cooling is lost [17].

Figure 14.4 shows a basic schematic of a SMES system [18]. Utility system feeds the power to the power conditioning and switching devices that provides energy to charge the coil, thus storing energy. When a voltage sag or momentary power outage occurs, the coil discharges through switching and conditioning devices, feeding conditioned power to the load. The cryogenic (refrigeration)

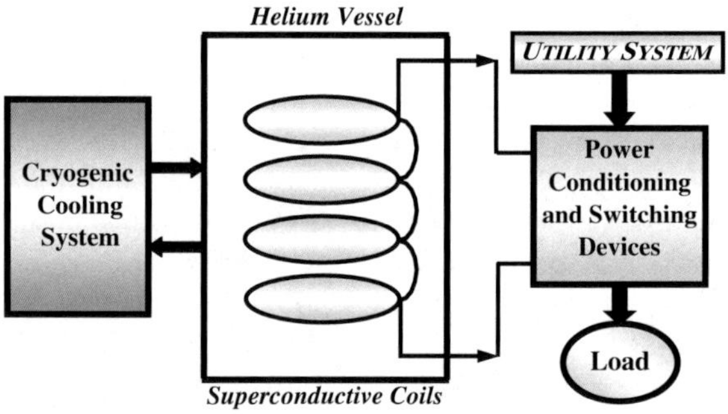

Fig. 14.4 Schematic diagram of the basic SMES system

system and helium vessel keep the conductor cold in order to maintain the coil in the superconducting state.

14.2.1 Advantages of SMES

There are several reasons for using superconducting magnetic energy storage instead of other energy storage methods. The most important advantages of SMES are that the time delay during charge and discharge is quite short. Power is available almost instantaneously and very high power output can be provided for a brief period of time. Other energy storage methods, such as pumped hydro or compressed air have a substantial time delay associated with the conversion of stored mechanical energy back into electricity. Thus if a customer's demand is immediate, SMES is a viable option. Another advantage is that the loss of power is less than other storage methods because the current encounters almost zero resistance. Additionally the main parts in a SMES are motionless, which results in high reliability. Also, SMES systems are environmentally friendly because superconductivity does not produce a chemical reaction. In addition, there are no toxins produced in the process.

The SMES is highly efficient at storing electricity (greater than 97% efficiency), and provide both real and reactive power. These systems have been in use for several years to improve industrial power quality and to provide a premium-quality service for individual customers vulnerable to voltage and power fluctuations. The SMES recharges within minutes and can repeat the charge/discharge sequence thousands of times without any degradation of the magnet [19]. Thus it can help to minimize the frequency deviations due to load variations [20]. However, the SMES is still an expensive device.

14.2.2 SMES for Load Frequency Control Application

A sudden application of a load results in an instantaneous mismatch between the demand and supply of electrical power because the generating plants are unable to change the inputs to the prime movers instantaneously. The immediate energy requirement is met by the kinetic energy of the generator rotor and speed falls. So system frequency changes though it becomes normal after a short period due to Automatic Generation Control. Again, sudden load rejections give rise to similar problems. The instantaneous surplus generation created by removal of load is absorbed in the kinetic energy of the generator rotors and the frequency changes. The problem of minimizing the deviation of frequency from normal value under such circumstances is known as the load frequency control problem. To be effective in load frequency control application, the energy storage system should be fast acting i.e., the time lag in switching from receiving (charging) mode to delivering (discharging) mode should be very small. For damping the swing caused by small load perturbations, the storage units for LFC application need to have only a small quantity of stored energy, though its power rating has to be high, since the stored energy has to be delivered within a short span of time.

14.3 Model System Considered for Simulation Analyses

The model system shown in Fig. 14.5 has been used in the simulation analyses in this chapter [13]. The model system consists of a wind farm (WF), a hydropower generator, SG1, two thermal power generators, SG2 and SG3, a nuclear power generator, SG4, and a load. The wind farm consists of five wind power generators (squirrel-cage induction machines, IGn, n = 1,2,…,5). SG1 and SG3 are operated under Load Frequency Control (LFC) mode, SG2 is under Governor Free (GF) control mode, and SG4 is under Load Limit (LL) operation [21]. LFC is used, in general, to control frequency fluctuations with a long period more than a few minutes, and GF is used to control fluctuations with a short period less than a minute. LL is used to output constant power. A SMES is connected to the wind farm terminal bus.

Q_{WF} and Q_{Load} are capacitor banks. Q_{WF} is used at the terminal of WF to compensate the reactive power demand of the wind generators at steady state. The value of the capacitor (0.45 pu) is chosen so that power factor of the wind power station during the rated operation without SMES installed becomes unity [13, 22]. Q_{Load} is used at the terminal of load to compensate the voltage drop by the impedance of transmission lines. The initial conditions and parameters of IG's and SG's are shown in Tables 14.1 and 14.2, respectively.

14.4 Governor and AVR Systems

In this study, different types of AVR and Governor Systems are considered for synchronous generators as explained below:

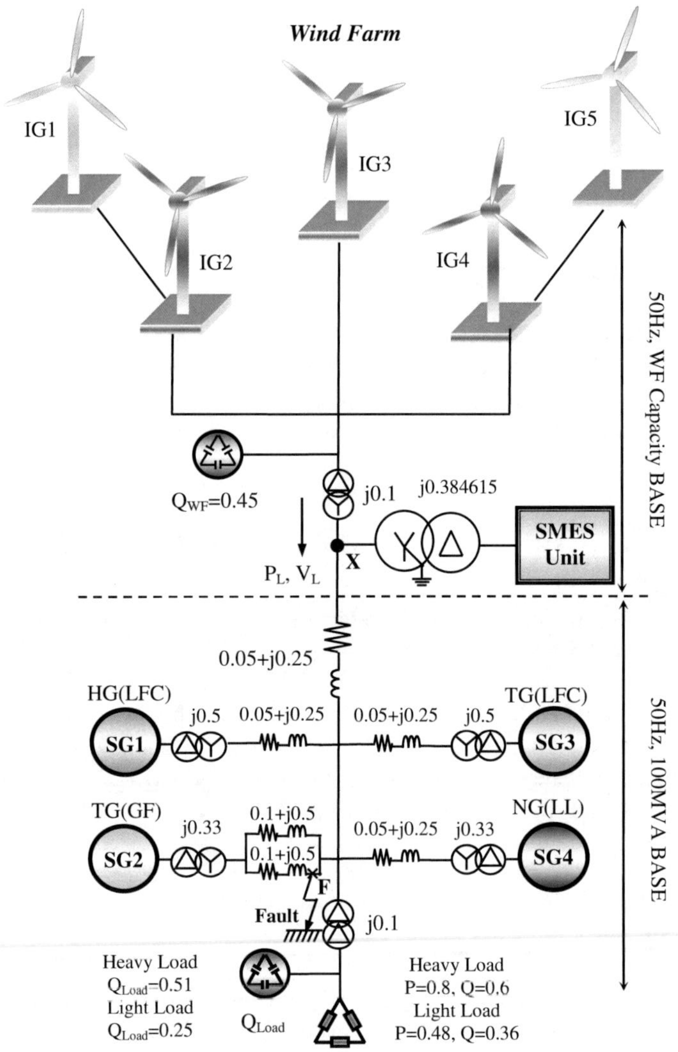

Fig. 14.5 Model system

Table 14.1 Initial conditions

	IG	SG1	SG2, SG3, SG4
P	0.1	1.00	1.00
V	1.00	1.05	1.05
Q	0.00	–	–
s(slip)	−1.733%	–	–

Table 14.2 Parameters of generators [13]

Wind generator (induction generator)	
	squirrel-cage type (IGn, n = 1,2…5)
MVA	2 (each)
R_1 (pu)	0.01
X_1 (pu)	0.10
X_m (pu)	3.5
R_{21} (pu)	0.035
X_{21} (pu)	0.03
R_{22} (pu)	0.014
X_{22} (pu)	0.098
2H s	1.5

Synchronous generators				
	Salient pole type (SG1)	Cylindrical type		
		SG2	SG3	SG4
MVA	20	30	20	30
Xd (pu)	1.2		2.11	
Xq (pu)	0.7		2.02	
H s	2.5		2.32	

14.4.1 Governor for Hydro, Thermal, and Nuclear Generators

The IEEE "non-elastic water column without surge tank" turbine model and "PID control including pilot and servo dynamics" speed-governing system [23] shown in Fig. 14.6 is used for synchronous generator, SG1. The IEEE generic turbine model and approximate mechanical-hydraulic speed-governing system [24] shown in Fig. 14.7 is used for synchronous generators, SG2, SG3, and SG4. In the governor models shown in Figs. 14.6 and 14.7, the values of P_{ref}, initial output, P_0, and turbine maximum output torque, $T_{m,max}$, are shown in Table 14.3, where, $\Delta\omega = \omega_{ref} - \omega$: the revolution speed deviation (pu), is set zero for SG1 and SG3 because these generators are operated under LFC to control frequency fluctuations with a relatively long period.

14.4.2 Automatic Voltage Regulator

Automatic Voltage Regulator (AVR) is used to keep the voltage of the synchronous generators constant. In the simulation analyses, IEEE alternator supplied rectifier excitation system (AC1A) [25] shown in Fig. 14.8 is used in the exciter model of all synchronous generators. Parameters of AVR model are shown in Table 14.4.

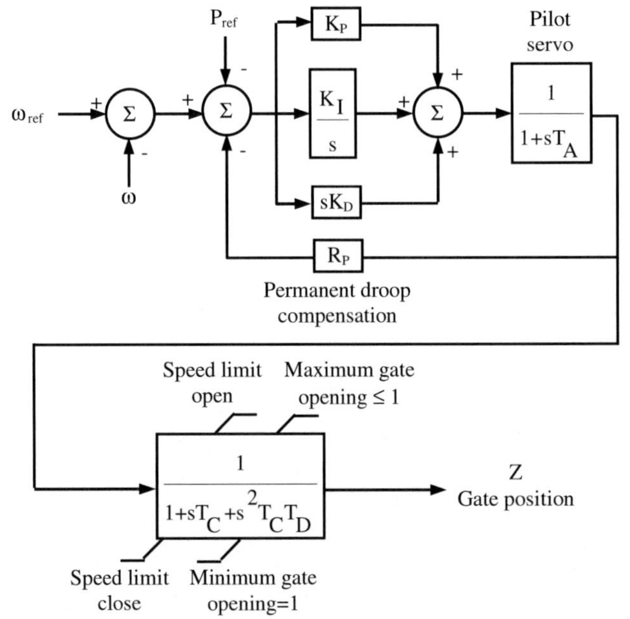

ω=speed, K_P=proportional gain [pu], K_I=integral gain[pu], K_D=derivative gain [pu], T_A=pilot servomotor time constant[s], T_C=gate servo gain[s], T_D=gate servomotor time constant[s], and R_p=permanent droop[pu]

Fig. 14.6 Hydro governor [23]

14.4.3 Load Frequency Control Model

In the Load Frequency Control (LFC), the control output signal is sent to LFC power plant when the frequency deviation is detected in the power system. Then, governor command signal and thus the output of LFC power plant is changed according to LFC signal. The frequency deviation is input into Low Pass Filter (LPF) to remove fluctuations with short period because the LFC is used to control frequency fluctuations with a long period. The LFC model used in this study is shown in Fig. 14.9, where, Tc: the LFC period = 200 s; ωc: the LFC frequency = 1/Tc = 0.005 Hz; ζ: the damping ratio = 1.

14.5 Method of Calculating Power System Frequency

In this study, the index of the smoothing effect is used in power system frequency analysis. Power system frequency fluctuation is occurred due to unbalance between supply and load power in power system [26, 27]. Then, the frequency fluctuation

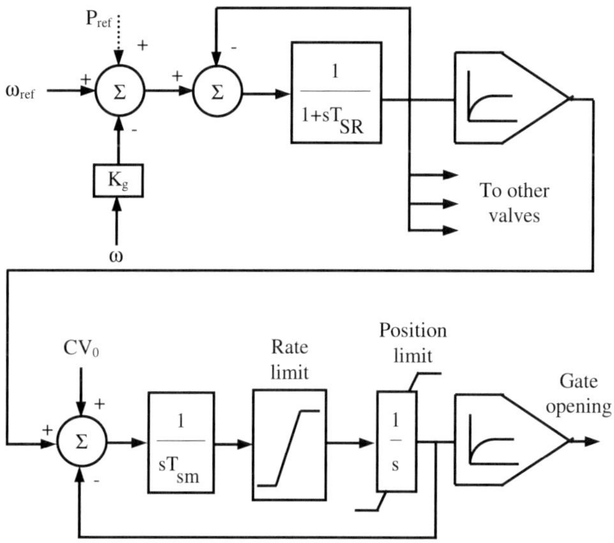

ω=speed, ω_{ref}=speed reference, K_g=inverse of regulation [1/pu], T_{sm}=gate servo time constant [s], T_{SR}=speed relay lag time constant[s], and CV_0=initial value of control valve flow area [pu]

Fig. 14.7 Thermal and nuclear governors [24]

Table 14.3 Values of P_{ref}, P_0 and $T_{m,MAX}$ [13]

Load (MVA)	Frequency control SG1(Hydro)	P_{ref}	P_0	Frequency control SG2(Thermal)	P_0	$T_{m,max}$
100 60	LFC	LFC signal	0.75	GF	0.8 0.4	No limit governor completely free
	SG3(Thermal)			SG4(Nuclear)		
100 60	LFC	LFC signal	0.70	LL	0.90	0.80

can be described by using two components, the rate of generator output variation, K_G (MW/Hz), and load variation, K_L (MW/Hz), respectively. They are representing the amount of power variation causing 1 (Hz) frequency fluctuation. When generator output variation, ΔG (MW), and load variation, ΔL (MW), are occurred, frequency fluctuation of the power system, ΔF (Hz), is expressed as follows:

$$\Delta F = \frac{\Delta G - \Delta L}{K_G + K_L} \tag{14.1}$$

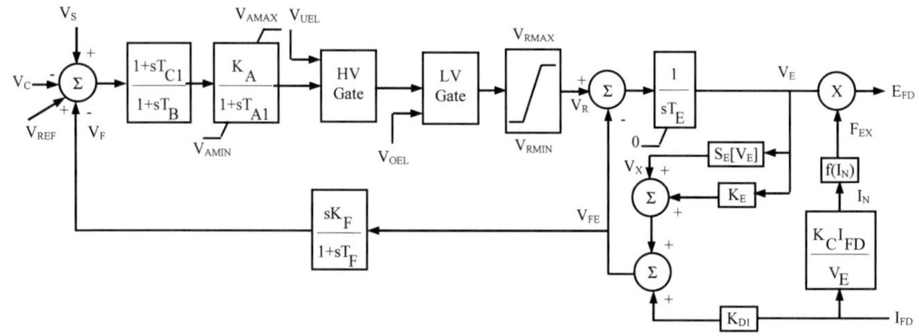

Fig. 14.8 AVR model system [25]

Table 14.4 Parameters of AVR

V_{REF} (pu)	1.05	K_C(pu)	0.20
K_A(pu)	400	K_D(pu)	0.38
T_A s	0.02	K_E(pu)	1.00
T_B s	0.00	T_E s	0.80
T_C s	0.00	K_F(pu)	0.03
V_{AMAX}(pu)	14.5	T_F s	1.00
V_{AMIN}(pu)	−14.5	$S_E V_{E1}$(pu)	0.10
V_{RMAX}(pu)	6.03	$S_E V_{E2}$(pu)	0.03
V_{RMIN}(pu)	−5.43	V_{E1}(pu)	4.18
V_{UEL}, V_{OEL}	Not applied	V_{E2}(pu)	3.14

Fig. 14.9 Load frequency control model

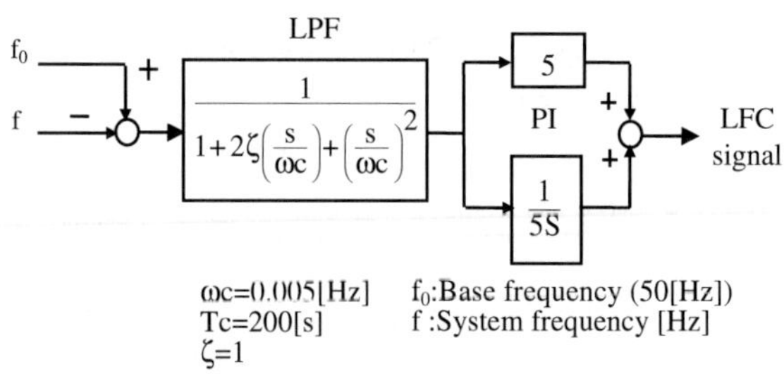

$$K = K_G + K_L \tag{14.2}$$

where, K is frequency characteristic constant.

In general, frequency characteristic is expressed as percentage K_G (expressed as $\%K_G$) for the total capacity of all generators and percentage K_L (expressed as $\%K_L$) for the total load. In general, it is known that $\%K_G$ and $\%K_L$ are almost constant and generally take a value of 8–15 and 2–6% (MW/Hz) , respectively. However, K_L and K_G change greatly during a day because the number of parallel generators changes depending on the amount of load during a day. And, when power imbalance ΔP is occurred in power system, frequency fluctuation $\Delta P/K$ cannot occur immediately due to the governor characteristic and generator

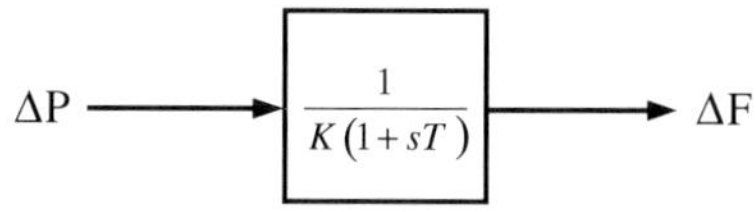

Fig. 14.10 Frequency calculation model

inertia. Normally, ΔF converges to a new steady state value in 2–3 s. In general, when ΔP is changing slowly, relationship between ΔP and ΔF can be expressed as follows:

$$\frac{\Delta F}{\Delta P} = \frac{1}{K(1 + sT)} \tag{14.3}$$

where, $\Delta P = \Delta G - \Delta L$.

Since changing load is not considered in this study, ΔL is “0”. Time constant, T (s), depending on the setting of generator governor and generator inertia, is generally 3–5 (s). In this study, power system capacity is assumed to be 100 (MW) and frequency characteristic K (MW/Hz) is selected to 8 (MW/Hz). This selection means that adjustability of the system frequency is weak, resulting a severe situation. Similarly, time constant T is selected to 3 s. In this study, frequency fluctuation in power system is evaluated by using Eq. 14.3. Therefore, frequency fluctuation, ΔF, is obtained as shown in Fig. 14.10.

14.5.1 Control System of SMES

The SMES system used is coupled (in Fig. 14.5) to the 66 kV line through a single step-down transformer (66/1.2 kV) with 0.384615 pu leakage reactance on the base value of 10 MVA, in this study. The proposed SMES [13] has the power rating and energy capacity of 2.6 MW and 312 MJ respectively, which will be explained later. Though SMES has virtually no resistance, the consideration of local LC resonance might be needed. However, as sub-synchronous resonance or shaft torsional oscillations are not the objective of this study, it is not considered here for simplicity. The control system of VSC used in this study is shown in Fig. 14.11. The SMES coil is charged or discharged by using the DC–DC chopper duty cycle (shown in Fig. 14.12). The parameters of PI controllers used in Figs. 14.11 and 14.13, which was determined by trial and error method are shown in Table 14.5.

14.5.2 Generation of Line Power Reference, P_{Lref}

IG line output power reference signal, P_{Lref}, is generated by the following way:

It is known from the results presented [28] that Low Pass Filter (LPF) method provides the best performance among the various reference generation methods

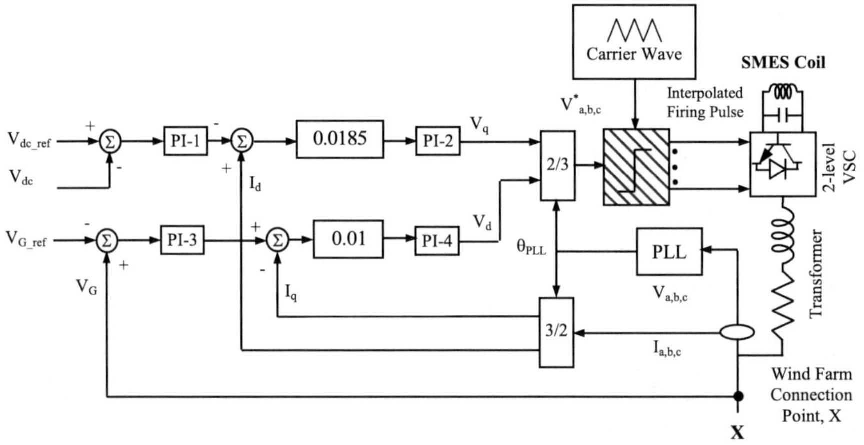

Fig. 14.11 Control system of the VSC [13]

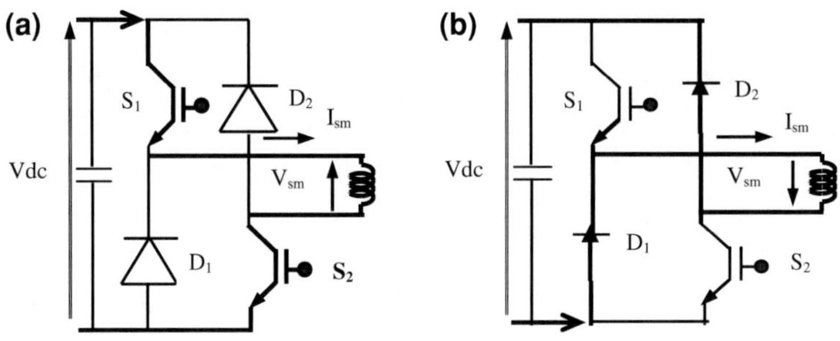

Fig. 14.12 The control concept of SMES charging and discharging mode

Fig. 14.13 Control system of two-quadrant DC–DC chopper

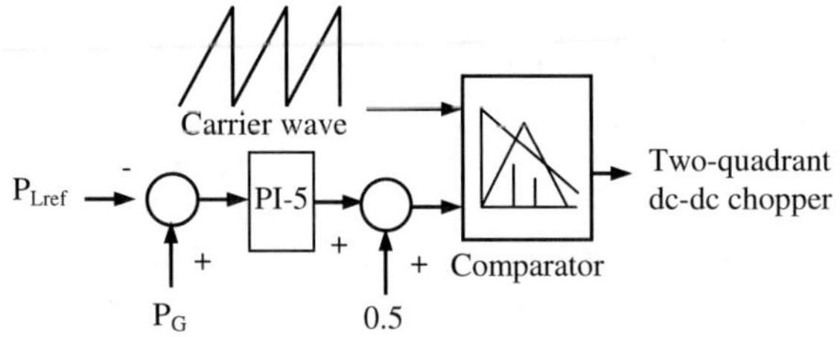

from the view points of the smoothing ability and energy storage capacity. Therefore, reference value of the transmission line power, P_{Lref}, is determined by using the LPF as shown in Fig. 14.14. The LPF suggests an increase or a decrease in the level of wind power output, which corresponds to charging or discharging of the stored energy.

Though it is very simple, it can be understood from Fig. 14.15 that reference value with enough smoothing effect can be obtained by using this type of LPF. Figure 14.15 shows an example how the time constant,T, affects the filtered wind

Table 14.5 Parameters of PI controllers

	PI-1	PI-2	PI-3	PI-4	PI-5
K_P	2.0	0.1	2.0	0.1	1.0
T_i	0.01	0.7	0.01	0.7	0.02

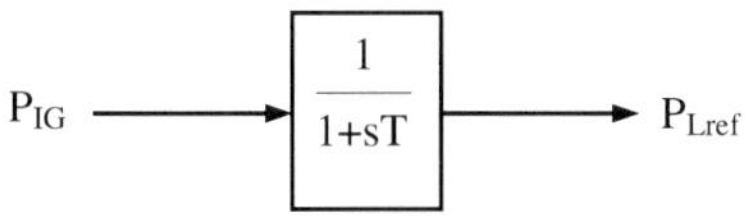

Fig. 14.14 Determination of reference line power

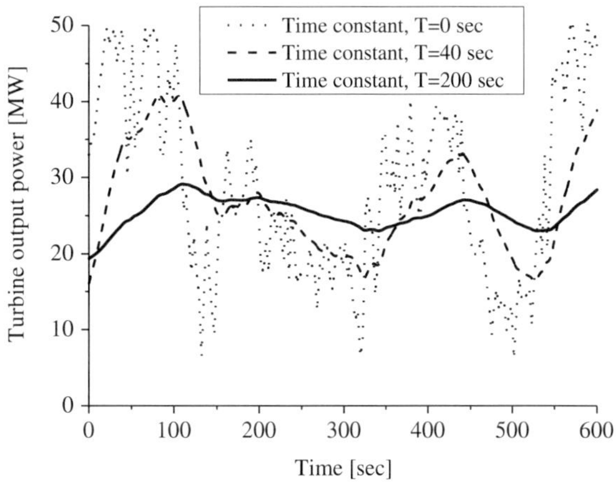

Fig. 14.15 Effect of time constant on the filtered wind power

power in practice. The values of T in the Fig. 14.15 correspond to energy storage system with different energy capacities. It is seen that the wind power fluctuation decreases as the LPF time constant increases. Therefore, if the transmission line power, P_{Lref}, is compensated according to the reference value, P_{Lref}, it is possible to decrease the system frequency fluctuation due to the wind generator output fluctuations.

The first-order passive low pass filter can be mathematically described as;

$$P_{\text{IG}} = P_{\text{Lref}} + T * P'_{\text{Lref}} \tag{14.4}$$

where T is the filtering time constant corresponding to energy storage capacity, P_{Lref} is the filter output function corresponding to the wind turbine output together with the storage unit, P'_{Lref} is the derivative of P_{Lref} and P_{IG} is the filter input function that corresponds to the wind turbine output without energy storage. When discrete data with a time step Δt are applied to a low pass filter and the derivative of P_{Lref} is expanded into a discrete form, Eq. 14.4 can be written for step k as

$$T\frac{P'_{\text{Lref,k}} - P'_{\text{Lref,k-1}}}{\Delta t} + P'_{\text{Lref,k}} = P_{IG,k} \tag{14.5}$$

Solving for $P_{\text{Lref, k}}$ gives

$$P_{\text{Lref},k} = \frac{T}{T + \Delta t} P_{\text{Lref},k-1} + \frac{\Delta t}{T + \Delta t} P_{\text{IG},k} \tag{14.6}$$

Defining a constant $\beta = \frac{T}{T+\Delta t}$, Eq. (14.6) can be rewritten as

$$P_{\text{Lref},k} = \beta P_{\text{Lref},k-1} + (1 - \beta) P_{\text{IG},k} \tag{14.7}$$

Now Eq. (14.7) has the form of an exponentially weighted moving average (EWMA) filter [29]. The subscript k corresponds to time, i.e. $t_k = t_0 + k\Delta t$, where Δt is the time step and t_0 is the starting point of the analysis.

With an EWMA filter the response of the energy storage system is

$$P_{\text{st},k} = P_{\text{IG},k} - P_{\text{Lref},k} \tag{14.8}$$

where $P_{\text{st,k}}$ is the power absorbed by the storage unit. Thus, the level of the stored energy in the system is in discrete form as

$$\mathrm{E}_k = \sum_{m=1}^{k} P_{\text{st},m} \Delta t \tag{14.9}$$

The energy storage capacity used for damping the fluctuations is then defined as

$$\mathrm{E}_{\text{storage}} = \max \mathrm{E}_k - \min \mathrm{E}_k, \quad \text{for } k = 1 \ldots \text{n} \tag{14.10}$$

where n is the total number of time points in the data sample.

14.6 Analysis of SMES Power Rating

Since a large number of wind turbine generators are going to be connected to power system in the near future, percentage of wind farm output to the total power system capacity is expected to be fairly large, and thus 10% (10 (MW)) wind power penetration is assumed in this study [13]. In this section, the relationship between SMES power rating and the smoothing ability is investigated by evaluating a wind farm output, P_{WF}, and reference value of transmission line power, P_{Lref}.

An interesting study has been performed in [14], where the SMES power rating to minimize the frequency fluctuation is determined by using single wind generator model which represents the aggregated wind farm. It is reported therein that SMES power rating of 55% of that of the wind farm is required to mitigate the frequency fluctuation of the grid. However, aggregated model of wind farm is not sufficient for the analysis because the output power of real wind farm is, in general, much more smooth than that of a single wind generator, and hence the required SMES power rating can be smaller than 55%, which is the result in [14]. Therefore, in this chapter the multiple wind generator-based wind farm model is used instead of a

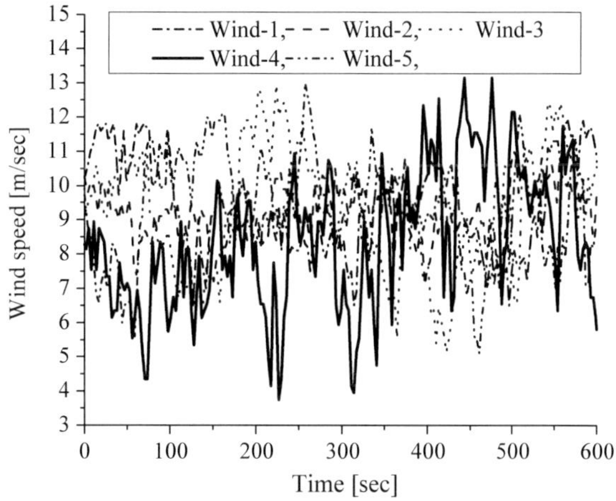

Fig. 14.16 Responses of wind speed data [Case-I & II]

single wind generator to determine the power rating of SMES unit more precisely. Then, the minimum energy storage capacity of SMES unit is also determined.

In this study, a wind farm of five wind generators with different wind speed patterns with relatively large fluctuations shown in Fig. 14.16 and Table 14.6, respectively is used in the analysis. In order to estimate a required power rating of the SMES, smoothing effect of the wind farm output is investigated by using P_{Lref} which is obtained through EWMA filter with considering several time constants. The effect of energy storage capacity on grid frequency fluctuations is also discussed.

SMES output is obtained as $P_{ot,k}$ of Eq. (14.8) in this analysis, and then a standard deviation of the SMES output, σ, is calculated. In addition, smoothing effect is evaluated by using frequency fluctuation, Δf. Power rating of the SMES required for smoothing wind farm output and EWMA time constant suitable for the reference value with enough smoothing effect are investigated by using σ and Δf in this analysis. Table 14.7 shows σ and maximum Δf for each EWMA time constant.

Figure 14.17 shows the maximum frequency fluctuation with respect to EWMA time constant. Table 14.7 shows that the frequency fluctuation decreases as EWMA time constant increases. Therefore, if the transmission line power is compensated according to the reference value, P_{Lref}, it is possible to decrease the system frequency fluctuation due to the wind farm output fluctuations. It is clear from Fig. 14.17 that the maximum frequency fluctuation is very small when EWMA time constant is over about 120 s.

Figure 14.18 shows standard deviation, σ, of the SMES output with respect to EWMA time constant. σ increases as EWMA time constant increases. However, as can be seen from the figure, the function of σ is not monotonous and it saturates where EWMA time constant is over about 120 s. Therefore, almost no improvement can be obtained by adopting longer EWMA time constant than 120 s and corresponding power rating of SMES. From these results, the reference value of

Table 14.6 Wind speed condition of each generator [13]

Wind data's name	Average wind speed (m/s)		Standard deviation of wind speed (m/s)	
Wind-1	Medium	9.44	Medium	1.39778
Wind-2	Medium	9.20	Medium	1.28063
Wind-3	Medium	9.30	Large	1.79351
Wind-4	Medium	8.27	Large	1.98949
Wind-5	Medium	9.42	Medium	1.55302

Table 14.7 Standard deviation of SMES output and maximum frequency fluctuation [13]

EWMA time constant (s)	10 MW wind farm	
	Δf (Hz)	σ (MW)
0	0.04320	0.00000
30	0.00651	0.65140
60	0.00419	0.75406
90	0.00317	0.84607
120	0.00253	0.86469
150	0.00209	0.86984
180	0.00178	0.87161
210	0.00154	0.87786
240	0.00135	0.88310
270	0.00120	0.88631
300	0.00108	0.88813

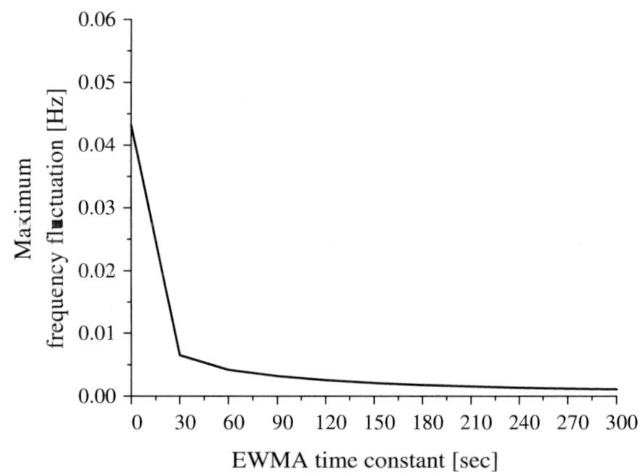

Fig. 14.17 Response of frequency fluctuation

transmission line power corresponding to σ for 120 s of EWMA time constant can be considered sufficient for the smoothing control. Consequently it can be said that, if 120 s time constant is adopted in EWMA filter, the suitable reference value with enough smoothing effect can be obtained.

If the power rating of SMES is determined based on the value of 3σ, approximately 99.7% of necessary smoothing effect can be achieved according to the characteristics of standard deviation as shown in Fig. 14.19. Therefore, if the

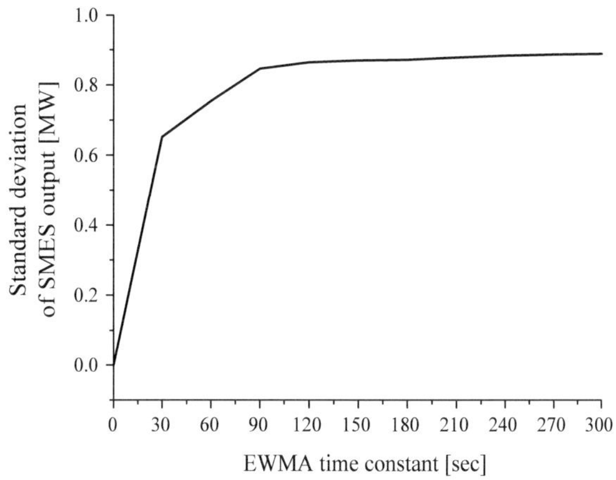

Fig. 14.18 Response of standard deviation of SMES output

SMES power rating is determined to 2.6 (MW) ($3\sigma = 3 \times 0.86469 = 2.59407 \approx 2.6$), it can be considered to be sufficient for the smoothing control of 10 (MW) wind farm. In the following simulation analyses, the SMES with 2.6 (MW) power rating is used with considering the same wind speed patterns and the effect of energy storage capacity of SMES on grid frequency fluctuations is investigated. Finally, the minimum energy storage capacity of SMES unit is determined.

14.7 Simulation Results

Simulation analyses are carried out to investigate the performance of the proposed controlled SMES [13]. The power capacity of SMES is 26% of that of the wind farm and several values are considered for its energy capacity in the analyses to determine its optimal value. The analyses have been performed by using PSCAD/EMTDC. Two cases are considered as given below:

Case-I, light load: The load is 60 MVA and all generators are in service except SG3. This case is more severe than Case-II from a viewpoint of system frequency control.

Case-II, heavy load: The load is 100 MVA and all generators are in service.

The real wind speed data shown in Fig. 14.16 is applied to each wind generator. The time step and simulation time have been chosen as 0.00001 and 600 s, respectively. Three energy capacities are considered for SMES in the analyses, which are 60, 90, and 120 s of the power rating (for example, the capacity is 2.6 MW $\times$ 120 s = 312 MJ in the case of 120 s). Figure 14.20 shows the responses of SMES energy storage level for the wind speed data of Fig. 14.16. Figures 14.21 and 14.22 show the responses of the line power in the cases of

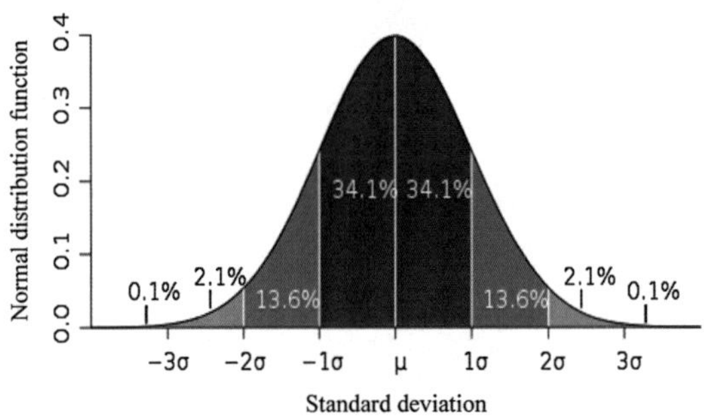

Fig. 14.19 Characteristics of standard deviation

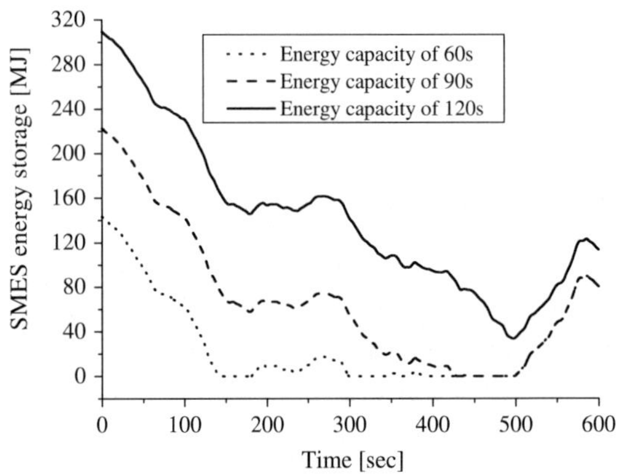

Fig. 14.20 Responses of SMES energy storage level [Case-I & II]

SMES energy capacity of 60 and 90 s for Case-I. It is clearly seen from the figures that the wind farm output cannot be smoothed in these cases. Also the system frequency cannot be maintained within the acceptable range as seen from Fig. 14.23. Similarly, it is seen from Fig. 14.24 that the system frequency cannot be maintained within the acceptable range also for Case-II.

As a result, 120 s can be expected to be optimal for the energy capacity of SMES unit, and detailed simulation analyses are performed using this value as shown in the following:

Figure 14.25 shows the wind farm output, which is fluctuating due to the wind speed variations in the case without SMES. But when SMES of 120 s energy capacity is installed, the line power can be smoothed effectively. Figures 14.26 and 14.27 show the output of hydro-power generator (SG1) and thermal power generator (SG3) respectively in the cases with and without SMES considered which are comparatively smooth. This is because these generators are operated under LFC to control the electric power fluctuations with long period. As SMES provides proper compensation for randomly varying wind farm output, SG1 and SG3 are generating comparatively less power to supply to the load as shown in Figs. 14.26 and 14.27 respectively for both cases. Figure 14.28 shows the thermal power generator (SG2) output. The response without SMES is fluctuating so much

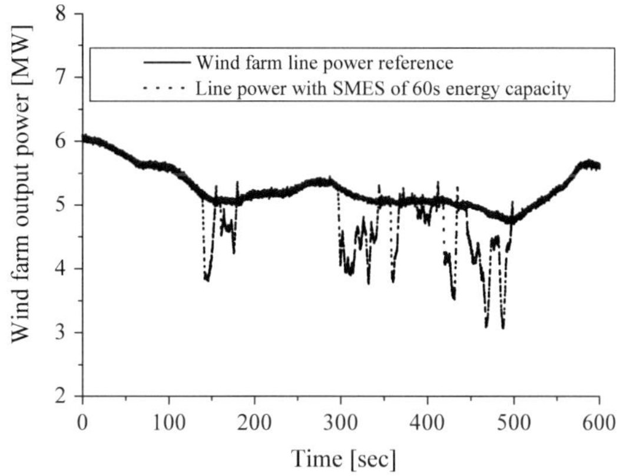

Fig. 14.21 Responses of wind farm line real power [Case-I]

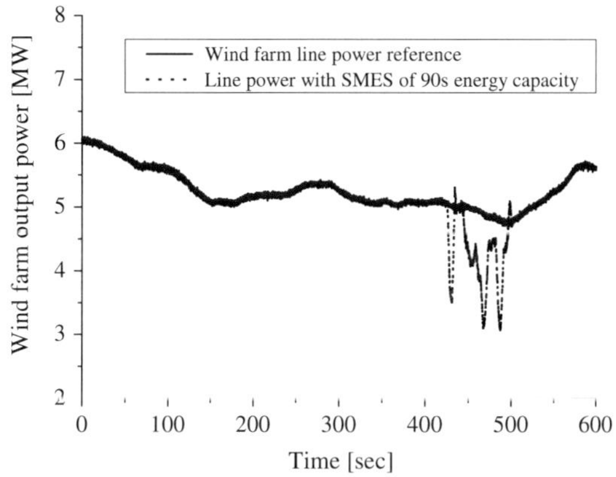

Fig. 14.22 Responses of wind farm line real power [Case-I]

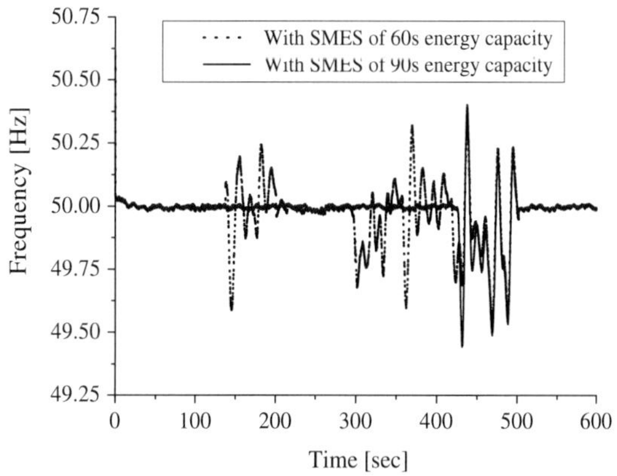

Fig. 14.23 Responses of power system frequency [Case-I]

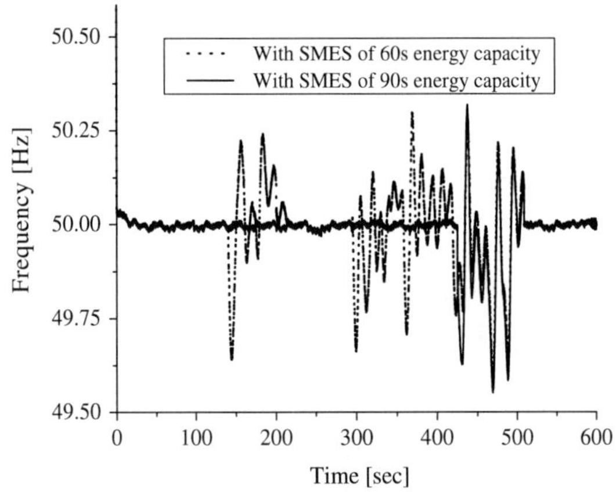

Fig. 14.24 Responses of power system frequency [Case-II]

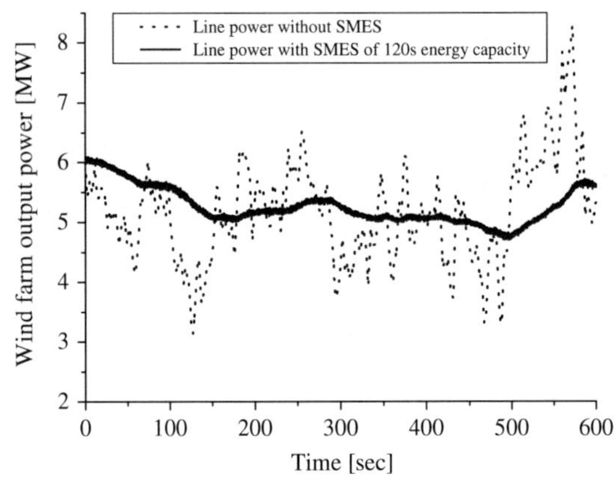

Fig. 14.25 Responses of wind farm line real power [Case-I & II]

because this generator is operated under GF to control the electric power fluctuations with short period.

However, in the case with considering SMES, the response does not vary so much because the grid power from the wind farm is smooth as shown in Fig. 14.25. Figure 14.29 shows the nuclear power generator (SG4) output, where the responses are maintained almost constant because this generator is operated under LL operation for both cases. Figure 14.30 shows the response of the SMES real power. It is seen that under the condition of randomly varying wind speed the SMES provide proper compensation of real power according to the variation of line power to maintain the grid frequency. Figures 14.31 and 14.32 show the power system frequency with and without using SMES for Case-I and II, respectively. When the power capacity of the wind farm is relatively large compared with that of the power system, the power system frequency cannot be maintained well by the frequency control of synchronous generators. But it can be maintained well to the rated value by the proposed controlled SMES. Moreover, frequency fluctuation without SMES is bigger in Case-I (light load) than in Case-II

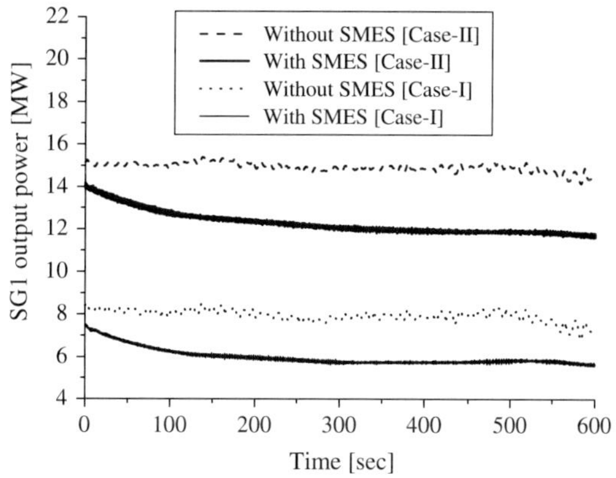

Fig. 14.26 Responses of hydro-power generator (SG1) outputs

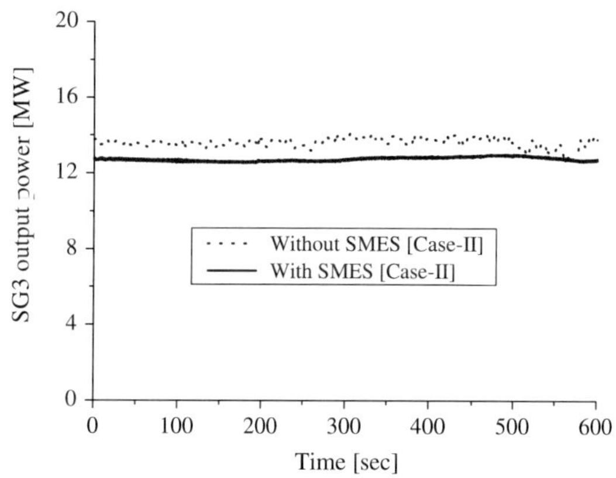

Fig. 14.27 Responses of thermal power generator (SG3) outputs

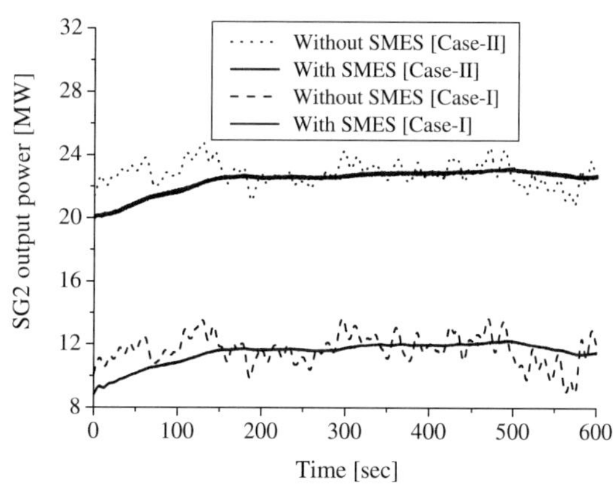

Fig. 14.28 Responses of thermal power generator (SG2) outputs

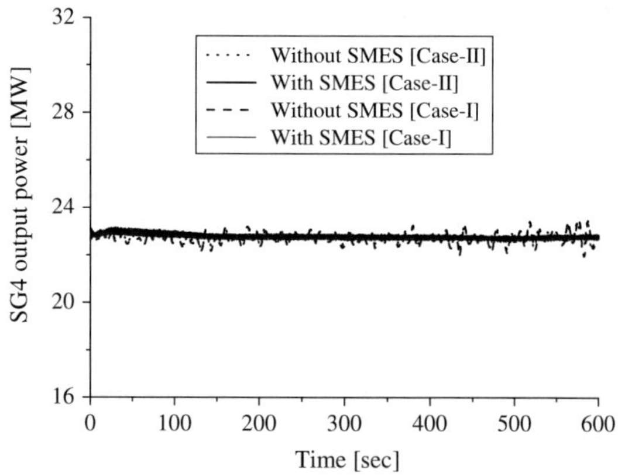

Fig. 14.29 Responses of nuclear power generator (SG4) outputs

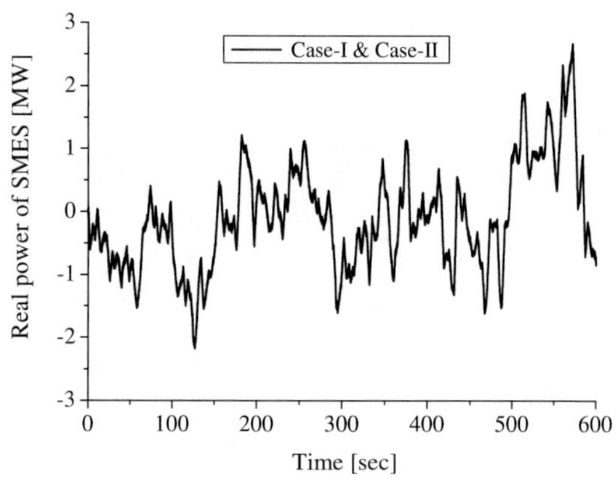

Fig. 14.30 Response SMES real power [Case-I & II]

(heavy load) as shown in Figs. 14.31 and 14.32, respectively. This is because one of the LFC synchronous generators must be stopped during the light load. Table 14.8 shows the maximum frequency fluctuation in the cases without and with SMES unit for each energy storage capacity.

Wind farm grid voltage can also be maintained constant by using the proposed SMES system as seen from Fig. 14.33. This fact indicates that the proposed controlled SMES can also decrease the voltage fluctuations.

From the simulation results shown in Figs. 14.25, 14.26, 14.27, 14.28, 14.29, 14.29, 14.30, 14.31, 14.32 and 14.33 and Table 14.8, it is seen that suitable reference value for the wind farm output can be obtained and then sufficient smoothing effect can be achieved by using the proposed SMES system.

Finally, it is concluded that the line power reference generation scheme using EWMA is very effective and the proposed SMES system can provide sufficient smoothing effect on the wind farm output as well as the grid system frequency fluctuations.

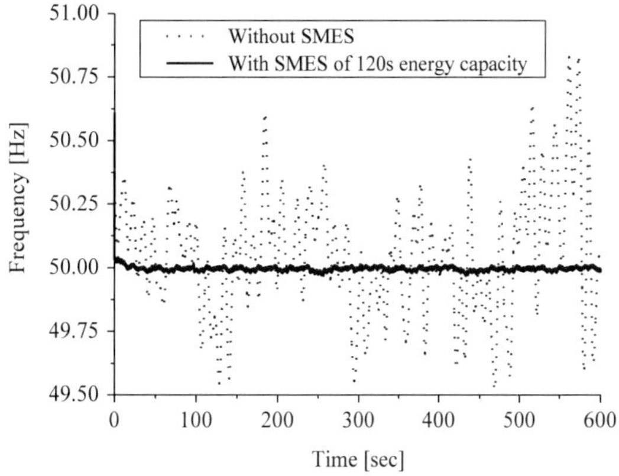

Fig. 14.31 Responses of power system frequency [Case-I]

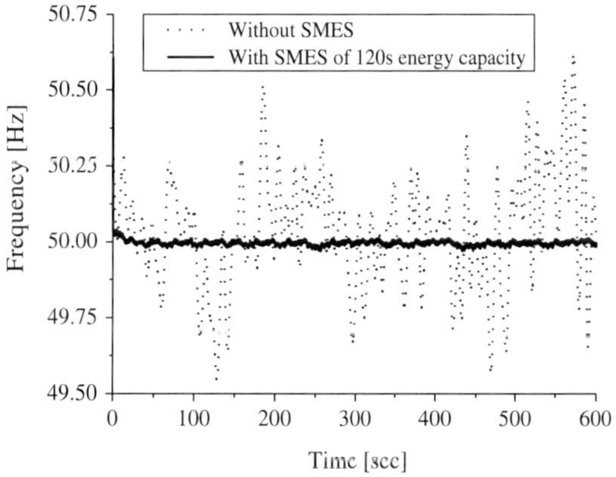

Fig. 14.32 Responses of power system frequency [Case-II]

Table 14.8 Maximum frequency fluctuation in each condition [13]

Load (MVA)	Wind farm capacity 10 MVA		
	Δf (Hz) without SMES	SMES energy capacity (s)	Δf (Hz) with SMES
60	0.83671	60	0.53416
		90	0.55683
		120	0.02408
100	0.62123	60	0.43312
		90	0.44866
		120	0.02275

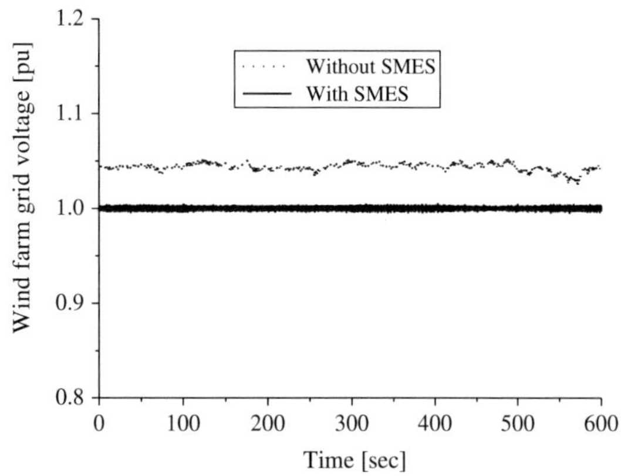

Fig. 14.33 Responses of wind farm terminal voltages

14.8 Conclusions

This chapter presents a sinusoidal PWM voltage-source converter and two-quadrant DC–DC chopper-controlled SMES for smoothing output power fluctuations of wind farm in order to maintain the grid frequency deviation within an acceptable range. The method of determining the power rating of the SMES is also presented. An EWMA filter is used to generate the reference value for the wind farm output. The effect of the smoothing control is evaluated using a power system model installed with the SMES unit, which has a power rating of 26% of the wind farm capacity and an energy capacity of 2 min multiplied by the power rating. These values of the SMES power rating and energy capacity are found to be optimum for the wind speed pattern obtained in Hokkaido, Japan, in which the speed fluctuation is very large compared with that in Europe. The simulation analyses show that, using the proposed SMES system, the wind farm output fluctuations can be decreased, and hence the frequency of the grid system, can be maintained within an acceptable range. Therefore, the integration of the proposed SMES system into a wind farm can be an effective means of mitigating the frequency fluctuations of the grid system.

References

1. http://en.wikipedia.org/wiki/Energy_in_the_United_States
2. Simoes MG, Farret FA (2004) Renewable energy system, design and analysis with induction generators. CRC Press, Washington, D.C
3. Heier S (1998) Grid integration of wind energy conversion systems. Wiley, Chicester, UK
4. http://www.olino.org/us/articles/category/wind-energy
5. World Wind Energy Association; Available online in http://www.wwindea.org/home/index.php
6. http://www.olino.org/us/articles/2009/12/03/technology-roadmap-wind-energy-iea

7. Yamazaki T, Takahashi R, Murata T, Tamura J, Fukushima T, Sasano E, Shinya K, Matstumoto T (2009) Smoothing control of wind generator output fluctuations by new pitch controller. IEEJ Trans Power Energy 129(7):880–888
8. Zhang L, Shen C, Crow ML, Dong L, Pekarek S, Atcitty S (2005) Performance indices for the dynamic performance of FACTS and FACTS with energy storage. Electr Power Compo Syst 33(3):299–314
9. Boenig HJ, Hauer JF (1985) Commissioning tests of the Bonneville power administration 30 MJ superconducting magnetic energy storage unit. IEEE Trans Power Apparatus Syst 104(2):302–309 PAS
10. Mitani Y, Tsuji K, Murakami Y (1988) Application of superconducting magnetic energy storage to improve power system dynamic performance. IEEE Trans Power Syst 3: 1418–1425
11. Banerjee S, Chatterjee JK, Tripathy SC (1990) Application of magnetic energy storage unit as load frequency stabilizer. IEEE Trans Energ Convers 5:46–51
12. Wu CJ, Lee YS (1991) Application of superconducting magnetic energy storage unit to improve the damping of synchronous generator. IEEE Trans Energ Convers 6(4):573–578
13. Sheikh MRI, Muyeen SM, Takahashi R, Murata T, Tamura J (2010) Smoothing Control of Wind Generator Output Fluctuations by PWM Voltage Source Converter and Chopper Controlled SMES. European Transactions on Electrical Power, 21(1):1–18, Published online in Wiley InterScience (http://www.interscience.wiley.com). DOI: 10.1002/etep.469
14. Asao T, Takahashi R, Murata T, Tamura J, Kubo M, Kuwayama A, Matsumoto T (2007) Smoothing control of wind power generator output by superconducting magnetic energy storage system. ICEMS, Seoul, Korea, pp 302–307
15. PSCAD/EMTDC Manual (1994) Manitoba HVDC Research Center
16. IEEE task force on benchmark models for digital simulation of FACTS, custom–power controllers, T&D committee, (2006) Detailed modeling of superconducting magnetic energy storage (SMES) system. IEEE Trans Power Delivery 21(2):699–710
17. Ali MH, Murata T, Tamura J (2008) Transient stability enhancement by fuzzy logic-controlled SMES considering coordination with optimal reclosing of circuit breakers. IEEE Trans Power Syst 23(2):631–640
18. http://www.doc.ic.ac.uk/ ~ matti/ise2grp/energystorage_report/node8.html
19. http://en.wikipedia.org/wiki/Superconducting_magnetic_energy_storage
20. Demiroren A, Yesil E (2004) Automatic generation control with fuzzy logic controllers in the power system including SMES units. Int J Electr Power Energ Syst 26:291–305
21. IEE of Japan, Standard Models of Electrical Power System.Technical Reports, 754:40 43
22. Sheikh MRI, Muyeen SM, Takahashi R, Murata T, Tamura J (2008) Wind generator stabilization by PWM voltage source converter and chopper controlled SMES. J Int Rev Autom Control (I.RE.A.CO) 1(3):311–320
23. Working group on prime mover, energy supply models for system dynamic performance studies (1992) Hydraulic turbine and turbine control models for system dynamic studies. IEEE Trans Power Syst 7(1):167–179
24. Working group on prime mover, energy supply models for system dynamic performance studies (1991) Dynamic models for fossil fuelled steam units on power system studies. IEEE Trans Power Syst 6(2):753–761
25. IEEE recommended practice for excitation system models for power system stability studies, IEEE Std. 421.5-1992
26. Koike T (1979) Electric power transmission and distribution. Youkendo. Co. Ltd, Tokyo, Japan
27. Sekine Y (1966) Power system engineering. Denkishoin. Co. Ltd, Tokyo
28. Sheikh MRI (2010) Stabilization of a grid-connected wind farm by using SMES. Ph.D. Thesis
29. Paatero JV, Lund PD (2005) Effect of energy storage on variations in wind power. Wind Energ 8:421–441

Part III
Offshore Trends

Chapter 15
Space-Based Observation of Offshore Strong Wind for Electric Power Generation

W. Timothy Liu and Xiaosu Xie

Abstract To optimize the deployment of offshore wind farms for electric power generation, the geographical and seasonal distributions of strong wind over global oceans were examined using nine years of equivalent neutral winds measured by a space-based scatterometer. The relation between scatterometer measurement and surface wind vector is explained. The dependence of wind strength on height and stability is examined. Near-shore locations of strong wind are identified.

15.1 Introduction

Offshore wind farms, together with better power transmission and storage technology, will lessen our dependence on fossil fuel. With the increasing demand of electric power and the need for reducing greenhouse gas emission, the importance of turning wind energy at sea into electric power has never been more evident. New technology has also enabled floating wind farms in the open seas to capture the higher wind energy and reduce the environmental impact. Knowledge of the distribution of wind strength at sea is needed to optimize the deployment of such wind farms [16, 18].

Just a few decades ago, almost all ocean wind measurements came from merchant ships. However, the quality and geographical distribution of these wind reports were uneven. Today, operational numerical weather prediction (NWP) also gives us wind information [2], but NWP depends on numerical models, which are limited by our knowledge of the physical processes and the availability of data.

W. T. Liu (✉) · X. Xie
Jet Propulsion Laboratory, California Institute of Technology,
4800 Oak Grove Dr., Pasadena, CA 91109, USA
e-mail: w.t.liu@jpl.nasa.gov

S. M. Muyeen (ed.), *Wind Energy Conversion Systems*,
Green Energy and Technology, DOI: 10.1007/978-1-4471-2201-2_15,

Recently, space-based scatterometers are giving us wind speed and direction with sufficient temporal and spatial sampling, night and day, under clear and cloudy conditions. The principles of wind retrievals are described in Sect. 15.2. The scatterometer measures ocean surface stress [19]; the relation between stress and wind depends on atmospheric stability (governed by vertical wind shear and density stratification); the stability effect is described in Sect. 15.4. The effective heights of various designs of the wind turbines, from the lower floating turbine that spins around a vertical axis to the higher anchored one that spins around a horizontal axis, are likely to be different. The dependence of power generation on turbine height has been recognized (e.g., [1]). Vertical wind shear that causes the dependence of wind strength on height is discussed in Sect. 15.3.

The climatology of scattorometer data in forms of mean wind (e.g., [21]), frequency of strong wind [22], and power density [17] has been produced. Global distribution of wind strength is discussed in Sect. 15.5. What is most relevant to power generation is not the distribution of wind strength in the open ocean, but the locations of strong winds near land where the power transmission problem is not prohibiting. Such strong winds are caused by the presence of land mass and will be discussed in Sect. 15.6.

15.2 Scatterometer

The capability of the space-based scatterometer in measuring wind vector at high spatial resolution is discussed by Liu [9] and Liu and Xie [12]. The scatterometer sends microwave pulses to the Earth's surface and measures the backscatter power. Over the ocean, the backscatter power is largely caused by small centimeter-scale waves on the surface, which are believed to be in equilibrium with stress (τ). Stress is the turbulent momentum transfer generated by vertical wind shear and buoyancy. Liu and Large [10] demonstrated, for the first time, the relation between measurements by a space-based scatterometer and surface stress measured on research ships. Although the scatterometer has been known to measure τ, it has largely been promoted as a wind-measuring instrument because the public is more familiar with wind than stress. The geophysical data product of the scatterometer is the equivalent neutral wind, U_N, at 10 m height [11], which is uniquely related to τ by definition, while the relation between τ and the actual winds at the reference level depends on atmosphere stability and ocean's surface current. U_N has been used as the actual wind, particularly in operational weather applications. The difference between U_N and the actual wind is assumed to be negligible because the marine atmosphere has near neutral stratification, and the magnitude of ocean current is small relative to wind speed over most ocean areas.

Because stress is small-scale turbulence generated by buoyancy (vertical density gradient) and wind shear, its magnitude should have strong spatial coherence with sea surface temperature and its direction should show influence by current. These small turbulences are driven by ocean processes and may not be fully represented in winds that are subjected to larger scale atmospheric factors, particularly over ocean fronts that have strong temperature gradient and current shears [13, 15]. Traditional concept of the increase of wind with stress may also break down under hurricane scale strong wind because of flow separation [19].

NASA launched a Ku-band scatterometer, QuikSCAT, in June 1999. Level-2 data at 12.5 km resolution are obtained from the Physical Oceanography Distributed Active Archive Center. Nine years of the data, from January 2000 to December 2008, organized in wind vector cells along satellite swath, are binned into uniform 1/8° grids over global oceans and used in this study.

15.3 Height Dependence

There is a long history of studying the wind profile in the atmospheric surface (constant flux) layer in terms of turbulent transfer. The flux-profile relation (also called similarity functions) of wind, as described by Liu et al. [14], is

$$\frac{U - U_s}{U_*} = 2.5\left(\ln\frac{Z}{Z_0} - \psi\right) = \frac{1}{\sqrt{C_D}} \tag{15.1}$$

where U_s is the surface current, $U_* = (\tau/\rho)^{1/2}$ is the friction velocity, ρ is the air density, Z_o is the roughness length, Ψ is the function of the stability parameter, and C_D is the drag coefficient. The stability parameter is the ratio of buoyancy to shear production of turbulence. The effect of sea state and surface waves (e.g., [3]) are not included explicitly in the relation. U_* and Z_o are estimated from the slope and zero intercept respectively of the logarithmic wind profile. The drag coefficient is an empirical coefficient in relating τ to ρU^2 [7, 8, 24] and is often expressed as a function of wind speed. An alternative to using the drag coefficient is to express Z_o as a function of U_*. For example, Liu and Tang [11] incorporated such a relation in solving the similarity function. They combined a smooth flow relation with Charnock's relation in rough flow to give

$$Z_o = 0.11\frac{\upsilon}{U_*} + 0.011\frac{U_*^2}{g} \tag{15.2}$$

where v is the kinematic viscosity and g is the acceleration due to gravity.

In general oceanographic applications, the surface current is assumed to be small compared with wind and the atmosphere is assumed to be nearly neutral. With the neglect of U_s and Ψ in Eq. 15.1, U becomes U_N by definition. The wind speed at a certain height z, U_z, relative to U_N at 10 m, $U_{10,}$ is given by

$$\frac{U_z}{U_{10}} = 1 + 2.5\sqrt{C_D}\ln\left(\frac{z}{10}\right) \tag{15.3}$$

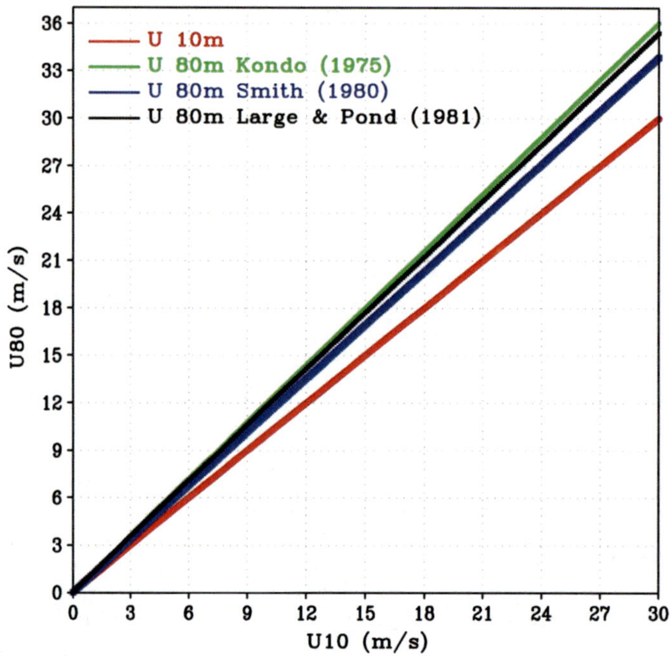

Fig. 15.1 Wind speed at 80 m height as a function of wind speed at 10 m under neutral stability for three formulations of drag coefficient

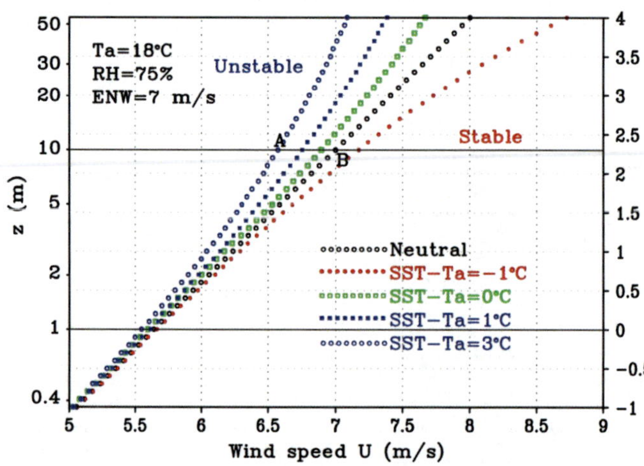

Fig. 15.2 Comparison of wind profiles under various stability conditions

and z is in meter. Figure 15.1 shows the variation of wind speed at 80 m as a function of wind speed at 10 m, under neutral conditions for three formulations of the drag coefficient. For example, the 80 m wind exceeds U_{10} of 10 m/s by 5% and

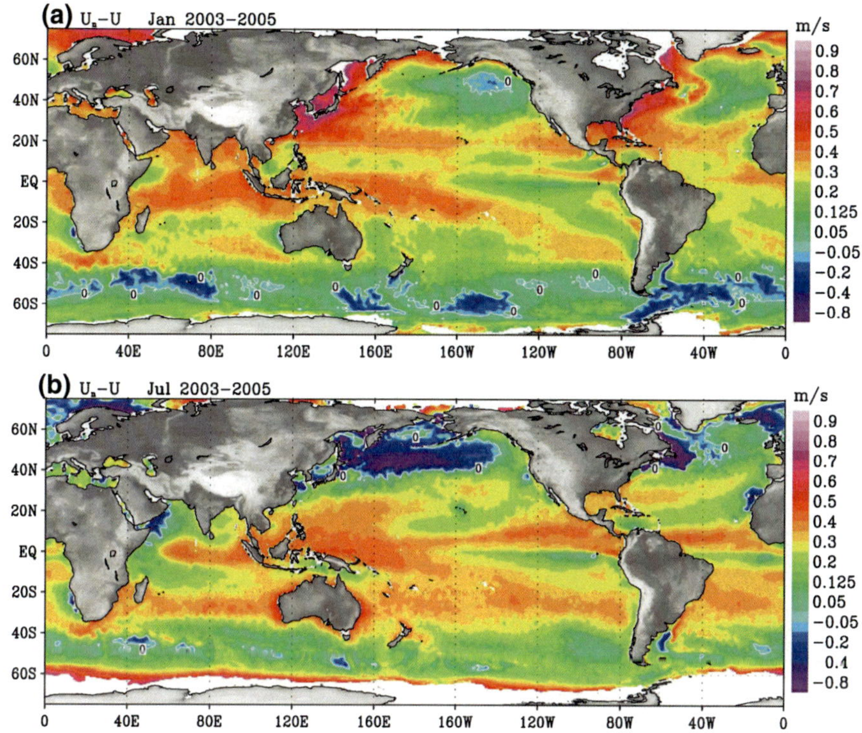

Fig. 15.3 Difference between equivalent neutral wind and actual wind at 10 m for (**a**) January and (**b**) July, averaged from 2003 to 2005

U_{10} of 30 m/s by 20%, according to the drag coefficient given by Kondo [7]. The horizontal axis wind turbines, which are anchored to ocean floor, may catch stronger winds with a higher structure, but the small vertical axis wind turbine, which are mostly floating, have lighter structure, and better air foils, may be more efficient and cost effective.

15.4 Stability Dependence

Typical wind profiles at various stabilities are shown in Fig. 15.2. At a given level, U_N is larger than the actual wind under unstable condition but lower under stable condition.

From Eq. 15.1 the difference between U_N and the actual wind U is

$$\delta U = U_N - U = 2.5 U_{*\Psi} \tag{15.4}$$

As described by Liu et al. [14] and Liu and Tang [11], the flux-profile relations for wind (Eq. 15.1), temperature, and humidity could be solved simultaneously for

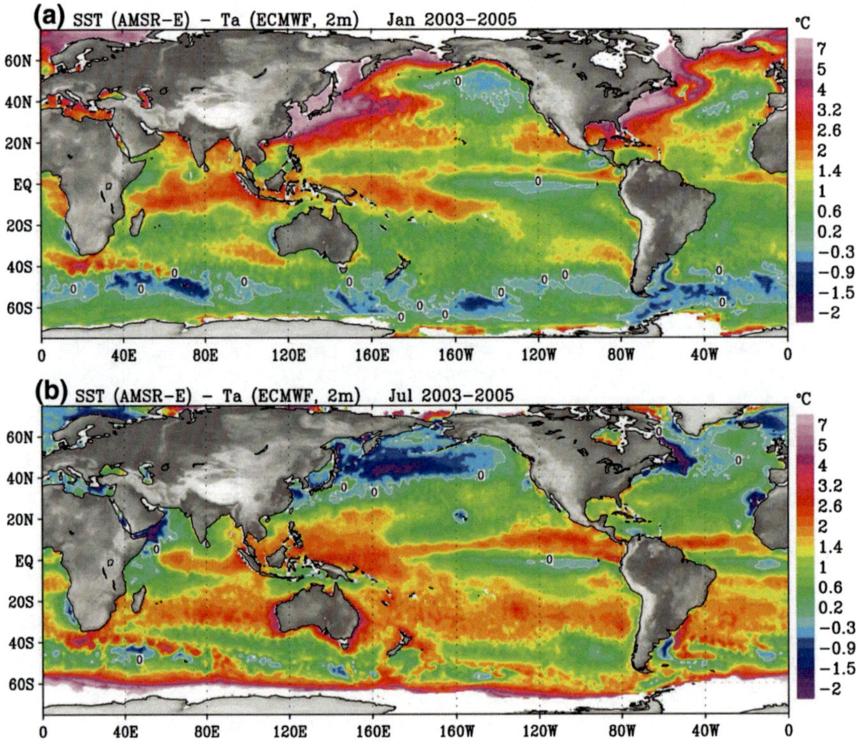

Fig. 15.4 Difference between sea surface temperature and air temperature (2 m) for (**a**) January and (**b**) July, averaged from 2003 to 2005

inputs of wind speed, temperature, and humidity at a certain level and the sea surface temperature to yield the fluxes of momentum (stress), heat, and water vapor. The value of Ψ is a by-product. Using U_{N} provided by QuikSCAT, sea surface temperature from AMSR-E, air temperature, and humidity from the reanalysis of the European Center for Medium-Range Weather Forecast, δU at 10 m averaged over a three-year period, for January and July, are computed and shown in Fig. 15.3. The distribution of stability effect shown in Fig. 15.3 closely follows the distribution of sea–air temperature difference shown in Fig. 15.4. The formulation of Ψ is based largely on experiment data on land, validated only with small amount of measurements over ocean. Although there have been many investigations to improve flux parameterization in the past few decades, there is no significant change in the formulation of Ψ.

Figures 15.3 and 15.4 show that U_N is higher than U by as much as 0.7 m/s in January over the western boundary currents in winter. It is also higher than U over the intertropical convergence zone, the South Pacific convergence zone, and the South Atlantic convergence zone. These are the regions with unstable atmosphere. U_N is lower than U in stable regions, such as over the circumpolar current, particularly in the Austral summer, and in the North Pacific and Atlantic,

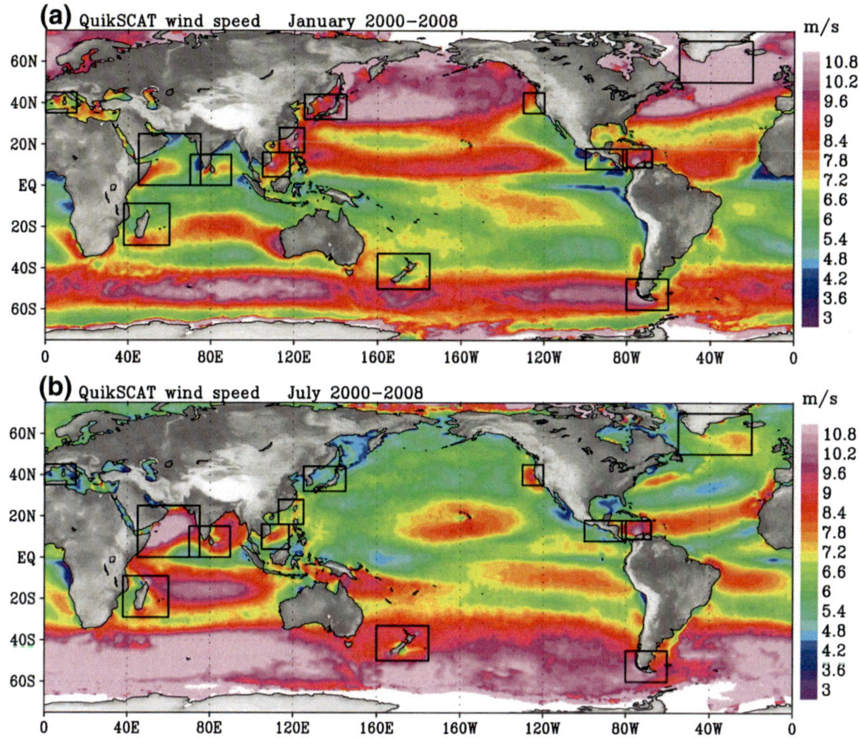

Fig. 15.5 Distributions of wind speed for (**a**) January, and (**b**) July, for a nine-year average from 2000 to 2008. Boxes indicate areas where regional wind characteristics are discussed in Sect. 15.6

particularly during Boreal summer. Figure 15.3 shows the inherent error of using the scatterometer measurements as the actual wind. In the unstable regions, higher sea surface temperature and atmospheric buoyancy generate turbulent momentum transport and increase stress magnitude. The higher wind stress affects atmospheric circulation aloft, as modified by atmospheric factors, and may be associated with high winds [15]. Atmospheric temperature varies at a much larger scale and mesoscale variations of stability is dominated by the small-scale sea surface temperature. The observed coherence between scatterometer measurements with sea surface temperature changes is an inherent characteristic of turbulent transfer [19]. Sea surface temperature has been suggested as an indicator of high wind occurrence [16, 22].

15.5 Climatological Distribution

The distributions of scatterometer winds U_N, as shown in Fig. 15.5, confirm the conventional knowledge—strongest winds over the mid-latitude storm tracks of the winter hemisphere, the relatively steady trade winds over the tropical oceans,

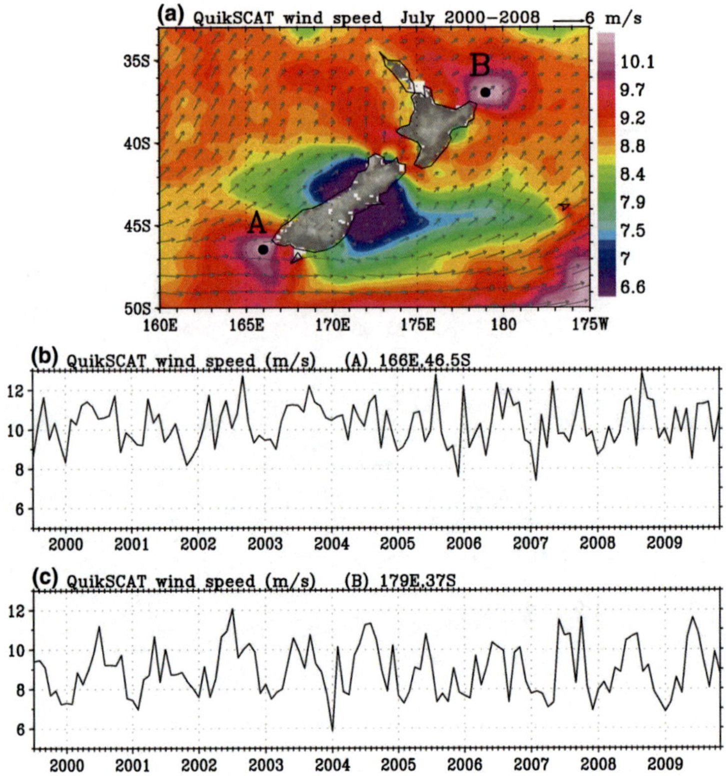

Fig. 15.6 (**a**) Wind speed (*color*) averaged for July from nine years of QuikSCAT data (2000–2008), with superimposition of QuikSCAT equivalent neutral wind vectors (*arrows*) for the same period, around New Zealand. (**b**) and (**c**) are time series of monthly mean wind speed at two selected locations marked in (**a**)

and the seasonal monsoons. At mid-latitude in the winter hemisphere, U_N is much larger than those in the tropics, making the display of the major features with the same color scale extremely difficult. The trade winds, particularly in the Western Pacific and Southern Indian Oceans are stronger in winter than summer, but the seasonal contrast is much less than those of the mid-latitude storm track. In the East China Sea, particularly through the Taiwan and Luzon Strait, the strong U_N is caused by the winter monsoon. In the Arabian Sea and Bay of Bengal, it is caused by the summer monsoon. The South China Sea is influenced by both summer and winter monsoons, the wind has two peaks. QuikSCAT data also reveal detailed wind structures not sufficiently identified before. The strong winds of transient tropical cyclones are not evident in U_N derived from the decadal ensemble. Nearshore regions of strong wind, marked by rectangles in Fig. 15.5, are discussed in Sect. 15.6.

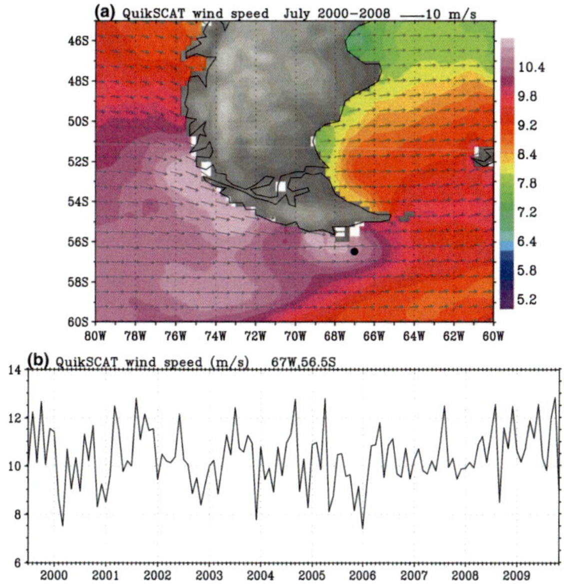

Fig. 15.7 Same as Fig. 15.6, except for a region around Tierra del Fuego

15.6 Regional Features

15.6.1 Aerodynamics

Regions with high U_N associated with the acceleration of strong prevailing winds when deflected by protruding landmass are ubiquitous. When the westerlies over the southern oceans blow around New Zealand (Fig. 15.6a) and Tierra del Fuego (Fig. 15.7a), they create wind shadows as well as strong jets. The winds are strong year-round except for slight weakening during a few Austral summers (Figs. 15.6c and 15.7b). Similarly strong winter jets are formed when the southeast trade wind is deflected at Cape d'Ambre and Cape Sainte Marie at the north and south tips of Madagascar (Fig. 15.8). Summer weakening is more prominent at the north tip. Asian summer monsoons are strengthened offshore of Somalia and Socotra Island (Fig. 15.9) and at Dandra Head in Sri Lanka (Fig. 15.10). Both summer and winter monsoons are strengthened at Phan Rang in Vietnam (Fig. 15.11), with double peaks in the time series (Fig. 15.11b). Strong winds are found near the north tip of Luzon Island of the Philippines, largely caused by the strengthening of the winter monsoons (Fig. 15.12). Strong winter monsoons are also found in the Taiwan Strait in the same figure. Strong wind is found downwind of Cape Mendocino in the United States (Fig. 15.13), with northerly winds year-round (strongest in summer). Similar features are found downwind Peninsula de La Guajira in Columbia (Fig. 15.14), with easterly winds year-round (weakest in fall). Very strong winter jets are found when the along shore wind coming south from the

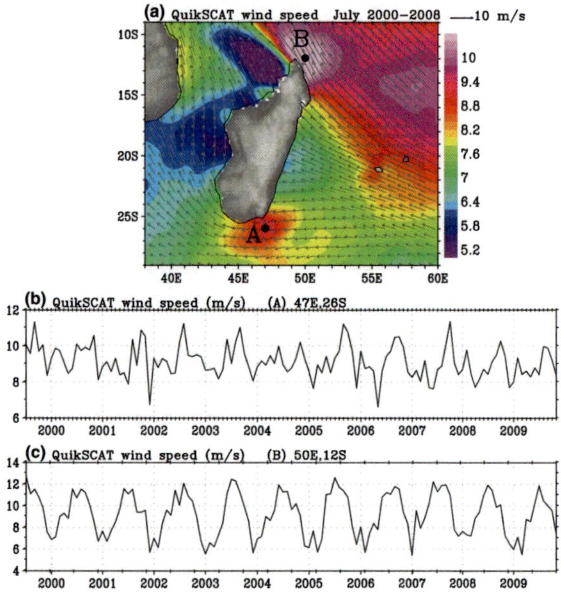

Fig. 15.8 Same as Fig. 15.6, except for a region around Madagascar

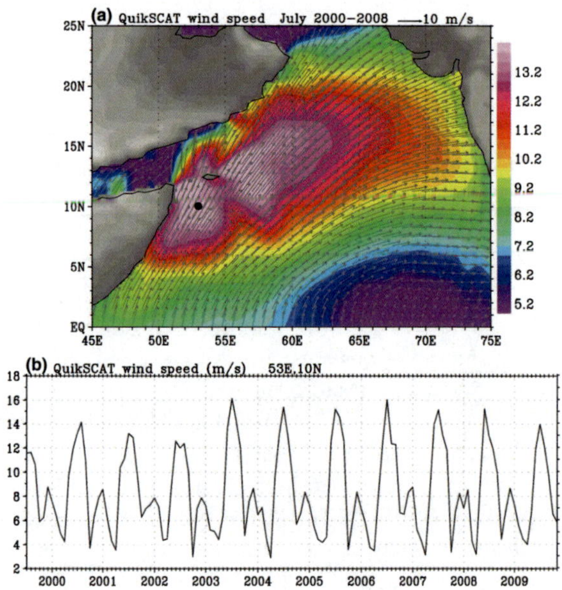

Fig. 15.9 Same as Fig. 15.6, except in the Indian Ocean of Somalia

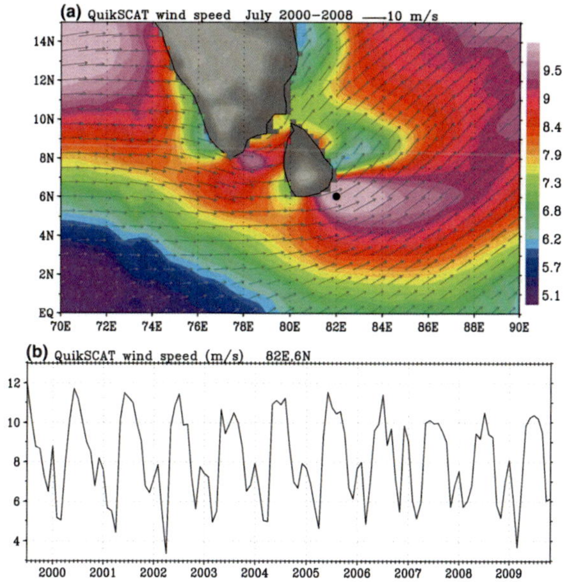

Fig. 15.10 Same as Fig. 15.6, except in the Indian Ocean around Sri Lanka

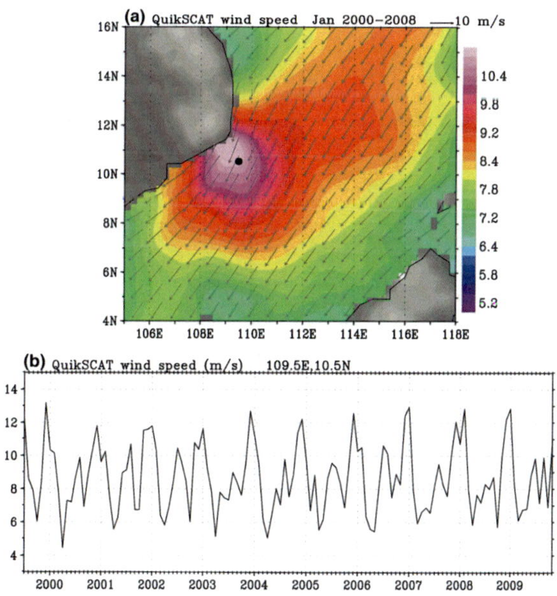

Fig. 15.11 Same as Fig. 15.6, except off the coast of Vietnam

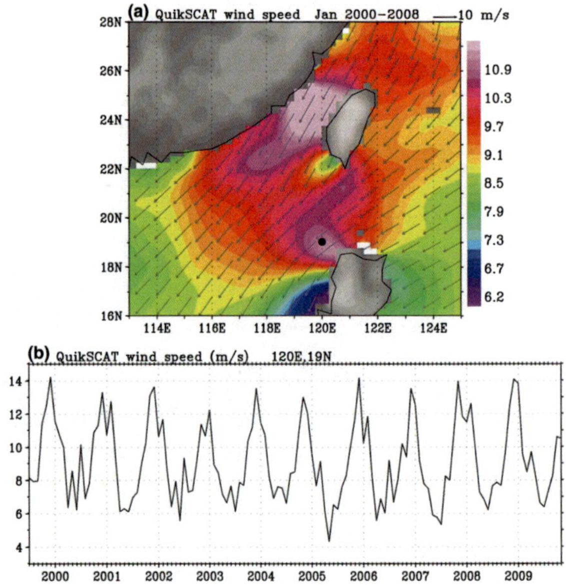

Fig. 15.12 Same as Fig. 15.6, except north of Luzon Island and around Taiwan

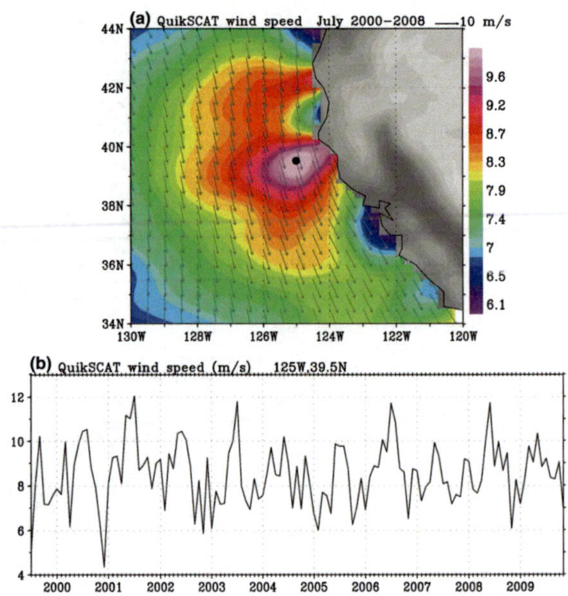

Fig. 15.13 Same as Fig. 15.6, except off the west coast of United States

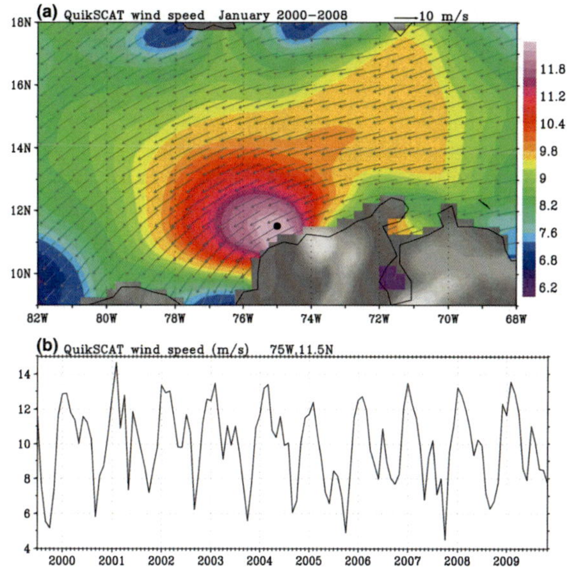

Fig. 15.14 Same as Fig. 15.6, except for a region in the Carribean

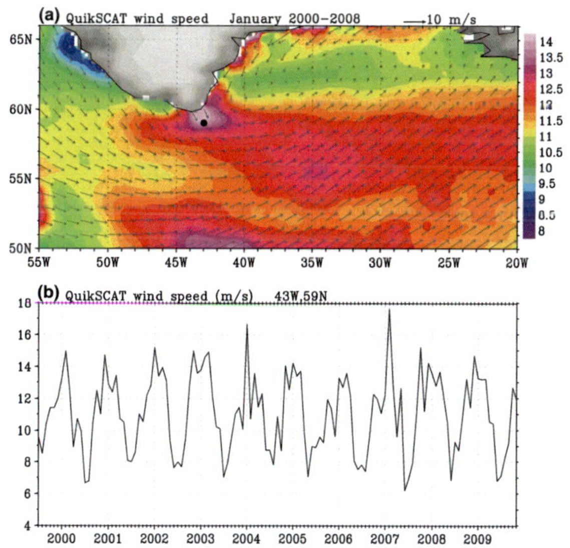

Fig. 15.15 Same as Fig. 15.6, except in the North Atlantic Ocean

Labrador sea along the West Greenland coast accelerates as it passes over Cape Farewell meeting with the wind blowing south along the East Greenland coast facing the Atlantic (Fig. 15.15).

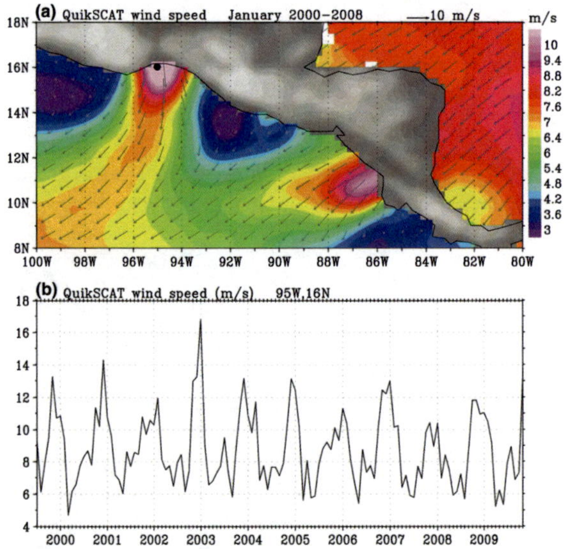

Fig. 15.16 Same as Fig. 15.6, except off the west coast of Central America

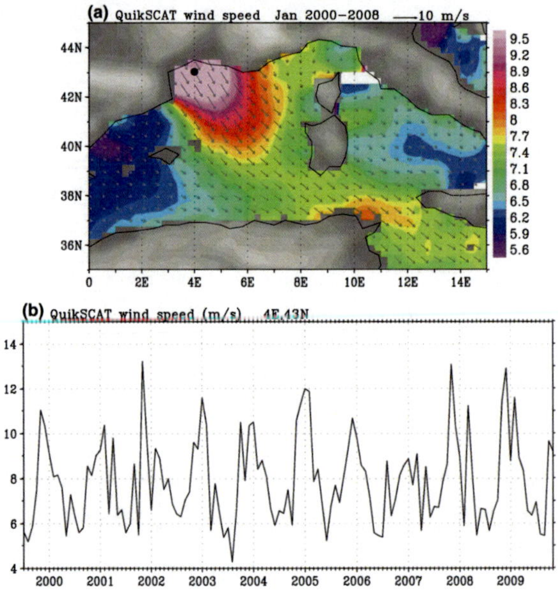

Fig. 15.17 Same as Fig. 15.6, except in the Mediterranean Sea

15.6.2 Land Topography

Strong U_N is also found (Fig. 15.5) when strong wind blows offshore, channeled by topography. There are many studies on the wind jets through the mountain gap

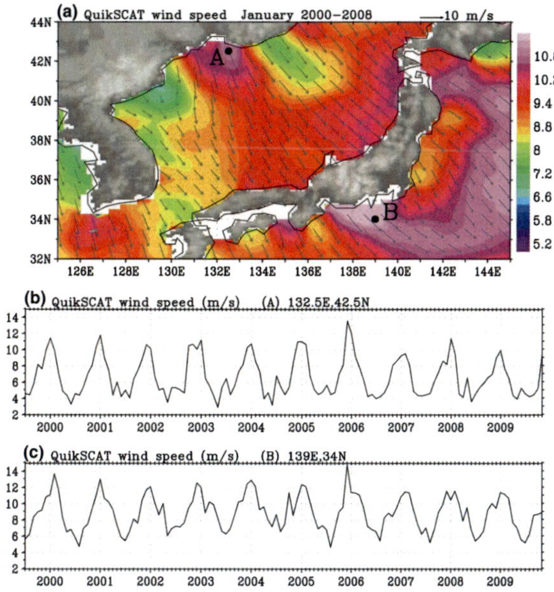

Fig. 15.18 Same as Fig. 15.6, except for a region around Japan

of Tehuantepec in Mexico and Papagayo of Costa Rica (e.g., [23, 25]). The jet is strongest when the northeast winds are strong in the Gulf of Mexico and the Caribbean Sea (Fig. 15.16). The well-known mistral blowing down to the Mediterranean from the gaps between Spain and France (e.g., [4]) is also obvious in Fig. 15.5 and Fig. 15.17 during winter. Another example is shown in Fig. 15.18, with strong wind blowing through the mountain gap south of Vladivostok causing cooling in the Sea of Japan [6]. These north/northwesterly winds then blow across the mountain of Japan and form strong wind jet west of Izu Islands [22]. Transient wind jets, such as the Santa Ana winds through the mountain passes and canyons of Southern California [5], do not significantly affect the long-term averages. Small-scale gap winds have also been identified using synthetic aperture radars, which usually require a priori knowledge of wind direction for wind speed retrieval (e.g., [20]).

15.7 Conclusions

We have gone from the old era of sailing and windmills as major wind applications to a new era of large-scale electric power generation. As a renewable energy, wind power has become an important factor, not only in reducing anthropogenic climate changes, but also in economy and employment. Although wind is not as intermittent as many people believe, it has strong spatial and temporal variations. Continuous monitoring of ocean surface wind from space would help in the optimal integration of wind into the power generation system.

A QuikSCAT follow-on mission is yet to be approved. The Advanced Scatterometer (ASCAT) operated by the Europeans has less coverage and spatial resolution. The scatterometers of India's Oceansat-2 and China's Haiyang-2 were launched in September 2009 and August 2011. The accuracy and data accessibility of the retrieved wind vector from these two missions have not been ascertained. Maintaining an accurate, consistent, and continuous ocean surface wind vector data set, with open accessibility, remains a technical and programmatic challenge.

Acknowledgment This study was performed at the Jet Propulsion Laboratory, California Institute of Technology under contract with the National Aeronautics and Space Administration (NASA). It was jointly supported by the Ocean Vector Winds and the Physical Oceanography Programs of NASA.

References

1. Barthelmie RJ (2001) Evaluating the impact of wind induced roughness change and tidal range on extrapolation of offshore vertical wind speed profiles. Wind Energ 4:99–105. doi:(10.1002/we.45)
2. Capps SB, Zender CS (2008) Observed and CAM3 GCM sea surface wind speed distributions: characterization, comparison, and bias reduction. J Clim 21:6569–6585
3. Donelan MA, Drenan WM, Katsaros KB (1997) The air-sea momentum flux in conditions of wind sea and swell. J Phys Oceanogr 27:2087–2099
4. Guénard V, Drobinski P, Caccia JL, Tedeschi G, Currier P (2006) Dynamics of the MAP IOP 15 severe Mistral event: Observations and high-resolution numerical simulations. Q J R Meteor Soc 132:757–777
5. Hu H, Liu WT (2003) Oceanic thermal and biological responses to santa ana winds. Geophys Res Lett 30(11):1596. doi:10.1029/2003GL017208
6. Kawamura H, Wu P (1998) Formation mechanism of Japan sea proper water in the flux center off vladivostok. J Geophys Res 103:21611–21622
7. Kondo J (1975) Airsea bulk transfer coefficients in diabatic conditions. Boundary-Layer Meteor 9:91–112
8. Large WG, Pond S (1981) Open ocean momentum flux measurements in moderate to strong winds. J Phys Oceanogr 11:324–336
9. Liu WT (2002) Progress in scatterometer application. J Oceanogr 58:121–136
10. Liu WT, Large WG (1981) Determination of surface stress by Seasat-SASS: a case study with JASIN data. J Phys Oceanogr 11:1603–1611
11. Liu WT, Tang W (1996) Equivalent neutral wind. *JPL Publication* 96–17, *Jet Propulsion Laboratory*, *Pasadena*, pp 16
12. Liu WT, Xie X, (2006) Measuring ocean surface wind from space. In: Gower J (ed) Remote sensing of the marine environment, manual of remote sensing, 3rd edn, vol 6, Chap 5. American society for photogrammetry and remote sensing, USA, pp 149–178
13. Liu WT, Xie X (2008) Ocean-atmosphere momentum coupling in the kuroshio extension observed from space. J Oceanogr 64:631–637
14. Liu WT, Katsaros KB, Businger JA (1979) Bulk parameterization of air-sea exchanges in heat and water vapor including the molecular constraints at the interface. J Atmos Sci 36:1722–1735
15. Liu WT, Xie X, Niiler PP (2007) Ocean-atmosphere interaction over agulhas extension meanders. J Clim 20(23):5784–5797
16. Liu WT, Tang W, Xie X (2008) Wind power distribution over the ocean. Geophys Res Lett 35:L13808. doi:10.1029/2008GL034172

17. Liu WT, Tang W, Xie X, Navalgund R, Xu K (2008) Power density of ocean surface wind-stress from international scatterometer tandem missions. Int J Remote Sens 29(21): 6109–6116
18. Liu WT, Tang W, Xie X (2010) Wind power at sea as observed from space. In: Muyeen SM (ed) Wind power, Chap 14. Intech, Vukovar, pp 341–352
19. Liu WT, Xie X, Tang W (2010b) Scatterometer's unique capability in measuring ocean surface stress. In: Barale V, Gower JFR, Alberotanza L (eds) Oceanography from space, Chap 6. Springer, Heidelberg, pp 93–111
20. Monaldo FM, Thompson DR, Pichel WG, Clemente-Colon P (2004) A systematic comparison of QuikSCAT and SAR ocean surface wind speeds. IEEE Trans Geosci Remote Sens 42:283–291
21. Risien CM, Chelton DB (2006) A satellite-derived climatology of global ocean winds. Remote Sens Environ 105:221–236
22. Sampe T, Xie S-P (2007) Mapping high sea winds from space: a global climatology. Bull Amer Meteor Soc 88:1965–1978
23. Sun F, Yu J-Y (2006) Impacts of central America gap winds on the SST annual cycle in the Eastern Pacific warm pool. Geophys Res Lett 33:L0670. doi:10.10.1029/2005GL024700
24. Smith SD (1980) Wind stress and heat flux over the ocean in gale force winds. J Phys Oceanogr 10:709–726
25. Xie S-P, Xu H, Kessler WS, Nonaka M (2005) Air-sea interaction over the eastern Pacific warm pool: gap winds, thermocline dome, and atmospheric convection. J Clim 18:5–20

Chapter 16
Power-Flow Control and Stability Enhancement of Four Parallel-Operated Offshore Wind Farms Using a Line-Commutated HVDC Link

Li Wang, Kuo-Hua Wang, Wei-Jen Lee and Zhe Chen

Abstract This chapter presents an effective control scheme using a line-commutated high-voltage direct-current (HVDC) link with a designed rectifier-current regulator (RCR) to simultaneously perform both power-fluctuation mitigation and damping improvement of four parallel-operated 80 MW offshore wind farms delivering generated power to a large utility grid. The proposed RCR of the HVDC link is designed by using modal control theory to contribute adequate damping to the studied four offshore wind farms under various wind speeds. A systematic analysis using a frequency-domain approach based on eigenvalue analysis and a time-domain scheme based on nonlinear model simulations is performed to demonstrate the effectiveness of the proposed control scheme. It can be concluded from the simulation results that the proposed HVDC link combined with the designed RCR can not only render adequate damping characteristics to the studied offshore wind farms under various wind speeds but also effectively mitigate power fluctuations of the offshore wind farms under wind-speed disturbance conditions (Wang et al., IEEE Trans Power Delivery 25(2):1190–1202, 2010).

L. Wang (✉) · K.-H. Wang
Department of Electrical Engineering, National Cheng Kung University, No. 1, University road, Tainan City 70101, Taiwan
e-mail: liwang@mail.ncku.edu.tw

W.-J. Lee
Energy Systems Research Center, The University of Texas at Arlington, Arlington, TX 76013, USA
e-mail: wlee@uta.edu

Z. Chen
Institute of Energy Technology, Aalborg University, Aalborg, Denmark
e-mail: zch@iet.aau.dk

S. M. Muyeen (ed.), *Wind Energy Conversion Systems*,
Green Energy and Technology, DOI: 10.1007/978-1-4471-2201-2_16,

16.1 Introduction

An offshore wind farm that is located a distance of less than 25 km away from seashore consists of several parallel-operated wind-turbine generators (WTGs). The use of several wind induction generators (IGs) and doubly-fed induction generators (DFIGs) connected directly to a power grid is the simplest way of running an offshore wind farm because wind IGs have the inherent advantages of cost effectiveness, less maintenance, and robustness, etc. However, operation of an offshore wind farm with a group of wind IGs requires large reactive power for magnetization, and the absorbed reactive-power of the offshore wind farm can severely affect the voltage magnitude of the bus to which the wind farms are connected under random wind-speed variations. This chapter proposes an effective control strategy using a line-commutated high-voltage direct-current (HVDC) link joined with a modal control designed RCR to perform both reactive-power compensation and stability enhancement of parallel-operated IG-based offshore wind farms and control the generated active power delivered to the onshore substation. The use of an HVDC link for the offshore wind farms has several advantages such as fast active-power modulation, effective reactive-power compensation, and less voltage drop on an onshore substation when comparing with the use of conventional long-distance AC transmission cables that contain higher inductive reactance and cause larger voltage drop.

To evaluate the performance of current controllers operated in the control loops of an HVDC link, the characteristics of using different types of current regulator were compared [2]. A novel current controller that was designed based on a linearized AC/DC system model and a pole-placement technique for an HVDC link was presented in [3]. The influence of the control loops of an HVDC link on dynamic stability of an HVDC-HVAC system was discussed in [4]. The design and implementation of an AC voltage dependent current order limiter at the Pacific HVDC Intertie was presented in [5] to demonstrate the consistent benefit of the proposed control scheme on enhancing stability and dynamic performance of the studied system. The potential of the fast growing offshore and onshore wind farms as well as the quick development of larger-capacity higher-efficiency WTGs in the whole world today can be obviously anticipated when an HVDC link is utilized to effectively control the generated power of the wind farms [6]. A time-domain model of an HVDC link using voltage-source converters (VSCs) for real-time simulations and its application to a wind energy conversion system were presented in [7]. An HVDC link combined with a capacitor commutated converter (CCC) model that was suitable for power flow and transient stability studies was proposed in [8].

The exploitation of an offshore wind farm was addressed in [9], and the proposed multiterminal VSC-HVDC system connected at the AC terminals of the WTGs enabled optimal acquisition and aggregation of wind power. The development of a linear continuous-time state model for analyzing small-signal dynamics of an HVDC transmission system was presented in [10], and a small-signal state-space model of the CIGRE benchmark HVDC transmission system was employed. The modeling and control design of a VSC-based HVDC using 12-pulse three-level converter topology was proposed in [11], and eigenvalue analysis, controller design, and transient simulations were performed. A linear time-invariant small-signal dynamic model for the main circuit of an HVDC link under synchronous rotating d-q reference frame by means of sampled-data modeling approach was proposed in [12], and the linearized model was validated by time-domain simulations.

The operation characteristics of a wind farm that was made up of several IG-based WTGs and modeled by an equivalent large IG incorporated VSC-HVDC were analyzed in [13]. Characteristics of a VSC-based HVDC link, an HVAC cable interconnection, and a synchronous generator under a faulted condition were compared, and the analyzed results showed that the VSC-based HVDC link demonstrated the ability to provide fault ride-through capability for the studied wind farm consisting of simple IG-based WTGs [14]. An improved model for the transient energy functions (TEFs) of integrated AC/DC power systems involved the omission of DC control dynamics was presented in [15]. The control requirements of a DFIG-based wind farm connected to a grid through a conventional thyristor-based HVDC link were investigated in [16]. The performance of a large IG-based offshore wind farm connected to a long-distance weak AC grid was studied, and the simulation results showed that the proposed HVDC link was able to supply the variable active power of the offshore wind farm to the weak grid and keep the AC voltage fluctuations at the point of common coupling (PCC) at an acceptable level [17].

A large 200 MW offshore wind farm consisting of 100 individual 2 MW WTGs connected to the grid through a multiterminal HVDC link with 25 VSCs was presented in [18]. A large 200 MW offshore wind farm consisting of 100 individual 2 MW WTGs connected to the grid through a multiterminal HVDC link with 25 current source inverters (CSIs)s was studied in [19]. A detailed technical–economic analysis of three transmission strategies (150, 400 kV AC, and VSC-HVDC) on benefits and drawbacks of onshore grid connection of three offshore wind farms (100, 200, and 500 MW) was evaluated and compared in [20]. The control system for a large offshore wind farm with an HVDC link using line-commutated converter (LCC) connection to the main onshore network using an aggregate system under operational and faulted conditions was proposed in [21].

A solution for integration of large DFIG-based offshore wind farms with a common collection bus controlled by a static compensator (STATCOM) into the main onshore grid using line-commutated HVDC link was presented in [22]. The control paradigm using the grid frequency control to regulate the rectifier firing angle or dc-link current to control the power flow of an offshore wind farm connected with line-commutated HVDC link was examined in [23]. A frequency controller in the converter to increase the ride-through capability of the VSC-HVDC-supplied industrial system was proposed in [24] to exploit the inertia energy of rotating masses in case of voltage disturbances. Three different frequency controllers and their effects on the voltage-disturbance ride-through capability of a VSC-HVDC-supplied industrial system were proposed in [25]. A new concept applicable to large power converters consisting of two series-connected twelve-pulse groups and a new type of converter control applicable to multilevel HVDC schemes with two or more 12-pulse groups per terminal were shown in [26]. Two types of FACTS devices, STATCOM and VSC-HVDC system, were presented in [27] to offer a realistic alternative to conventional network reinforcement to solve a range of operational problems in power systems.

This chapter presents damping controller design of rectifier, steady-state eigenvalue analysis and transient time-domain simulations of four parallel-operated IG-based offshore wind farms connected to an onshore substation through an HVDC link. For steady-state analysis, system eigenvalues, eigenvalue sensitivity, and design of RCR of the HVDC link are performed. For time-domain simulations, transient responses of the studied offshore wind farms with and without the designed RCR subject to different wind-speed disturbances are also carried out. This chapter is organized as follows. Section 16.3 introduces the configuration and the employed mathematical models for the offshore wind farms with an HVDC link in this chapter. Section 16.4 demonstrates the design procedure for the RCR of the HVDC link using pole-placement technique. Section 16.5 describes the results of eigenvalue analysis under various wind speeds and eigenvalue sensitivity with respect to the parameters of the designed RCR. Section 16.6 compares dynamic responses of the studied offshore wind farm with and without the designed RCR under wind-speed disturbance conditions. Specific important conclusions of this chapter are drawn in Sect. 16.6.

16.2 Configuration of the Studied System

Figure 16.1 shows the configuration of the studied four 80 MW offshore wind farms connected to an onshore power grid through an HVDC link. The four 80 MW offshore wind farms are represented by four equivalent wind IGs (IG_1–IG_4) connected to a common AC bus. Each equivalent 80 MW IG that is driven by an equivalent variable-speed wind turbine through an equivalent gearbox (GB) is obtained by aggregating forty 2 MW IGs together.

The output of each equivalent 80 MW IG is connected to the common AC bus through an excitation capacitor bank C and a step-up transformer with an equivalent impedance of $R_T + jX_T$. The HVDC link consisting of an AC-to-DC converter (or rectifier), a T-equivalent DC line, and a DC-to-AC inverter delivers the generated power of the four offshore wind farms to the onshore power grid through a step-down transformer and an AC transmission line. The equivalent impedance of $R_L + jX_L$ in Fig. 16.1 is used to represent the combined impedance of the step-down transformer and the AC transmission line. The employed system parameters are listed in Appendix for conciseness and the mathematical models for the

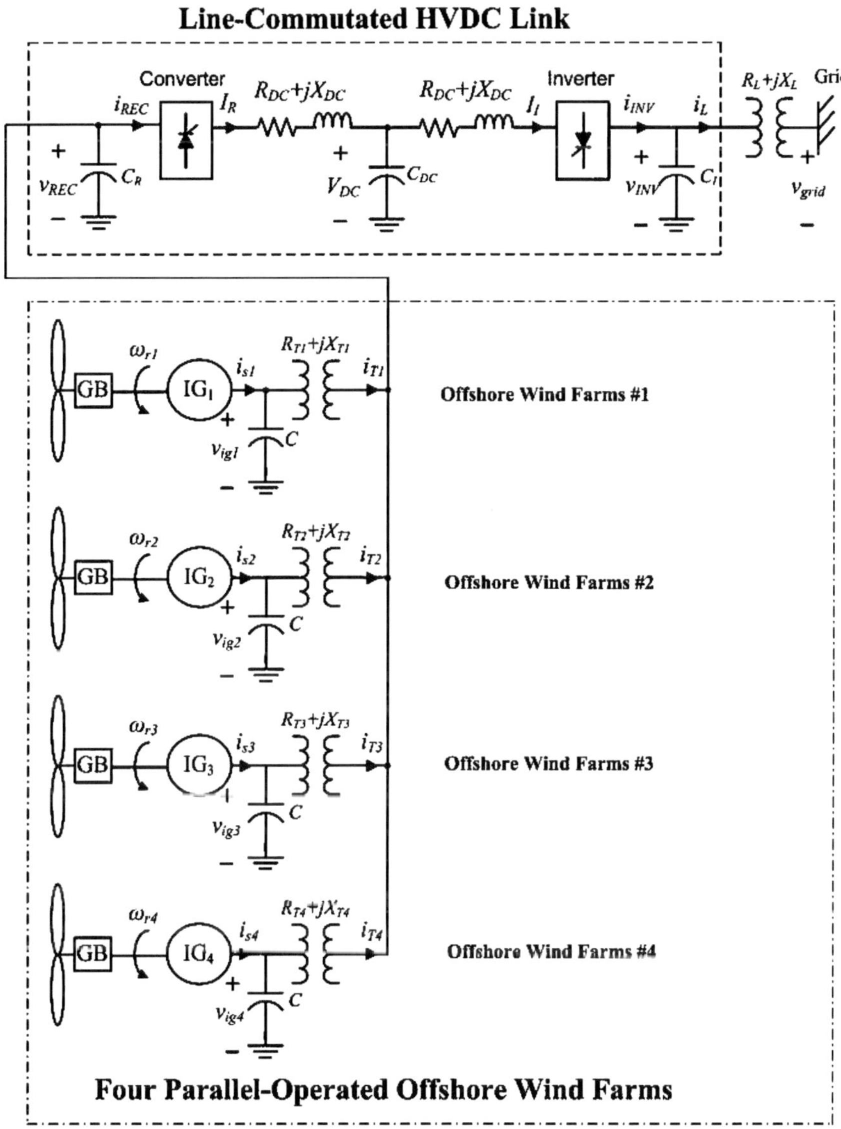

Fig. 16.1 Configuration of the studied four parallel-operated offshore wind farms with a line-commutated HVDC link (©2010 IEEE. Reprinted from *IEEE Trans. Power Delivery*, vol. 25, no. 2, April 2010)

subsystems shown in Fig. 16.1 are described as below. The equations in the following subsections are expressed in per-unit or MKS quantities except that the time variable t is in second.

16.2.1 Wind Speed Model

The wind speed is modeled as the algebraic sum of base wind speed, gust wind speed, ramp wind speed, and noise wind speed [28]. The base wind speed can be expressed by

$$V_{\mathrm{WB}} = K_{\mathrm{B}} \tag{16.1}$$

where K_{B} is a constant. It is assumed that V_{WB} is always present in the study of wind-turbine systems. The gust wind speed can be described by

$$V_{\mathrm{WG}} = \begin{cases} 0, & t < T_{1\mathrm{G}} \\ V_{\cos}, & T_{1\mathrm{G}} < t < T_{1\mathrm{G}} + T_{\mathrm{G}} \\ 0, & t > T_{1\mathrm{G}} + T_{\mathrm{G}} \end{cases} \tag{16.2}$$

where

$$V_{\cos} = \frac{MAXG}{2}\left[1 - \cos 2\pi\left(\frac{t}{T_{\mathrm{G}}} - \frac{T_{1\mathrm{G}}}{T_{\mathrm{G}}}\right)\right], \tag{16.3}$$

T_{G}, $T_{1\mathrm{G}}$, and $MAXG$ are the period, the starting time, and the peak of the gust wind speed, respectively. The gust wind is the usual (1 – cosine) gust used in wind studies. The ramp wind speed can be depicted by

$$V_{\mathrm{WR}} = \begin{cases} 0, & t < T_{1\mathrm{R}} \\ V_{\mathrm{ramp}}, & T_{1\mathrm{R}} < t < T_{2\mathrm{R}} \\ 0, & t > T_{2\mathrm{R}} \end{cases} \tag{16.4}$$

where

$$V_{\mathrm{ramp}} = \mathrm{MAXR}\left(1 - \frac{t - T_{2\mathrm{R}}}{T_{1\mathrm{R}} - T_{2\mathrm{R}}}\right) \tag{16.5}$$

$MAXR$, $T_{1\mathrm{R}}$, and $T_{2\mathrm{R}}$ are the maximum, the starting time, and the maximum time of the ramp wind speed, respectively. The noise wind speed can be expressed by

$$V_{\mathrm{WN}} = 2\sum_{i=1}^{N} \sqrt{S_{\mathrm{V}}(\omega_i)\Delta\omega}\cos(\omega_i t + \varphi_i) \tag{16.6}$$

where $\omega_{\mathrm{i}} = (i - 0.5)\Delta\omega$, φ_{i} is a random variable with uniform probability density on the interval $[0, 2\pi]$,

$$S_V(\omega_i) = \frac{2K_N F^2|\omega_i|}{\pi^2\left[1 + (F\omega_i/\mu\pi)^2\right]^{4/3}} \tag{16.7}$$

is the spectral density function, K_{N} is the surface drag coefficient, F is the turbulence scale, and μ is the mean speed of wind at the reference height. Various researchers used $N = 50$ and $\Delta\omega = 0.5 - 2.0$ rad/s to obtain excellent results. According to the above four wind speeds, the employed wind speed model in this chapter can be defined as

$$V_W = V_{WB} + V_{WG} + V_{WR} + V_{WN} \tag{16.8}$$

16.2.2 Wind Turbine Model

The mechanical power produced by a wind turbine is given by

$$P_W = \frac{1}{2}\rho A_r V_W^3 C_p(\lambda, \beta) \tag{16.9}$$

where P_W is the mechanical power in W, ρ is the air density in kg/m^3, A_r is the blade impact area in m^2, V_W is the wind velocity in m/s as depicted in (16.8), and C_p is the power coefficient [29] that can be expressed by

$$C_p(\psi_k, \beta) = \left[c_1\left(\frac{c_2}{\psi_k} - c_3\beta - c_4\beta^{c_5} - c_6\right)\right]\exp\left[-\frac{c_7}{\psi_k}\right] \tag{16.10}$$

in which

$$\frac{1}{\psi_k} = \frac{1}{\lambda + c_8\beta} - \frac{c_9}{\beta^3 + 1} \tag{16.11}$$

$$\lambda = \frac{R_B\omega_B}{V_W} \tag{16.12}$$

where ω_B and R_B are respectively the blade angular velocity in rad/s and blade radius in m, λ is the tip speed ratio, β is the blade pitch angle in degrees, and c_1–c_9 are the constant coefficients for power coefficient C_p.

16.2.3 Mass-Spring-Damper Model

Figure 16.2 shows the simplified equivalent mass-spring-damper model of a three-blade wind turbine coupled to the rotor of the studied wind IG [30]. The masses of gearbox and IG are combined together. The per-unit mechanical equations of each blade are described by

$$(2H_{Bk})p(\omega_{Bk}) = T_{Wk} - D_{Bk}\omega_{Bk} - d_{BkH}(\omega_{Bk} - \omega_H) - K_{BkH}(\theta_{Bk} - \theta_H) \tag{16.13}$$

$$p(\theta_{Bk}) = \omega_b(\omega_{Bk} - \omega_{Bk0}) \tag{16.14}$$

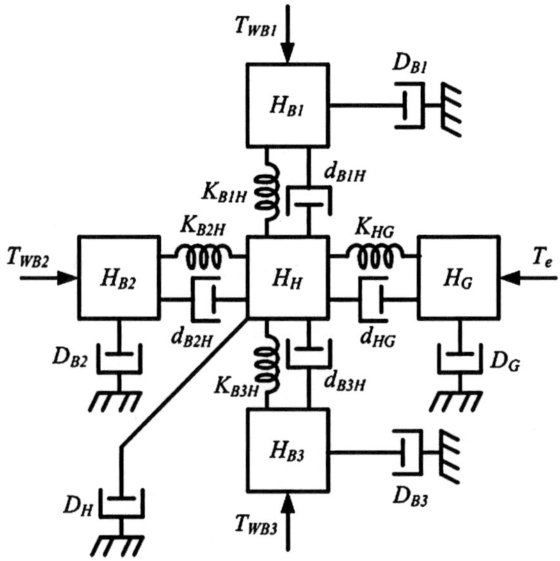

Fig. 16.2 Simplified equivalent mass-spring-damper model of a three-blade wind turbine coupled to the shaft of an induction generator (©2010 IEEE. Reprinted from *IEEE Trans. Power Delivery*, vol. 25, no. 2, April 2010)

where ω and θ are respectively the per-unit angular speed and angle displacement of each mass, T_W is the per-unit input mechanical torque from wind, D is the per-unit self damping coefficient, d is the per-unit mutual damping coefficient, K is the per-unit spring constant or stiffness, ω_{Bko} is the per-unit initial angular speed of the k-th blade, and $k = 1$, 2, and 3 refers to the quantities of blades 1, 2, and 3, respectively. The per-unit mechanical equations of the hub are given by

$$(2H_H)p(\omega_H) = \sum_{k=1}^{3}[K_{BkH}(\theta_{Bk} - \theta_H) + d_{BkH}(\omega_{Bk} - \omega_H)] - D_H\omega_H - d_{HG}(\omega_H - \omega_G) - K_{HG}(\theta_H - \theta_G) \tag{16.15}$$

$$p(\theta_H) = \omega_b(\omega_H - \omega_{H0}) \tag{16.16}$$

where ω_{H0} is the per-unit initial angular speed of the hub. The per-unit mechanical-torque equations of the IG are determined by

$$(2H_G)p(\omega_G) = D_G\omega_G + d_{HG}(\omega_H - \omega_G) + K_{HG}(\theta_H - \theta_G) - T_e \tag{16.17}$$

$$p(\theta_G) = \omega_b(\omega_G - \omega_{G0}) \tag{16.18}$$

where ω_{G0} is the per-unit initial angular speed of the IG, and T_e is the per-unit electromagnetic torque of the IG and it can be expressed by [31]

$$T_e = X_m(i_{dr}i_{qs} - i_{qr}i_{ds}) \tag{16.19}$$

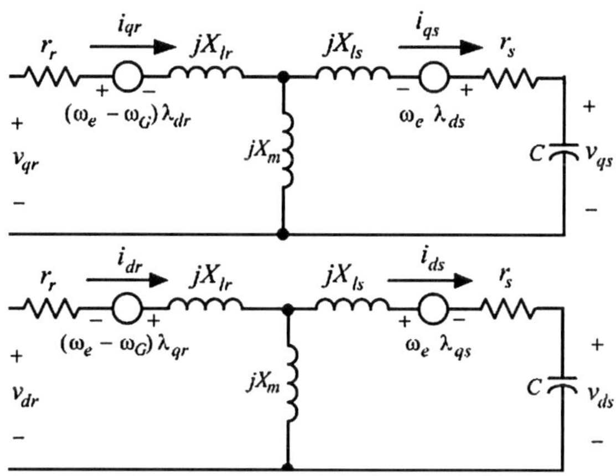

Fig. 16.3 d-q axis equivalent circuit of the studied IG (©2010 IEEE. Reprinted from *IEEE Trans. Power Delivery*, vol. 25, no. 2, April 2010)

16.2.4 Induction Generator Model

Figure 16.3 shows the *d-q* axis equivalent circuit of an IG with an excitation capacitor bank *C* under a synchronously rotating reference frame [32, 33]. The per-unit voltage-current equations of the studied IG can be described by

$$X_{ss}p(i_{qs}) - X_m p(i_{qr}) = (-r_s i_{qs} - \omega_e X_{ss} i_{ds} + \omega_e X_m i_{dr} - v_{qs})\omega_b \quad (16.20)$$

$$X_{ss}p(i_{ds}) - X_m p(i_{dr}) = (-r_s i_{ds} + \omega_e X_{ss} i_{qs} - \omega_e X_m i_{qr} - v_{ds})\omega_b \quad (16.21)$$

$$X_{rr}p(i_{qr}) - X_m p(i_{qs}) = \left[-r_r i_{qr} + (\omega_e - \omega_G)X_m i_{ds} - (\omega_e - \omega_G)X_{rr} i_{dr}\right]\omega_b \quad (16.22)$$

$$X_{rr}p(i_{dr}) - X_m p(i_{ds}) = \left[-r_r i_{dr} - (\omega_e - \omega_G)X_m i_{qs} + (\omega_e - \omega_G)X_{rr} i_{qr}\right]\omega_b \quad (16.23)$$

where X_m is the per-unit mutual reactance between stator and rotor windings, $X_{ss} = X_{ls} + X_m$ and $X_{rr} = X_{lr} + X_m$ are respectively the per-unit self (leakage) reactances of stator and the rotor windings of the studied IG, and X_{ls} and X_{lr} are respectively the per-unit leakage reactances of stator and the rotor windings of the studied IG.

16.2.5 Excitation Capacitor Bank Model

The per-unit voltage-current equations of the excitation capacitor bank *C* shown in Fig. 16.3 can be expressed by

$$i_{qC} = (C/\omega_b)p(v_{qs}) + \omega_e C v_{ds} = i_{qs} - i_{qT} \quad (16.24)$$

$$i_{\mathrm{sdC}} = (C/\omega_b)p(v_{\mathrm{ds}}) - \omega_e C v_{\mathrm{qs}} = i_{\mathrm{ds}} - i_{\mathrm{dT}} \tag{16.25}$$

where i_{qC} and i_{dC} are respectively the per-unit q- and d-axis excitation capacitor currents while i_{qT} and i_{dT} are the per-unit q- and d-axis currents flowing through the step-up transformer, respectively.

16.2.6 Step-up Transformer, AC Line, and Power Grid Models

The per-unit voltage-current equations from the output terminals of the inverter to the power grid through the step-up transformer and the AC transmission line are given by

$$v_{\mathrm{q,INV}} = v_{\mathrm{q,grid}} + R_{\mathrm{L}} i_{\mathrm{qL}} + (X_L/\omega_b)p(i_{\mathrm{qL}}) + \omega_e X_L i_{\mathrm{dL}} \tag{16.26}$$

$$v_{\mathrm{d,INV}} = v_{\mathrm{d,grid}} + R_{\mathrm{L}} i_{\mathrm{dL}} + (X_{\mathrm{L}}/\omega_{\mathrm{b}})p(i_{\mathrm{dL}}) - \omega_e X_{\mathrm{L}} i_{\mathrm{qL}} \tag{16.27}$$

where i_{qL} and i_{dL} are respectively the per-unit q- and d-axis line currents, $v_{\mathrm{q,grid}}$ and $v_{\mathrm{d,grid}}$ are respectively the per-unit q- and d-axis grid voltages, and $v_{\mathrm{q,INV}}$ and $v_{\mathrm{d,INV}}$ are the per-unit q- and d-axis output voltages of the inverter, respectively.

16.2.7 Line-Commutated HVDC Link Model

The line-commutated HVDC link model consists of an AC-to-DC converter, a DC line, and a DC-to-AC inverter [34–39]. The base values for the AC and DC quantities should be properly selected such that the pu values of DC quantities remain unchanged when they are converted to the synchronous reference frame of the AC system [35, 37, 39]. The pu output DC voltage and current of the rectifier can be properly converted with reference to the common reference frame of the dq-axis of the AC system according to the phasor diagram shown in Fig. 16.4 [37–39].

$$V_{\mathrm{qR}} = V_{\mathrm{q}} \cos\delta_{\mathrm{R}} + V_{\mathrm{d}} \sin\delta_{\mathrm{R}} \tag{16.28}$$

$$V_{\mathrm{dR}} = -V_{\mathrm{q}} \sin\delta_{\mathrm{R}} + V_{\mathrm{q}} \cos\delta_{\mathrm{R}} \tag{16.29}$$

$$I_{\mathrm{qR}} = I_{\mathrm{q}} \cos\delta_{\mathrm{R}} + I_{\mathrm{d}} \sin\delta_{\mathrm{R}} \tag{16.30}$$

$$I_{\mathrm{dR}} = -I_{\mathrm{q}} \sin\delta_{\mathrm{R}} + I_{\mathrm{q}} \cos\delta_{\mathrm{R}} \tag{16.31}$$

where δ_{R} is the angle by which q_{R}-axis lags the q-axis as shown in Fig. 16.4. Since the reference direction of the q_{R}-axis with respect to the q-axis is arbitrary, the operation of the converter can be easily expressed when $V_{\mathrm{dR}} = 0$ is properly chosen. Under this particular selection for the reference frame, we have

Fig. 16.4 Phasor diagram for transformation of *d-q* and d_R-q_R quantities (©2010 IEEE/. Reprinted from *IEEE Trans. Power Delivery*, vol. 25, no. 2, April 2010)

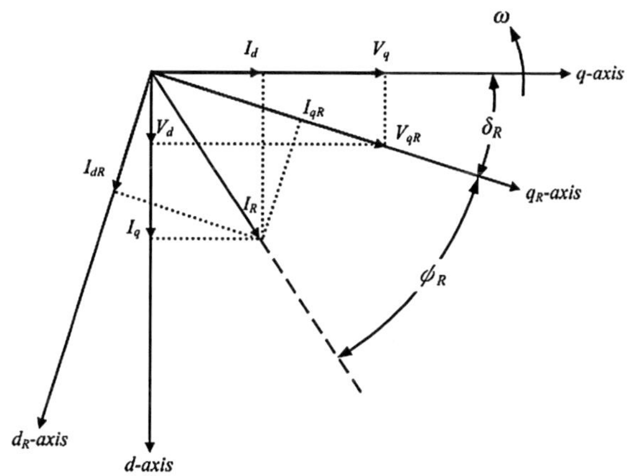

$$\cos\delta_{\mathrm{R}} = V_{\mathrm{q}}/V_{\mathrm{qR}} \tag{16.32}$$

$$\sin\delta_{\mathrm{R}} = V_{\mathrm{d}}/V_{\mathrm{qR}} \tag{16.33}$$

$$I_{\mathrm{q}} = I_{\mathrm{qR}}\cos\delta_{\mathrm{R}} - -I_{\mathrm{dR}}\sin\delta_{\mathrm{R}} \tag{16.34}$$

$$I_{\mathrm{d}} = I_{\mathrm{qR}}\sin\delta_{\mathrm{R}} + I_{\mathrm{qR}}\cos\delta_{\mathrm{R}} \tag{16.35}$$

The pu d_{R}- and q_{R}-axis components shown in Fig. 16.4 can be expressed by

$$I_{\mathrm{qR}} - I_{\mathrm{R}}\sin\psi_{\mathrm{R}} \tag{16.36}$$

$$I_{\mathrm{dR}} = I_{\mathrm{R}}\cos\psi_{\mathrm{R}} \tag{16.37}$$

The DC line is represented by a T-equivalent circuit that has two identical series impedances of $(R_{\mathrm{DC}} + jX_{\mathrm{DC}})$ placed on both sides of a lumped charging capacitor C_{DC}. The per-unit voltage-current equations of the rectifier and inverter can be written as, respectively [37–39],

$$V_{\mathrm{R}} = V_{\mathrm{qR}}\cos\alpha_{\mathrm{R}} - (\pi/6)X_{\mathrm{CR}}I_{\mathrm{R}} \tag{16.38}$$

$$V_{\mathrm{I}} = V_{\mathrm{qI}}\cos\gamma_{\mathrm{I}} - (\pi/6)X_{\mathrm{CI}}I_{\mathrm{I}} \tag{16.39}$$

where α_{R} is the delay angle of the converter, γ_{I} is the extinction angle of the inverter, X_{CR} and X_{CI} are respectively the per-unit commutation choke reactance of the converter and inverter, V_{qR} and V_{qI} are respectively the per-unit AC-side voltage of the converter and inverter, $V_{\mathrm{qR}} = [(V_{\mathrm{q,REC}})^2 + (V_{\mathrm{d,REC}})^2]^{1/2}$, and $V_{\mathrm{qI}} = [(V_{\mathrm{q,INV}})^2 + (V_{\mathrm{d,INV}})^2]^{1/2}$.

Figures 16.5 and 16.6 show the block diagrams of the rectifier-current regulator (RCR) of the converter and the inverter-current regulator (ICR) of the inverter

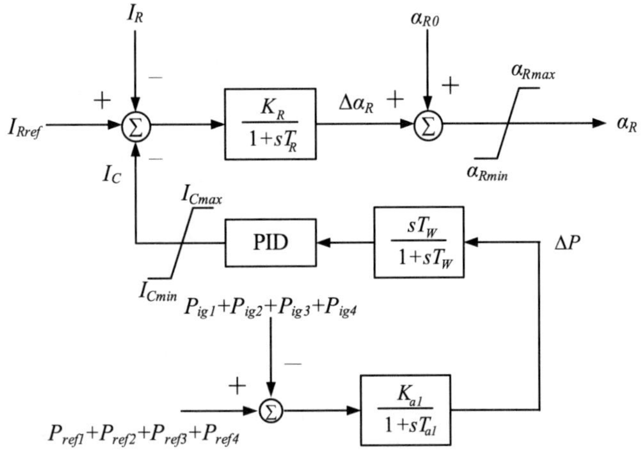

Fig. 16.5 Block diagram of the rectifier-current regulator (RCR) of the converter including the designed PID RCR (©2010 IEEE. Reprinted from *IEEE Trans. Power Delivery*, vol. 25, no. 2, April 2010)

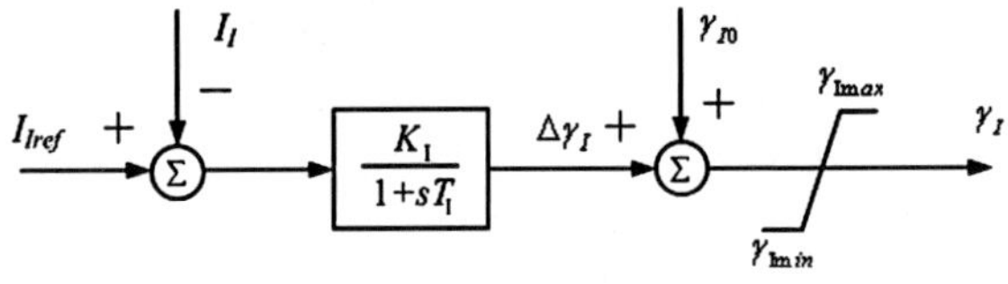

Fig. 16.6 Block diagram of the inverter-current regulator (ICR) of the inverter (©2010 IEEE. Reprinted from *IEEE Trans. Power Delivery*, vol. 25, no. 2, April 2010)

[37–39], respectively. The per-unit equations for the RCR and the ICR can be expressed by, respectively,

$$p(\alpha_R) = (K_R/T_R)(I_{Rref} - I_R + I_C) - (1/T_R)\alpha_R \tag{16.40}$$

$$p(\gamma_I) = (K_I/T_I)(I_{Iref} - I_I) - (1/T_I)\gamma_I \tag{16.41}$$

where the per-unit damping control signal I_C is generated by the PID controller that will be designed in the next section.

16.3 Design of a PID RCR Using Modal Control Theory

This section presents a unified approach based on modal control theory to design a PID RCR for the proposed HVDC link. To have a clear observation of the damping effect contributed by the proposed PID RCR, the eigenvalues to be assigned on the complex plane are properly selected. The design procedures and results are given as below [37–39].

16.3.1 Linearized System

The nonlinear system equations developed in the previous section are first linearized around a nominal operating point to obtain a set of linearized dynamic equations of the form:

$$
\begin{aligned}
p\mathbf{X} &= \mathbf{AX} + \mathbf{BU} + \mathbf{VW} \\
\mathbf{Y} &= \mathbf{CX} + \mathbf{DU}
\end{aligned}
\tag{16.42}
$$

where $\mathbf{X}$ is the state vector, $\mathbf{Y}$ is the output vector, $\mathbf{U}$ is the external or compensated input vector, $\mathbf{W}$ is the disturbance input vector while $\mathbf{A}$, $\mathbf{B}$, $\mathbf{C}$, and $\mathbf{D}$ are all constant matrices of appropriate dimensions. To design a damping controller for the proposed RCR, the variation term $\mathbf{V}$ and the external input $\mathbf{D}$ in (16.42) are properly neglected by letting $\mathbf{D} = \mathbf{V} = 0$. The state vector $\mathbf{X}$ can be partitioned into four substate vectors as $\mathbf{X} = [\mathbf{X}_{IG}, \mathbf{X}_{LC}, \mathbf{X}_{WT}, \mathbf{X}_{HVDC}]^T$, where $\mathbf{X}_{IG}$, $\mathbf{X}_{LC}$, $\mathbf{X}_{WT}$, and $\mathbf{X}_{HVDC}$ are referred to the state vectors of the four wind IGs, the excitation capacitor banks and the transmission line, the mechanical systems of four wind turbines, and the HVDC link, respectively.

Since the wind speed of the studied four offshore wind farms seldom reaches the rated wind speed of 15 m/s of the WTGs, V_W of 12 m/s is properly selected as the operating point for the PID RCR design in the next subsection. The eigenvalues of the studied four offshore wind farms with the proposed HVDC link under V_W of 12 m/s are listed in the second column of Table 16.1. The eigenvalues listed in Table 16.1 with ( )$\times$ n, $n = 3$ or 4 represent the repeated eigenvalues or modes of the studied four offshore wind farms since the studied four offshore wind farms have identical WTG-IG sets and parameters. An examination of these eigenvalues listed in Table 16.1 reveals that the damping of both Λ_{11} and Λ_{12} can be improved by adding a damping controller to the HVDC link. The PID RCR will be designed by using modal control theory in the next subsection.

16.3.2 Design of a PID RCR

Figure 16.5 showed the control block diagram of the RCR of the converter including the designed PID RCR. It is seen that the PID damping controller employs ΔP as a feedback signal to generate a damping control signal I_C in order that the damping characteristics of the poorly damped modes listed in Table 16.1 can be improved. The block diagram for determining ΔP was given in Fig. 16.5, where the reference values for active-power of four IGs (P_{ref1}–P_{ref4}) can be determined from the rotor speeds of the four wind IGs.

Hence, the output signal in (16.42) is

$$
\mathbf{Y} = \Delta P \tag{16.43}
$$

Table 16.1 Eigenvalues (rad/s) of the studied system under wind speed of 12 m/s

	Four offshore wind farms with the HVDC link only	Four offshore wind farms with the HVDC link and the designed PID RCR
Λ_1	$-887.11 \pm j11336$	$-886.88 \pm j11336$
Λ_2	$-870.88 \pm j10779$	$-870.52 \pm j10779$
Λ_3	$-535.17 \pm j7355.2$	$-535.17 \pm j7355.2$
Λ_4	$(-728 \pm j5653.7) \times 3$	$(-728 \pm j5653.7) \times 3$
Λ_5	$-58.878 \pm j1590.1$	$-58.911 \pm j1590.3$
Λ_6	$-71.556 \pm j1245.8$	$-71.581 \pm j1245.6$
Λ_7	$(-728.29 \pm j4899.8) \times 3$	$(-728.29 \pm j4899.8) \times 3$
Λ_8	$-110.97 \pm j581.83$	$-110.92 \pm j581.78$
Λ_9	$(-13.226 \pm j538.8) \times 4$	$(-13.226 \pm j538.8) \times 4$
Λ_{10}	$(-69.316 \pm j359.14) \times 3$	$(-69.316 \pm j359.14) \times 3$
Λ_{11}	$-54.32 \pm j45.735$	$-55.0 \pm j45.0$*
Λ_{12}	$-1.5245 \pm j44.32$	$-2.0 \pm j43.0$*
Λ_{13}	-10.0	$-14.173 \pm j35.967$
Λ_{14}	$(-128.89, -117.54) \times 3$	$(-128.89, -117.54) \times 3$
Λ_{15}	$(-7.0228 \pm j46.429) \times 3$	$(-7.0228 \pm j46.429) \times 3$
Λ_{16}	-1.4865	-1.4865
Λ_{17}	$(-0.80404 \pm j133.44) \times 4$	$(-0.80404 \pm j133.44) \times 4$
Λ_{18}	$(-0.80402 \pm j133.44) \times 4$	$(-0.80402 \pm j133.44) \times 4$

*denotes exactly assigned eigenvalues

and

$$\mathbf{U} = I_C \tag{16.44}$$

is the input vector. The transfer function $\mathbf{H}(s)$ of the proposed PID RCR in s domain is given by

$$H(s) = \frac{\mathbf{U}(s)}{\mathbf{Y}(s)} = \frac{I_C(s)}{\Delta P(s)} = \frac{sT_{\mathrm{Wa}}}{1 + sT_{\mathrm{Wa}}}\left(K_P + \frac{K_I}{s} + sK_D\right) \tag{16.45}$$

where T_{Wa} is the time constant of a washout term, and K_{P}, K_{I}, and K_{D} are gains of the damping controller. The four unknowns for the PID damping controller can be determined by using modal control theory and the first step is to obtain the state equations in s domain by taking Laplace transformation from (16.42) to (16.44). The second step is to derive an algebraic equation of the feedback system using eigenvalue analysis in s domain and the unknown parameters can be solved. The input signal in s domain can be expressed by

$$\mathbf{U}(s) = \mathbf{H}(s)\Delta P(s) = \mathbf{H}(s)\mathbf{Y}(s) = \mathbf{H}(s)\mathbf{C}\mathbf{X}(s) \tag{16.46}$$

Combining (16.42), (16.43), (16.45) and (16.46), we have

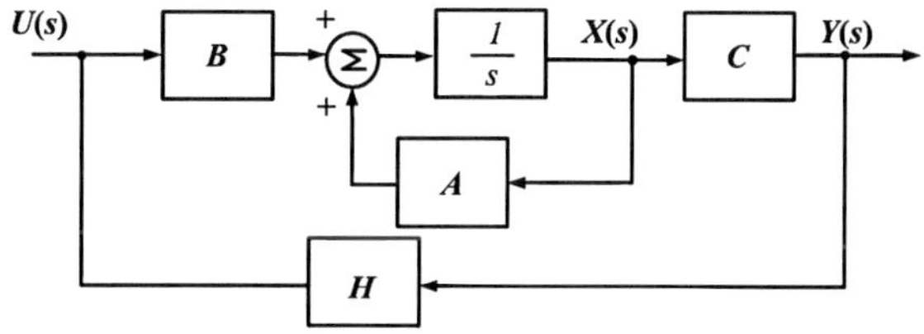

Fig. 16.7 Block diagram of the studied system with the designed damping controller (©2010 IEEE. Reprinted from *IEEE Trans. Power Delivery*, vol. 25, no. 2, April 2010)

$$s\mathbf{X}(s) = \{\mathbf{A} + \mathbf{B}[\mathbf{H}(s)\mathbf{C}]\}\mathbf{X}(s) \tag{16.47}$$

The block diagram of the studied system with the designed damping controller is given in Fig. 16.7. The characteristic equation of the closed-loop system with the designed PID damping controller is given by

$$\det\{s\mathbf{I} - [\mathbf{A} + \mathbf{BH}(s)\mathbf{C}]\} = 0 \tag{16.48}$$

Since the studied system with the proposed PID RCR constitutes a single-input single-output (SISO) system, (16.48) is a scalar equation. The parameters of the PID controller can be determined by substituting two pairs of prespecified complex-conjugated eigenvalues corresponding to the assigned poles into (16.48). Four simultaneous algebraic equations with the four unknowns, K_P, K_I, K_D, and T_{Wa}, can then be solved to obtain a unique solution. The design results are presented as follows.

Eigenvalues of four offshore wind farms with the HVDC link

$$\Lambda_{11} = -54.32 \pm j45.735 \quad \Lambda_{12} = -1.5245 \pm j44.32$$

Prespecified eigenvalues

$$\Lambda_{11} = -55.0 \pm j45.0 \quad \Lambda_{12} = -2.0 \pm j43.0$$

Parameters of the PID RCR

$$K_P = 3.0,\ K_I = 114.1,\ K_D = 0.06,\ \text{and}\ T_{Wa} = 0.01\ \text{s}$$

The system eigenvalues of the studied four offshore wind farms with the designed PID RCR are listed in the third column of Table 16.1. It can be obviously found from Table 16.1 that the eigenvalues of both modes Λ_{11} and Λ_{12} have been exactly assigned on desired locations of the complex plane.

16.4 Steady-State Analysis

This section examines the effectiveness of the proposed HVDC link joined with the designed PID RCR on dynamic-stability enhancement of the studied four offshore wind farms. The variations of the designed PID RCR's parameters on the closed-loop eigenvalues are also carried out using relative sensitivity coefficients (RSCs).

16.4.1 Steady-State Operating Conditions under Various Wind Speeds

Table 16.2 lists the computed steady-state operating conditions of the studied four offshore wind farms. Since four wind farms have identical parameters, the calculated results for a single wind farm are listed in Table 16.2. These steady-state results are useful for determining the eigenvalues of the studied system under different wind speeds in the next subsection.

16.4.2 Dynamic Stability under Various Wind Speeds

Table 16.3 lists the calculated eigenvalues of the studied four offshore wind farms with the designed PID RCR when wind speed increases from 4 m/s (the cut-in wind speed of wind turbine) to 24 m/s (the cut-off wind speed of wind turbine). The eigenvalue results listed in Table 16.3 are analyzed as below according to different wind-speed ranges. The root-loci plots of these eigenvalues when wind speed varies from 4 to 24 m/s are also shown in Fig. 16.8.

(a) When $V_W \leq 12$ m/s, the damping of both Λ_{12} and Λ_{16} gets worse when wind speed becomes lower. All system modes are located on the left half of the complex plane. Some mechanical modes are close to the imaginary axis of the complex plane but they are nearly fixed on the complex plane.
(b) When $V_W > 12$ m/s, the damping of Λ_{12} initially drops at V_W of 14 m/s and then increases to a higher value at V_W of 15 m/s. When $V_W > 15$ m/s, the damping of Λ_{12} gets worse as wind speed becomes higher. The damping of Λ_{16} gets better before V_W reaches 18 m/s. When V_W is higher than 18 m/s, the damping of Λ_{16} becomes worse. Again, all system eigenvalues are positioned on the left half of the complex plane and some mechanical modes are nearly fixed on the complex plane.

16.4.3 Eigenvalue Sensitivity of PID RCR's Parameters

Whenever there is a small change in one of the system parameter ζ, the system matrix $\mathbf{A}$ in (16.42) and the associated system eigenvalues will vary accordingly. The eigenvalue sensitivity can be determined by

Table 16.2 Computed steady-state operating conditions under different values of wind speed V_W (m/s)

	Single wind farm						HVDC link				Power grid			
V_W	v_G	i_G	ω_G (rpm)	P_G	Q_G	PF_G	V_{DC}	I_R	α_R (°)	γ_I (°)	v_{grid}	i_{grid}	P_{grid}	Q_{grid}
4	1.247	0.626	1802.081	−0.007	−0.781	−0.009	0.121	−0.600	84.159	83.430	1.0	0.429	−0.140	−0.591
6	1.358	0.687	1810.551	0.074	−0.930	−0.080	0.214	0.654	75.049	77.400	1.0	1.349	0.190	−1.898
8	1.488	0.772	1821.470	0.217	−1.127	−0.189	0.532	0.790	56.885	58.164	1.0	1.509	0.724	−2.007
10	1.554	0.848	1835.214	0.412	−1.252	−0.313	0.769	1.000	40.577	45.349	1.0	1.758	1.314	−2.112
12	1.569	0.930	1852.090	0.637	−1.314	−0.436	0.910	1.294	25.843	35.240	1.0	2.125	2.007	−2.237
14	1.670	1.046	1861.843	0.862	−1.519	−0.494	0.964	1.629	24.357	32.414	1.0	2.585	2.609	−2.561
15*	1.663	1.093	1870.228	0.973	−1.534	−0.536	1.040	1.785	12.556	9.868	1.0	2.721	3.287	−2.002
16	1.698	1.143	1874.325	1.075	−1.616	−0.554	1.052	1.941	13.065	7.039	1.0	2.939	3.589	−2.095
18	1.733	1.230	1883.203	1.255	−1.723	−0.589	1.069	2.216	10.906	1.196	1.0	3.402	4.080	−2.549
20	1.764	1.296	1889.140	1.392	−1.814	−0.509	1.075	2.432	11.363	0.983	1.0	3.714	4.443	−2.801
22	1.783	1.339	1892.875	1.481	−1.873	−0.520	1.079	2.572	11.729	0.973	1.0	3.839	4.695	−2.726
24	1.763	1.365	1898.201	1.529	−1.858	−0.635	1.081	2.645	5.066	0.799	1.0	3.943	4.813	−2.816

*denotes the rated wind speed on wind turbine

Table 16.3 Eigenvalues (rad/s) of the closed-loop system with the designed PID RCR under different wind speeds

V_W (m/s)	4	6	8	10	12**	14
Λ_1	$-742.09 \pm j11488$	$-783.94 \pm j11324$	$-814.12 \pm j11331$	$-846.88 \pm j11335$	$-886.88 \pm j11336$	$-932.68 \pm j11282$
Λ_2	$-746.75 \pm j10584$	$-762.6 \pm j10753$	$-788.07 \pm j10759$	$-822.82 \pm j10769$	$-870.52 \pm j10779$	$-918.83 \pm j10834$
Λ_3	$-865.75 \pm j6930$	$-718.85 \pm j7384.7$	$-666.63 \pm j7396.4$	$-603.91 \pm j7391.1$	$-535.17 \pm j7355.2$	$-472.75 \pm j7273.8$
Λ_4	$(-728 \pm j5653.7)\times 3$	$(-728 \pm j5653.7)\times 3$	$(-728 \pm j5653.7)\times 3$	$(-728 \pm j5653.7)\times 3$	$(-728 \pm j5653.7)\times 3$	$(-728 \pm j5653.7)\times 3$
Λ_5	$-26.488 \pm j1538.5$	$-31.344 \pm j1541.6$	$-47.614 \pm j1556.7$	$-55.862 \pm j1575.4$	$-58.911 \pm j1590.3$	$-60.823 \pm j1587.7$
Λ_6	$-41.548 \pm j1304.5$	$-51.149 \pm j1257.4$	$-52.559 \pm j1254.1$	$-59.664 \pm j1248.9$	$-71.581 \pm j1245.6$	$-84.578 \pm j1240$
Λ_7	$(-728 \pm j4899.8)\times 3$	$(-728 \pm j4899.8)\times 3$	$(-728 \pm j4899.8)\times 3$	$(-728 \pm j4899.8)\times 3$	$(-728 \pm j4899.8)\times 3$	$(-728 \pm j4899.8)\times 3$
Λ_8	$-105.31 \pm j504.7$	$-96.218 \pm j551.82$	$-91.46 \pm j560.15$	$-96.829 \pm j572.53$	$-110.92 \pm j581.78$	$-124.41 \pm j588.64$
Λ_9	$(-13.23 \pm j538.8)\times 4$	$(-13.23 \pm j538.8)\times 4$	$(-13.23 \pm j538.8)\times 4$	$(-13.23 \pm j538.8)\times 4$	$(-13.23 \pm j538.8)\times 4$	$(-13.23 \pm j538.8)\times 4$
Λ_{10}	$(-69 \pm j358.67)\times 3$	$(-69.1 \pm j358.74)\times 3$	$(-69.2 \pm j358.83)\times 3$	$(-69.3 \pm j358.97)\times 3$	$(-69.3 \pm j359.14)\times 3$	$(-69.4 \pm j359.28)\times 3$
Λ_{11}	$-35.324 \pm j18.733$	$-37.757 \pm j27.312$	$-40.889 \pm j43.05$	$-50.369 \pm j48.031$	$-55 \pm j45.0$	$-56.137 \pm j47.591$
Λ_{12}	$-0.33224 \pm j43.448$	$-0.39971 \pm j44.319$	$-0.62452 \pm j43.876$	$-1.0264 \pm j43.361$	$-2.0 \pm j43.0$	$-1.8021 \pm j43.007$
Λ_{13}	$-4.1146 \pm j27.94$	$-9.1093 \pm j42.478$	$-21.169 \pm j45.016$	$-19.406 \pm j40.63$	$-14.173 \pm j35.967$	$-13.065 \pm j34.24$
Λ_{14}	$(-126.5 \pm j17.24)\times 3$	$(-125.5 \pm j14.79)\times 3$	$(-124.2 \pm j11.15)\times 3$	$(-123.4 \pm j5.911)\times 3$	$(-128.9, -117.54)\times 3$	$(-131.1, -114.62)\times 3$
Λ_{15}	$(-4.316 \pm j45.73)\times 3$	$(-5.166 \pm j45.95)\times 3$	$(-6.285 \pm j46.23)\times 3$	$(-6.91 \pm j46.39)\times 3$	$(-7.023 \pm j46.42)\times 3$	$(-7.296 \pm j46.49)\times 3$
Λ_{16}	-0.94805	-1.12	-1.3411	-1.4636	-1.4865	-1.5389
Λ_{17}	$(-0.80 \pm j133.44)\times 4$	$(-0.80 \pm j133.44)\times 4$	$(-0.80 \pm j133.44)\times 4$	$(-0.80 \pm j133.44)\times 4$	$(-0.80 \pm j133.44)\times 4$	$(-0.80 \pm j133.44)\times 4$
Λ_{18}	$(-0.80 \pm j133.44)\times 4$	$(-0.80 \pm j133.44)\times 4$	$(-0.80 \pm j133.44)\times 4$	$(-0.80 \pm j133.44)\times 4$	$(-0.80 \pm j133.44)\times 4$	$(-0.80 \pm j133.44)\times 4$
V_W (m/s)	15	16	18	20	22	24
Λ_1	$-946.22 \pm j11105$	$-956.42 \pm j11063$	$-973.69 \pm j10949$	$-1176.7 \pm j10732$	$-1206.8 \pm j10665$	$-1308.3 \pm j10605$
Λ_2	$-933.49 \pm j10675$	$-938.19 \pm j10748$	$-956.63 \pm j10660$	$-816.98 \pm j10725$	$-843.3 \pm j10666$	$-798.8 \pm j10612$
Λ_3	$-505.93 \pm j6322.8$	$-485.77 \pm j6329.3$	$-497.27 \pm j6260.2$	$-442.45 \pm j6318.5$	$-462.96 \pm j5953$	$-442.76 \pm j5967.2$
Λ_4	$(-728 \pm j5280.5)\times 3$	$(-728 \pm j5318.7)\times 3$	$(-728 \pm j5082.9)\times 3$	$(-728 \pm j4906.8)\times 3$	$(-728 \pm j4748.4)\times 3$	$(-728 \pm j4604.9)\times 3$
Λ_5	$-57.962 \pm j1577.7$	$-57.374 \pm j1574$	$-57.662 \pm j1568.3$	$-52.166 \pm j1560.8$	$-52.964 \pm j1561$	$-50.628 \pm j1558.2$
Λ_6	$-85.171 \pm j1199.8$	$-87.809 \pm j1199.3$	$-88.496 \pm j1173.3$	$-93.711 \pm j1142.3$	$-95.749 \pm j1126.1$	$-98.32 \pm j1104.1$
Λ_7	$(-728 \pm j4526.7)\times 3$	$(-728 \pm j4564.9)\times 3$	$(-728 \pm j4329.1)\times 3$	$(-728 \pm j4153)\times 3$	$(-728 \pm j3994.6)\times 3$	$(-728 \pm j3851.2)\times 3$
Λ_8	$-123.8 \pm j537.18$	$-124.84 \pm j545.6$	$-125.48 \pm j511.1$	$-128.57 \pm j490.79$	$-132.51 \pm j466.33$	$-135.26 \pm j446.68$
Λ_9	$(-13.2 \pm j538.8)\times 4$	$(-13.2 \pm j538.8)\times 4$	$(-13.2 \pm j538.8)\times 4$	$(-13.2 \pm j538.8)\times 4$	$(-13.2 \pm j538.8)\times 4$	$(-13.2 \pm j538.8)\times 4$

(continued)

Table 16.3 (continued)

V_W (m/s)	4	6	8	10	12**	14
Λ_{10}	$(-69.4 \pm j359.29)\times 3$	$(-69.4 \pm j359.3)\times 3$	$(-69.5 \pm j359.33)\times 3$	$(-69.61 \pm j359.4)\times 3$	$(-69.6 \pm j59.48)\times 3$	$(-69.6 \pm j359.51)\times 3$
Λ_{11}	$-55.884 \pm j47.3$	$-54.449 \pm j50.666$	$-55.122 \pm j52.62$	$-48.479 \pm j58.329$	$-49.764 \pm j58.014$	$-46.722 \pm j60.835$
Λ_{12}	$-2.2829 \pm j42.627$	$-1.2789 \pm j42.56$	$-1.6914 \pm j42.263$	$-0.65071 \pm j42.897$	$-1.2275 \pm j42.459$	$-1.0692 \pm j42.696$
Λ_{13}	$-13.521 \pm j35.054$	$-17.308 \pm j38.073$	$-17.634 \pm j36.918$	$-25.29 \pm j38.508$	$-22.731 \pm j35.82$	$-25.143 \pm j35.407$
Λ_{14}	$(-131.3, -113.09)\times 3$	$(-131.4, -111.64)\times 3$	$(-131.4, -109.72)\times 3$	$(-131.1, -110.16)\times 3$	$(-130.2, -111.8)\times 3$	$(-129.58, -112.9)\times 3$
Λ_{15}	$(-7.89 \pm j46.615)\times 3$	$(-8.48 \pm j46.735)\times 3$	$(-9.32 \pm j46.895)\times 3$	$(-9.27 \pm j46.873)\times 3$	$(-8.90 \pm j46.78)\times 3$	$(-8.7 \pm j46.739)\times 3$
Λ_{16}	-1.6513	-1.76	-1.9131	-1.9038	-1.8352	-1.801
Λ_{17}	$(-0.80 \pm j133.44)\times 4$	$(-0.80 \pm j133.44)\times 4$	$(-0.80 \pm j133.44)\times 4$	$(-0.80 \pm j133.44)\times 4$	$(-0.80 \pm j133.44)\times 4$	$(-0.80 \pm j133.44)\times 4$
Λ_{18}	$(-0.80 \pm j133.44)\times 4$	$(-0.80 \pm j133.44)\times 4$	$(-0.80 \pm j133.44)\times 4$	$(-0.80 \pm j133.44)\times 4$	$(-0.80 \pm j133.44)\times 4$	$(-0.80 \pm j133.44)\times 4$

**denotes the operating condition at which the PID RCR is designed

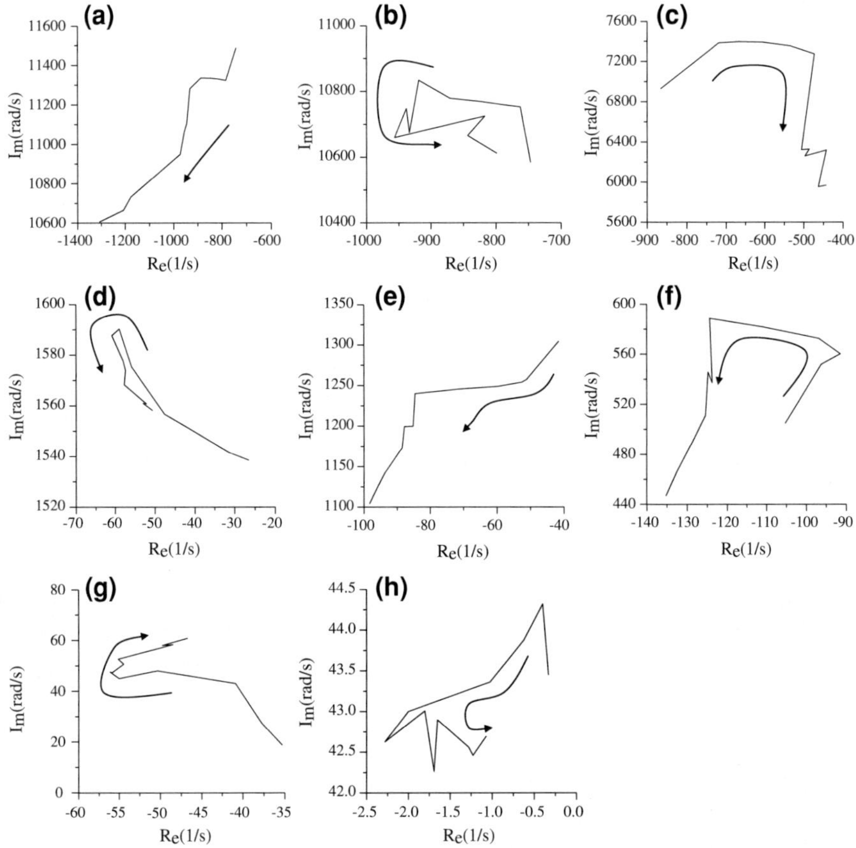

Fig. 16.8 Root loci of the eigenvalues listed in Table 16.3 when the wind speed is varied from 4 to 24 m/s. **a** Λ_1. **b** Λ_2. **c** Λ_3. **d** Λ_5. **e** Λ_6. **f** Λ_8. **g** Λ_{11}. **h** Λ_{12} (©2010 IEEE. Reprinted from *IEEE Trans. Power Delivery*, vol. 25, no. 2, April 2010)

$$\frac{\partial \Lambda_i}{\partial \zeta} = \frac{\mathbf{V}_i^T \frac{\partial \mathbf{A}}{\partial \zeta} \mathbf{U}_i}{\mathbf{V}_i^T \mathbf{U}_i} \tag{16.49}$$

where $\mathbf{U}_i$ ($\mathbf{V}_i$) is the eigenvector of $\mathbf{A}$ ($\mathbf{A}^{\mathrm{T}}$) associated with the *i*th eigenvalue Λ_i. Since the parameter ζ in (16.42) may range from a very small value to very large one, it is more meaningful to determine the fractional variation of the parameter. The relative sensitivity coefficient (RSC) is defined by [40]

$$RSC = \frac{\partial \Lambda_i}{\partial \zeta} \zeta \approx \frac{\Delta \Lambda_i}{\Delta \zeta / \zeta} \tag{16.50}$$

Table 16.4 lists the RSCs of the closed-loop eigenvalues with respect to the four PID RCR's parameters (K_{P}, K_{I}, K_{D}, and T_{Wa}). It is clearly found from the RSC results listed in Table 16.4 that only Λ_{11}–Λ_{13} are more sensitive to the variations of K_{I}, K_{D}, and T_{Wa} while all RSCs are very small. Hence, a small variation of four

Table 16.4 Relative sensitivity coefficients of the closed-loop eigenvalues with respect to the designed PID RCR's four parameters

	K_P	K_I	K_D	T_{Wa}
Λ_1	$8.4212\times10^{-8} \pm j4.0636\times10^{-7}$	$0.0045316 \pm j0.001315$	$-18.925 \pm j50.202$	$4.8595\times10^{-5} \pm j2.3346\times10^{-4}$
Λ_2	$2.8789\times10^{-7} \pm j3.0584\times10^{-7}$	$-0.0035473 \pm j0.002837$	$-27.492 \pm j40.707$	$0.00016527 \pm j1.759\times10^{-4}$
Λ_3	$4.1272\times10^{-13} \pm j5.1888\times10^{-12}$	$-3.8386\times10^{-8} \pm j2.5877\times10^{-10}$	$1.864\times10^{-5} \pm j2.8248\times10^{-4}$	$2.3313\times10^{-10} \pm j2.982\times10^{-9}$
Λ_4	$(-1.01\times10^{-6} \pm j0.7377\times10^{-6})\times3$	$(-0.010655 \pm j0.002003)\times3$	$(3.8131 \pm j16.827)\times3$	$(-0.0006047 \pm j3.8684\times10^{-3})\times3$
Λ_5	$-1.1447\times10^{-32} \pm j9.4828\times10^{-33}$	$9.475\times10^{-30} \pm j6.8173\times10^{-30}$	$7.4815\times10^{-26} \pm j5.8876\times10^{-26}$	$-4.3004\times10^{-29} \pm j8.0965\times10^{-28}$
Λ_6	$-1.3931e-006 \pm j\ 7.4901e-006$	$0.0094296 \pm j0.00$[illegible]1991	$0.81871 \pm j11.832$	$-0.00076629 \pm j0.0043117$
Λ_7	$(-1.147\times10^{-33} \pm j1.0143\times10^{-33})\times3$	$(5.726\times10^{-30} \pm j2.13$[illegible]$8\times10^{-30})\times3$	$(-4.16\times10^{-26} \pm j5.1837\times10^{-26})\times3$	$(2.257\times10^{-28} \pm j2.2668\times10^{-29})\times3$
Λ_8	$7.8924\times10^{-6} \pm j1.3811\times10^{-5}$	$0.0071592 \pm j0.0051235$	$-4.3567 \pm j3.4858$	$0.0046803 \pm j0.0078825$
Λ_9	$(-8.79\times10^{-11} \pm j1.845\times10^{-11})\times4$	$(-8.778\times10^{-9} \pm j4.7607\times10^{-8})\times4$	$(2.5767\times10^{-5} \pm j4.1002\times10^{-6})\times4$	$(-5.071\times10^{-8} \pm j9.6649\times10^{-9})\times4$
Λ_{10}	$(4.245\times10^{-30} \pm j9.9488\times10^{-30})\times3$	$(5.749\times10^{-27} \pm j2.4771\times10^{-27})\times3$	$(-8.88\times10^{-25} \pm j3.8812\times10^{-24})\times3$	$(-1.696\times10^{-26} \pm j4.399\times10^{-26})\times3$
Λ_{11}	$0.23866 \pm j0.037942$	$-11.419 \pm j12.826$	$50.842 \pm j1219.3$	$154.79 \pm j8.9866$
Λ_{12}	$0.02053 \pm j0.026057$	$-1.1615 \pm j0.83067$	$-33.396 \pm j51.607$	$7.9373 \pm j16.997$
Λ_{13}	$-0.25757 \pm j0.24787$	$12.566 \pm j5.7509$	$28.738 \pm j533.48$	$-113.63 \pm j187.55$
Λ_{14}	$(-3.1599\times10^{-27}, 4.929\times10^{-28})\times3$	$(3.5094\times10^{-25}, 5.4748\times10^{-26})\times3$	$(-3.687\times10^{-23}, 5.7467\times10^{-24})\times3$	$(-5.1503\times10^{-24}, 3.2012\times10^{-24})\times3$
Λ_{15}	$(-2.66\times10^{-25} \pm j5.4743\times10^{-26})\times3$	$(4.165\times10^{-24} \pm j8.9611\times10^{-24})\times3$	$(3.492\times10^{-22} \pm j1.2631\times10^{-22})\times3$	$(-1.070\times10^{-22} \pm j5.842\times10^{-23})\times3$
Λ_{16}	1.3484×10^{-25}	-2.0616×10^{-24}	8.3773×10^{-23}	-1.3905×10^{-22}
Λ_{17}	$(-1.59\times10^{-16} \pm j7.8594\times10^{-17})\times4$	$(1.061\times10^{-14} \pm j2.1244\times10^{-14})\times4$	$(2.826\times10^{-12} \pm j1.4337\times10^{-12})\times4$	$(-8.79\times10^{-14} \pm j5.1769\times10^{-14})\times4$
Λ_{18}	$(-2.09\times10^{-29} \pm j7.7051\times10^{-31})\times4$	$(5.684\times10^{-28} \pm j1.0401\times10^{-28})\times4$	$(6.994\times10^{-27} \pm j5.1892\times10^{-26})\times4$	$(3.747\times10^{-27} \pm j5.8302\times10^{-27})\times4$

parameters of the designed PID RCR will not cause severe variations on the closed-loop eigenvalues. The closed-loop system can maintain stable operation subject to a small variation of the PID RCR's parameters.

16.4.4 Summary of the Analyzed Results

The major constraints for selecting the assigned eigenvalues in this chapter are analyzed as below.

(a) The obtained parameters of the PID RCR should be reasonable. For example, the time constant of the washout term must be positive and the gains of the PID RCR should be as small as possible.
(b) The eigenvalues of the closed-loop system including the designed PID RCR must be completely located on the left half of the complex plane under different operating conditions.
(c) The eigenvalues of the closed-loop system should be insensitive to the variation of all parameters of the designed PID RCR.
(d) According to the closed-loop eigenvalues with the designed PID RCR at the selected operating point listed in the third column of Table 16.1, all eigenvalue results of the closed-loop system under various wind speeds listed in Table 16.3, the designed parameters of the PID RCR, and all RSCs of the closed-loop eigenvalues with respect to the parameters of the PID RCR listed in Table 16.4, it can be obviously concluded that all system eigenvalues have adequate damping under various wind speeds and the parameters of the designed PID RCR are also appropriate.
(e) The eigenvalues of the closed-loop system including the designed PID RCR must be completely located on the left half of the complex plane under different operating conditions.
(f) The eigenvalues of the closed-loop system should be insensitive to the variation of all parameters of the designed PID RCR.

16.5 Dynamic Simulations under Various Wind-Speed Disturbance Conditions

This section employs the nonlinear model developed in Sect. 16.3 to compare the damping effect contributed by the designed PID RCR on dynamic behaviors of the studied four offshore wind farms subject to different wind-speed disturbances. Since the wind speed of the studied four offshore wind farms seldom reaches WTG's rated wind speed of 15 m/s, base wind speed of 12 m/s is properly selected as the nominal operating point in this section.

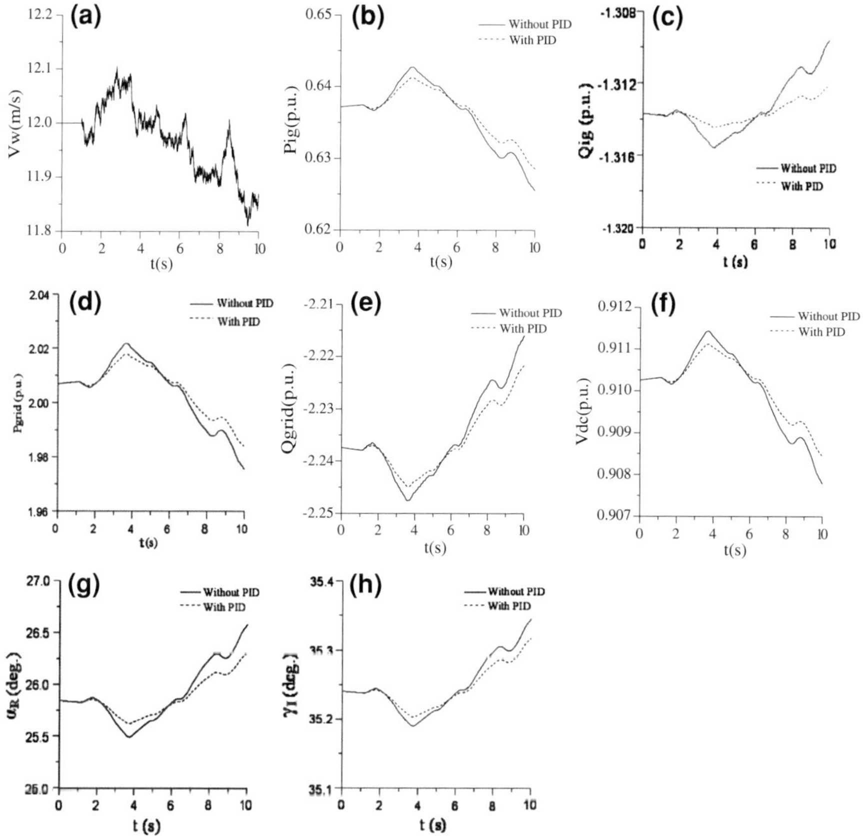

Fig. 16.9 Dynamic responses of the studied system with and without the designed PID RCR when the studied four equivalent 80 MW wind IGs are subject to the same wind-speed disturbance. (©2010 IEEE. Reprinted from *IEEE Trans. Power Delivery*, vol. 25, no. 2, April 2010). **a** V_W. **b** P_G . **c** Q_G. **d** P_{grid}. **e** Q_{grid}. **f** V_{DC}. **g** α_R. **h** γ_I

16.5.1 Identical Wind-Speed Disturbance Applied to Four Wind IGs

Figure 16.9 shows the comparative dynamic responses of the studied four offshore wind farms with and without the designed PID RCR when the same wind-speed disturbance is applied to the blades of the four equivalent 80 MW wind IGs. Because the four identical wind IGs are subject to the same wind-speed disturbance, only the active-power and reactive-power responses of one IG are shown in Fig. 16.9. The simulated wind-speed conditions are described as below.

(a) When 0 s $< t <$ 1.0 s, the four offshore wind farms operates under a base wind speed of V_W = 12 m/s.

(b) When 1.0 s $< t <$ 4 s, the wind speed initially drops, then gradually rises to 12.1 m/s, and finally step drops to 12 m/s.
(c) When 4 s $< t <$ 20 s, the wind speed gradually drops toward the final value of 11.8 m/s but some sudden rises can also be observed during the time interval.

It is clearly seen from the comparative transient responses shown in Fig. 16.9 that the designed PID RCR can effectively mitigate the variations of all quantities of the studied offshore wind farm such as active powers of each wind IG and grid, reactive powers of each wind IG and grid, dc voltage of the HVDC line, firing angle of the converter, and extinction angle of the inverter. Since the variations of the studied offshore wind farm are obviously reduced by the employment of the designed PID RCR, the fluctuating amplitude of all quantities of the studied four offshore wind farms subject to random wind-speed disturbances can be effectively decreased and the dynamic stability of the four offshore wind farms can be effectively enhanced.

16.5.2 Different Wind-Speed Disturbances Applied to Four Wind IGs

Different wind-speed disturbances of the studied four offshore wind farms with and without the PID RCR are simulated and compared. The comparative dynamic responses of the studied four offshore wind farms are shown in Fig. 16.10. Table 16.5 also lists the maximum errors between the transient responses of the system with and without the designed PID RCR. The simulated wind-speed conditions for the four wind IGs are analyzed as below.

(a) Four wind speeds V_{W1}–V_{W4} shown in Figs. 16.10a, d, g, and j are applied to the blades of wind IG_1–IG_4, respectively.
(b) For $0 < t < 1$ s, the initial wind speeds for the four wind IGs are identical base wind speeds of $V_W = 12$ m/s.
(c) When 1 s $< t <$ 10 s, both V_{W1} and V_{W4} have the same variations and their wind speeds gradually drop toward 11.5 m/s while both V_{W2} and V_{W3} gradually rise toward 13.5 and 12.5 m/s, respectively.
(d) The selected wind speeds can be used to examine identical wind speeds and different wind speeds on power fluctuations of the studied four offshore wind farms. This selection is also a practical consideration that different offshore wind farms are located at various sites with different wind directions and wind speeds. Moreover, several offshore wind farms could have different distances away from one another, and the wind-speed characteristics of various offshore wind farms could be quite different.
(e) The generated power of various offshore wind farms subject to different and identical wind-speed characteristics can be integrated together to deliver to the onshore substation through the HVDC link. The active-power fluctuations of the integrated four offshore wind farms can be effectively reduced by the designed PID RCR.

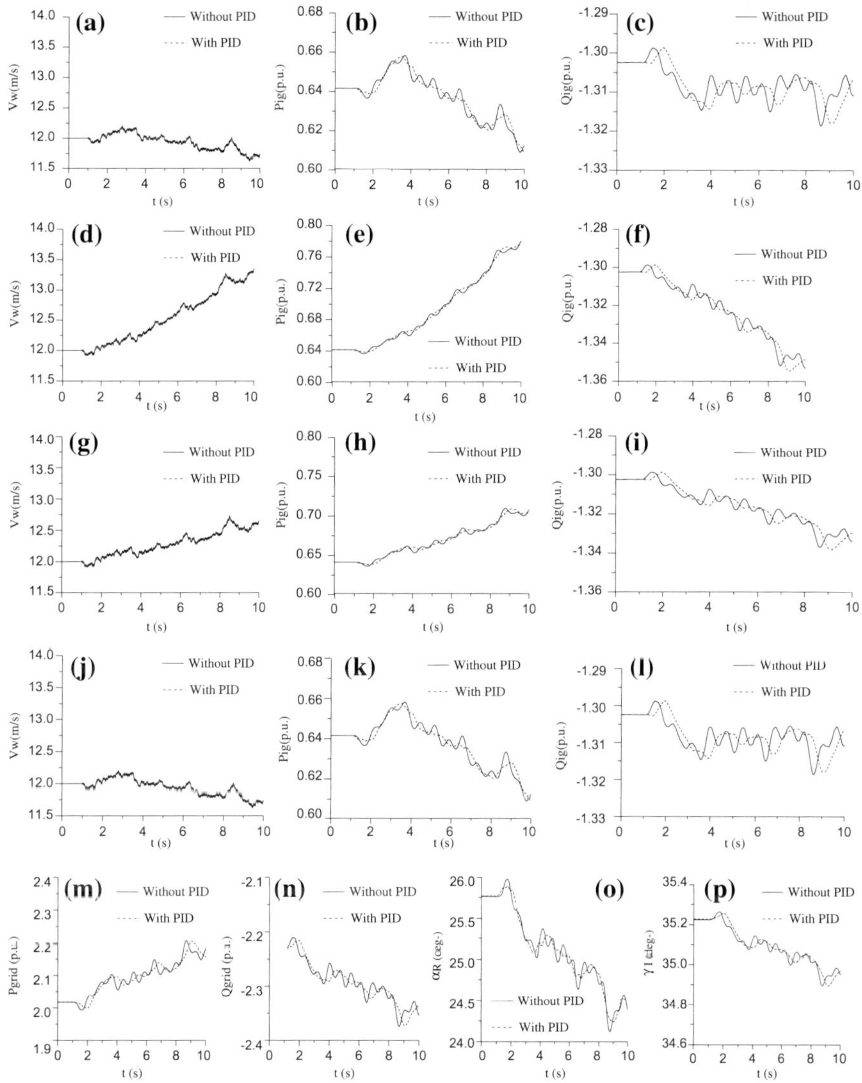

Fig. 16.10 Transient responses of the studied system with and without the designed PID RCR when the studied four equivalent wind IGs are subject to different wind-speed disturbances (©2010 IEEE. Reprinted from *IEEE Trans. Power Delivery*, vol. 25, no. 2, April 2010). **a** V_{W1}. **b** P_{G1}. **c** Q_{G1}. **d** V_{W2}. **e** P_{G2}. **f** Q_{G2}. **g** V_{W3}. **h** P_{G3}. **i** Q_{G3}. **j** V_{W4}. **k** P_{G4}. **l** Q_{G4}. **m** P_{grid}. **n** Q_{grid}. **o** α_R. **p** γ_I

It can be clearly seen from the comparative dynamic waveforms shown in Fig. 16.10 and the maximum errors listed in Table 16.5 that, again, the designed PID RCR can effectively mitigate the variations of all quantities of the studied four offshore wind farms even if different wind speeds are simultaneously applied to the

Table 16.5 Comparative maximum values and deviations (per-unit) of the quantities shown in Fig 16.10 when the HVDC link is with and without the designed PID RCR

	P_{G1}	P_{G2}	P_{G3}	P_{G4}	Q_{G1}	Q_{G2}	Q_{G3}	Q_{G4}	P_{gird}	Q_{gird}	α_R	γ_I
Operating point	0.641583	0.641583	0.641583	0.641583	−1.30243	−1.30243	−1.30243	−1.30243	2.017	−2.22973	25.76048	35.22375
Maximum value (With PID RCR)	0.643391	0.671216	0.659122	0.643391	−1.3076	−1.31439	−1.31142	−1.3076	2.073153	−2.27274	25.26331	35.12683
Maximum value (Without PID RCR)	0.644265	0.672067	0.659984	0.644265	−1.30701	−1.31381	−1.31083	−1.30701	2.073163	−2.27112	25.21551	35.11474
Maximum deviation (With PID RCR)	0.001807	0.029633	0.017539	0.001807	−0.00517	−0.01196	−0.00899	−0.00517	0.056153	−0.04301	−0.49717	−0.09691
Maximum deviation (Without PID RCR)	0.002682	0.030484	0.0184	0.002682	−0.00459	−0.01138	−0.00841	−0.00459	0.056163	−0.04139	−0.54497	−0.10901

blades of the four wind IGs, respectively. The dynamic results shown in Fig. 16.10 show that the employment of the designed PID RCR has the ability to reduce random fluctuations of the studied four offshore wind farms. The amplitude of power fluctuations of the studied four offshore wind farms with an HVDC link subject to different random wind-speed disturbances can be effectively mitigated and the dynamic stability of the four offshore wind farms can be effectively improved provided that the designed PID RCR is employed to join with the proposed line-commutated HVDC link.

16.6 Conclusion

This chapter has presented both damping enhancement and mitigation of power fluctuations of four parallel-operated offshore wind farms using a line-commutated HVDC link joined with a modal control designed PID RCR. Pole-assignment technique has been effectively applied to the design of the PID RCR to exactly locate the specified poles of the studied four offshore wind farms on the desired locations of the complex plane to achieve damping improvement. Steady-state operating conditions and eigenvalue analysis of the studied offshore wind farms under various wind speeds and relative sensitivity coefficients of the closed-loop eigenvalues with respect to four parameters of the PID RCR have been both performed to validate the effectiveness and advantages of the proposed HVDC joined with designed PID RCR on damping enhancement of the offshore wind farms. Comparative dynamic responses of the studied offshore wind farms subject to wind-speed disturbances have demonstrated the effectiveness of the proposed HVDC link joined with the designed damping PID RCR on mitigating power fluctuations of the four parallel-operated offshore wind farms.

Acknowledgements This work is supported by the National Science Council of Taiwan under Grant NSC 96-2221-E-006-313-MY3, Grant NSC 97-2918-I-006-010, Grant NSC 99-3113-P-006-007, and Grant NSC 100-3113-E-006-014.

Appendix

System Parameters

Reference [1]

$$V_b = 161\,\text{kV}, S_b = 80\,\text{MW (One of four offshore wind farms)}, \omega_b = 2\pi f_b,$$
$$f_b = 50\text{Hz}$$

(a) Single 2 MW wind induction generator (IG) (per-unit) $V_b = 690\,\text{V}$, $S_b = 2\text{MW}$, $\omega_b = 2\pi f, f_b = 50\text{Hz}\, r_s = 0.00488, X_{ss} = 0.09241$, $r_r = 0.00549$, $X_{rr} = 0.09955, X_m = 3.95279, H_G = 3.5\text{s}$

(b) Excitation capacitor bank and AC transmission line (per-unit) $X_C = 0.375$, $R_T = 0.01, X_T = 0.04, R_L = 0.02, X_L = 0.08$

(c) Line-commutated HVDC link (per-unit) $R_{DC} = 0.05, X_{DC} = 0.2$, $C_{DC} = 0.6$, $C_I = 0.3$, $C_R = 0.6$, $T_{a1} = 0.05s$, $K_{a1} = 0.1, T_R = 0.1s$, $K_R = 1.0$, $T_I = 0.1s$, $K_I = 1.0$, $I_{Cmax} = 0.1$, $I_{Cmin} = -0.1$, $\alpha_{Rmax} = 35°$, $\alpha_{Rmin} = 15°, \gamma_{Imax} = 45°$, $\gamma_{Imin} = 25°$

(d) Wind-turbine characteristics and coefficients (Fig. 16.11, Table 16.6).

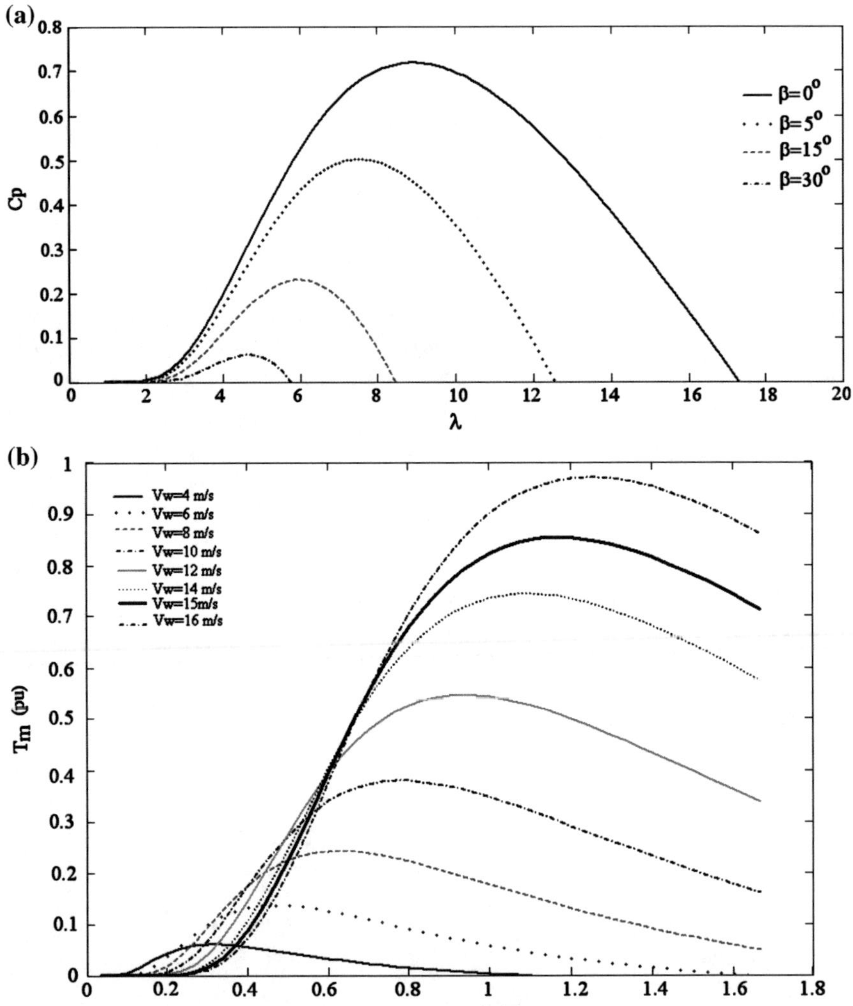

Fig. 16.11 Characteristics of the employed wind turbine model for simulations (©2010 IEEE. Reprinted from *IEEE Trans. Power Delivery*, vol. 25, no. 2, April 2010) (**a**) C_P versus λ (**b**)T_m versus ω_r

Table 16.6 Coefficients c_1–c_9 employed in wind-turbine simulations

	c_1	c_2	c_3	c_4	c_5	c_6	c_7	c_8	c_9
Constant speed	0.44	125	0	0	0	6.94	16.5	0	−0.02
Variable speed	0.73	151	0.58	0.02	2.14	13.2	18.4	−0.02	−0.003

References

1. Wang L, Wang K-H, Lee W-J, Chen Z (2010) Power-flow control and stability enhancement of four parallel-operated offshore wind farms using a line-commutated HVDC link. IEEE Trans Power Delivery 25(2):1190–1202
2. Narendra KG, Khorasani K, Sood VK, Pate RV (1998) Intelligent current controller for an HVDC transmission link. IEEE Trans Power Syst 13:1076–1083
3. Jovcic D, Pahalawaththa N, Zavahir M (1999) Novel current controller design for elimination of dominant oscillatory mode on an HVDC line. IEEE Trans Power Delivery 14:543–548
4. Jovcic D, Pahalawaththa N, Zavahir M (1999) Stability analysis of HVDC control loops. IEE Proc Gener Transm Distrib 146: 143–148
5. Bunch R, Kosterev D (2000) Design and implementation of AC voltage dependent current order limiter at pacific HVDC intertie. IEEE Trans Power Delivery 15:293–299
6. Kirby NM, Luckett MJ, Xu L (2001) HVDC transmission for large offshore wind farms. AC-DC Power Trans 485:162–168
7. Chan KH, Parle JA, Johnson N, Acha E (2001) Real-time implementation of a HVDC-VSC application in a scaled-down wind energy conversion system (WECS). AC-DC Power Trans 485:169–174
8. Gomes S, Martins N, Jonsson T, Menzies D, Ljungqvist R (2002) Modeling capacitor commutated converters in power system stability studies. IEEE Trans Power Syst 17: 371–377
9. Lu W, Ooi BT (2003) Optimal acquisition and aggregation of offshore wind power by multiterminal voltage-source HVDC. IEEE Trans Power Delivery 18:201–206
10. Osauskas C, Wood A (2003) Small-signal dynamic modeling of HVDC systems. IEEE Trans Power Delivery 18:220–225
11. Padiyar KR, Prabhu N (2004) Modelling control design and analysis of VSC based HVDC transmission systems. In proceedings of the IEEE 2004 international conference on power system technology, vol 1. pp 774–779
12. Yang X, Chen C (2004) HVDC dynamic modelling for small signal analysis. IEE Proc Gener Transm Distrib 151: 740–746
13. Yin M, Li G, Zhou M, Liu Y (2005) Analysis and control of wind farm incorporated VSC-HVDC in unbalanced conditions. In: Proceedings of the 2005 IEEE/PES transaction and distribution conference, pp 1–6
14. Reidy A, Watson R (2005) Comparison of VSC based HVDC and HVAC interconnections to a large offshore wind farm. In: Proceedings IEEE power engineering society general meeting, vol 1. pp 1–8
15. Fernandopulle N, Alden RTH (2005) Integration of HVDC control dynamics into transientenergy functions. Can J Eelectr Comput Eng 30
16. Xiang D, Ran L, Bumby JR, Tavner PJ, Yang S (2006) Coordinated control of an HVDC link and doubly fed induction generators in a large offshore wind farm. IEEE Trans Power Delivery 21:463–470
17. Koutiva XI, Vrionis TD, Vovos NA, Giannakopoulos GB (2006) Optimal integration of an offshore wind farm to a weak AC grid. IEEE Trans Power Delivery 21:987–994
18. Jovcic D (2006) Interconnecting offshore wind farms using multiterminal VSC-based HVDC. In: Proceedings of the IEEE power engineering society, pp 1–7

19. Jovcic D, Milanovic JV (2006) Offshore wind farm based on variable frequency mini-grids with multiterminal DC interconnection. AC and DC power transactions, pp 215–219
20. Bresesti P, Kling WL, Hendriks RL, Vailati R (2007) HVDC connection of offshore wind farms to the transmission system. IEEE Trans Energy Convers 22:37–43
21. Bozhko SV, Giménez RB, Li R, Clare JC, Asher GM (2007) Control of offshore DFIG-based wind farm grid with line-commutated HVDC connection. IEEE Trans Energy Convers 22:71–78
22. Bozhko S, Asher G, Li R, Clare J, Yao L (2008) Large offshore DFIG-based wind farm with line-commutated HVDC connection to the main grid: engineering studies. IEEE Trans Energy Convers 23:119–127
23. Li R, Bozhko S, Asher G (2008) Frequency control design for offshore wind farm grid with LCC-HVDC link connection. IEEE Trans Power Electron 23:1085–1092
24. Du C, Agneholm E, Olsson G (2008) Use of VSC-HVDC for industrial systems having onsite generation with frequency control. IEEE Trans Power Delivery 23:2233–2240
25. Du C, Agneholm E, Olsson G (2008) Comparison of different frequency controllers for a VSC-HVDC supplied system. IEEE Trans Power Delivery 23:2224–2232
26. Murray NJ, Arrillaga J, Liu YH, Watson NR (2008) Flexible reactive power control in multigroup current-sourced HVDC interconnections. IEEE Trans Power Delivery 23: 2160–2167
27. Jaén A, Acha E, Gómez Expósito A (2008) Voltage source converter modeling for power system state estimation: STATCOM and VSC-HVDC. IEEE Trans Power Syst 23:1552–1559
28. Anderson PM, Bose A (1983) Stability simulation of wind turbine system. IEEE Trans Power Apparatus Syst 102:3791–3795
29. Heier S, Waddington R (1998) Grid integration of wind energy conversion systems. Wiley, New York
30. CIGRE (2000) Modeling new forms of generation and storage. TF.01.10, Fifth draft
31. Krause PC (1987) Analysis of electric machinery. McGraw-Hill, New York
32. Anderson PM, Fouad AA (1977) Power system control and stability. The Iowa University Press, AMES,IA
33. Kundur P (1994) Power system stability and control. McGraw-Hill, New York
34. Kimbark EW (1971) Direct current transmission. Wiley, New York
35. Peterson HA, Krause PC (1966) A direct- and quadrature-axis representation of a parallel AC and DC power system. IEEE Trans Power Apparatus Syst 86:210–224
36. Rahim AHMA, El-Amin IM (1986) Stabilization of a high voltage AC/DC power system I, Evaluation of control strategies. IEEE Trans Power Syst 1:128–136
37. Hsu Y-Y, Wang L (1988) Damping of a parallel ac-dc power system using PID power system stabilizers and rectifier current regulators. IEEE Trans Energy Convers 3:540–549
38. Hsu Y-Y, Wang L (1989) Modal control design of an HV DC system for the damping of subsynchronous oscillations. IEE proceedings, part C, vol 136. pp 76–86
39. Wang L (1993) A comparative study of damping schemes on damping generator oscillations. IEEE Trans Power Syst 8:613–619
40. Wang L, Mau S-J, Chuko C-C (1993) Suppression of common torsional mode interactions using shunt reactor controllers. IEEE Trans Energy Convers 8:539–545

Chapter 17
Fault Ride-Through of HVDC Connected Large Offshore Wind Farms

Lie Xu and Liangzhong Yao

Abstract Voltage source converter-based HVDC systems have been proposed for connecting large offshore wind farms. Fault ride-through during onshore ac fault is one of the main grid code requirements and technical challenges. Due to the reduction of transmitted grid power during such faults, the wind farm generated power must be either reduced or damped to ensure continuous system operation. Three different strategies, i.e., telecommunication based, offshore frequency modulation, and DC damping resistor are investigated.

17.1 Large Offshore Wind Farm Integration Using HVDC

Large offshore wind farms have attracted a lot of attention in the last few years and the European wind energy association estimates that 120 GW of offshore wind power will be installed in the next two decades, amounting to 10% of the installed generation capacity. The integration of such large offshore wind farms to the grid over distances of tens, and sometimes hundreds, of kilometers is one of the main challenges facing developers and system operators [1, 2]. It has been well accepted

L. Xu (✉)
School of Electronics, Electrical Engineering and Computer Science,
Queen's University of Belfast, Belfast, BT9 5AH, UK
e-mail: lie.xu@ieee.org

L. Yao
State Grid Electric Power Research Institute,
No 8, NARI Road, Nanjing, 210003 China

S. M. Muyeen (ed.), *Wind Energy Conversion Systems*,
Green Energy and Technology, DOI: 10.1007/978-1-4471-2201-2_17,

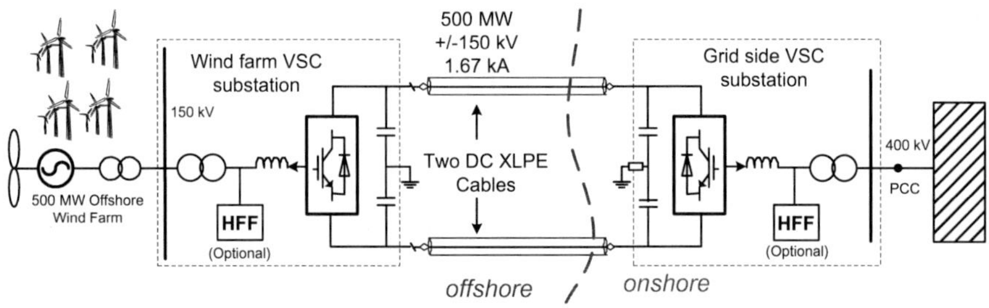

Fig. 17.1 Single-line diagram of wind farm integration using VSC-based HVDC transmission

that HVDC transmission technology will become attractive and likely to be the only feasible option for connecting large remote offshore wind farms [3–5].

HVDC technologies based on both the line commutated converter (LCC) [4–8] and the voltage source converter (VSC) [4, 9–12] have been proposed for wind farm integration. It has been identified that VSC scheme is superior to LCC systems in terms of independent reactive power control, no need for external voltage source, less harmonic pollution, and fast system control. Apart from the common point-to-point system,a multi-terminal VSC-based HVDC system which contains an offshore DC grid has also been proposed for connecting large wind farms [13–16].

Similar to ac connection, offshore wind farms connected using HVDC system are also required to meet the grid code requirement at the onshore ac connection point. Among the grid code requirements, fault ride-through (FRT) is one of the main challenges when HVDC connection is used. The key issue is how to balance the generated and transmitted power immediately after grid fault while maintaining the offshore ac network considering the fact that the wind farms and the faulty ac system are decoupled by the DC connectors.

17.2 System Outline and Operation Principles

Figure 17.1 shows the single-line diagram of a point-to-point VSC-based HVDC system for integrating offshore wind farms. The wind farm considered in this example is rated at 500 MW and consists of wind turbines with a power electronics converter interface. The HVDC system contains a wind farm VSC substation (WFVSC), a grid side VSC substation (GSVSC), and a pair of dc XLPE cables. The high frequency filter (HFF) shown in Fig. 17.1 might not be required if the converters generate near-sinusoidal ac waveforms due to the use of a multi-level converter structure. Based on the point-to-point system shown in Fig. 17.1, a multi-terminal HVDC system can be envisaged. Figure 17.2 shows a “typical” four-terminal system which contains an offshore DC grid and connects two 500 MW offshore wind farms to two ac networks. For the following studies,the two-terminal system shown in Fig. 17.1 will be used as the bench mark but the adopted methodology can also be applied to multi-terminal systems.

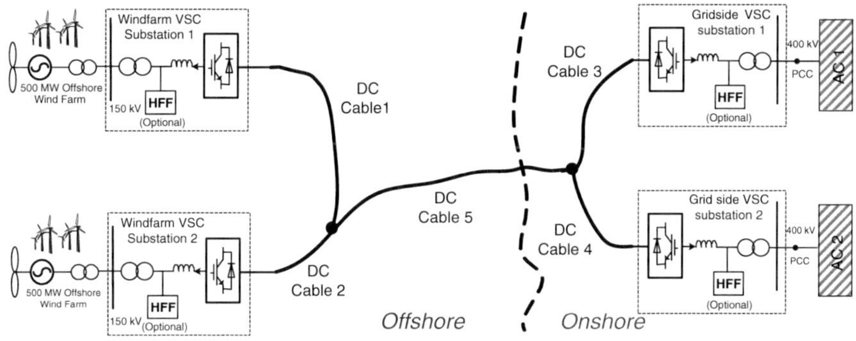

Fig. 17.2 Single-line diagram of wind farm integration using a four-terminal HVDC system

The wind turbines in each offshore wind farm are considered to be connected together using a local ac network. The WFVSC collects energy from the offshore wind farm and then converts ac into dc. The WFVSC also controls the ac voltage and frequency of the wind farm ac network. The power transmitted through the dc cables is then converted back into ac via the GSVSC. The GSVSC can also provide network support, e.g., reactive power / ac voltage control, power oscillation damping etc., to the onshore ac network [17].

Under normal operation, the dc voltage of the HVDC system must be maintained at a set value with limited variation range. Abnormal dc link voltage can cause the system to trip and disrupt its operation. Furthermore, a constant dc voltage indicates balanced active power flow between the wind farm input (WFVSC) and the grid output (GSVSC). To achieve this balance, the GSVSC is controlled as the dc voltage regulator ensuring that the generated wind energy is fully transmitted to the onshore grid network.

17.2.1 GSVSC Modeling and Control

For the GSVSC, Fig. 17.3a, b show its equivalent ac and dc circuits respectively where V_s and V_c refer to the source and converter generated voltages. L and R are the inductance and equivalent resistance of the AC coupling inductor, and I_s is the converter ac output current. C is the capacitance of the main dc capacitor and V_{dc} is the dc link voltage. Referring to Fig. 17.3a, in the synchronous d-q reference frame rotating at the supply frequency of ω_s where the d-axis is fixed to the source voltage V_s and usually obtained by a phase-locked loop, the system on the ac side can be expressed as:

$$\frac{d}{dt}\begin{bmatrix} i_{sd} \\ i_{sq} \end{bmatrix} = [A]\begin{bmatrix} i_{sd} \\ i_{sq} \end{bmatrix} + \frac{1}{L}\begin{bmatrix} -v_{sd} + v_{cd} \\ -v_{sq} + v_{cq} \end{bmatrix} \tag{17.1}$$

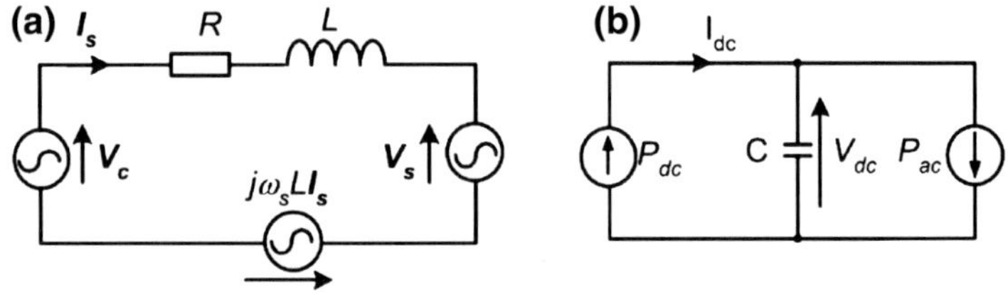

Fig. 17.3 Equivalent circuit of the GSVSC. **a** AC. **b** DC

where $[A] = \begin{bmatrix} -R/L & \omega_s \\ -\omega_s & -R/L \end{bmatrix}$.

Using the power balancing equation, the dc side system as shown in Fig. 17.3b is expressed as:

$$P_{ac} = \frac{3}{2}\left(v_{cd} \cdot i_{sd} + v_{cq} \cdot i_{sq}\right), \quad P_{dc} = I_{dc} \cdot V_{dc} \tag{17.2a}$$

$$I_{dc} \cdot V_{dc} = \frac{3}{2}\left(v_{cd} \cdot i_{sd} + v_{cq} \cdot i_{sq}\right) + C\frac{dV_{dc}}{dt} \cdot V_{dc} \tag{17.2b}$$

Under Pulse Width Modulation (PWM) control, the amplitude of the converter output fundamental voltage is controlled by the modulation index as

$$V_c = M \cdot V_{dc}/2 \tag{17.3}$$

Assuming an ideal converter model, the ac and dc side systems showing in (17.1) and (17.2a, 17.2b) can be expressed as:

$$\begin{aligned} \frac{d}{dt}\begin{bmatrix} i_{sd} \\ i_{sq} \end{bmatrix} &= [A]\begin{bmatrix} i_{sd} \\ i_{sq} \end{bmatrix} - \frac{1}{L}\begin{bmatrix} v_{sd} \\ v_{sq} \end{bmatrix} + \frac{V_{dc}}{2L}\begin{bmatrix} M_d \\ M_q \end{bmatrix} \\ \frac{dV_{dc}}{dt} &= \frac{1}{C} \cdot I_{dc} - \frac{3}{4C}\left(M_d \cdot i_{sd} + M_q \cdot i_{sq}\right) \end{aligned} \tag{17.4}$$

where the d- and q-axis modulation indices are defined as
$M_d = 2 \cdot v_{cd}/V_{dc}$ and $M_q = 2 \cdot v_{cq}/V_{dc}$.

Thus the main requirement of the GSVSC controller is to control i_{sd}, i_{sq} and V_{dc} [18]. Figure 17.4 shows the block diagram of a typical control system for GSVSC. As shown, an outer dc/ac voltage loop and an inner current loop are used where V_G refers to the grid ac voltage.

17.2.2 WFVSC Modeling and Control

The WFVSC can be modeled in a similar way as for the GSVSC [16]. By considering the wind farm as a controlled current source, Fig. 17.5a, b show the schematic diagram and the ac equivalent circuit of the WFVSC in the d-q

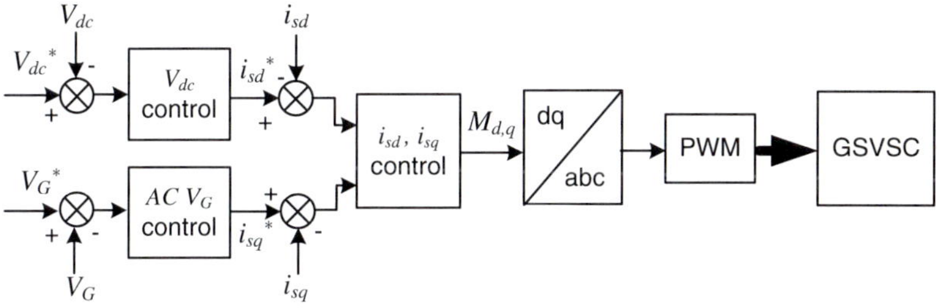

Fig. 17.4 Block diagram for GSVSC controller

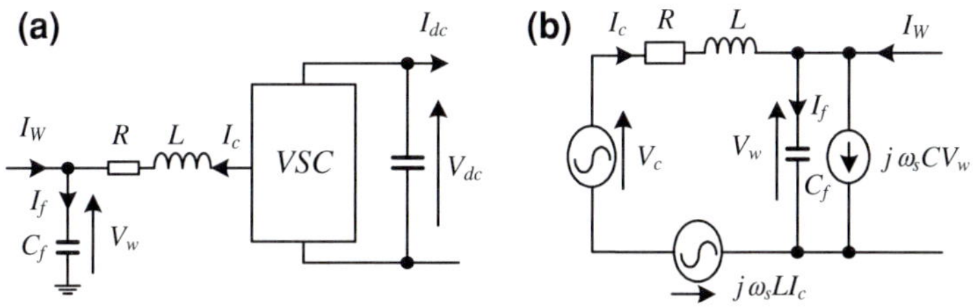

Fig. 17.5 Equivalent circuit of the WFVSC. **a** Schematic diagram. **b** ac equivalent circuit

synchronous reference frame. V_w is the wind farm ac voltage. I_c and I_w are the converter output and total input currents respectively whereas I_f is the current flowing into the filter. At the fundamental frequency, the high frequency ac filter is dominated by its capacitance. Thus it is represented by a simple capacitor C_f as shown in Fig. 17.5.

As there is no synchronous generator in the offshore wind farm network, the frequency and phase of the ac voltage are determined solely by the control of the WFVSC, and therefore can be set directly by the controller. Based on Fig. 17.5b, the system on the ac side can be expressed in the synchronous d-q reference frame as: [16]

$$\begin{aligned} V_c &= RI_c + L\frac{dI_c}{dt} + j\omega_s LI_c + V_w \\ I_f &= I_c + I_w - j\omega_s C_f V_w = C_f\frac{dV_w}{dt} \end{aligned} \tag{17.5}$$

Separating (17.5) into d-q components yields:

$$\begin{aligned} \frac{d}{dt}\begin{bmatrix} i_{cd} \\ i_{cq} \end{bmatrix} &= \begin{bmatrix} -R/L & \omega_s \\ -\omega_s & R/L \end{bmatrix}\begin{bmatrix} i_{cd} \\ i_{cq} \end{bmatrix} - \frac{1}{L}\begin{bmatrix} v_{wd} \\ v_{wq} \end{bmatrix} + \frac{V_{dc}}{2L}\begin{bmatrix} M_{cd} \\ M_{cq} \end{bmatrix} \\ \frac{d}{dt}\begin{bmatrix} v_{wd} \\ v_{wq} \end{bmatrix} &= \begin{bmatrix} 0 & \omega_s \\ -\omega_s & 0 \end{bmatrix}\begin{bmatrix} v_{wd} \\ v_{wq} \end{bmatrix} + \frac{1}{C_f}\begin{bmatrix} i_{cd} + i_{wd} \\ i_{cq} + i_{wq} \end{bmatrix} \end{aligned} \tag{17.6}$$

where $M_{cd} = 2 \cdot v_{cd}/V_{dc}$ and $M_{cq} = 2 \cdot v_{cq}/V_{dc}$.

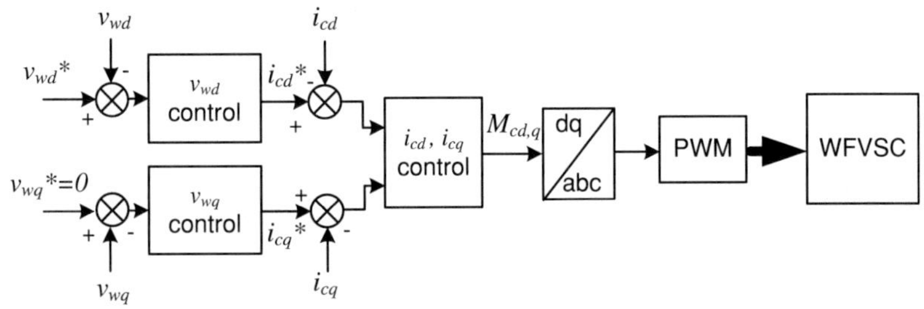

Fig. 17.6 Schematic diagram of the control system for the WFVSC

WFVSC operation requires i_{cd} and i_{cq} to follow varying reference points. In addition, the wind farm side ac voltages v_{wd} and v_{wq} need to be controlled to the set points. Similar to the GSVSC, the WFVSC system control involves two control loops, i.e., an outer ac voltage loop and an inner current loop as Fig. 17.6 shows. As shown, v_{wd}^* refers to the reference ac voltage for the wind farm ac network.

An alternative control method for the WFVSC is for it to resemble an infinite voltage source with constant frequency, voltage amplitude, and phase angle. Thus, as in the case when a wind farm is connected to an infinite ac system, the power generated by the wind farm is automatically absorbed by the source resembled by the WFVSC and then transmitted to the dc side. The control system for the WFVSC does not need to identify whether the power is active or reactive and no extra frequency or active power control loops are required [4]. The schematic diagram of such a control strategy for the WFVSC is shown in Fig. 17.7 where it is controlled and operated as a voltage source with constant ac frequency (f) and phase angle (θ). The only feedback control loop used is the ac voltage control via the modulation index M_c of the converter. This is similar to a synchronous generator but with improved frequency control capability as the output ac frequency is set directly by the output of the WFVSC.

17.3 System Fault Ride-Through During Onshore AC Fault

To ensure FRT of the HVDC system during close fault on the onshore ac network, the following two requirements must be met:

1. Generated and transmitted/consumed active power must be balanced during fault. During an onshore ac fault, active power transmitted by the GSVSC to the grid may be significantly reduced due to the reduced ac voltage. Consequently, the total generated wind power could exceed the maximum power being transmitted to the onshore network. This could result in dc over-voltage and necessiate the shutdown of the whole system. It is thus important that either the wind farm generated active power is being reduced or active power is being

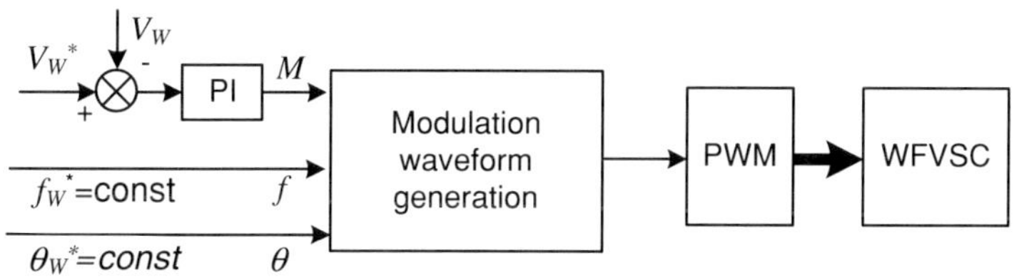

Fig. 17.7 Simplified control blocks for WFVSC controlled as an infinite source

damped (consumed) in a separate device such that the dc voltage of the HVDC system can be maintained.
2. Offshore AC systems need to be maintained with limited frequency and voltage variations such that when onshore fault is cleared, the system can go back to normal operation quickly.

Under normal operation, the mean value of the dc voltage is well controlled by the GSVSC and the variation range is limited. During grid fault, the dc voltage could exceed its normal variation range and thus abnormal dc voltage can be used as an indication of fault condition on the grid side.

17.3.1 GSVSC

During onshore ac fault, the GSVSC is no longer able to control the dc voltage due to the reduction of its terminal ac voltage. Its dc voltage controller will push it into current limit operation automatically. However, in order to avoid the unnecessary interaction between GSVSC's dc voltage controller and WFVSC's control system, the GSVSC's dc voltage controller is bypassed and is switched to current limit operation upon the detection of abnormal dc voltage. The schematic diagram shown in Fig. 17.8a illustrates this approach.

After fault clearance, the onshore ac voltage recovers quickly. Consequently, the active power output from the GSVSC to the grid increases rapidly as the controller is unlikely to change the converter current under such fast transient. This could result in a rapid reduction of the dc voltage, and its value could drop to a very low level if both the converter current and the increase in ac voltage were high. In order to limit the impact of the rapid increase in ac voltage and active power on system control and operation during recovery, a voltage dependent current limit is employed as schematically shown in Fig. 17.8b [12]. As shown, the lower the fault ac voltage is which indicates a possible larger step change in ac voltage during recovery, the lower the current limit is set. Thus the increase of active power output and the resulted converter dc voltage drop can be limited during recovery.

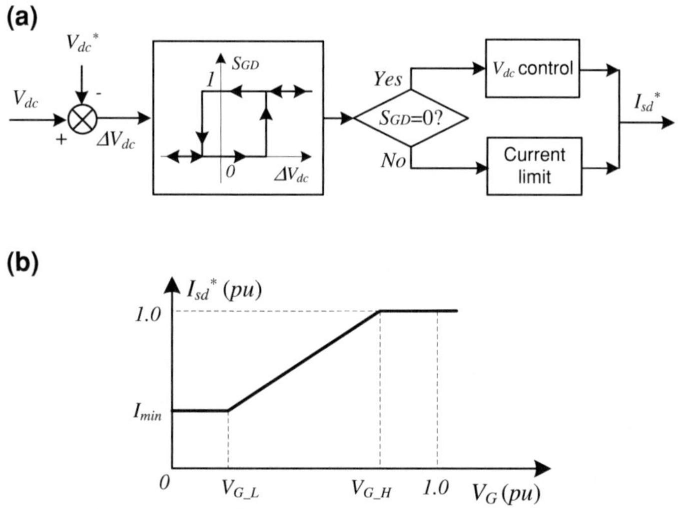

Fig. 17.8 **a** GSVSC control switch during grid fault, **b** Voltage dependent current limit

17.3.2 WFVSC

Since the GSVSC is now operated at current limit mode, the WFVSC must regulate the dc voltage. In order to maintain the dc voltage, the active power generated by the wind farm must match the total transmitted/consumed power by the system. In the following studies, the WFVSC is assumed to be controlled as an infinite voltage source shown in Fig. 17.7. A number of options exist: [19].

Option 1: Fast telecommunication between the WFVSC and individual wind turbine. Once dc over-voltage is detected, signals are sent to each wind turbine using telecommunications such that the active power output from the wind turbines are reduced immediately. The main issue for this option in practical systems is that small signal delay could result in excessive dc over-voltage.

Option 2: Offshore wind farm network frequency modulation. In this approach, the offshore ac system frequency is increased using the WFVSC during fault, and the active power from the wind turbines is automatically reduced upon the detection of abnormal ac frequency. While this approach does not require extra hardware, its reliability would be one of the main concerns. In addition, the wind turbines must be specially designed to respond to fast frequency variation.

Option 3: DC damping resistor on the GSVSC substation. A damping resistor is placed on the dc side of the GSVSC and is switched in to consume the surplus power upon the detection of dc over-voltage. Thus the wind farm and WFVSC can operate as normal with no disruption. The scheme is very reliable and there is no need for wind turbines to adopt special designs. However, this method incurs extra hardware cost as the damping resistor and switch have to be rated properly.

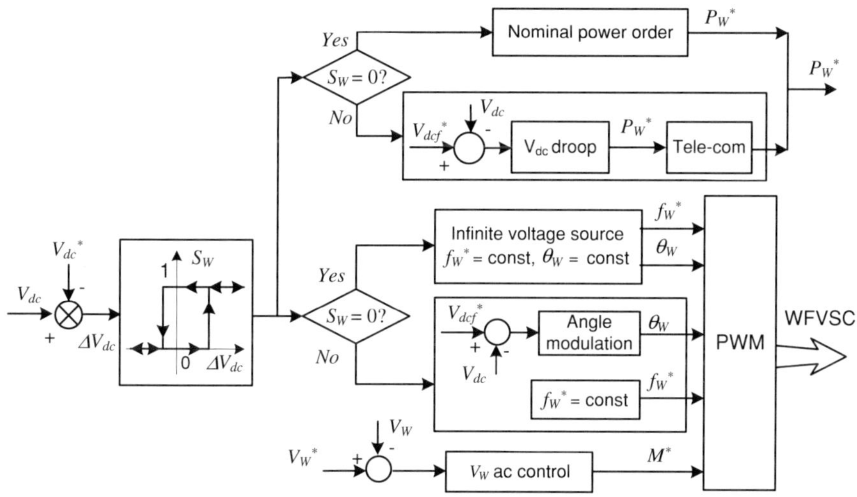

Fig. 17.9 WFVSC and wind farm control during grid side fault (Option 1)

17.3.2.1 Option 1: Communication based

The schematic diagram of the control process under Option 1 is shown in Fig. 17.9. As shown, when the dc over-voltage reaches a certain threshold, the control flag S_W is set to 1 and the WFVSC is switched from infinite voltage source to dc voltage control by regulating its phase angle in order to reduce the power import from the wind farm. The dc voltage reference during fault V_{dcf}^{*} can be set slightly above its nominal value (say +3%). At the same time a reduced power order signal is sent to individual wind turbine using fast telecommunication, to temporarily over-ride the nominal power order. The wind farm power output is reduced in accordance with the new power order and consequently the dc voltage of the VSC-HVDC system is regulated. The ac frequency/voltage of the wind farm network stays at its nominal value. When the fault is cleared the onshore ac voltage recovers rapidly. Consequently, the power output from the GSVSC to the grid is also significantly increased and this causes the dc voltage to drop. The voltage drop is detected by both converters and the GSVSC and WFVSC return to dc voltage control and infinite voltage source modes, respectively. The imposed wind farm power limit is also removed. The wind generator will speed up during the fault due to the reduced electric power output. However, this is unlikely to be a major concern due to the variable speed wind turbine operation.

17.3.2.2 Option2: offshore frequency modulation

Instead of using telecommunication, Option 2 uses the offshore ac network for transmitting the requirement for power reduction between the WFVSC and

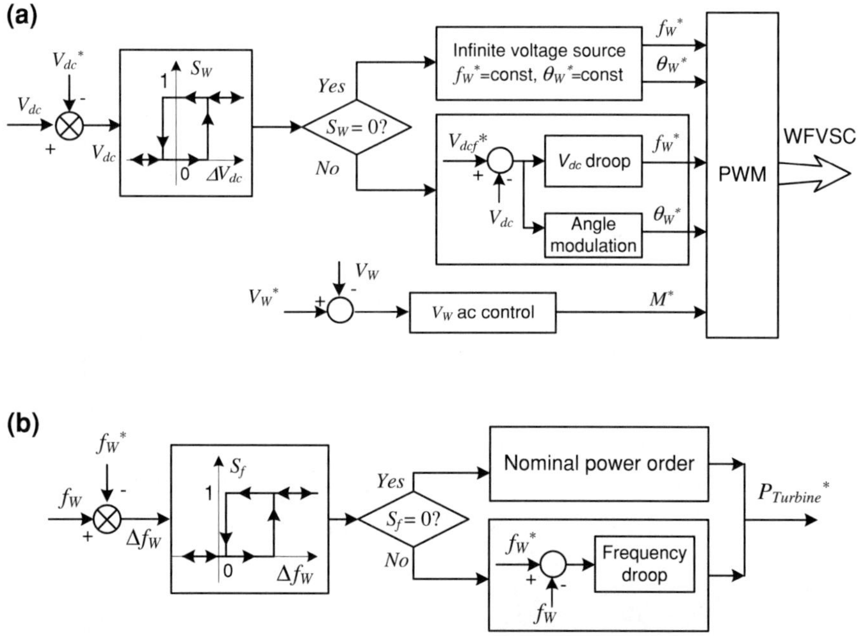

Fig. 17.10 System control for the WFVSC and wind farm for Option 2. **a** WFVSC control switch during grid fault (Option 2). **b** Wind turbine frequency control during fault

individual wind turbine. AC frequency, rate of ac frequency change, rate of ac voltage change, etc., can all be used as the indication of fault. However, the study here only utilizes the ac frequency.

Figure 17.10a shows the control diagram for the WFVSC. As can be seen, once abnormal dc voltage is detected, the controller is switched from infinite voltage source to frequency modulation. Thus the frequency reference for the WFVSC output is increased according to the pre-set droop characteristic. Again, an extra dc voltage control loop regulating the phase angle of the WFVSC output is also employed. As usual, the ac voltage V_w is controlled through the modulation index. Once the fault is cleared and the dc voltage back to its nominal value, the WFVSC is switched back to infinite voltage source mode with constant frequency and phase angle.

As ac frequency plays no direct part in active and reactive power in wind turbines with power electronics interface, it is essential that an extra outer frequency loop is added to the nominal power controller at each turbine as Fig. 17.10b shows. As seen, upon detecting abnormal frequency at each turbine, the nominal wind turbine active power order, e.g. from maximum power point tracking curve, is replaced immediately by the output from the frequency droop controller. Thus the wind farm output active power is reduced immediately.

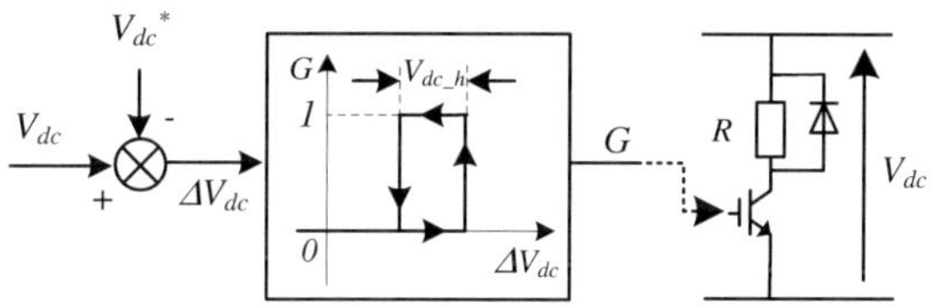

Fig. 17.11 Control of dc damping resistor for Option 3

17.3.2.3 DC Damping Resistor

The principle of this method is to dissipate excessive energy in the system into dc resistors during onshore ac fault. The dc resistor can be connected to the dc terminals of the GSVSC via a controllable power device such as an IGBT valve. If the dc damping resistor is rated at the full system rating, the power output from the wind farm through WFVSC will not be affected and their normal operation undisrupted even during a close ac fault to the GSVSC terminal. A simple hysteresis control can be used to control the power switch as schematically shown in Fig. 17.11. As seen, by repeatedly switching the dc damping resistor in and out during the ac fault period, the dc voltage can be maintained within the control band of V_{dc_h}. The method requires no coordination between the WFVSC and individual wind turbines and provides a reliable and secure way of operation. The drawback for this method is the extra cost of the dc damping resistor and switching devices which all have to be rated at the full system rating.

17.4 Case Studies

Simulation studies of the HVDC system for integrating a 500 MW wind farm based on doubly fed induction generators have been performed. The system comprises a 500 MW wind farm and a 500 MW/300 kV HVDC system based on conventional 2-level converter switched at 2 kHz. The main dc capacitor on each side is 100 μF and the length of the dc cables is 100 km. The GSVSC is connected to an ac source with a short circuit ratio of five. One HFF rated at 75 MVar is used at each end and is connected permanently to the respective networks as shown in Fig. 17.1. The wind farm is simulated as one lumped 500 MW wind generator model with an inertia constant of 3 s. During the simulation, a 100 ms solid 3-phase to ground fault is applied to the primary side of the GSVSC coupling transformer at 0.5 s.

17.4.1 Option 1

Figure 17.12 shows the simulated results with Option 1 during the 100 ms ac fault. Prior to the fault, the system operates around its full capacity of 500 MW. Upon

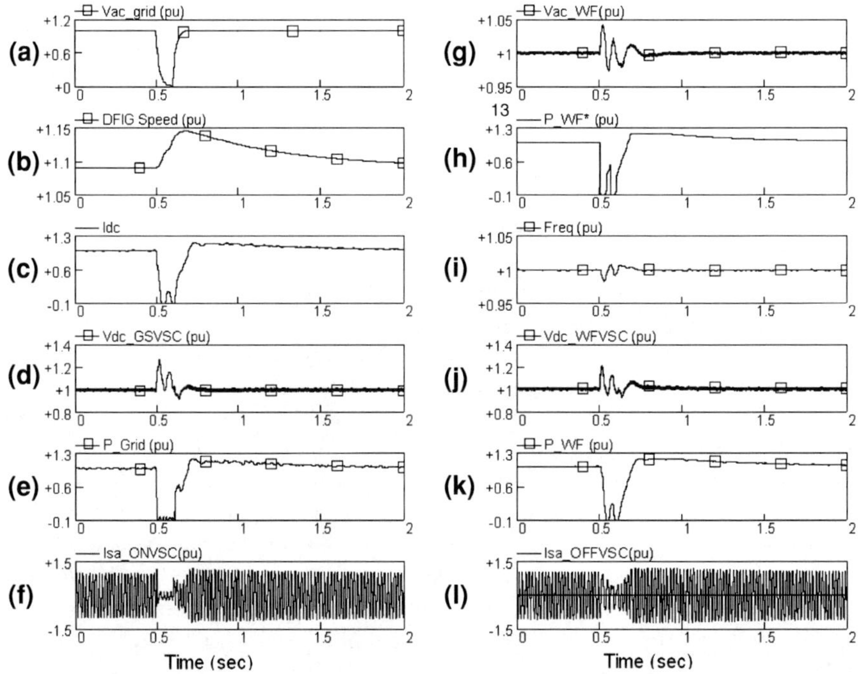

Fig. 17.12 Simulation results with grid side ac fault with Option 1: **a** Grid ac voltage, **b** Wind turbine speed **c** Converter dc current, **d** GSVSC dc voltage **e** GSVSC active power output, **f** GSVSC phase current **g** Wind farm ac voltage **h** Wind farm power reference **i** Wind farm ac frequency **j** WFVSC dc voltage **k** Active power from the wind farm **l** WFVSC phase current

the ac fault, the rapid reduction of the power export by the GSVSC causes the dc voltage to rise quickly, as can be seen from Fig. 17.12d, e, j. When the dc over-voltage reaches 8%, the WFVSC switches to dc voltage control and generates the active power order for the wind farm which is transmitted to the turbines using telecommunication. Only a very small telecom delay of 0.5 ms is used in the simulation. As can be seen in Fig. 17.12h, k, the wind farm output active power is reduced and so is the dc voltage. During the fault, the GSVSC operates in current limit mode. The maximum dc over-voltages for this study are around 27% and 22% for the GSVSC and WFVSC, respectively. Due to the reduced electric power output during the fault the wind turbine speeds up as shown in Fig. 17.12d.

When the fault is cleared at 0.6 s, the dc voltage drops and the WFVSC and the GSVSC switch back to infinite ac voltage source mode and dc voltage control mode respectively. A 20% over-load capability is assumed for the whole system. As can be seen, the accumulated energy in the turbine during the fault period is transferred back to the grid and the system eventually back to normal operation.

As can be seen, FRT is achieved although considerable dc over-voltage is observed. System operation with larger communication delay was found to be more problematic.

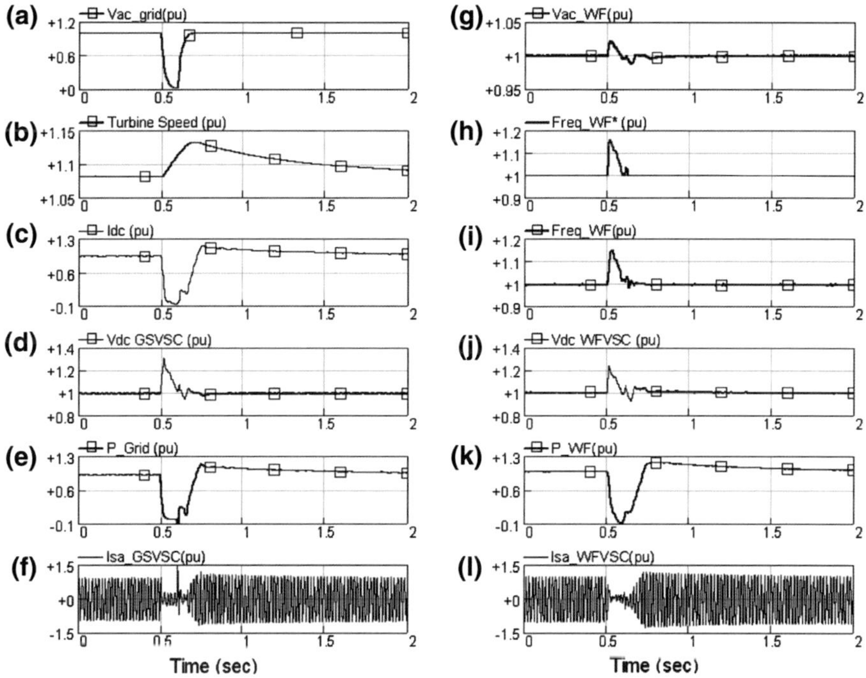

Fig. 17.13 Simulation results with grid side ac fault with Option 2 **a** Grid ac voltage, **b** Wind turbine speed, **c** Converter dc current, **d** GSVSC dc voltage, **e** GSVSC active power output, **f** GSVSC phase current, **g** Wind farm ac voltage, **h** Wind farm ac frequency reference, **i** Wind farm ac frequency, **j** WFVSC dc voltage, **k** Active power from the wind farm, **l** WFVSC phase current

17.4.2 Option 2

Under the same operating condition, Fig. 17.13 shows the system response with Option 2. Again, the ac fault caused the dc voltages on both converters to increase quickly and exceed the upper threshold of 1.08 pu. The control mode for the GSVSC was then switched to current limit. Meanwhile, the WFVSC was switched to frequency modulation and dc voltage control mode. The wind farm ac frequency started to rise due to the control action by the WFVSC as can be seen in Fig. 17.13h, i. The abnormal ac frequency was detected by the wind turbines and once the threshold of 1.03 pu is exceeded, their output active power is reduced. Consequently, the dc over-voltage is reduced. The maximum dc over-voltages for this study are about 32% and 23% for the GSVSC and WFVSC respectively. The maximum ac frequency increase on the wind farm is about 16%.

When the fault is cleared, the dc voltage on both converters drops and causes the GSVSC and the WFVSC to switch back to dc voltage control and infinite voltage source mode respectively. The wind farm ac frequency also returns to its

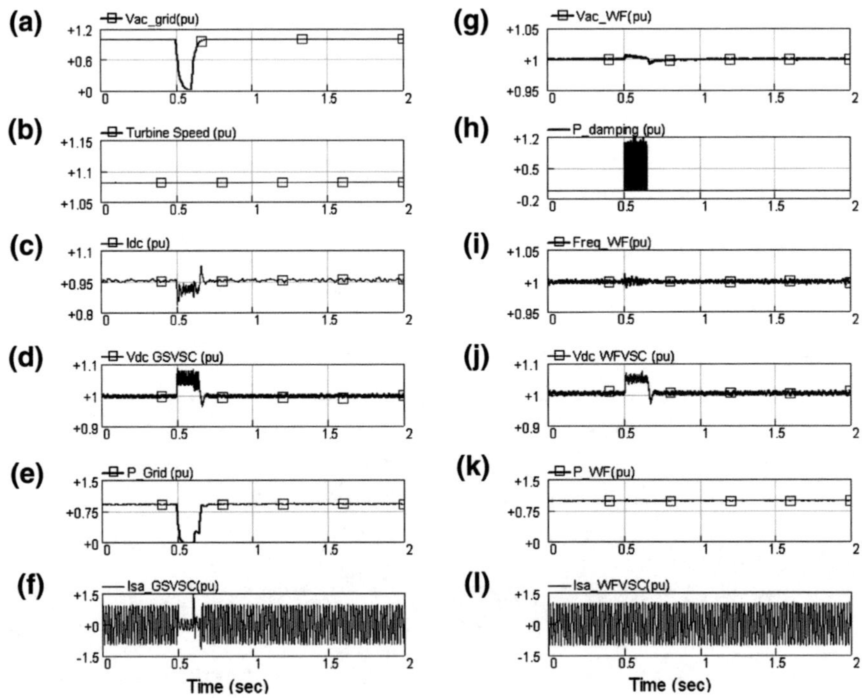

Fig. 17.14 Simulation results with grid side ac fault with Option 3: **a** Grid ac voltage, **b** Wind turbine speed, **c** Converter dc current, **d** GSVSC dc voltage, **e** GSVSC active power, **f** GSVSC phase current, **g** Wind farm ac voltage, **h** dc damping resistor current, **i** Wind farm ac frequency, **j** WFVSC dc voltage, **k** Active power from the wind farm, **l** WFVSC phase current

nominal value. Again, FRT is achieved with this method although higher dc over-voltage is noticed.

17.4.3 Option 3

In this case, a dc damping resistor of 180 Ω was connected to the GSVSC's dc terminal via a controllable switch such that the full 500 MW could be consumed by the dc resistor under the most critical condition. The dc voltage was fed through a low pass filter with a time constant of 0.0015 s before feeding to the hysteresis controller with the band being set at 1.05 pu and 1.04 pu respectively. As can be seen in Fig. 17.14d, h, j, when ac fault occurs, the energy generated by the wind farm is consumed by the dc damping resistor and the dc over-voltage is limited. The WFVSC and the turbines operate unaffected with little voltage and frequency variation for the offshore ac network. As the turbines continue to generate power, their speeds do not increase as previously seen with Options 1 and 2.

Once the fault is cleared, the dc voltage drops and the dc damping resistor automatically comes out of operation. The system can return to normal operation very quickly and the transition is very smooth and satisfactory.

Although the case studies for the four-terminal HVDC system shown in Fig. 17.2 have not been provided, the system model and control strategies described in this chapter can be directly applied to a multi-terminal HVDC system. The detailed models and case study results of a multi-terminal system can be found in [16] and [17].

17.5 Conclusions

Fault ride-through of VSC-based HVDC systems used for connecting large offshore wind farms is one of the main grid code requirements and technical challenges. To ensure safe operation of the HVDC system, the total generated and transmitted/consumed power must be balanced under all conditions. Three different strategies, i.e., telecommunication based, offshore frequency modulation, and dc damping resistor, for ensuring satisfactory fault ride-through are investigated. The dc damping resistor option is the simplest and most reliable among the three methods. It requires no coordination between the VSC converters and the wind turbines, and also results in least disturbance to the offshore network.

References

1. Eriksson E, Halvarsson P, Wensky D, Hausler M (2003) System approach on designing an offshore windpower grid connection. In: Proceedings of the fourth international workshop on large-scale integration of wind power and transmission networks for offshore wind farms, Sweden
2. Bolik SM (2003) Grid requirements challenges for wind turbines. In: Proceedings of the fourth international workshop on large-scale integration of wind power and transmission net works for offshore wind farms, Sweden
3. Ackermann T (2002) Transmission systems for offshore wind farms. IEEE Power Eng Rev 22(12):23–27
4. Xu L, Andersen BR (2006) Grid connection of large offshore wind farms using HVDC. Wind Energy 9(4):371–382
5. Kirby N, Xu L, Lucket M, Siepmann M (2002) HVDC Transmission for large offshore windfarms. IEEE Power Eng J 16:135–141
6. Xiang D, Ran L, Bumby JR, Tavner P, Yang S (2006) Coordinated control of an HVDC link and doubly fed induction generators in a large offshore wind farm. IEEE Trans Power Deliv 21(1):463–471
7. Bozhko S, Blasco-Gimenez R, Li R, Clare J, Asher G (2007) Control of offshore DFIG-based wind farm grid with line-commutated HVDC connection. IEEE Trans Energy Convers 22(1):71–78
8. Foster S, Xu L, Fox B (2008) Control of an LCC HVDC system for connecting large offshore wind farms with special consideration of grid fault. In: Proceedings of the IEEE PES general meeting, Pittsburgh, USA

9. Sobrink KH, Sorensen PL, Christensen P, Sandersen N, Eriksson K, Holmberg P (1999) Dc feeder for connection of a wind farm. In: Proceedings of Cigre symposium, Malaysia
10. Skytt AK, Holmberg P, Juhlin LE (2001) HVDC Light for connection of wind farms. In: Proceedings of the second international workshop on transmission networks for offshore wind farms, Sweden
11. Koutiva XI, Vrionis TD, Vovos NA, Giannakopoulos GB (2006) Optimal integration of an offshore wind farm to a weak grid. IEEE Trans Power Deliv 21(2):987–994
12. Xu L, Yao L, Sasse C (2007) Grid integration of Large DFIG based wind farms using VSC transmission. IEEE Trans Power Syst 22(3):976–984
13. Lu W, Ooi BT (2003) Optimal acquisition and aggregation of offshore wind power by multiterminal voltage-source HVDC. IEEE Trans Power Deliv 18(1):201–206
14. Jiao L, Joos G, Abbey C, Zhou F, Ooi BT (2004) Multi-terminal dc (MTDC) systems for wind farms powered by doubly-fed induction generators (DFIGs). In: Proceedings of the IEEE power electronics specialist conference, Aachen, Germany
15. Lu W, Ooi BT (2002) Multiterminal LVDC system for optimal acquisition of power in wind-farm using induction generators. IEEE Trans Power Electron 17(4):558–563
16. Xu L, Yao L, Bazargan M, Williams BW (2008) Control and operation of multi-terminal dc systems for integrating large offshore wind farms. In: Proceedings of the seventh international workshop on large-scale integration of wind power and transmission networks for offshore wind farms, Spain
17. Xu L, Wang Y, Yao L, Rasolonjanahary JL (2009) Multi-terminal HVDC system for large offshore wind farm integration and transmission network support. In: Proceedings of the eighth international workshop on large-scale integration of wind power and transmission networks for offshore wind farms, Germany
18. Xu L, Andersen BR, Cartwright P (2005) VSC Transmission system operating under unbalanced network conditions—analysis and control design. IEEE Trans Power Deliv 20(1):427–434
19. Xu L, Yao L, Bazargan M (2009) Fault ride through of large offshore wind farms using HVDC transmission. In: Proceedings of power tech conference, Romania

Chapter 18
Connection of Off-Shore Wind Farms Using Diode Based HVDC Links

R. Blasco-Gimenez, S. Añó-Villalba, J. Rodríguez-D′Derlée, S. Bernal-Perez and F. Morant

Abstract The development of off-shore wind farms located at a large distance from the coastline imposes a series of technical challenges. At distances larger than 50–70 km and installed powers larger than 500 MW, the use of HVDC links based on line commutated converters (LCC-HVDC) is the most advantageous solution for the connection to the on-shore transmission grid. At the same time, the market share of wind turbines with fully rated converters is increasing. Currently, manufacturers are offering either direct-drive, single-stage or double-stage gearbox multi-megawatt wind turbines for off-shore applications. The additional fault-ride through and control capabilities of wind turbines with fully rated converters can be

Based on "Distributed Voltage and Frequency Control of Offshore Wind Farms Connected With a Diode-Based HVdc Link", by R. Blasco-Gimenez, S. Añó-Villalba, J. Rodríguez-D'Derlée, F. Morant, and S. Bernal-Perez which appeared in IEEE Transactions on Power Electronics, vol. 25, no. 12, pp. 3095–3105. © 2010 IEEE.

R. Blasco-Gimenez (✉) · J. Rodríguez-D′Derlée · F. Morant
Dept. de Ingenieria de Sistemas y Automatica-DISA,
Universidad Politecnica de Valencia, Camino de Vera s/n,
46022 Valencia Spain
e-mail: rblasco@isa.upv.es

S. Añó-Villalba · S. Bernal-Perez
Dept. de Ingenieria Electrica, Universidad Politecnica de Valencia,
Camino de Vera s/n, 46022 Valencia Spain
e-mail: sanyo@die.upv.es

J. Rodríguez-D′Derlée
e-mail: jdderlee@gmail.com

S. Bernal-Perez
e-mail: sbernal@die.upv.es

F. Morant
e-mail: fmorant@isa.upv.es

S. M. Muyeen (ed.), *Wind Energy Conversion Systems*,
Green Energy and Technology, DOI: 10.1007/978-1-4471-2201-2_18,

exploited to control off-shore ac-grid voltage and frequency. Moreover, the wind turbine front-end converters can also be used to control the power transmited through the HVDC link. At this stage, the use of a controlled rectifier is no longer needed, and a diode-based rectifier can be used at the off-shore side of the HVDC link. This chapter shows how an adequate control system on each wind turbine can be used to enable the diode-based HVDC rectifier to be operated in voltage or current control mode, in a similar way as standard HVDC links. Moreover, the proposed control strategy includes adequate protection and fault ride-through capability against most common grid disturbances.

18.1 Introduction

Most of the currently installed wind generation capacity is based on Doubly Fed Induction Generator (DFIG) based wind turbines, as they allow the use of fractional power converters. However, there is a strong trend on the development and installation of wind turbines based on fully rated converters. Table 18.1 shows a list of major manufacturers currently offering wind turbines with fully rated converters.

The main reasons for this trend are more stringent grid codes, together with falling power electronic prices, with additional technological advantages such as weight reduction when using Permanent Magnet Synchronous Generators (PMSG), reduced stage gearboxes or direct connection to the rotor.

The use of full power inverters allows for much greater control flexibility. For example, this kind of wind turbines can be controlled as traditional power plants, in order to provide grid support capability, while exhibiting a much faster speed of response.

These additional control capabilities are specially relevant for HVDC-connected off-shore wind farms, where an integrated control of wind turbines and HVDC converter can lead to substantial installation, operation and maintenance savings.

This chapter introduces an integrated control strategy for a large off-shore wind farm and its corresponding HVDC rectifier which allows the substitution of the traditional thyristor based rectifier for a more reliable and inexpensive diode based rectifier.

18.1.1 HVDC Connection of Large Off-Shore Wind Farms

When large off-shore wind farms are installed further than 50–70 km from the coastline, the most economical solution for the connection of such wind farms is the use of HVDC links [12, 15]. The technical alternatives for such HVDC connection consist on the use of Voltage Source Converters (VSC) [23] or Line Commutated Converters (LCC). VSC-HVDC rectifiers exhibit desirable characteristics for off-shore applications, such as the use of smaller filters and the capability of operation without a relatively strong ac-grid. Therefore, there is no need for additional equipment to provide a strong off-shore ac-grid voltage

Table 18.1 Major manufacturers offering full-rated converter wind turbines

Manufacturer	Turbine model	Rated power (MW)	Rotor diameter (m)	Generator type	Drive-train
Enercon	E126	7.5	127	Separately excited SG	direct-drive
Gamesa	G128-4.5	4.5	128	PMSG	2-stage
GE wind	4.0-110	4.0	110	PMSG	direct-drive
Siemens	SWT-3.6	3.6	107	Squirel cage	3-stage
Vestas	V112-3	3.0	112	PMSG	4-stage

regulation. These two characteristics allow for significant reductions in weight and volume, which are extremely important for off-shore applications.

However, VSC-HVDC converters exhibit higher losses than their LCC-HVDC counterparts. Moreover, the voltage and power ratings of LCC-HVDC links are still higher than those achievable with VSC-HVDC links.

On the other hand LCC-HVDC rectifiers require a relatively strong off-shore ac-grid voltage, relatively large filters and transformers with tap-changers for optimal rectifier operation. All these elements add to the weight of the off-shore rectifier station. Nevertheless, the smaller converter losses and higher operating voltages and power ratings still suggest that LCC-HVDC links have some economic advantage over SVC-HVDC links [15].

18.1.2 Diode-Based HVDC Links

Several researchers have proposed the use of diode-based HVDC rectifiers for links with unidirectional power flow [3, 10, 17]. The reported advantages over thyristor-based HVDC rectifiers being smaller conduction losses, smaller installation cost and higher reliability. To overcome the lack of control of diode-based rectifiers, generator excitation regulation was proposed in order to control the rectifier side HVDC voltage. The power transmitted by the HVDC link being regulated by on-shore inverter current control. This mode of operation leads to lower efficiency and higher chance of commutation failure.

The short circuit current of a diode rectifier HVDC link during faults is only limited by the transient inductance of the generator. Generators can be designed with an adequate transient inductance, however, relatively long recovery times (around 1.5 s) have been reported [17].

The aforementioned drawbacks have been the important deterrents to the use of diode rectifier HVDC links. Moreover, in submarine lines linking two relatively large networks, there is need for bidirectional power flow and large cable cost offset possible savings obtained from the use of diode rectifiers.

However, installation and maintenance costs of off-shore installations are much higher than their on-shore counterparts. Therefore, reduction in equipment weight and maintenance requirements have a large impact on overall HVDC rectifier station costs.

18.1.3 Proposed Solution

As previously mentioned, the use of wind turbines with fully rated, paralleled converters [6, 7] offers the possibility of tight control of the off-shore ac-grid voltage and frequency. Several alternatives for the control of inverter-based islanded grids have been proposed for distributed generation and microgrids [1, 8, 11, 14, 19, 21] , generally using standard $P/f, Q/V$ droop control.

In the present work, the characteristics of the off-shore ac-grid are known to a great extent. Therefore, from the analysis of the network dynamics, a $P/V, Q/f$ control strategy has been developed, similar to that in [21], leading to a relatively straight forward decoupling of P and Q sharing from grid dynamics and from frequency and voltage regulation.

The present work is an extension of that in [2] and will introduce an integrated control strategy for the distributed regulation of both off-shore ac-grid voltage and frequency in islanded and connected modes of operation. The proposed control strategy allows the joint off-shore wind farm and HVDC diode rectifier to operate either in voltage control mode or in current control mode, in a similar way as standard fully controlled LCC-HVDC links. Moreover, a new protection strategy is introduced so the behaviour of the overall system is comparable to that of fully controlled rectifiers in the event of on-shore faults.

18.2 Overall System Description and Modeling

18.2.1 General Overview

Figure 18.1 shows the diagram of the proposed off-shore wind farm, consisting of a number of SG-based wind turbines with fully rated converters, with their corresponding medium voltage transformers T_W which connect them to the variable voltage off-shore ac-grid. The 12-pulse HVDC diode-based rectifier station is connected to the off-shore ac-grid by means of the transformer without a tap changer (T_R). Harmonic filtering and reactive power compensation for the diode rectifier and transformer is carried out by the capacitor and filter bank (C_F, Z_F). The on-shore inverter is a standard 12-pulse thyristor-based converter.

The specific details of the different components are covered below.

18.2.2 Wind turbine

Each aggregated wind turbine i $(i = 1, \ldots, n)$ corresponds to a direct-drive permanent magnet synchronous generator with mechanical and electrical parameters extrapolated from [20].

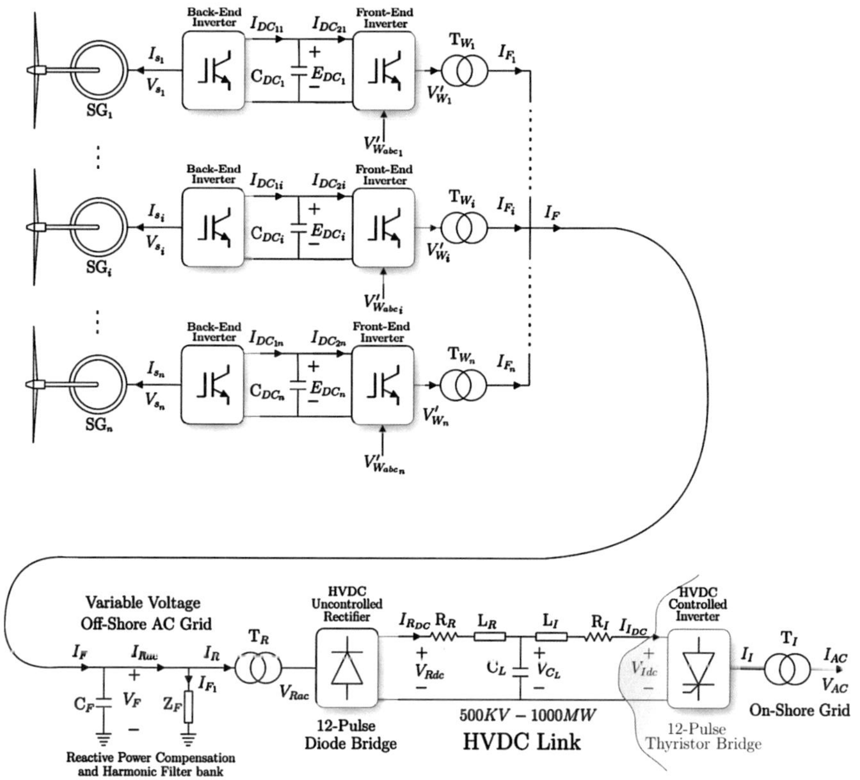

Fig. 18.1 SG-Based off-shore wind farm with HVDC connection

18.2.2.1 Wind Turbine Rotor

The wind turbine rotor converts the kinetic wind power into mechanical-rotational power. The relationship between wind speed ($V_{\text{wind},i}$) and the mechanical wind turbine rotor output power (P_{mi}) is:

$$P_{mi} = \frac{1}{2}\rho A_i V_{\text{wind},\,i}^3 C_{Pi} \tag{18.1}$$

where ρ is the air density, A_i is the area covered by the wind turbine rotor of radius R and C_{Pi} is the power coefficient that depends on the wind rotor design. For the present dynamic studies, the following expression of C_{Pi} is used:

$$C_{Pi} = 0.5176\left(\frac{116}{\lambda_{0i}} - 0.4\beta_i - 5\right)e^{\frac{-21}{\lambda_{0i}}} + 0.0068\lambda_i \tag{18.2}$$

where:

$$\frac{1}{\lambda_{0i}} = \frac{1}{\lambda_i + 0.08\beta_i} - \frac{0.035}{\beta_i^3 + 1} \tag{18.3}$$

and λ_i is the tip-speed ratio defined by:

$$\lambda_i = \frac{\omega_{Ti} R}{V_{\mathrm{wind},i}} \tag{18.4}$$

where ω_{Ti} is the wind turbine rotor angular speed. This is a modified version of the general equation proposed for C_P in [9]. The wind turbine rotor radius R is 60 m and the rated power is 5 MW. The torque is obtained as follows:

$$T_{Ti} = \frac{P_{mi}}{\omega_{Ti}} \tag{18.5}$$

18.2.2.2 Mechanical Drive Train

The mechanical dynamics of the drive train is described by the following two-mass representation:

$$T_{Ti} - D_{Ti}\frac{\mathrm{d}\theta_{Ti}}{\mathrm{d}t} - k_i(\theta_{Ti} - \theta_{ri}) = J_{Ti}\frac{d^2\theta_{Ti}}{\mathrm{d}t^2} \tag{18.6}$$

$$T_{Gi} - D_{Gi}\frac{\mathrm{d}\theta_{ri}}{\mathrm{d}t} + k_i(\theta_{Ti} - \theta_{ri}) = J_{Gi}\frac{d^2\theta_{ri}}{\mathrm{d}t^2} \tag{18.7}$$

where the subscripts T and G stand for wind turbine rotor and generator rotor, respectively. J is the moment of inertia, D is the damping constant and k is the shaft stiffness. T is the torque and θ is the mass angle. The angular speeds are:

$$\omega_{Ti} = \frac{\mathrm{d}\theta_{Ti}}{\mathrm{d}t} \tag{18.8}$$

$$\omega_{ri} = \frac{\mathrm{d}\theta_{ri}}{\mathrm{d}t} \tag{18.9}$$

Table 18.2 shows the values of the mechanical parameters used for the wind turbine model.

18.2.2.3 Permanent Magnet Synchronous Generator

A permanent magnet synchronous generator is used to convert the mechanical-rotational power into electrical power. The dynamics of the generator in a synchronous-frame (d-q axis) rotating at ω_{ri} can be written as:

Table 18.2 Mechanical parameters

Parameter	Value
Wind turbine inertia J_T (kg.m^2)	10×10^6
Generator rotor inertia J_G (kg.m^2)	100×10^3
Shaft stiffness k (N.m/rad)	1.6×10^9
Wind turbine damping D_T (N.m/rad/s)	20
Generator rotor damping D_G (N.m/rad/s)	100

Table 18.3 PMSG characteristics

Generator parameters	Value
Base L–L voltage (kV)	2
Rated frequency (Hz)	20
R_s (mohm)	13.6
L_{sd} (mH)	5.09
L_{sq} (mH)	6.37
λ_m (Wb)	9.31
pole pairs p	80

$$V_{sdi} = R_s I_{sdi} + L_{sd} \frac{dI_{sdi}}{dt} - \omega_{ri} \lambda_{Gqi} \tag{18.10}$$

$$V_{sqi} = R_s I_{sqi} + L_{sq} \frac{dI_{sqi}}{dt} + \omega_{ri} \lambda_{Gdi} \tag{18.11}$$

where R_s is the stator resistance, and L_{sd} and L_{sq} are the d-q stator leakage inductances, respectively. λ_{Gdi} and λ_{Gdi} are the flux linkages defined by the following equations:

$$\lambda_{Gdi} = L_{sdi} I_{sdi} + \lambda_m \tag{18.12}$$

$$\lambda_{Gqi} = L_{sqi} I_{sqi} \tag{18.13}$$

where λ_m is the flux linkage due to the rotor permanent magnets. The characteristics of the 5 MVA generator are shown in Table 18.3. The following equation gives the internal electromechanical torque of the generator:

$$T_{Gi} = 3p \left(\lambda_{Gdi} I_{sqi} - \lambda_{Gqi} I_{sdi} \right) \tag{18.14}$$

where p is the number of generator pole pairs.

18.2.2.4 Back-to-Back Converter

The full-scale back-to-back converter extracts the energy from the generator and delivers it to the off-shore ac-grid. The dc-link voltage (E_{DCi}) dynamics can be expressed as:

$$I_{DC1i} - I_{DC2i} = C_{DCi}\frac{dE_{DCi}}{dt} \tag{18.15}$$

The rated value of E_{DCi} is 5.4 kV and the dc-link capacitance C_{DCi} is set to store 115 kJ at rated voltage. Neglecting losses in the back-to-back converter, the following power balance equations are obtained:

$$-3\left(V_{sdi}I_{sdi} + V_{sqi}I_{sqi}\right) = E_{DCi}I_{DC1i} \tag{18.16}$$

$$E_{DCi}I_{DC2i} = 3\left(V_{Wdi}I_{Fdi} + V_{Wqi}I_{Fqi}\right) \tag{18.17}$$

If the front-end converter current I_{Wi} is scaled, the result is an aggregated wind turbine; for example to obtain a 100 MW aggregated wind turbine the current is multiplied by a factor $\frac{100}{5}$.

18.2.2.5 Transformer

The front-end converter voltage is stepped up by the wind turbine transformer T_{Wi} which is modelled neglecting its shunt branches. The dynamics of this simplified model in a synchronous-frame rotating at ω_F, can be written as:

$$V_{Wdi} = R_{T_{Wi}}I_{Fdi} + L_{T_{Wi}}\frac{dI_{Fdi}}{dt} - L_{T_{Wi}}I_{Fqi}\omega_F + V_{Fd} \tag{18.18}$$

$$V_{Wqi} = R_{T_{Wi}}I_{Fqi} + L_{T_{Wi}}\frac{dI_{Fqi}}{dt} + L_{T_{Wi}}I_{Fdi}\omega_F + V_{Fq} \tag{18.19}$$

where $R_{T_{Wi}}$ and $L_{T_{Wi}}$ values are 0.005 and 0.06 pu, respectively.

18.2.3 Off-Shore Ac-Grid and Diode-Based HVDC Link

The off-shore ac-grid connects the wind turbines to the off-shore rectifier transformers through submarine cables. The parameters of the cable have been neglected as its serial and parallel impedances can be lumped into the wind transformer (T_w) and the rectifier capacitor (C_F).

The main components of the HVDC transmission system are shown in Fig. 18.1. The system configuration and parameters (except when noted) are those from the CIGRE benchmark model for HVDC system studies [22].

The off-shore rectifier is a 12-pulse diode bridge, and the on-shore inverter is a 12-pulse controlled line commutated thyristor bridge. The values of the off-shore rectifier harmonic and reactive power compensation banks have been modified with respect to the original CIGRE benchmark model (Table 18.4).

The filter has been retuned as the wind turbines offer better voltage regulation than that obtained from the original Thevenin equivalent of the rectifier ac-grid. Therefore, with wind farm voltage regulation, the full load off-shore ac-grid voltage is closer to its base value, and the values of the filter components had to be redesigned.

Table 18.4 Parameters of the off-shore rectifier filter bank

Component	Original value	New value
$C_F(\mu F)$	3.342	2.856
$C_{a1}(\mu F)$	6.685	5.714
$C_{a2}(\mu F)$	74.28	63.49
$R_{a1}(\Omega)$	29.76	34.82
$R_{a2}(\Omega)$	261.87	306.4
L_a (mH)	136.4	159.6
$C_b(\mu F)$	6.685	5.714
$R_b(\Omega)$	83.32	97.49
L_b (mH)	13.6	15.91

For the dynamic analysis, the base value for the off-shore ac-grid has been assumed to be 345 kV. Clearly, this value is not realistic, as typically, the wind turbines would be connected to a medium voltage grid (33–66 kV) and the wind farm substation would rise that voltage to a level suitable for HVDC rectification and transmission. Generally, a single transformer stage is not sufficient to rise the off-shore ac-grid voltage (33–66 kV) to the high voltage required by the HVDC rectifier. However, only two transformer stages have been assumed for the present dynamic study. This simplification is valid for dynamic studies, provided that the additional reactance and resistance introduced by the extra transformer are taken into account.

18.3 Integrated Wind Farm and HVDC Control

18.3.1 Overall Control Strategy

The use of an uncontrolled HVDC rectifier places important restrictions both on the control of the HVDC link and of the off-shore ac-grid. Therefore, all the different control tasks have to be performed by the wind turbine converters.

The wind turbine control system should be able to perform standard control duties, such as wind turbine torque and speed control, converter dc-link voltage (E_{DCi}) control and converter current control. Additionally, the wind turbines should provide ac-grid voltage and frequency control, HVDC-link current limitation, optimal wind power tracking and adequate fault protection. Wind turbine speed control is based on standard pitch control, which will prevent turbine overspeeding.

The synchronous generator is controlled using standard vector control, with synchronous frame PI current controllers. Usually, the front-end active current (I_{Fdi}) is used to control the dc-link voltage (E_{DCi}), whereas the synchronous generator torque current (I_{sqi}) is used for optimum power tracking. However, in our case, the wind turbine converter dc-link voltage is controlled by changing the reference of the synchronous generator torque current (I_{sqi}) [5]. This choice of control action allows E_{DCi} to be controlled without relying on the off-shore ac-grid, contributing decisively to the individual wind turbine black-start capability. On the other hand, this control alternative leads to a somehow more complex plant, but allows both front-end active and reactive currents to be used for off-shore ac-grid control.

In this way, when the wind farm is operating in islanding mode, the active and reactive power delivered by each wind turbine can be used to control the voltage and frequency of the off-shore ac-grid. Depending on the grid dynamics, standard P/f and Q/V droop control can be used. In our case, the off-shore ac-grid dynamics are well known and are dominated by the HVDC rectifier capacitor and filter banks. Therefore, it has been found that a *P/V* and *Q/f* control is more suitable in this particular case.

Once the HVDC link is conducting, the HVDC diode rectifier acts as a voltage clamp on the off-shore ac-grid voltage. Therefore, the power reference of each wind turbine is no longer used to control the ac-grid voltage, and instead is used to set an individual wind turbine optimal power reference depending on the particular wind condition.

Current limit during on-shore or HVDC cable faults is also provided by the individual wind turbines. A novel distributed protection algorithm is also introduced, based on standard Voltage Dependant Current Order Limit (VDCOL). With this kind of algorithm, the current limit of each wind turbine depends on the actual off-shore ac-grid voltage, preventing damaging overcurrents.

The following sections offer a detailed description of the aforementioned control strategy.

18.3.2 Wind Turbine Control

The proposed wind turbine control strategy consists of two main tasks. Namely a speed control to prevent the wind turbine from overspeeding and a generator side converter control to achieve back-to-back dc-link voltage (E_{DCi}) regulation.

18.3.2.1 Wind Turbine Speed Control

For high wind speeds the extracted wind power has to be limited via blade pitching. Additionally, the wind turbine might not be able to deliver active power to the grid (i.e. when the HVDC link is disconnected or during faults). A speed controller is used to prevent overspeeding when the wind turbine cannot deliver all the available wind power.

Considering the model of the mechanical drive train described in (18.6) and (18.7), and only for control purposes, it is possible to simplify the analysis assuming a rigid shaft, i.e. $\theta_{Ti} = \theta_{Gi}$, thus the new model will be equivalent to one-mass model and can be written as:

$$T_{Ti} + T_{Gi} - (D_{Ti} + D_{Gi})\frac{\mathrm{d}\theta_{Ti}}{\mathrm{d}t} = (J_{Ti} + J_{Gi})\frac{\mathrm{d}^2\theta_{Ti}}{\mathrm{d}t^2} \tag{18.20}$$

Defining a decoupling input $U_{Ti} = T_{Ti} + T_{Gi}$, a moment of inertia $J = J_{Ti} + J_{Gi}$ and damping constant as $D_i = D_{Ti} + D_{Gi}$, Eq. 18.20 turns to:

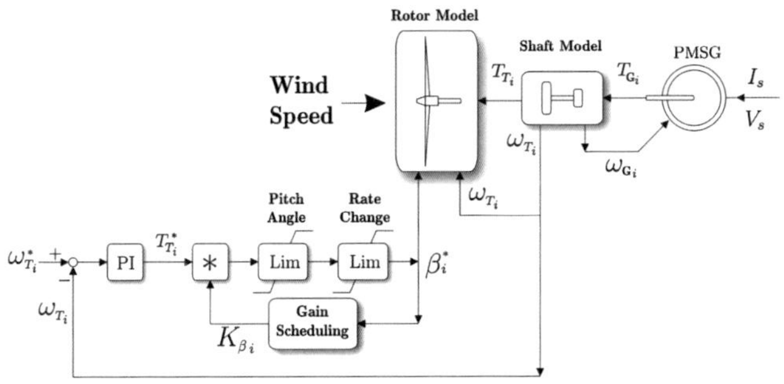

Fig. 18.2 Wind turbine speed control

$$J\frac{d\omega_{Ti}}{dt} + D\omega_{Ti} = U_{Ti} \tag{18.21}$$

Note T_{Gi} is set indirectly by the E_{DCi} control strategy covered in Sect. 18.3.2.3 and cannot be used for speed control. On the other hand, the wind turbine torque T_{Ti} can be modified by means of pitch control. For operation at rated power and rated speed, the linearised function relating the pitch angle reference to the torque reference has been obtained from the wind turbine aerodynamic characteristics $\beta_i^* = f(T_{Gi}^*)$.

Therefore, the individual wind turbine speed ω_{Ti} can be controlled to follow desired speed references with simple PI controllers, with the aforementioned linearised expression acting as a nonlinear gain. The proposed ω_{Ti} control loop is depicted in Fig. 18.2.

Moreover, limits on the pitch angle range (0–30°) and its maximum rate of change (14 deg/s) are included in the model. The rate of change limitation is particularly relevant during grid faults because it will determine the wind turbine maximum transient speed.

18.3.2.2 PMSG Current Control

Permanent magnet synchronous generator current control has been carried out by defining:

$$V_{sdi} = -\omega_{ri} L_{sqi} I_{sqi} + u_{di} \tag{18.22}$$

$$V_{sqi} = \omega_{ri} L_{sdi} I_{sdi} + \omega_{ri}\lambda_m + u_{qi} \tag{18.23}$$

Therefore, the dynamics defined by (18.10) and (18.11) become:

$$u_{di} = R_s I_{sdi} + L_{sdi}\frac{dI_{sdi}}{dt} \tag{18.24}$$

$$u_{qi} = R_s I_{sqi} + L_{sqi}\frac{dI_{sqi}}{dt} \tag{18.25}$$

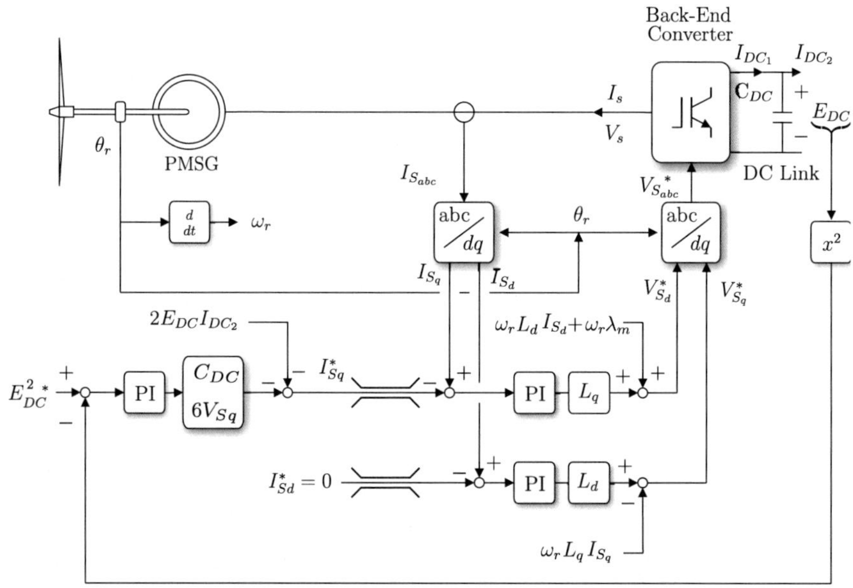

Fig. 18.3 Wind turbine control system

Table 18.5 PMSG current controller parameters

Parameter	Value
K_{Pd}	0.8157
T_{Id}	1.594×10^{-2} s
K_{Pq}	1.0362
T_{Iq}	1.255×10^{-2} s

The resulting first order systems are easily controlled by synchronous-frame PI controllers (Fig. 18.3). The controller parameters are shown in Table 18.5.

The field current reference is set to zero ($I^*_{sdi} = 0$). Therefore the electromagnetic torque will be proportional to I_{sqi}:

$$T_{Gi} = 3p\lambda_m I_{sqi} \tag{18.26}$$

18.3.2.3 Back-to-Back Converter DC-Link Voltage Control

From the power balance between AC and DC sides of the back-end converter (18.16) and neglecting losses, we have:

$$-3\left(I_{sdi}V_{sdi} + I_{sqi}V_{sqi}\right) = E_{DCi}I_{DC1i} \tag{18.27}$$

Table 18.6 Wind turbine dc-link voltage (E_{DC}) controller parameters

Parameter	Value
K_P	16.0771
T_I	8.086×10^{-3} s

Substituting $I_{DC1i} = C_{DCi}\frac{\mathrm{d}}{\mathrm{d}t}(E_{DCi}) + I_{DC2i}$, and assuming $I_{sdi} = 0$, we have:

$$\frac{\mathrm{d}}{\mathrm{d}t}\left(E_{DCi}^2\right) = -\frac{1}{C_{DCi}}\left(6I_{sqi}V_{sqi} + 2I_{DC2i}E_{DCi}\right) \tag{18.28}$$

Using simple decoupling techniques, the plant to be controlled becomes a simple integrator and the PMSG torque current reference can be calculated as:

$$I_{sqi}^* = -\frac{C_{DCi}}{6V_{sqi}}u_i - \frac{I_{DC2i}E_{DCi}}{3V_{sqi}} = -\frac{C_{DCi}}{6V_{sqi}}(u_i + 2I_{DC2i}E_{DCi}) \tag{18.29}$$

where u_i is the output of the E_{DCi}^2 PI controller. The generator torque current reference I_{sqi}^* is limited so the generator never operates in the motoring region. The E_{DCi} control loop is depicted in Fig. 18.3, and its parameters are shown in Table 18.6.

Figure 18.4 shows the performance of the speed and back-to-back converter dc-link control against large load variations. Initially, the system is operated at rated wind and power. At $t = 0.1$ s, the load is suddenly disconnected. After disconnection, the PMSG torque current I_{sq} is rapidly reduced and the wind turbine speed (ω_T) steadily increases. At this point, pitch control will act to bring the turbine back to rated speed. At $t = 5.1$ s, rated power operation is restored, leading to decreased turbine speed, as the pitch angle returns to zero. The maximum speed excursion clearly depends on the maximum pitch angle rate. Note the generator oscillations due to the direct-drive relatively low damping.

It is worth stressing that the wind turbine dc-link voltage E_{DCi} remains within acceptable limits in spite of the large load transients applied to the system. Therefore, in the following sections, a constant dc-link value E_{DCi} is assumed.

18.3.3 Wind Turbine Grid Integration

For control purposes, the wind farm is modelled as a total of five aggregated wind turbines of different powers. The total power considered is 1 GW and each aggregated model corresponds to a number of 5 MW synchronous generators, modelled as explained in Sect. 18.2.

Figure 18.5 shows a simplified model of the off-shore wind farm depicted in Fig. 18.1. The HVDC link is modelled using a *T*-equivalent of both the dc smoothing reactors and dc transmission line. The on-shore inverter station is modelled by a variable voltage dc source, since the on-shore inverter would determine the voltage V_{Idc} when controlled on a constant voltage or minimum$-\gamma$

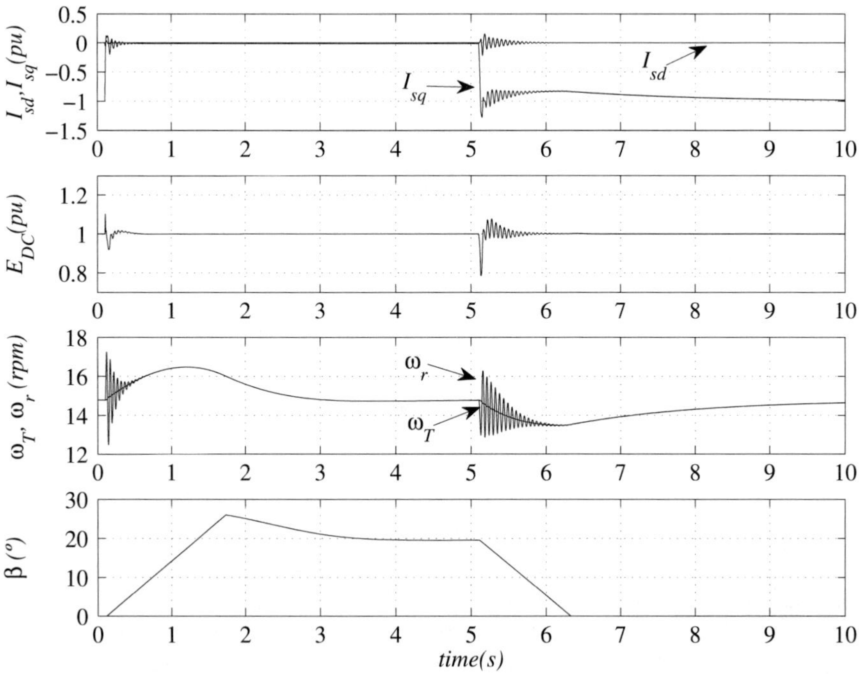

Fig. 18.4 Wind turbine control performance against large load variations

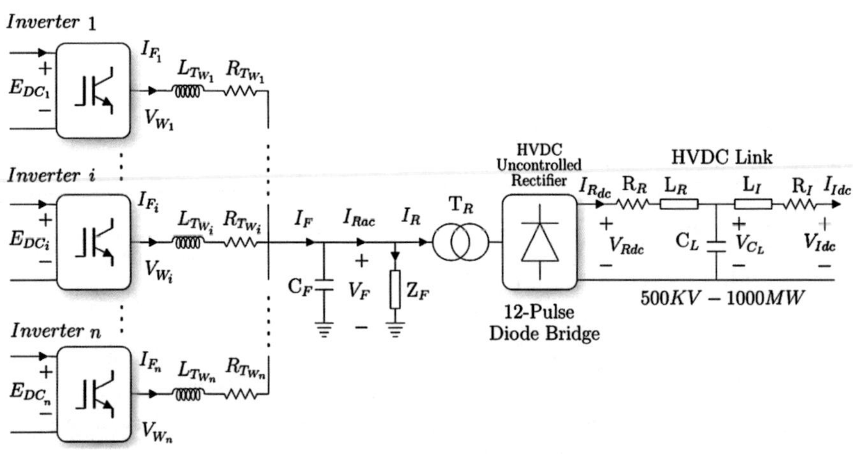

Fig. 18.5 Wind farm simplified model

strategy [4, 16]. Note the voltage drop caused by the on-shore inverter overlap angle and transformer leakage inductance in T_I can be lumped into R_I.

For control purposes, the simplified analysis of the off-shore ac-grid shown in Fig. 18.5 has been carried out assuming that the off-shore ac-line impedances

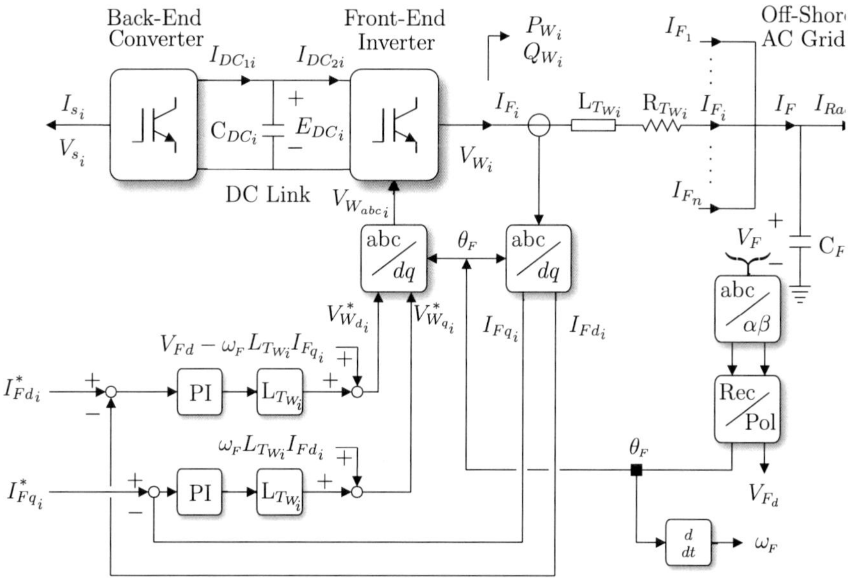

Fig. 18.6 Off-shore I_{Fdi} and I_{Fqi} current control loops

are negligible when compared to wind turbine transformer leakage reactance and neglecting the wind turbine transformer shunt branches.

18.3.3.1 Front-End Converter Current Control

The dynamics of the simplified model in Fig. 18.5 in a synchronous frame rotating at ω_F and oriented on V_F, i.e. $V_{Fq} = 0$, can be written as:

$$\frac{\mathrm{d}}{\mathrm{d}t} I_{Fdi} = -\frac{R_{T_{Wi}}}{L_{T_{Wi}}} I_{Fdi} + \omega_F I_{Fqi} + \frac{1}{L_{T_{Wi}}} V_{Wdi} - \frac{1}{L_{T_{Wi}}} V_{Fd} \tag{18.30}$$

$$\frac{\mathrm{d}}{\mathrm{d}t} I_{Fqi} = -\omega_F I_{Fdi} - \frac{R_{T_{Wi}}}{L_{T_{Wi}}} I_{Fqi} + \frac{1}{L_{T_{Wi}}} V_{Wqi} \tag{18.31}$$

$$\frac{\mathrm{d}}{\mathrm{d}t} V_{Fd} = \frac{1}{C_F} \sum_{i=1}^{n} I_{Fdi} - \frac{1}{C_F} I_{Racd} \tag{18.32}$$

$$\omega_F V_{Fd} = \frac{1}{C_F} \sum_{i=1}^{n} I_{Fqi} - \frac{1}{C_F} I_{Racq} \tag{18.33}$$

A set of decoupling inputs for each wind turbine is defined as follows:

$$u_{Fdi} = L_{T_{Wi}} \omega_F I_{Fqi} + V_{Wdi} - V_{Fd} \tag{18.34}$$

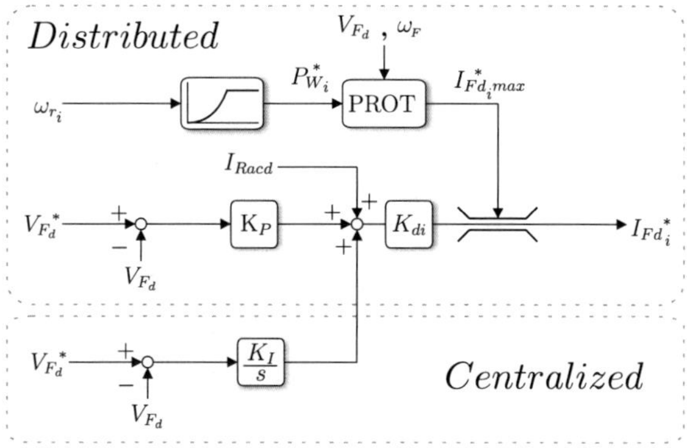

Fig. 18.7 Off-shore ac-grid voltage control

$$u_{Fqi} = -L_{T_W i}\omega_F I_{Fdi} + V_{Wqi} \tag{18.35}$$

Therefore, the individual wind turbine d-q currents (I_{Fdi} and I_{Fqi}) can be controlled to follow desired step references with simple PI controllers. The I_{Fdi} and I_{Fqi} control loops of each wind turbine are depicted in Fig. 18.6. The parameters used for the current controllers are $K_P = 583.8 \times 10^{-6}$ and $K_I = 0.048$.

Distributed Voltage and Frequency Control

Assuming sufficiently fast current control loops, $I_{Fdi} = I^*_{Fdi}$ and $I_{Fqi} = I^*_{Fqi}$. Hence, the system dynamics will be:

$$\frac{d}{dt}V_{Fd} = \frac{1}{C_F}\sum_{i=1}^{n} I^*_{Fdi} - \frac{1}{C_F}I_{Racd} \tag{18.36}$$

$$\omega_F V_{Fd} = \frac{1}{C_F}\sum_{i=1}^{n} I^*_{Fqi} - \frac{1}{C_F}I_{Racq} \tag{18.37}$$

Therefore, the overall front-end inverter active current ($\sum_{i=1}^{n} I^*_{Fdi}$) can be used to control the off-shore ac-grid voltage V_{Fd}, whereas the overall reactive current ($\sum_{i=1}^{n} I^*_{Fqi}$) can control the frequency ω_F. Note this is the opposite to that commonly used in power systems, where the active power is used to control the frequency and the reactive power to control the voltage. In our case, the grid topology and load (capacitor) characteristics are well known, and their dynamics determine the coupling between $\sum_{i=1}^{n} P_{Wi}$, $\sum_{i=1}^{n} Q_{Wi}$, V_{Fd} and ω_F.

Distributed Off-Shore Ac-Grid Voltage Control

Figure 18.7 shows the off-shore ac voltage (V_{Fd}) control loops. The distributed part is local to each wind turbine, whereas the integral part is centralized, i.e. common to the complete wind farm. The output of each control loop with

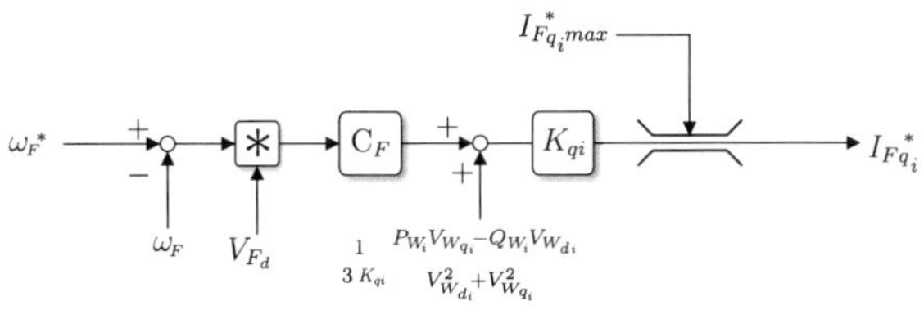

Fig. 18.8 Frequency control loop

remote integrator is multiplied by a constant K_{di} proportional to the rated power of each wind turbine so that $\sum_{i=1}^{n} K_{di} = 1$.

The required remote measurement of I_{Racd} can be avoided either by not using the feedforward term, with the corresponding performance degradation, or by estimating I_{Racd} from local variables.

The voltage control loop is designed to have a 20 Hz closed loop bandwidth $(K_P = 583.8 \times 10^{-6}, K_I = 0.048)$. Therefore, communication delays on the centralised integrator in the range of 5–10 ms can be easily tolerated.

Distributed Off-Shore Ac-Grid Frequency Control

As previously mentioned, (18.37) implies that the wind turbine front-end reactive current can be used to control the off-shore ac-grid frequency. Moreover, as (18.37) is just an algebraic equation, the off-shore ac-grid frequency can be controlled by setting:

$$I_{Fq}^{*} = \sum_{i=1}^{n} I_{Fqi}^{*} = V_{Fd} C_F \omega_F + I_{Racq} \tag{18.38}$$

$$I_{Fqi}^{*} = K_{qi} I_{Fq}^{*} \tag{18.39}$$

where K_{qi} represents the reactive current contribution factor of each converter, and $\sum_{i=1}^{n} K_{qi} = 1$. However, such an open loop scheme is very sensitive to C_F estimation errors and does require the use of a remote measurement of I_{Racq}.

Alternatively, I_{Racq} can be estimated by using local measurements:

$$\hat{I}_{Racq} = \frac{1}{3K_{qi}} \frac{V_{Wqi} P_{Wi} - V_{Wdi} Q_{Wi}}{V_{Wdi}^2 + V_{Wqi}^2} - C_F \omega_F V_{Fd} \tag{18.40}$$

Substituting (18.40) in (18.37), we have:

$$I_{Fq}^{*} = C_F V_{Fd} \left(\omega_F^{*} - \omega_F\right) + \frac{1}{3K_{qi}} \frac{V_{Wqi} P_{Wi} - V_{Wdi} Q_{Wi}}{V_{Wdi}^2 + V_{Wqi}^2} \tag{18.41}$$

The corresponding block diagram is shown in Fig. 18.8. Note the frequency control loop corresponds to a simple proportional control with a feedforward compensation term. Moreover, it is clear that the frequency control system is more

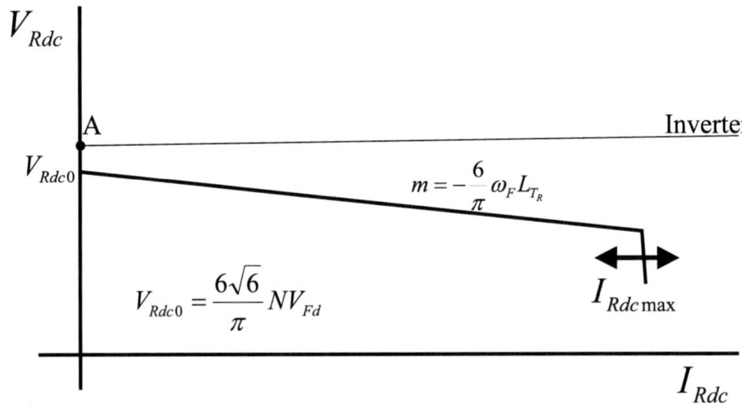

Fig. 18.9 HVDC-link steady state operation characterstic (mode A: islanded operation)

robust to C_F estimation errors, as C_F now appears within the control loop and can be considered as a loop gain, rather than a plant parameter.

Note all the variables used in Fig. 18.8 are local to each wind turbine. Moreover, the control design based on the analysis of the off-shore grid dynamics has led to a scheme similar to traditional droop control.

18.3.4 Integrated Wind Farm and HVDC Control

18.3.4.1 General Description and Modes of Operation

Using the previously explained distributed control system, the off-shore ac-grid voltage is controlled by the wind turbines. When the HVDC diode rectifier is conducting, the relationship between the HVDC rectifier dc voltage (V_{Rdc}) and thc grid voltage (V_{Fd}) is [13]:

$$V_{Rdc} = \frac{6\sqrt{6}}{\pi} NV_{Fd} - \frac{6}{\pi}\omega_F L_{T_R} I_{Rdc} \tag{18.42}$$

where N is the rectifier transformer turns ratio, V_{Fd} represents line to neutral rms values, and L_{T_R} is the rectifier transformer leakage inductance.

Usually, there will be a saturation limit on the overall active current delivered by the wind turbines $I_{Fd\max} = \sum_i I_{Fdi\max}$. Assuming no losses on the off-shore ac-grid, $I_{Fd\max} = I_{Rd\max}$ in steady state. Hence, the relationship between $I_{Fd\max}$ and $I_{Rdc\max}$ is:

$$I_{Rdc\max} = \frac{\sqrt{6}\pi V_{Rdc}}{12NV_{Rdc} - 6\sqrt{6}\omega_F L_{T_R} I_{Fd\max}} I_{Fd\max} \tag{18.43}$$

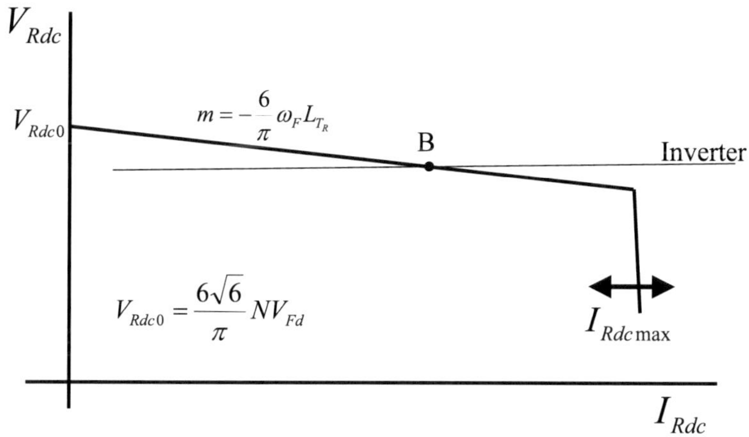

Fig. 18.10 HVDC-link steady state operation characterstic (operation mode B)

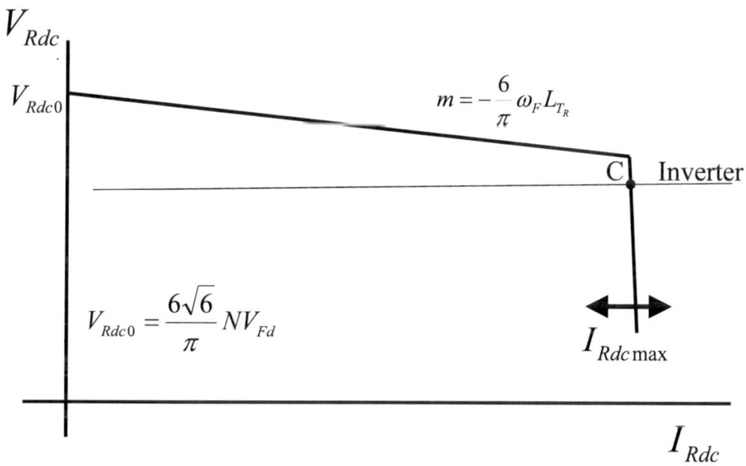

Fig. 18.11 HVDC-link steady state operation characterstic (operation mode C)

Equations 18.42 and 18.43 define the steady state operation characteristic of the HVDC rectifier (Figs. 18.9, 18.10, 18.11). Clearly, this characteristic can be modified by changing the set point of the off-shore ac-grid voltage (V_{Fd}^*) and/or by modifying the wind farm overall active current limit $(I_{Fd\max})$.

The intersection between the HVDC diode rectifier and that of the inverter will therefore define the HVDC link steady state operating point. Figures 18.9, 18.10, 18.11 show the operating points for different off-shore ac-grid voltage values (V_{Fd}). These figures depict the HVDC rectifier steady state characteristic, as per (18.42) and (18.43). It is assumed that the inverter would be operating in voltage or in minimum-γ control. The inverter steady state characteristic is represented by an almost horizontal line. Clearly, the slope of this line would be slightly positive for inverter voltage control and slightly negative for minimum-γ control.

A. Islanded operation (rectifier voltage control and inverter voltage control)

Figure 18.9 shows the operation with a low value of V_{Fd} such that $V_{Rdc0} = \frac{6\sqrt{6}}{\pi} N V_{Fd} < V_{Rdc}$ (operation mode A). Therefore, the on-shore ac-grid voltage is not high enough for the HVDC rectifier to conduct. Hence, the off-shore wind farm will be operating in islanded mode. The off-shore ac-grid voltage and frequency are controlled by the wind farm front-end converters.

B. Connected operation (rectifier voltage control and inverter voltage control)

Figure 18.10 shows the operation of the HVDC link with an intermediate value of V_{Fd} (operation mode B). Now the off-shore ac-grid voltage is large enough for the HVDC diode rectifier to conduct. The active current delivered by the front-end converters have not yet reached its limit $I_{Fd\max}$. This fact implies that some of the the voltage control loops in Fig. 18.7 are not saturated and that the wind farm still keeps control of the off-shore ac-grid voltage.

C. Connected operation (rectifier current control and inverter voltage control)

If the off-shore ac-grid voltage reference V_{Fd}^* is increased beyond a certain point (Fig. 18.11), then all the voltage control loops in Fig. 18.7 will be saturated (operation mode C). At this point, the HVDC link current will be determined by (18.43). Additionally, as the voltage control loops are saturated, the steady state off-shore ac-grid voltage will be determined by the inverter side dc voltage (V_{Idc}) and the HVDC link current (I_{Rdc}):

$$V_{Fd} = \frac{\sqrt{6}}{36} \frac{\pi}{N} \left(V_{Idc} + \left(R_R + R_I + \frac{6}{\pi} L_{T_R} \omega_F \right) I_{Rdc} \right) \tag{18.44}$$

Therefore, the off-shore ac-grid voltage will be indirectly controlled by the inverter.

As shown in Figs. 18.10 and 18.11, the transition from operation mode B to operation mode C can be achieved by either rising V_{Fd}^* or by decreasing the front-end converter active current limits $I_{Fdi\max}$(which, in turn, determines $I_{Rdc\max}$).

At this stage, the active current limit on each front-end converter ($I_{Fdi\max}^*$ in Fig. 18.7) can be set to follow the wind turbine optimal characteristic [18]:

$$I_{Fdi\max}^* = k_2 \frac{P_{Wi}^*}{V_{Fd}} = k_2 \frac{K_{opti} \omega_{ri}^3}{V_{Fd}} \tag{18.45}$$

where ω_{ri} is the wind turbine speed and K_{opti} is an optimal constant that depends on the physical characteristics of the wind turbine rotor and the air density.

18.3.4.2 Self-Start Operation

The proposed control system also allows for self-start operation. As a first step, the on-shore inverter must energise the HVDC line to its rated voltage.

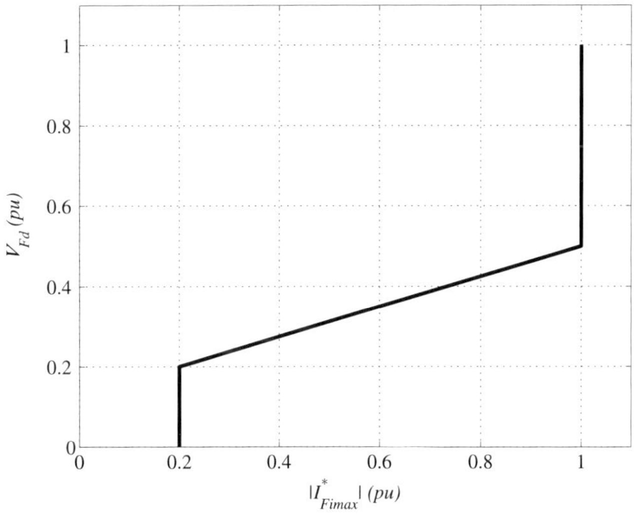

Fig. 18.12 VDCOL protection characteristic

At the rectifier side, the proposed control strategy allows all the wind turbines to have grid-forming capability. Therefore, provided that the wind resource is enough to compensate for system losses, the turbine can start rotating and produce enough energy to reach a constant dc-link value (E_{DCi}).

When a constant E_{DCi} has been obtained, the front-end inverters can start operation at reduced V_{Fd} voltage (operation mode A). The voltage V_{Fd} can now be slowly increased until the integrated wind farm and HVDC diode rectifier are in current control mode (operation mode C). At this point, optimum power tracking can be carried out as per (18.45).

Before reaching operation mode C, the wind turbines do not operate at its maximum capability for a given wind speed. At this stage, pitch control will ensure that individual wind turbines do not exceed their maximum operational speed.

18.3.5 Protection Using a VDCOL

Similarly to standard practice in thyristor-based HVDC links, a VDCOL has been included (Fig. 18.7). The VDCOL characteristic is shown in Fig. (18.12). A limit is imposed on $|I^*_{Fi}|$ depending on the measured grid voltage V_{Fd}. When used for VDCOL operation, the measured voltage V_{Fd} is rate limited, with a larger limit value for decreasing $V_{Fd}(10 \times 10^3$ kV/s) than for increasing V_{Fd} (100 kV/s). The difference in rate limits is consistent with standard practice to reduce oscillations and possible instability during fault recovery [13]. As the VDCOL for each wind turbine can be calculated locally, this scheme represents a simple distributed strategy for the protection of the HVDC link.

The limits for the front-end converter currents I^*_{Fdi} and I^*_{Fqi} in Figs. 18.7 and 18.8 would be:

$$I^*_{Fqi\,\max} = \left|I^*_{Fi}\right|_{\max} \tag{18.46}$$

$$I^*_{Fdi\,\max} = \min\left(\sqrt{\left|I^*_{Fi}\right|^2_{\max} - I^2_{Fqi}}, k_2 \frac{K_{opti}\omega^3_{ri}}{V_{Fd}}\right) \tag{18.47}$$

The limit on I^*_{Fdi} is the smaller of two quantities, namely $\sqrt{\left|I^*_{Fi}\right|^2_{\max} - I^2_{Fqi}}$ obtained from the VDCOL characteristic and $k_2 \frac{K_{opti}\omega^3_{ri}}{V_{Fd}}$ which represents the maximum power that can be extracted for a particular wind speed (18.45).

Alternative protection schemes are possible, as $I^*_{Fdi\,\max}$ and $I^*_{Fqi\,\max}$ (or even V^*_F) can be modified using more elaborated schemes. Moreover, other variables such as the ac-grid frequency ω_F could also be used as signals for the distributed detection of the disturbance. However, the proposed VDCOL has been chosen to show that the wind farm together with the diode rectifier can be operated in a similar way as traditional thyristor-based HVDC rectifiers.

18.4 System Performance

The control strategies previously described have been validated using PSCAD simulations. Several scenarios have been considered to check voltage and frequency control during islanded operation, adequate start-up procedure, response to power generation variations and response during different transient conditions (on-shore faults, generation disconnection, HVDC rectifier ac-breaker tripping and reconnection and capacitor bank switching). The parameters of the HVDC link have been obtained from the CIGRE benchmark model [22], using a diode rectifier instead of a thyristor rectifier and modelling the on-shore inverter station by the means of a variable voltage DC source (Fig. 18.5) [4, 16]. The wind farm has been modelled using a total of five equivalent wind turbines of different rated power ($S_{R1} = 390$ MVA, $S_{R2} = 300$ MVA, $S_{R3} = 200$ MVA, $S_{R4} = 100$ MVA and $S_{R5} = 10$ MVA). The sharing coefficients are $K_{di} = K_{qi} = S_{Ri}/S_{RT}$, with the wind farm rated power being $S_{RT} = 1000$ MVA, which is also the base power for the whole system.

A switching frequency of 1 kHz has been assumed for the wind turbine front-end converters. Therefore, the front-end converter current loop bandwidth has been designed to be around 180 Hz. The off-shore ac-grid voltage control loop bandwidth is 20 Hz.

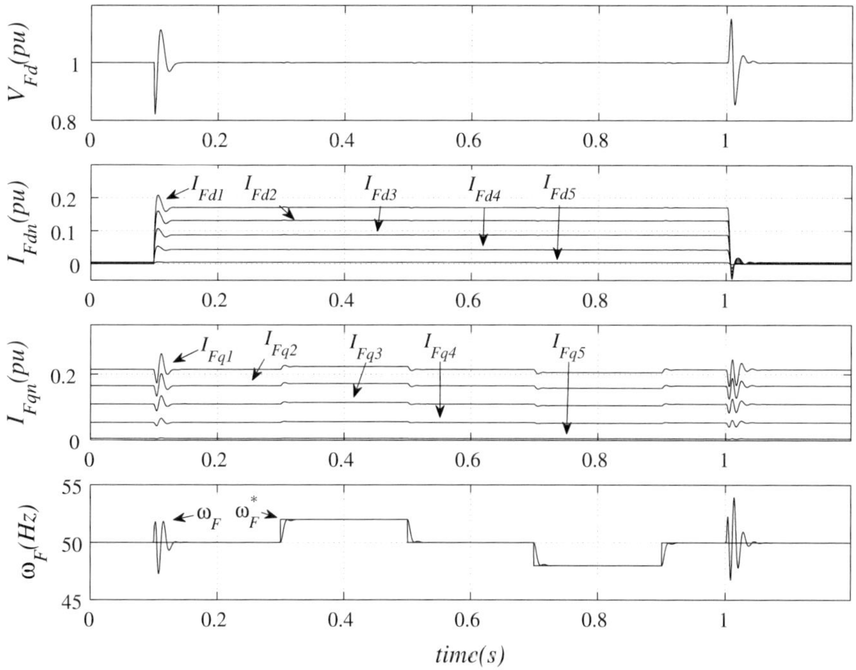

Fig. 18.13 Off-shore ac-grid control during islanded operation (operation mode A)

18.4.1 Islanded Operation (Operation Mode A)

The performance of the proposed voltage and frequency control systems has been initially tested with the off-shore grid operating in islanded mode, i.e. the HVDC diode rectifier not conducting. Figure 18.13 shows the response of the proposed control system to load changes and to frequency reference changes during islanded operation. Initially, the off-shore grid is operated at no-load and rated voltage and frequency. At $t = 0.1$ s, a 0.5 pu resistive load is connected. After a transient lasting approximately 40 ms, the system returns to the steady state, showing an excellent sharing of active and reactive power among the different turbines. At $t = 1$ s, the resistive load is disconnected. The minimum voltage during load connection is 0.85 pu and the maximum voltage during load disconnection is 1.15 pu. Note the sudden connection and disconnection of a 500 MW local load on the off-shore grid represents an extreme scenario, unlikely to happen in an actual system. Nevertheless the good response to such a load variation proves the good voltage regulation and load sharing capability of the proposed control strategy.

During the previously mentioned load transient, a frequency reference change of ± 2 Hz is carried out in order to show the performance of the frequency control loop. The frequency reference is reached in 10 ms.

18.4.1.1 Self- Start Operation

As shown in Fig. 18.6, the wind turbine front-end converter current control loops are oriented on V_{Fd}. Therefore, when the off-shore ac-grid is not operational $(V_{Fd} = 0)$, an adequate start-up procedure is required for wind turbine connection.

Figure 18.14 shows the start-up transient, including the HVDC-link energisation. At $t = 0$ s, the on-shore inverter starts operating, rising the HVDC-link voltage from 0 to its rated value. At $t = 0.3$ s, after rated HVDC-link voltage has been reached, the off-shore ac-grid voltage and frequency control loops are enabled and the V_{Fd} demand is increased linearly from 0 to 1.1 pu. Frequency reference is kept constant at 50 Hz. From $t = 0.3$ s to $t = 1.7$ s, the HVDC rectifier is not conducting and the off-shore ac-grid is effectively operated in islanded mode, with a fixed frequency reference and a variable voltage reference.

Note the reactive current components I_{Fqi} are relatively large when the HVDC diode rectifier is not conducting. Clearly, when the diode rectifier is not conducting, the capacitor and filter banks are overcompensating and the wind turbines must balance the excess reactive power produced by the capacitor banks.

When V_{Fd} reaches a value of 0.87 pu (at $t = 1.7$ s), the off-shore rectifier starts conducting and currents I_{Fdi} and I_{Rdc} increase. As current I_{Rdc} increases, the V_{Rdc} ripple increases due to the higher harmonic content at higher currents. When I_{Fdi} reaches a value of 1 pu, the voltage control loop in Fig. 18.7 saturates and the off-shore grid voltage (V_{Fd}) does no longer follow its reference. Therefore, Fig. 18.14 shows a smooth transition between islanded (mode A) and connected (mode B) operation at $t = 1.7$ s, as well as a smooth transition between voltage and current control mode (mode C) operation at $t = 1.9$ s. The ac-grid frequency remains close to 50 Hz during the transient.

The transient in Fig. 18.14 shows reliable steady state operation at rated power and stand alone operation of the wind farm with good voltage and frequency regulation when the HVDC diode rectifier is not conducting and excellent active and reactive power sharing during the complete transient. It is worth stressing that all wind turbine front-end converters are controlled as grid-forming inverters and therefore all of them contribute to the control of the off-shore grid during start-up.

18.4.1.2 Self-Start Operation with Power Limits in Some Wind Turbines

The connection transient shown in Fig. 18.14 assumes that rated active power is available from all the wind turbines. However, depending on the wind conditions, wind turbines might not be able to deliver full-rated active power. To consider this situation, the active power delivered by the 390 MVA wind turbine is limited to

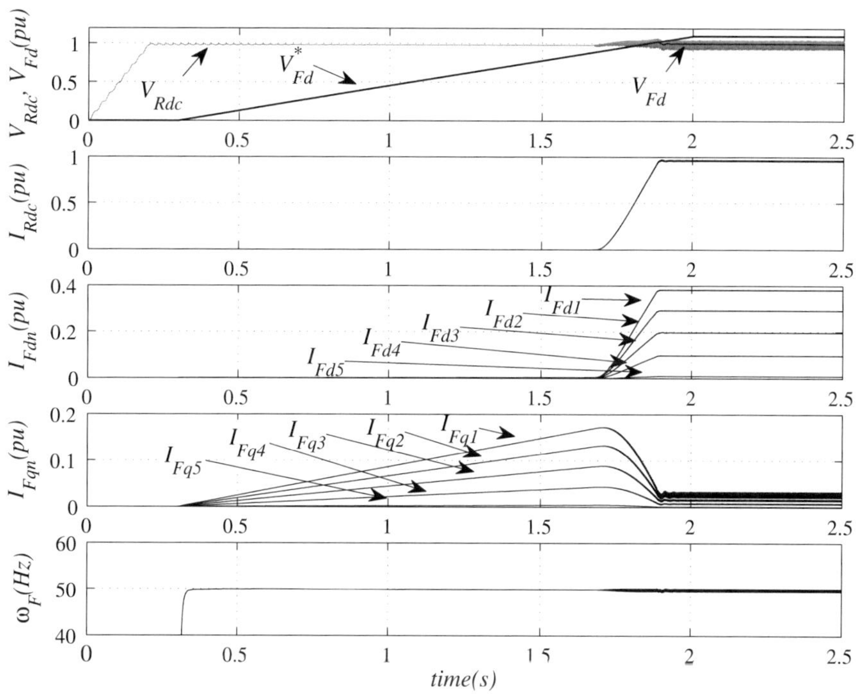

Fig. 18.14 HVDC link and off-shore ac-grid start-up operation

30% its rated value. The connection of the wind farm with this active power constraint is shown in Fig. 18.15.

The behaviour of the system is exactly the same as the unconstrained case up to 0.16 s. At this point, I_{Fd1} saturates and stops contributing to the off-shore ac-grid voltage control. Nevertheless, the rest of the wind turbines would still be keeping V_{Fd} very close to its reference. Notice the increase slope on I_{Fd2} to I_{Fd5} to compensate for I_{Fd1} saturation. At $t = 0.24$ s, all the wind turbines saturate and the system can no longer track V_{Fd}^*.

During the transient there are no constraints on reactive power components I_{Fqi} and therefore reactive power is shared amongst all wind turbines. Note that, in contrast with Fig. 18.14, the reactive current components do not reach a zero value after the transient. This effect is caused again by the overcompensation of capacitor and filter banks when the HVDC diode rectifier is not operating at full power.

It is worth stressing that, from $t = 0.16$ s to $t = 0.24$ s, some of the wind turbines are controlling the ac-grid voltage, while some others are saturated due to insufficient wind resource. As seen in Fig. 18.15, the proposed distributed control algorithm does handle this situation automatically.

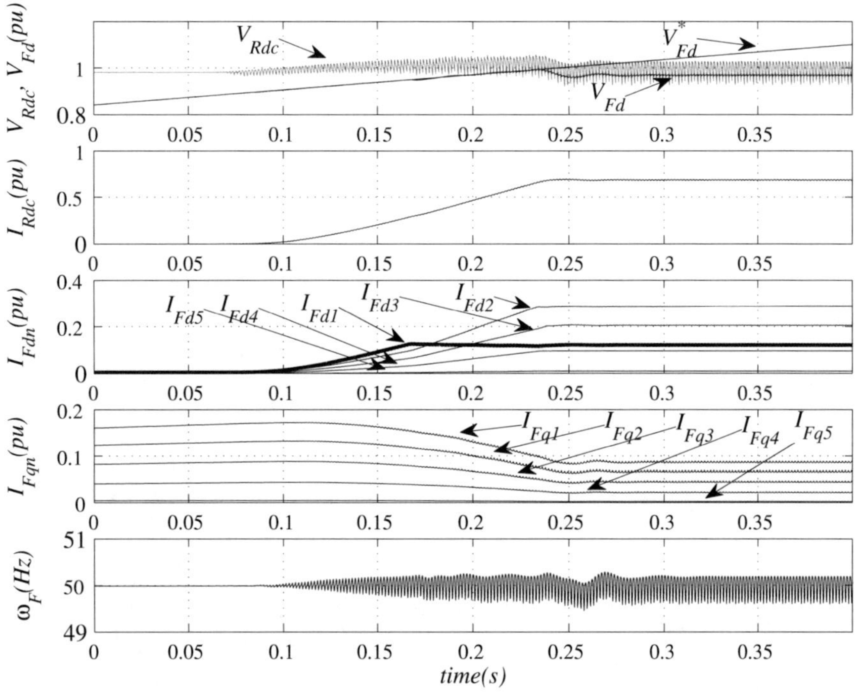

Fig. 18.15 Wind farm connection with power limits on some wind turbines

18.4.2 Connected Operation

18.4.2.1 Frequency Control

The performance of the frequency control loop is shown in Fig. 18.16, while the system is operating in current control mode at rated power. At $t = 0.1$ s the frequency demand rises to 52 Hz (1.04 pu) and then is changed back to 50 Hz (at $t = 0.3$ s). Similarly, a reduction on frequency demand (to 48 Hz) is carried out at $t = 0.5$ s. In both cases, the actual frequency reaches its reference value in around 12 ms. The wind turbine active currents I_{Fdi} remain constant during the transients. On the other hand I_{Fqi} show minor variations, due to the frequency dependant nature of the reactive power of the capacitor bank, filters and leakage reactances.

18.4.2.2 Power Tracking

Figure 18.17 shows the response of the system to changes in the 390 MVA wind turbine power reference. The value of $I^*_{Fd1\,\max}$ in Fig. 18.7 has been reduced from 1 pu (390 MW) to 0.51 pu (200 MW) in 0.1 s, kept constant for 0.2 s, ramped down to 0.1 pu, kept constant again for 0.2 s and then ramped up to 1 pu to

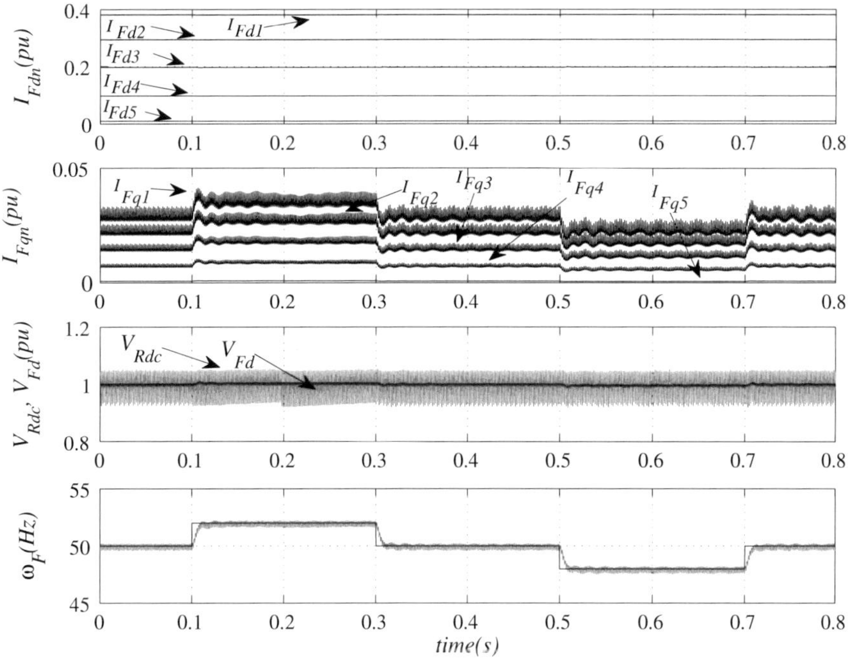

Fig. 18.16 Response to frequency demand changes at rated power

simulate power variations for changing wind conditions. Note a 351 MW power transient in 0.2 s might not be realistic, as power generated from a wind farm of this size does not vary so rapidly. Therefore, the response to realistic power changes would always be better than that shown in Fig. 18.17.

It should be pointed out that the wind turbine front-end reactive currents (I_{Fq1} to I_{Fq5}) increase as I_{Fd1} decreases, as the capacitor and filter banks are again overcompensating the reactive power absorbed by the rectifier transformer leakage reactance. On the other hand, the HVDC-link and off-shore ac-grid voltages decrease slightly along with I_{Fd1} due to smaller voltage drop on the HVDC-link resistance and rectifier transformer leakage reactance. During the whole transient, the frequency of the off-shore ac-grid follows its reference value.

18.4.3 Transient Performance

18.4.3.1 Capacitor Bank Switching

The reactive power absorbed by the leakage inductances of the transformers and that due to the overlap angle of the HVDC rectifier are functions of the generated power. Although, as shown in the previous figures, the wind turbine front-end

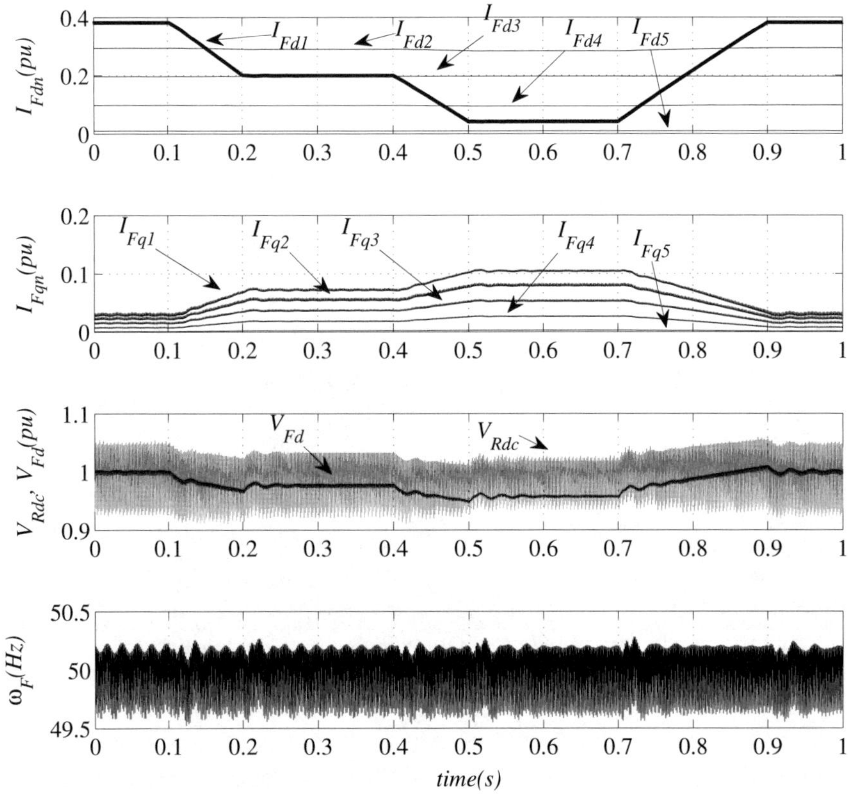

Fig. 18.17 Response to changes in wind farm power set point

converters can compensate for reactive power variations, the capacitor banks would usually be switched on and off depending on the required overall reactive power.

Figure 18.18 shows the response of the system to the connection of a 100 MVA capacitor bank at $t = 0.05$ s and subsequent disconnection at $t = 0.25$ s, when the system is being operated at rated power. The connection transient lasts for about 50 ms. When the capacitor bank is connected, the control system reacts by absorbing the additional reactive power generated by the capacitor bank. Note the adequate reactive power sharing among the different wind turbines.

During the connection transient, the off-shore ac-grid voltage reaches a maximum value of 1.1 pu. However, during capacitor bank disconnection, the transient is almost negligible. This difference in behaviour is due to the fact that all the front-end converters are initially operating at rated current, therefore, when they are required to provide additional reactive current, the front-end current references are limited by (18.46) and (18.47). Note the front-end inverter active current limits are reduced to provide enough transient reactive current capability for the system to maintain rated frequency. When the transient is over ($t = 0.1$ s), rated power operation is resumed.

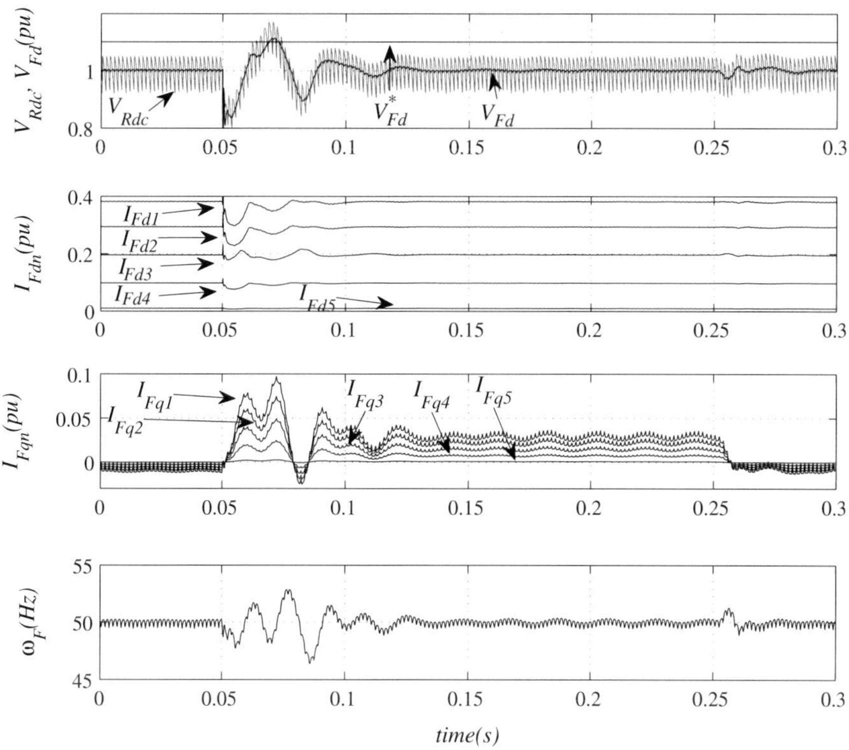

Fig. 18.18 Response to the switching of a 100 MVA capacitor bank

It is worth stressing that the controller parameters have not been modified during this transient, showing therefore, good robustness to relatively large sudden changes of the system overall capacitance.

18.4.3.2 Fault Ride-Through Operation

On-Shore Ac-Grid Voltage Sag

Figure 18.19 shows the response of the system to an 80% voltage sag at the on-shore inverter terminals. The fault has a duration of 500 ms with a 700 ms recovery time to 100% of the pre-fault voltage. As the HVDC-link voltage reduces, the off-shore ac-voltage V_{Fd} will reduce almost proportionally. This behaviour is caused by the HVDC diode rectifier voltage clamp on V_{Fd}. Therefore, the VDCOL mechanism on each independent wind turbine will sense the ac-grid voltage reduction and therefore will decrease the wind turbine active and reactive currents limits ($I_{Fdi\,\max}$ and $I_{Fqi\,\max}$). When all reactive current components I_{Fqi} reach their saturation limits, the frequency control loop saturates and the frequency

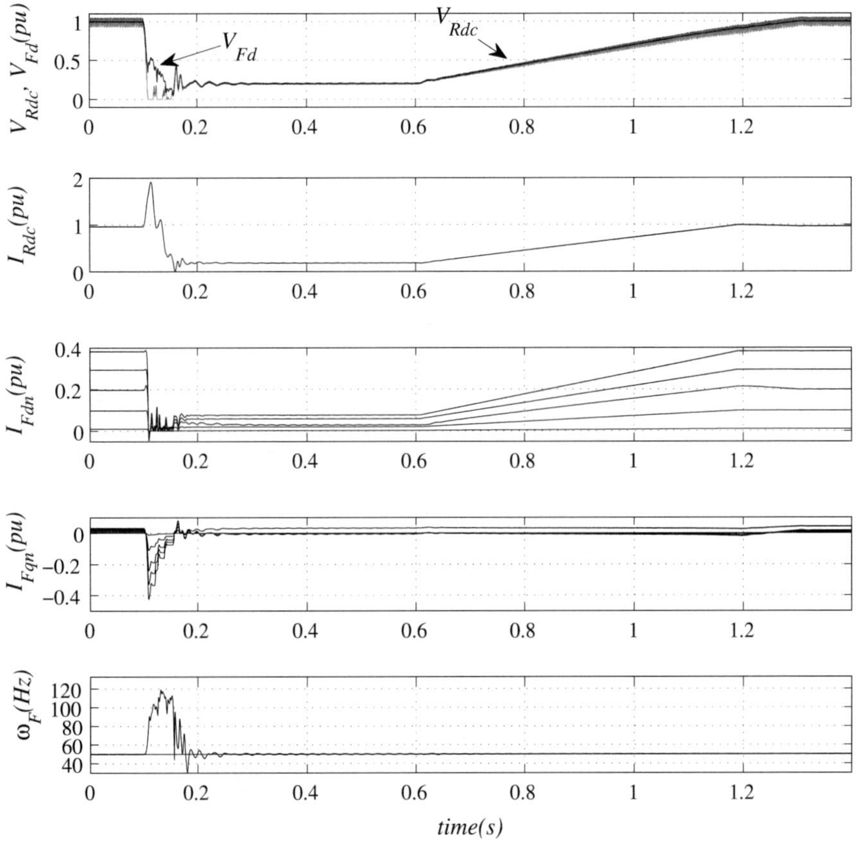

Fig. 18.19 Response to an 80% voltage sag at on-shore inverter terminals

of the off-shore ac-grid reaches a maximum of 120 Hz during less than 100 ms. After 100 ms I_{Fqi} are no longer saturated and frequency control is regained. Full power delivery is resumed as the voltage V_{Rdc} recovers its pre-fault value. It is worth stressing that the peak HVDC-link current is below 2 pu, which is comparable to the values obtained with fully controlled rectifiers.

Note the wind turbine active currents (I_{Fd1} to I_{Fd5}) are positive during the complete transient, except from few milliseconds right after the fault ($t = 0.1$ s). Therefore, during this period of time, active power will flow into the wind turbine front-end converters. The maximum power flowing into the wind turbine front end converters is 60 MW (0.06 pu) and the negative power flow lasts for less than 3 ms. Therefore, the overall energy flowing into the converters is at most 180 kJ and could be easily absorbed by the dynamic breaking protection of the wind turbines.

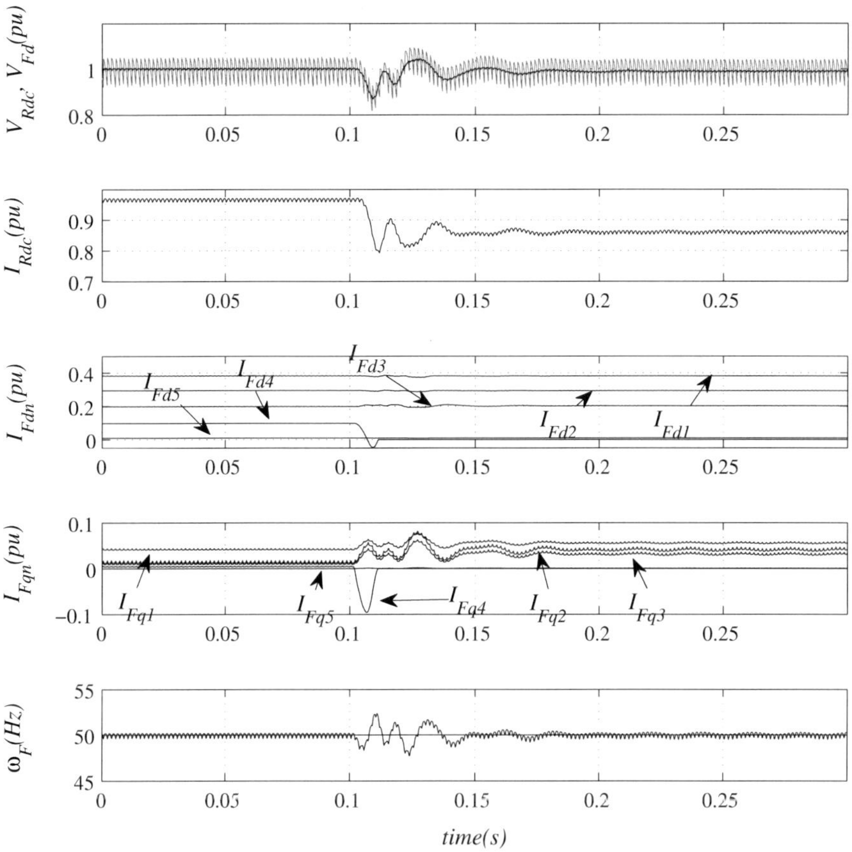

Fig. 18.20 Response to a sudden disconnection of 10% of the wind turbines

Disconnection of a Substantial Number of Wind Turbines

The response to a sudden disconnection of 10% of the wind turbines is shown in Fig. 18.20. At $t = 0.1$ s wind turbine 4 (100 MVA) breaker is opened. After a small transient, lasting less than 50 ms, the off-shore ac-grid voltage and current return to their pre-fault values, while the HVDC current (I_{Rdc}) reflects the 10% reduction in generating power. Again, the remaining wind turbines have to provide the reactive power required to keep the off-shore ac-grid frequency at 50 Hz. As in previous transients, the wind turbines have to absorb the reactive power excess compensation provided by the capacitor and filter banks.

HVDC Rectifier Breaker Trip and Reclosure

Figure 18.21 shows the response of the control system to a sudden trip of the HVDC rectifier ac-breaker and its subsequent reclosure. At $t = 20$ ms the HVDC

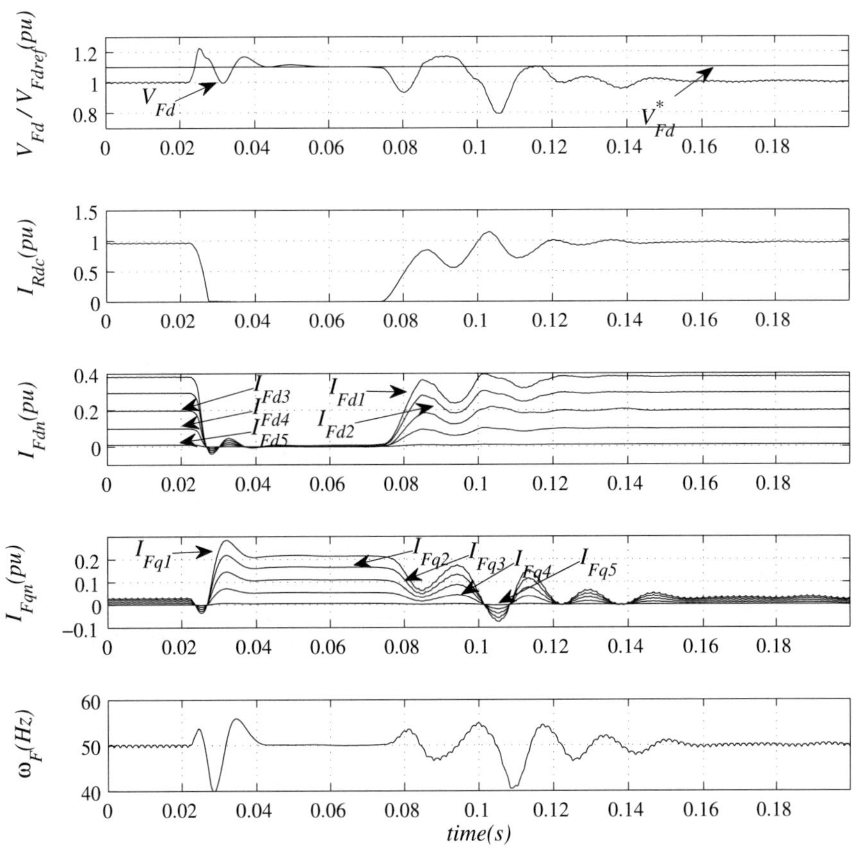

Fig. 18.21 Response to a HVDC rectifier ac-breaker trip and reclosure

rectifier ac-breaker trips, disconnecting the rectifier and the transformer banks from the off-shore ac-grid. As a consequence of the disconnection, the off-shore ac-grid voltage (V_{Fd}) is no longer limited by the HVDC diode rectifier and increases up to 1.22 pu. After approximately 40 ms the control system drives V_{Fd} to its reference value (1.1 pu). Note the decrease of the HVDC-link current I_{Rdc} and the front-end active currents. At the same time, the front-end reactive current increases to compensate for the capacitor bank overcompensation, now that the HVDC rectifier is not conducting.

At $t = 70$ ms the HVDC breaker is reclosed and, after a transient lasting approximately 50 ms, normal operation is resumed with rated active power transmission. The off-shore ac-grid voltage V_{Fd} is again being determined by the HVDC-link inverter. During the whole transient, both active and reactive currents are adequately shared among all the front-end inverters.

18.5 Discussion and Conclusions

The integrated control of wind turbines and off-shore ac-grid, and the diode-based HVDC link allows for substantial installation and operational cost reduction.

The proposed control strategy includes the use of the Synchronous Generator torque current to control the wind turbine dc-link E_{DCi} voltage. In this way, both active and reactive currents of the front-end converters can be used for off-shore ac-grid control.

A strategy for the distributed control of the off-shore ac-grid voltage and frequency has been presented, which allows for black-start operation and shows an excellent performance when in islanded operation. Moreover, the presented control strategy allows for smooth transition between islanded and connected modes of operation. The distributed control strategy can perform adequately in the presence of limited wind resource at the wind turbines.

In connected mode of operation, the joint wind farm and HVDC diode rectifier can be used to either set the voltage or the current at the dc-terminals of the HVDC rectifier. This behaviour is similar to industrial practice in LCC-HVDC links. Moreover, a strategy for optimum power tracking has been developed with the system operating in current control mode.

The distributed control systems has been designed so most of the variables used for the different control loops are local to each wind turbine. The remote variables used in the control system are the feedforward terms I_{Racd}, I_{Racq} and the output of the integrator in Fig. 18.7. The aforementioned feedforward terms can be neglected with a slight performance penalty. On the other hand, the frequency contents of the output of the integrator in Fig. 18.7 are such that relatively slow communications can be used.

A new distributed fault protection strategy has also been implemented, similar to standard VDCOL. The proposed protection system keeps all the devices within their operational limits in the event of a fault, with recovery times comparable to those achieved with controlled HVDC rectifiers. Moreover, the control system reacts adequately to substantial generation disconnection, capacitor bank switching and HVDC rectifier ac-breaker tripping and reclosure.

The presented technical feasibility study shows the advantages of integrated design of wind turbine and HVDC-link control, allowing, in this case, important savings by using diode-based HVDC rectifiers.

Acknowledgments The authors would like to thank the Spanish Ministry of Education and the European Union FEDER funds for their support through research grants DPI2007-64730 and DPI2010-16714.

References

1. Balaguer-Alvarez I, Lei Q, Yang S, Supatti U, Peng Z (2010) Control for grid-connected and intentional islanding operations of distributed power generation. IEEE Trans Ind Electron PP(99):1. doi:10.1109/TIE.2010.2049709

2. Blasco-Gimenez R, Añó-Villalba S, Rodríguez-D'Derlé J, Morant F, Bernal-Perez S (2010) Distributed voltage and frequency control of offshore wind farms connected with a diode-based HVdc link. IEEE Trans Power Electron 25(12):3095–3105. doi:10.1109/TPEL.2010.2086491
3. Bowles J (1981) Multiterminal HVDC transmission systems incorporating diode rectifier stations. IEEE Trans Power Apparatus Syst PAS-100(4):1674–1678. doi:10.1109/TPAS.1981.316562
4. Bozhko S, Blasco-Gimenez R, Li R, Clare J, Asher G (2007) Control of offshore DFIG-based wind farm grid with line-commutated HVDC connection. IEEE Trans Energy Convers 22(1):71–78. doi:10.1109/TEC.2006.889544
5. Cardenas R, Pena R, Asher G, Clare J, Blasco-Gimenez R (2004) Control strategies for power smoothing using a flywheel driven by a sensorless vector-controlled induction machine operating in a wide speed range. IEEE Trans Ind Electron 51(3):603–614. doi:10.1109/TIE.2004.825345
6. Chandorkar M, Divan D, Adapa R (1993) Control of parallel connected inverters in standalone AC supply systems. IEEE Trans Ind Appl 29(1):136–143. doi:10.1109/28.195899
7. Chen Z, Guerrero JM, Blaabjerg F (2009) A review of the state-of-the-art of power electronics for wind turbines. IEEE Trans Power Electron 24(8):1859–1875
8. Green T, Prodanovic M (2007) Control of inverter-based micro-grids. Electr Power Syst Res 77(9):1204–1213. doi:10.1016/j.epsr.2006.08.017
9. Heier S, Waddington R (2006) Grid integration of wind energy conversion systems. Wiley, Chichester
10. Hungsasutra S, Mathur R (1989) Unit connected generator with diode valve rectifier scheme. IEEE Trans Power Sys 4(2):538–543
11. Kanellos F, Hatziargyriou N (2008) Control of variable speed wind turbines in islanded mode of operation. IEEE Trans Energy Convers 23(2):535–543. doi:10.1109/TEC.2008.921553
12. Kirby N, Xu L, Luckett M, Siepmann W (2002) HVDC transmission for large offshore wind farms. Power Eng J [see also Power Engineer] 16(3):135–141
13. Kundur P (1994) Power system stability and control. McGraw-Hill Professional, New York
14. Lasseter R, Piagi P (2004) Microgrid: a conceptual solution. In: Proceedings of 2004 IEEE 35th annual power electronics specialists conference, PESC 04, Aachen
15. Lazaridis LP (2005) Economic comparison of HVAC and HVDC solutions for large offshore wind farms under special consideration of reliability. Master's Thesis, Royal Institute of Technology Stockholm
16. Li R, Bozhko S, Asher G (2008) Frequency control design for offshore wind farm grid with LCC-HVDC link connection. IEEE Trans Power Electron 23(3):1085–1092. doi:10.1109/TPEL.2008.921193
17. Machida T, Ishikawa I, Okada E, Karasawa E (1978) Control and protection of HVDC systems with diode valve converter. Electr Eng Jpn 98:62–70. doi:10.1002/eej.4390980109
18. Pena R, Clare J, Asher G (1996) A doubly fed induction generator using back-to-back PWM converters supplying an isolated load from a variable speed wind turbine. IEE Proc Electr Power Appl 143(5):380–387. doi:10.1049/ip-epa:19960454
19. Pogaku N, Prodanovic M, Green T (2007) Modeling, analysis and testing of autonomous operation of an inverter-based microgrid. IEEE Trans Power Elect 22(2):613–625. doi:10.1109/TPEL.2006.890003
20. Polinder H, Pijl FVD, Vilder GD, Tavner P (2006) Comparison of direct-drive and geared generator concepts for wind turbines. IEEE Trans Energy Convers 21(3):725–733. doi:10.1109/TEC.2006.875476 http://dro.dur.ac.uk/6704
21. Sao C, Lehn P (2008) Control and power management of converter fed microgrids. IEEE Trans Power Syst 23(3):1088–1098. doi:10.1109/TPWRS.2008.922232
22. Szechtman M, Wess T, Thio CV (1991) First benchmark model for HVDC control studies. Electra 135(4):54–73
23. Weimers L (1998) HVDC light: a new technology for a better environment. IEEE Power Eng Rev 18(8):19–20. doi:10.1109/MPER.1998.691411

Chapter 19
Wind Farm with HVDC Delivery in Inertial and Primary Frequency Response

Lingling Fan, Zhixin Miao and Dale Osborn

Abstract In this chapter, investigation of wind farms with line commutated converter (LCC) HVDC delivery in system frequency response participation is carried out. While LCC-based high power HVDC is a viable choice to deliver large-scale wind power, the consequent responsibility of such wind energy systems in frequency response should be assumed. A coordination control strategy for wind farms with HVDC delivery for participating in inertial response and primary frequency control is discussed in this chapter. The coordination philosophy is to feedback the grid frequency and its derivative and adjust the delivery power of the HVDC link according to the feedback signals. The feedback loop employing the derivative of the grid frequency aims to improve the inertial response while the feedback loop employing the grid frequency deviation introduces a droop at the rectifier control loop. When the grid frequency is too high or too low, active power flow through the HVDC link will be ramped down or up. In turn, the wind generation will increase or decrease the blade angles to reduce or increase the captured wind power through pitch control. A case study demonstrates the effectiveness of the inertial enhancement and frequency droop in HVDC control. Simulation results in TSAT are given.

L. Fan (✉) · Z. Miao
University of South Florida, 4202 E. Fowler Ave, Tampa, FL 33620, USA
e-mail: fll@ieee.org

Z. Miao
e-mail: Zhixin.miao@ieee.org

D. Osborn
Midwest ISO, 1125 Energy Park Drive, St Paul, MN 55108, USA
e-mail: DOsborn@midwestiso.org

S. M. Muyeen (ed.), *Wind Energy Conversion Systems*,
Green Energy and Technology, DOI: 10.1007/978-1-4471-2201-2_19,

19.1 Introduction

Wind energy, one of the most promising renewable energy sources, has been harvested for decades. With the advances in high power electronic converters, a single wind turbine can now reach a rating of 2–5 MW, about a hundred time increase in power ratings from the wind turbines in the 1980s. The power rating of a wind farm can reach hundreds of MW, e.g., the Buffalo Ridge wind farm in southwest Minnesota has a power rating of about 600 MW. With such large power ratings, integration of wind farms usually occurs at a transmission level with 100 kV and above voltage level. In May 2008, the U.S. Department of Energy (DOE) issued a report [1] entitled "20% Wind Energy by 2030," which found that the nation possesses affordable wind energy resources far in excess of those needed to enable a 20% scenario.

The most urgent problem is how to transport wind energy-based electricity to millions of users. In the U.S., some of the richest wind resources are far from the metropolitan areas. For example, the midwest region (North Dakota, South Dakota, Minnesota and Iowa) has abundant wind power resources, and wind farms have been installed in a rapid pace. The major load centers are in the East. Moreover, the current AC system in U.S. is already congested [2]. Therefore, an enhanced transmission system is the solution.

The Eastern Wind Integration and Transmission Study (EWITS) [3, 4] has studied transmission scenarios with high wind penetrations (20 or 30%). One important finding is that the huge difference of the locational marginal prices (LMP) of generation between the Midwest and the East results in an economic benefit that can justify the transmission system expansion. Given the recent BP oil spill, deep sea oil drilling will be restricted in the future, and gas prices will be even greater for the most expensive gas turbines, which determine the LMPs. Therefore, building transmission systems to deliver wind power is economical.

One concern is, which transmission system should be used—high voltage DC (HVDC) or high voltage AC? The answer from the industry is both. Given the large-scale wind power in the Midwest, ITC proposed the "Green Express Project," building a 765 kV AC transmission system as a backbone system to deliver wind power from the Dakotas and western Minnesota and Iowa toward the east and the load centers in that direction. In EWITS, an HVDC transmission overlay is proposed to deliver wind power from the Midwest to the load centers on the East Coast, while high voltage AC collector systems are used to feed the HVDC transmission. Whether to use AC or DC depends on distance and needs. AC transmission systems benefit loads along the transmission paths. Power can be tapped or diverted along the path. DC transmission systems are more suitable for long distance point-to-point electricity transportation. Compared to AC transmission, DC transmission is more efficient since there is no reactive consumption in the lines. However, converters at the terminals are costly. For long distance transmission over the break-even distance (500–800 km), DC lines are more cost effective [5].

Besides the economic reason, there are other justifications for using HVDC. The technology of HVDC to deliver electric power generated from hydro or fossil fuel is mature, and there are applications of HVDC around the world [6]. Recent years have seen a surge of HVDC applications in China and India for long distance bulk power transmission [7–9]. Further, HVDC systems have excellent controllability due to their power electronic devices. HVDC allows power flow control and provides voltage and stability control [10]. This becomes very important for wind power delivery. HVDC delivery is also suitable for offshore wind power delivery. Undersea cables need reactive power compensation due to the high charges when they are powered by AC voltages, and it is difficult to install reactive power devices under the sea. A DC grid is preferred for undersea electricity deployment. Europe has proposed to build an undersea HVDC grid to transport wind power [11]. Such a grid fits the scope of Smart Grid due to its capability of power flow control [12].

The coordination between the wind farm maximum power extracting controller and the HVDC wind farm side converter has been addressed in [13–15]. All of the above mentioned papers only address the issue of wind farm and HVDC rectifier coordination. However, the issue not addressed is: When the AC system has a change in load or generation, synchronous generators will act by first releasing the kinetic energy (inertial response) from their rotors and then changing the prime mover power (primary frequency response). It is desirable to have the wind generators act the same way as the synchronous generators. With inertial response contribution from wind, the frequency deviation of the system during transients will be less significant. With primary frequency control, the frequency deviation at steady state will be less significant and the wind farm can share the active power demand along with the synchronous generators. How can a wind farm with HVDC delivery participate in inertial response and primary frequency control? This is the focus of this chapter.

There are two types of HVDC systems to deliver wind power: line commutated converter (LCC) based and voltage source converter (VSC) based HVDC systems. This paper focuses on LCC-based HVDC. The major advantage of LCC-based HVDC is its high power transfer capability compared with VSC-based HVDC. Hence LCC-based HVDC is suitable to deliver large-scale wind power to load centers for a long distance. In the 2008 DOE report [1], 20% of wind power penetration is studied in Eastern Interconnection. The majority of wind farms are located in Midwest while the load centers are far away. To deliver such large-scale wind power to the grid, LCC-based HVDC systems are proposed in JCSP report [16].

Research has been carried out on inertial response and active power sharing or primary frequency response for doubly fed induction generator (DFIG)-based wind farms directly interconnecting to the AC grids [17–23]. References [17–19] point out that DFIG-based wind farms have negligible contribution to inertia response without additional control. However, further study has shown that wind turbine generators have enough kinetic energy to provide inertia support [21]. A supplementary control loop using the derivative of the system frequency as the input signal can be introduced to provide additional inertia [17, 20] and the improvements on the inertia control loop are reported in [24, 25]. To provide primary frequency control, a feedback loop with the system frequency deviation as the input signal is introduced in the wind generator electrical control system [20, 22].

These approaches, using the derivative and deviation of the system frequency as the input signals for inertia contribution and frequency control, have also been applied in full converter permanent magnet synchronous generator (PMSG) based wind energy systems [26].

The desired operation of a wind farm is to get the maximum output power for a certain wind speed. Under the above circumstances, the wind farms will have no reserve to supply more active power when the grid frequency is low. In order to have improved frequency response, the wind farms should operate with reserves. Wind farms have pitch controllers to increase or decrease the captured wind power. This feature can help wind farms to participate in power sharing when the system frequency is higher or lower than the nominal value. In a Denmark wind farm, the amount of total power production is reduced by a predefined set point. The production of the wind farm can be briefly increased and decreased according to the power system requirement [27] (Chap. 7). In [28, 22], wind farms have been operated under deloaded maximum power exporting conditions to carry out system operator requests. Primary frequency control and supervisory wind farm control are superimposed on the reference power signal for rotor side converter (RSC) control loops. A 20% reserve margin is assumed in [28, 22]. Operation at deloaded maximum power extraction is also adopted in [29].

Further, the reserve allocation provided by each DFIG wind turbine is distributed according to the available wind speed. Fuzzy logic technique is adopted in [30] to provide primary frequency control without the measurement of wind speed. In order to have wind farms with HVDC delivery provide inertia contribution and frequency control, the HVDC needs supplementary controls and also coordination with the wind farm. Power transferred through an LCC-HVDC link is controlled by the firing angle of the rectifier converter. Hence it is reasonable to introduce feedback loops with grid frequency derivation and grid frequency deviation as the input signals.

The next question is: how much can the wind farm with HVDC delivery contribute to inertial response, and how much should HVDC power transfer or the wind farm exporting be adjusted?

The inertia response contribution will be determined by the control loop gain [24, 25], which is determined by the kinetic energy the wind turbines possess [25]. The steady state frequency response will be determined by the droop gain of the frequency control loop. The concept is similar to the frequency droop in the speed governor of a synchronous generator. If a frequency-power droop is introduced, the wind farms with HVDC delivery can successfully participate in primary frequency response when the system has a load or generation change. The prime mover (the wind turbine) will change its output through its pitch controller.

In [31], a simple system with a wind farm with HVDC delivery participating in primary frequency control via., a supplementary frequency droop control is presented.

In this chapter, work in [32] will be presented, including the detailed modeling, primary frequency control of LCC-HVDC, and the inertia contribution from wind farms with HVDC delivery. Wind farms are assumed to operate with enough reserve to provide frequency control.

The rest of the chapter is organized as follows. Section 19.2 presents the LCC-HVDC's average model and conventional controls. Section 19.3 presents the inertial response enhancement through HVDC and the frequency-power droop added to the HVDC rectifier. Section 19.4 presents the coordination in wind generation. Section 19.5 presents simulation results by TSAT [33] and discussion. Section 19.6 concludes the chapter.

19.2 LCC-HVDC Average Model and Conventional Control

For a monopole, 12-pulse inverter of HVDC link, the expressions for the AC/DC voltage, current and power considering ignition delay α are given by:

$$\begin{cases} V_{\mathrm{dr}} = \dfrac{3\sqrt{6}}{\pi} V_{\mathrm{ac}} \cos\alpha \\ I_{\mathrm{dc}} = \dfrac{\sqrt{6}}{\pi} I_{\mathrm{ac}} \\ V_{\mathrm{dr}} = I_{\mathrm{dc}} R + V_{\mathrm{di}} \\ P_{\mathrm{dc}} = V_{\mathrm{dr}} I_{\mathrm{dc}} \end{cases} \tag{19.1}$$

where the leakage inductance of the converter transformer is neglected, V_{ac} is the rms value of the bus voltage, I_{ac} is the rms value of the bus current, R is the total resistance of the dc transmission line, V_{dr} and V_{di} are voltages at the dc link rectifier and inverter sides, I_{dc} is the current through the dc link, and P_{dc} is the dc link power flow at the rectifier side.

The power through the dc link is given by:

$$P_{\mathrm{dc}} = 3 V_{\mathrm{ac}} I_{\mathrm{ac}} \cos\alpha \tag{19.2}$$

From (19.2), it is seen that the larger the firing angle, the lower will be the delivered power. Hence in order to increase the power delivered through a HVDC link, the firing angle should be reduced. Since the DC power is proportional to the DC current, a feedback control can be implemented to adjust the firing angle based on the DC current measurement.

The widely used control scheme for the HVDC link is the current control as shown in Fig. 19.1 [34], where the current reference is determined by the power reference divided by the measured DC voltage. The measured DC current is then compared with the current reference and the error is passed through a proportional integral (PI) controller to generate a firing angle reference.

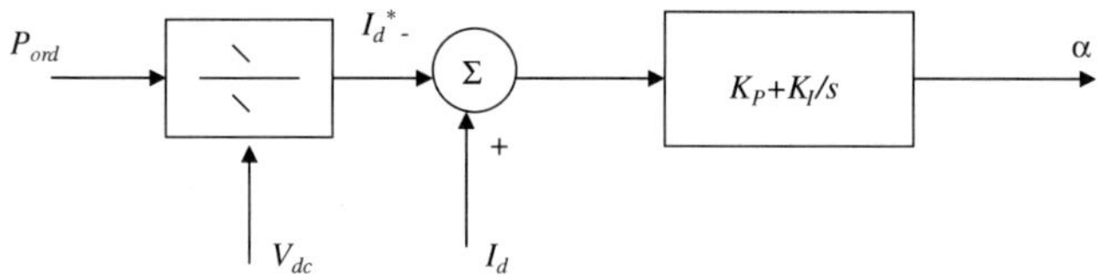

Fig. 19.1 Current control scheme at an HVDC's rectifier

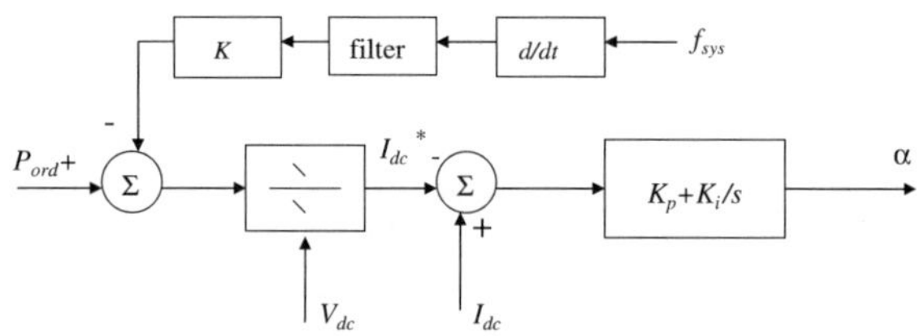

Fig. 19.2 Control loop for HVDC rectifier with enhanced inertia loop [40] © [2010] IEEE

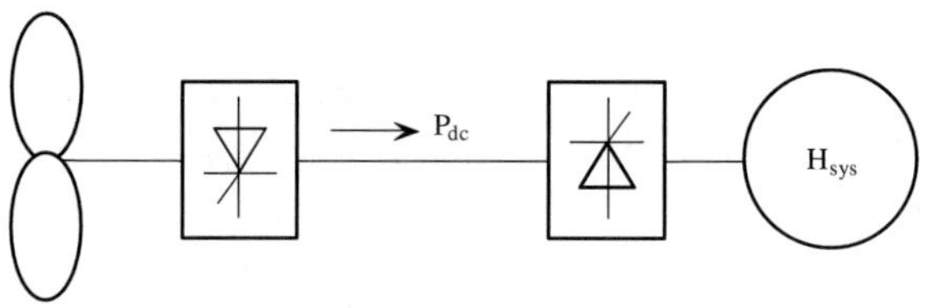

Fig. 19.3 A wind farm with HVDC delivery connected to a system with aggregated inertia H_{sys}

19.3 Inertial Response Enhancement and Frequency Droop Control via HVDC

19.3.1 Inertial Response Enhancement

In order to have the HVDC converter provide electrical inertia, the control scheme shown in Fig. 19.2 is proposed. The derivative of the system frequency is obtained and the power reference is modified. To get rid of impulses like noise measurements due to abrupt wind changes or other system variations, a low pass filter is used along with the derivative function block [29]. In this paper, the low pass filter has a transfer function of 1/(1 + 0.01 s). The bandwidth of the low pass filter is about 16 Hz. The time frame of the frequency response study is around 30–100 s. The electromechanical mode, which determines the frequency dynamic response, has a much lower bandwidth. Therefore the high frequency noise can be filtered out but the concerned frequency responses will be captured.

The need for the inertial enhancement loop can be explained using the simple system shown in Fig. 19.3 where a wind farm with HVDC delivery is connected to a system with an aggregate inertia H_{sys}.

The dynamics of the system frequency f_{sys} can be written as:

$$2H_{\text{sys}}\frac{\mathrm{d}f_{\text{sys}}}{\mathrm{d}t} = P_{\text{m,sys}} - P_{\text{e,sys}} + P_{\text{dc}} \tag{19.3}$$

where $P_{\text{m,sys}}$ is the equivalent prime mover power in the system and $P_{\text{e,sys}}$ is the equivalent generation output of the system. Assuming that there is no power loss in the HVDC converters, the power fed from the HVDC is equal to P_{dc}. With the inertial enhancement control, P_{dc} will have the following dynamics:

$$P_{\text{dc}} = P_{\text{ord}} - K\frac{\mathrm{d}f_{\text{sys}}}{\mathrm{d}t} \tag{19.4}$$

Equations 19.3 and 19.4 lead to the following dynamics of the system frequency:

$$2\left(H + \frac{K}{2}\right)\frac{\mathrm{d}f_{\text{sys}}}{\mathrm{d}t} = P_{\text{m,sys}} - P_{\text{e,sys}} + P_{\text{ord}}. \tag{19.5}$$

From (19.5), it can be observed that through inertial enhancement control, the overall inertia of the system with wind generation will be improved. In other words, the wind generation system with HVDC delivery will now be contributing to the overall system inertia. A larger inertia means a smaller transient frequency shift during disturbances.

In the mean time, due to the additional contribution to the inertia, the damping of the swing mode will be decreased. This can be explained by a simple two-order linearized swing equation with the equivalent rotor angle δ as the state variable and also by considering the damping of the system. If the mechanical power is assumed to be constant compared with the electric power, then the linearized electric power can be expressed as

$$\Delta P_{\text{e, sys}} - D\frac{\mathrm{d}}{\Delta}\delta\mathrm{d}t + T_{3}\Delta\delta \tag{19.6}$$

where D is the damping coefficient and T_{S} is the synchronizing coefficient. Then, the linearized swing equation can be written as

$$\frac{2H}{\omega_e}\frac{\mathrm{d}^2\Delta\delta}{\mathrm{d}t^2} + D\frac{\mathrm{d}\Delta\delta}{\mathrm{d}t} + T_{\text{S}}\Delta\delta = 0. \tag{19.7}$$

It is found that with an increasing inertia, the eigenvalue moves right toward the right half plane. Hence it can be reasoned that with an inertia control loop, the system will have a poor damping. This phenomenon can be observed in the simulation results.

19.3.2 Frequency Droop Control

In order to let wind farms share active power, a frequency droop is introduced to the HVDC rectifier control loop (Fig. 19.4). The idea is the same as the frequency droop in the turbine governor in a synchronous generator.

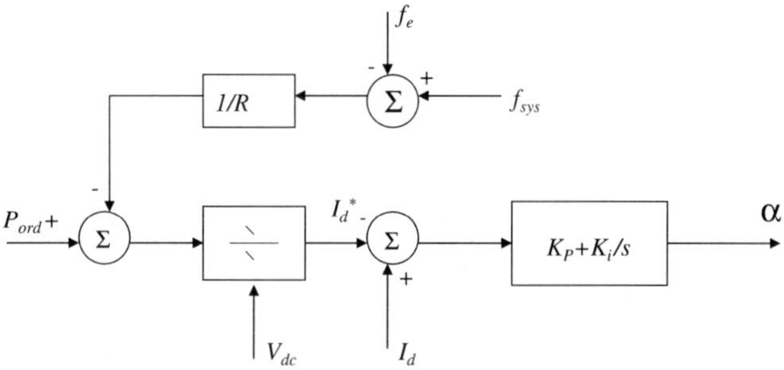

Fig. 19.4 Control loop for HVDC rectifier [40] © [2010] IEEE

By introducing a frequency droop loop into the HVDC rectifier control loop, the total change in system frequency will be reduced as

$$\Delta f = \frac{-\Delta P}{(1/R_1) + (1/R_1) + \cdots + (1/R_n) + (1/R_{\mathrm{HVDC}})} \tag{19.8}$$

where n is the number of synchronous generators. If the system has a high penetration of wind power, it will be necessary to have frequency droop in the HVDC loop. Otherwise the frequency variation due to the load change will be too large.

With the addition of the droop control, the power through the DC link can be expressed as:

$$P_{\mathrm{dc}} = P_{\mathrm{ord}} - \frac{1}{R_{\mathrm{HVDC}}}\Delta f. \tag{19.9}$$

Moreover, the equivalent swing equation can be written as:

$$2H\frac{\mathrm{d}f_{\mathrm{sys}}}{\mathrm{d}t} + \frac{1}{R_{\mathrm{HVDC}}}\Delta f = P_{\mathrm{m,\,sys}} - P_{\mathrm{e,\,sys}} + P_{\mathrm{ord}}. \tag{19.10}$$

From (19.10), it is obvious that the introduction of the droop control enhances the damping of the electromechanical swing mode.

19.4 Coordination in Wind Generation

When there is no HVDC interface, DFIG-based variable-speed wind generators use supplementary torque or power control loop around RSC for inertia and frequency regulation [17, 20]. With HVDC interface and supplementary inertia and frequency regulation loops at HVDC rectifier side, no additional control loops will be needed for DFIGs. This is because the purpose of these control loops is to adjust the output power from the wind generators. Whether the control loops are applied at HVDC converters or DFIG converters, the ultimate purposes are same.

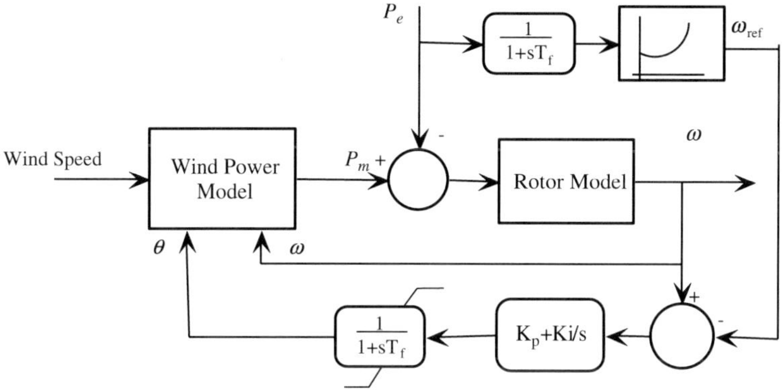

Fig. 19.5 Block diagram of wind turbine control

To coordinate with the inertia and frequency regulation loops in the HVDC interface, the mechanical power of the wind turbine should match the delivered power to HVDC. The following paragraphs will discuss the relationship among the mechanical power, the blade angle and the rotor speed. Insights into the coordination in wind generation will also be given.

The commercial multi-megawatt variable-speed wind turbine (1.5 MW) developed by GE is used in this study and the block diagram of the wind turbine control is shown in Fig. 19.5, where θ is the blade angle and ω is the rotating speed of the wind generator. Pitch controllers use the rotating speed of the wind turbines (ω) as the input signal (Fig. 19.5) [35]. If there is a difference between the mechanical power and the delivered power, the rotating speed will change. Pitch controllers sense the speed change and regulate the blade angle accordingly.

When the available wind power is above the equipment rating, the blades are pitched to limit the mechanical power (P_{m}) delivered to the shaft to the equipment rating (1.0 pu). When the available wind power is less than rated, the blades are set at minimum pitch to maximize the mechanical power.

The extracted wind power can be expressed as [35]:

$$P_{\mathrm{w}} = \frac{\rho}{2} A_{\mathrm{r}} V_{\mathrm{w}}^{3} C_{\mathrm{p}}(\lambda, \theta) \tag{19.11}$$

where ρ is the air density in kg/m^3, A_{r} is the area swept by the rotor blades in m^2, V_{w} is the wind speed in m/sec, and C_{p} is the power coefficient (function of λ-tip ratio ($\lambda = V_{\mathrm{tip}}/V_{\mathrm{w}}$), and θ-pitch angle).

A changing power output impacts the speed reference and hence the rotating speed of wind generators. This will in turn change the extracted wind power. The reference speed (ω_{ref}) is generated for maximum power tracking based on the measured electric power (P_{e}). When P_{e} is greater than 75% of the rated power, ω_{ref} will be set at 1.2 pu. When P_{e} is less than 75% of the rated power, ω_{ref} is given by

$$\omega_{\mathrm{ref}} = 0.67 P_{\mathrm{e}}^{2} + 1.43 P_{\mathrm{e}} + 0.51. \tag{19.12}$$

The power setting of the HVDC line Pord in Figs. 19.1, 19.2, and 19.4 depends on the wind speed. Normally, there are four operation modes:

(1) When the wind speed is less than the cut-in speed, the output of the wind farm is zero. The power order of the HVDC line is set to zero.
(2) When the wind speed is greater than the cut-in speed and less than 120% of the nominal wind speed, the wind farm outputs the maximum power. The power setting of the HVDC line also follows such principle.
(3) When the wind speed is greater than 120% of the nominal wind speed and less than the cutoff speed, the wind farm exports constant power. When the wind speed changes, the pitch control adjusts the blade angle so that the output is kept constant. In turn, the power setting of the HVDC line is set to the same value.
(4) When the wind speed is greater than the cutoff speed, the wind farm shuts down. The power setting of the HVDC line is set to zero.

Besides the wind speeds, the power setting of the HVDC line will also be changed should the system have a load or generation change. HVDC control makes the power delivered from the wind farm decrease or increase. The mechanical power from the wind turbines will match the change. Pitch controllers in wind farms have the ability to adjust the blade angle and change the extracted wind power. If wind farms are not operated at the maximum power point, then, when the power delivered through the HVDC increases, the pitch controller should reduce blade angles to extract more wind power.

Following a drop in the system frequency, the HVDC increases its delivered power. The increased electric power from the wind generator results in an increased speed reference. Hence the rotating speed increases while the blade pitch angle decreases and the extracted wind power increases to match the electric power.

In order to provide improved frequency response to the power system, the wind farm is operated with reserve. Further, wind farms suffer the risk of stall for inertial control [36].

When the electrical power increases to support the grid, the rotating speed of the wind turbine decreases for the first several seconds. The pitch controllers deal with the mechanical systems and their responses are relatively slower than the electrical responses. This causes the captured wind power decrease too much and thus exasperates the speed decline.

Hence the inertial control must provide enough margin above stall. When the wind speed is low, the inertial control has limited usage. Therefore in the test system study that follows, the assumption of a high wind speed is made to demonstrate the effect of the inertial and primary frequency control.

19.5 Case Study

In the study system shown in Fig. 19.6, a wind farm (330 turbines and 1.5 MW capacity per turbine. Total rated capacity: 550 MVA, 495 MW) is connected to a two-area four synchronous generator system via an LCC-based HVDC link. At the

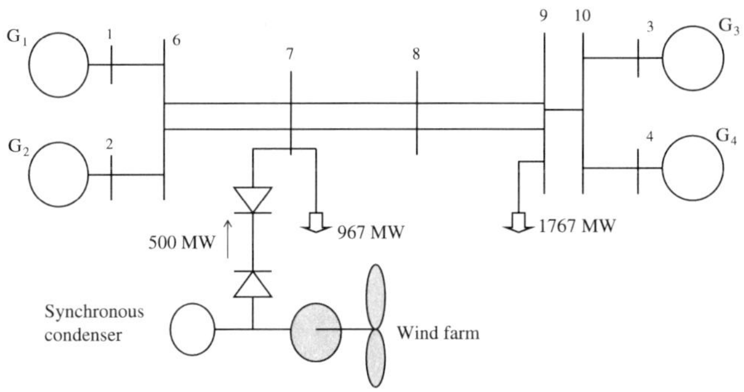

Fig. 19.6 The test system

rectifier side of the HVDC link, a synchronous condenser is installed to provide reactive power. The four synchronous generators are equipped with IEEE Type 2 speed governors [33].

Since the focus of the paper is on inertial and primary frequency responses, no automatic generator control (AGC) is included in these generators. Moreover, AGC is considered to have a much slower response than the inertial and primary frequency response. This is because the control centers of power systems usually update area control errors (ACEs), the input signal of AGC every two minutes [37]. The current-source DFIG wind farm model developed by GE [35] is used in this paper. The wind speed is assumed to be constant (14 m/s) which is above the synchronous speed. The wind farm is operated with a 10% margin ($P_e = 445.5$ MW). Hence the wind farm can increase or decrease its output power. The inertial and primary frequency control will be modeled and tested.

In this study, we assume that the wind speed does not change for a period of time because the focus is on inertial and primary frequency responses. Therefore, the power setting of the HVDC line (P^*_{ord}) remains constant throughout the simulation period.

In the AC system, 600 MW generation is tripped. Without any inertial enhancement and frequency droop included in the HVDC rectifier, the system has a frequency drop of 1.2764 Hz $\left(\Sigma \frac{1}{R_i} = \frac{600\,\text{MW}}{1.2764\,\text{Hz}} = 400\,\frac{\text{MW}}{\text{Hz}}\right)$. During the transient period, the frequency of Generator 1 can drop to 58.5 Hz (See Fig. 19.7-no control). The two-area study system is based on the two-area case in TSAT User Manual [33]. Parameters of the synchronous generators in two-area system are listed in Table 19.1 (all pu data on generator MVA base).

19.5.1 Discussion on Inertial Enhancement Results

The gain of the inertial enhancement control will impact the transient response of the system frequency. With a larger gain, more inertia will be contributed to the system resulting in a lower system frequency deviation during the transient period.

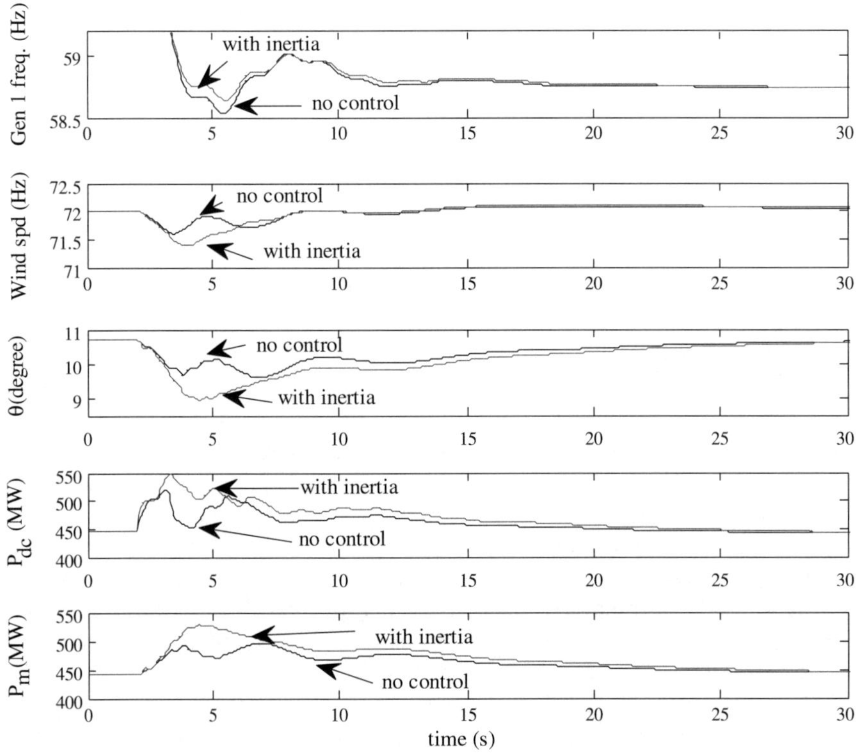

Fig. 19.7 Comparison of the dynamic responses of the system frequency, wind speed, blade angle θ, P_{dc}, and the mechanical power from wind turbines P_m

The inertia used in this study is $H = 4.94$ pu for each turbine. For the aggregate wind farm which is equivalent to a single DFIG with a capacity of 550 MVA, the inertia is 4.94 pu as well. In the study, the base power is chosen to be 100 MW.

Hence the inertia of the wind farm is 4.94*550/100 = 27.17 pu. The gain should be at most $2H$ (54.34 pu). Since the nominal frequency is 60 Hz and the base power is 100 MW, the maximum gain will be $K = 54.34$ pu*100 MW/60 Hz = 90.57 MW/Hz. In this study, K is chosen to be 50 MW/Hz and the simulation results are shown in Fig. 19.7.

Inertial enhancement control will reduce the frequency deviation during the transient period. The dynamic responses of the frequencies of the synchronous generator 1, rotor speeds of the wind generator, and the blade angles are shown in Fig. 19.7. The dynamics of the HVDC power P_{dc} and the mechanical power of the wind turbine P_m are shown in Fig. 19.7. It is observed that an additional 100 MW can be supplied to the HVDC link during the transient period due to the inertial enhancement control. As a result, the frequency deviation during transient period is reduced by about 10%.

Table 19.1 Parameters of the synchronous generators in per unit and the HVDC link

Line resistance (ohm) R	15.0	H	6.5	X''_d	0.25
Line reactance (mH) (L)	1000	D	0	X''_q	0.25
Capacitance (μF) ($C/2$)	10.0	X_d	1.8	T'_{d0}	8.0
Number of bridges	2	X_q	1.7	T'_{q0}	0.4
Transformer ratio	0.6	X'_d	0.3	T''_{d0}	0.03
Commutating reactance (ohm)	15.0	X'_q	0.55	T''_{q0}	0.05
Desired rectifier power (MW)	500	MVA Rating	900	$1/R$	8.0
Desired rectifier firing angle	15.0°				
Desired inverter current (A)	−1,000				
Desired inverter extinction angle	18.0°				

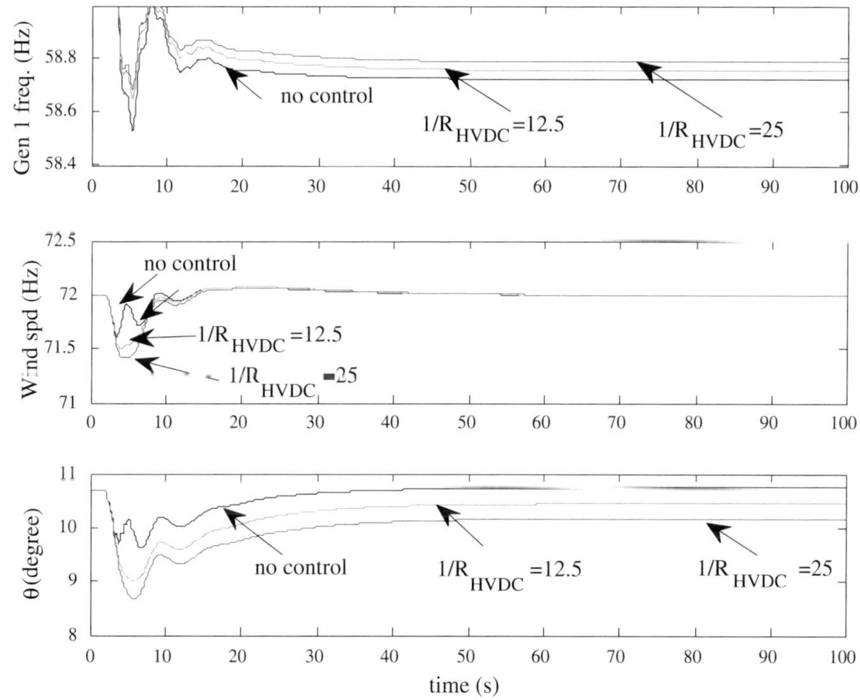

Fig. 19.8 Dynamic responses of the speed of Generator 1, the rotating speed of wind turbine, and blade angle

For the initial 2–3 s following the 600 MW generation trip, it is observed that the wind speed drops. During this time interval, the system frequency drops while the power from the wind turbines P_m has a slower response and can be assumed constant. The electric power from the wind farm P_e increases since $P_e = P_{dc}$ and the DC power P_{dc} increases due to the inertial enhancement control. Thus the wind

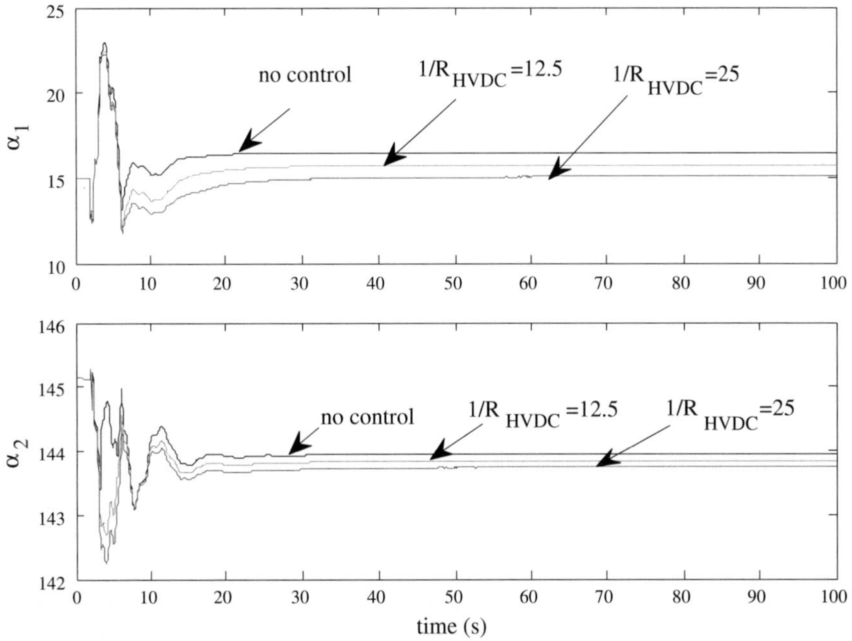

Fig. 19.9 HVDC converter firing angles at the rectifier side and at the inverter side

speed drops and the kinetic energy released from the wind turbine blades supplies the temporary increase in demand. From the simulation results in Fig. 19.7, it is found that the power through the HVDC link can be changed instantaneously. However the increase of mechanical power is slower. After the initial period, the pitch controller will work and the pitch angle will be reduced. The mechanical power Pm from the wind turbines will be adjusted according to the relationship among wind power, blade pitch angle, and the rotating speed relationship. The lower the pitch angle, the higher will be the wind power extracted.

19.5.2 Frequency Droop

With a frequency droop introduced in the HVDC rectifier control loop, the system frequency response at steady state will be improved. Two droop values are tested in the simulationI. Simulation results are shown in 19.8, Figs. 19.9 and 19.10. The computed results agree with the simulation results well.

Figure 19.8 shows the dynamic responses of the synchronous generator speed, the rotating speed of the wind farm and the blade angle. Figure19.9 shows the dynamic responses of the firing angles at the rectifier and the inverter sides. Due to the droop control, it is found that the deviation of the synchronous generator speed

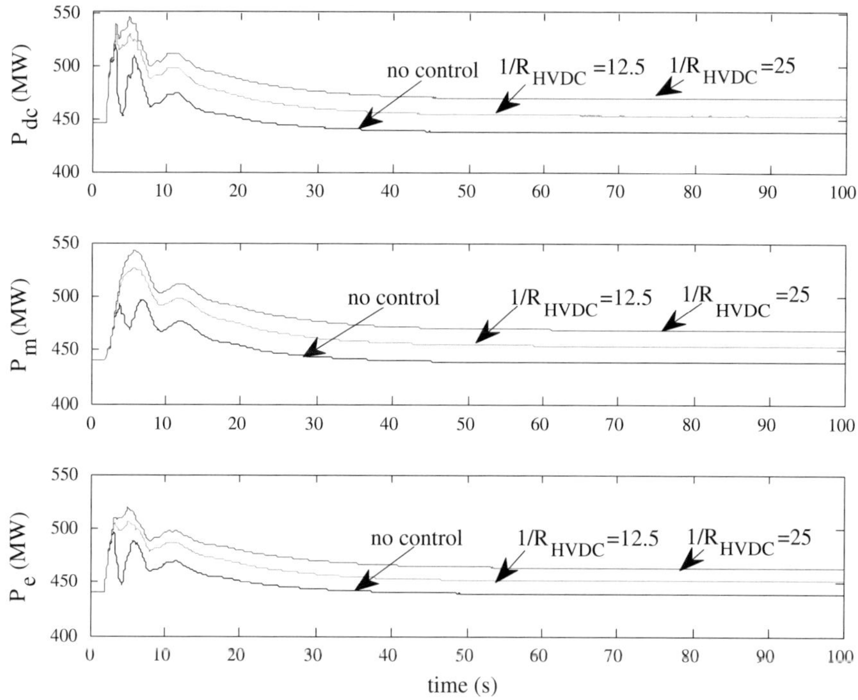

Fig. 19.10 HVDC power, wind turbine mechanical power, and wind farm output power

reduces. The rotating speed of the wind farm stays constant at 72 Hz due to the setting point of speed. The blade angle reduces to extract more wind power. The higher the value of $1/R_{\text{HVDC}}$, the higher will be the participation of the wind farm with HVDC in active power sharing, and smaller will be the frequency deviation. Meanwhile, the higher the value of $1/R_{\text{HVDC}}$, the higher will be the power sent through the HVDC line and the lower the rectifier side firing angle α_1.

The HVDC power, wind turbine mechanical power and wind farm output power are plotted in Fig. 19.10. It is found that the HVDC power and wind farm output electrical power have instantaneous changes once there is a generation trip in the system. The mechanical power is slower to respond following the generator trip.

19.5.3 With Both Inertial Enhancement and Frequency Droop

Finally, both inertial enhancement and frequency droop will be applied to the HVDC converters. The purpose is to have wind farms with HVDC delivery not only contribute to the system inertia but also contribute to load sharing. A comparison of the system responses without any of the controls and with both controls are shown in Fig. 19.11.

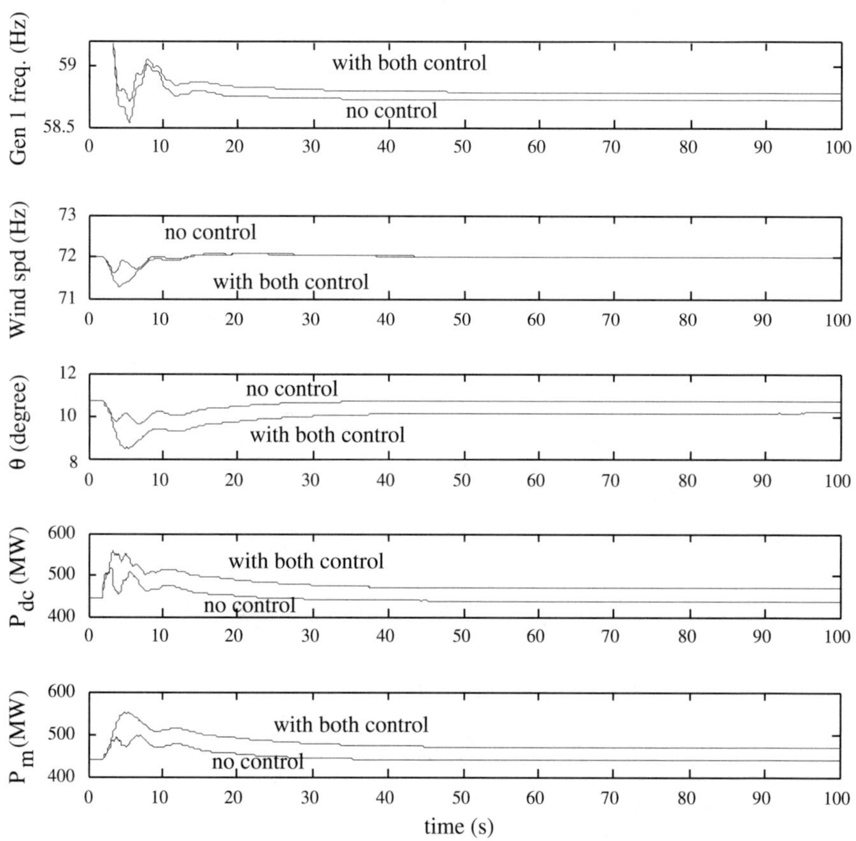

Fig. 19.11 Dynamic responses of the frequency of Generator 1, wind generator, blade angle, P_{dc} and P_m. **a** no control; **b** with both inertial enhancement $K = 50$ MW/Hz and frequency droop $1/R_{HVDC} = 25$ MW/Hz

It is found that with both inertial enhancement and frequency droop, both the transient and steady state deviation of the system frequency will be reduced by allowing the wind farm with HVDC delivery to release kinetic energy during the transient period and to participate in active power sharing.

19.5.4 Application in VSC-Based HVDC

The basic concept presented in this paper to improve frequency responses is to modulate the power transferred through the DC link by using feedback signals such as the deviation of the system frequency or the derivative of the system frequency.

This can also be done in VSC-based HVDC system. The only difference is in the inner control method and the parameter that is controlled. In LCC-based HVDC, it is the rectifier angle that needs to be controlled. In VSC-based HVDC, it is the three-phase voltage magnitude and angle of the rectifier converter that will be controlled. The grid side converter is controlled to have a constant DC link voltage and a constant AC side voltage output. Additional literature on VSC-based HVDC power control can be found in [38,39].

19.6 Conclusion

This chapter presents a method to help wind farms with HVDC delivery improve the inertial response and primary frequency control or active power sharing during system load or generation change. A case study demonstrates the effectiveness of the inertial enhancement and frequency droop control in the HVDC rectifier control loop. With both controls, wind farms can improve the inertial response, reduce the deviation of the system frequency both during transient period and steady state, and share the change in the active power of the AC system.

References

1. U.S. Department of Energy, 20% Wind Energy by 2030 (2008) Increasing wind energy's contribution to U.S. electricity supply. http://www1.eere.energy.gov/windandhydro/pdfs/41869.pdf
2. U.S. Department of Energy, National Transmission Grid Study (2002) URL: http://certs.lbl.gov/ntgs/main-print.pdf
3. National Renewable Energy Laboratory, Eastern Wind Integration and Transmission Study (2010) http://www.nrel.gov/wind/systemsintegration/pdfs/2010/ewits executive summary.pdf
4. Corbus D, Lew D, Jordan G, Winters W, Hull FV, Manobianco J, Zavadil B, (2009) Up with wind. IEEE Power Energy Mag 7(6):36–46
5. Padiyar KR (1990) HVDC power transmission system. Wiley, New York
6. Long W, Nilsson S (2007) HVDC transmission: yesterday and today. IEEE Power Energy Mag 5(2):22–31
7. Szechtman M, Sarma Maruvada P, Nayak RN (2007) 800-kV HVDC on the horizon. IEEE Power Energy Mag 5(2):61–69
8. Mukhopadhyay S (2007) Towards electricity for all. IEEE Power Energy Mag 5(5):71–78
9. Garrity T (2003) Shaping the future of global energy delivery. IEEE Power Energy Mag 1(5):26–30
10. Reed G, Paserba J, Salavantis P (2003) The FACTS on resolving transmission gridlock. IEEE Power Energy Mag 1(5):41–46
11. Blau J (2010) Europe plans a North Sea grid. IEEE Spectrum 47(3):12–13
12. Liserre M, Sauter T, Hung JY (2010) Future energy systems: integrating renewable energsources into the smart power grid through industrial electronics. IEEE Ind Electron Mag 4(1):18–37

13. Xiang D, Ran L, Bumby JR, Tavner PJ, Yang S (2006) Coordinated control of an HVDC link and doubly fed induction generators in a large offshore wind farm. IEEE Trans Power Del 21(1):463–471
14. Li R, Bozhko S, Asher G (2008) Frequency control design for offshore wind farm grid with LCC-HVDC link connection. IEEE Trans Power Electron 23(3):1085–1092
15. Bozhko SV, Blasco-Gimenez R, Li R, Clare JC, Asher GM (2007) Control of offshore DFIG-based windfarm grid with line-commutated HVDC connection. IEEE Trans Energy Convers 22(1):71–78
16. Joint coordinated system plan 2008 (2008) Midwest ISO, PJM, SPP and TVA, Technical Report
17. Ekanayake J, Jenkins N (2004) Comparison of the response of doubly fed and fixed-speed induction generator wind turbines to changes in network frequency. IEEE Trans Energy Convers 19(4):800–802
18. Lalor G, Mullane A, O'Malley M (2005) Frequency control and wind turbine technologies. IEEE Trans Power Syst 20(4):1905–1913
19. Mullane A, O'Malley M (2005) The inertia response of induction machine based wind turbines. IEEE Trans Power Syst 20(4):1496–1503
20. Morren J, Haan S, Kling WL, Ferreira JA (2006) Wind turbines emulating inertia and supporting primary frequency control. IEEE Trans Power Syst 21(1):433–434
21. Keung P-K, Li HBP, Ooi BT (2009) Kinetic energy of wind turbine generators for system frequency support. IEEE Trans Power Syst 24(1):279–287
22. Almeida RG, de Lopes JAP (2007) Participation of doubly fed induction wind generators in system frequency regulation. IEEE Trans Power Syst 22(3):944–950
23. Ullah N, Thiringer T, Karlsson D (2008) Temporary primary frequency control support by variable speed wind turbines—potential and applications. IEEE Trans Power Syst 23(2):601–612
24. Mauricio J, Marano A, Gomez-Exposito A, Ramos JM (2009) Frequency regulation contribution through variable-speed wind energy conversion system. IEEE Trans Power Syst 24(1):173–180
25. Kayikci M, Milanovic J (2009) Dynamic contribution of DFIG-based wind plants to system frequency disturbances. IEEE Trans Power Syst 24(2):859–867
26. Conroy J, Watson R (2008) Frequency response capability of full converter wind turbine generators in comparison to conventional generation. IEEE Trans Power Syst 23(2):649–656
27. Ackermann T (2004) Wind power in power systems. Wiley, Chichester
28. de ARG, Castronuovo ED, Lopes JAP (2006) Optimum generation control in winparks when carrying out system operator requests. IEEE Trans Power Syst 21(2):718–725
29. Chang-Chien L-R, Hung C-M, Yin Y-C (2008) Dynamic reserve allocation for system contingency by DFIG wind farms. IEEE Trans Power Syst 23(2):729–736
30. Mokadem M, Courtecuisse V, Saudemont C, Robyns B, Deuse J (2009) Fuzzy logic supervsor-based primary frequency control experiments of a variable-speed wind generator. IEEE Trans Power Syst 24(1):407–417
31. Fan L, Miao Z, Osborn D (2009) Wind farms with HVDC delivery in load frequency control. IEEE Trans Power Syst 24(4):1894–1895
32. Miao Z, Fan L, Osborn D, Yuvarajan S (2010) Wind farms with HVDC delivery in inertial response and load frequency control. IEEE Trans Energy Convers 25(4):1171–1178
33. Transient security assessment tool (TSAT) user manual (2007) Powertech Labs Inc., Surrey, Canada, Technical Report
34. Kundur P (1994) Power system stability and control. McGraw-Hill Professional, New York
35. Miller N, Price W, Sanchez-Gasca J, (2005) Modeling of GE wind turbine-generators for grid studies. General Electric International, Inc Schenectady, New York, Technical Report
36. Clark K, Miller NW, Sanchez-Gasca JJ, (2009) Modeling of GE wind turbine-generators for grid studies. GE-power systems energy consulting, General Electric International, Inc., Schenectady, NY, Technical Report http://www.gepower.com/prod serv/products/utility

software/en/downloads/09100 Modeling of GE Wind Turbine-Generators for Grid Studies.pdf
37. Wood AJ, Wollenberg BF (1996) Power generation, operation, and control, 2nd edn. Wiley-Interscience, New York
38. Xu L, Yao L, Sasse C (2007) Grid integration of large DFIG-based wind farms using VSC transmission. IEEE Trans Power Syst 22(3):976–984
39. Feltes C, Wrede H, Koch F, Erlich I (2009) Enhanced fault ride-through method for wind farms connected to the grid through VSC-based HVDC transmission. IEEE Trans Power Syst 24(3):1537–1546

Chapter 20
HOTT Power Controller With Bi-Directional Converter (HPB)

Mohammad Lutfur Rahman, Shunsuke Oka and Yasuyuki Shirai

Abstract The innovative renewable energy conversion system called "hybrid offshore-wind and tidal turbine power controller with bi-directional converter (HPB)" was proposed. The system was used to evaluate a tidal turbine (induction machine) with the ability to change the AC frequency reference to a bi-directional converter for optimal performance. The tidal power generation system consists of an induction generator/motor connected to a DC capacitor link through an insulated-gate bi-polar–transistor (IGBT) bi-directional converter. Using PWM control of the bi-directional converter, the speed of the rotating magnetic field (control by the AC voltage frequency) fed to the induction machine can be controlled. The generation power can be controlled quickly by only changing the frequency reference for the PWM converter.

20.1 Introduction

The development of economy, the improvement of living standard, people pay more and more attention to survive the living environment improvement, environmental protection consciousness unceasingly strengthens, need some clean, renewable new energy. Development and utilization of various renewable energy

---

M. L. Rahman (✉) · S. Oka · Y. Shirai
Graduate School of Energy Science, Department of Energy Science and Technology, Kyoto University, 53/4 BUET Staff Quarter, Kyoto 606-8501, Japan
e-mail: lutfur@pe.energy.kyoto-u.ac.jp

S. Oka
e-mail: oka@pe.energy.kyoto-u.ac.jp

Y. Shirai
e-mail: shirai@energy.kyoto-u.ac.jp

S. M. Muyeen (ed.), *Wind Energy Conversion Systems*,
Green Energy and Technology, DOI: 10.1007/978-1-4471-2201-2_20,

represented by offshore-wind become the strategic choice of human existence problem solving. Wind power generation technology is currently the most mature technology and one of the most widely used renewable power generation technology, many countries focus on developing new energy technology.

Electric power generation using renewable energy sources has been sought for economic and environmental reasons, mainly since the 1970s [1–3]. In this regard, great efforts have been carried out to overcome the drawbacks of induction generators by applying power electronic converters and machine control techniques because induction machines best match the requirements of variable-speed generation systems.

In this proposed system, the impact of bi-directional converter AC voltage frequency change on the hybrid power system performance was investigated. The results and conclusions of this analysis will be applied to the future development of an HPB system [4–6].

To achieve this purpose, the HPB generator was connected to the power system through the AC-DC-AC converter, and tidal power can be controlled quickly by just changing the frequency reference for the PWM converter to clarify the ability of the proposed system [7–9].

The reference frequency for the PWM of the bi-directional inverter was manually changed from 60 to 50 Hz and back to 60 Hz. It was assumed that the servo-motor (tidal) was controlled to maintain a stable rotating speed (1,100 rpm).

20.2 Proposed HPB Model System

20.2.1 Model Setup

Figure 20.1a shows how the hybrid offshore-wind and tidal turbine (HOTT) system will be set up offshore, and how it will function. A wind and tidal turbine experimental model plays an important task in the hybrid turbine modeling, particularly for analyzing the interaction between the tidal and offshore-wind power systems, which are connected on the DC side.

Figure 20.1a–c shows a photo and schematic view of the small laboratory-based hybrid power system model that is designed and fabricated. The system has two types of generation, the tidal motor/generator and the offshore-wind turbine generator. The tidal turbine (induction machine) can act as either a motor or generator, depending on the need. The tidal generator provides smooth output power, whereas the output power of a wind turbine depends on the wind velocity.

Figure 20.1a–d shows how the HOTT system is set up in the laboratory and how it will function. These figures also show the conceptual schematic of the proposed HOTT system connected to the power system, and the detailed circuit configuration. The AC power generated by the wind and tidal turbine generators is converted into DC power. It is converted again into AC power through the maximum power point tracking (MPPT) inverter.

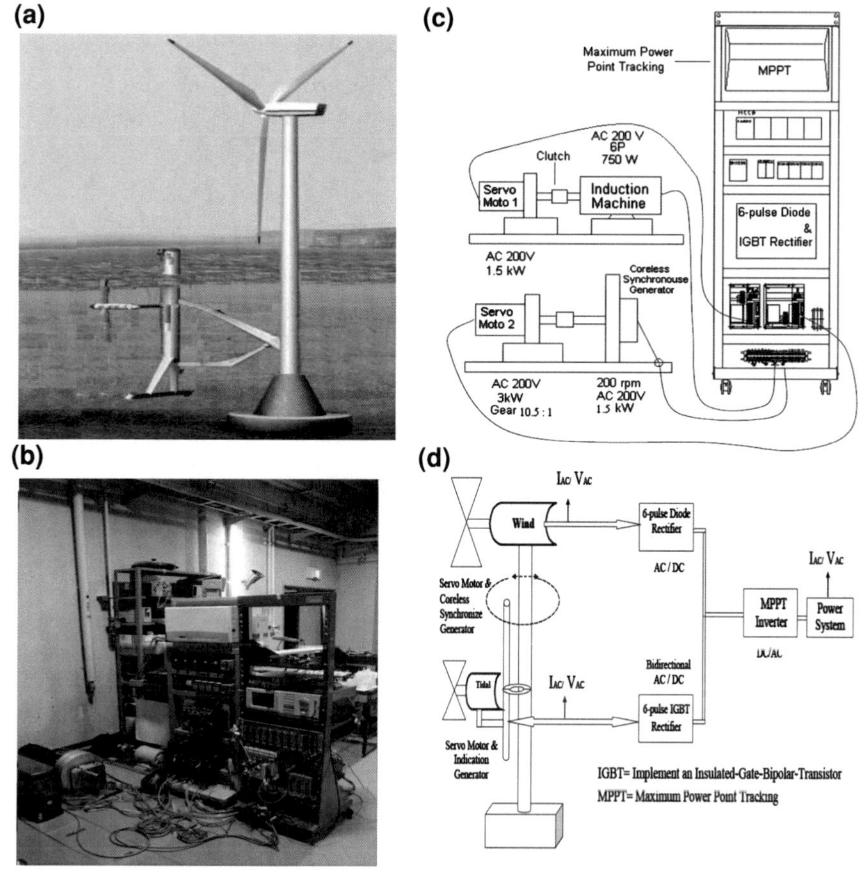

Fig. 20.1 **a** HOTT conceptual image ([9], ©IEEE 2010), **b** Photo of laboratory scale prototype model of hybrid offshore-wind and tidal turbine system with flywheel ([9], © IEEE 2010), **c** Schematic of prototype model of hybrid offshore-wind and tidal turbine system with flywheel ([9], © IEEE 2010), **d** Schematic of prototype model of hybrid offshore-wind and tidal turbine system with flywheel

20.2.2 *Offshore-Wind Turbine*

Figure 20.2 shows an experimental model of the offshore-wind turbine generator system. It consists of a coreless synchronous generator and a servo-motor. The offshore-wind turbine is simulated by the servo-motor. In this model system with the small servo-motor, the rated rotating speed is 2,500 rpm and the gear ratio is 10.5:1. In the real system, the wind turbine would have a slower rotating speed without the step-down gear. The rotating speed or the torque of the servo-motor is controlled by a computer. The electrical energy depends on the rotations per

Fig. 20.2 Offshore-wind turbine generator experimental model ([9], © IEEE 2010)

Table 20.1 Rating of main components (offshore-wind servo-motor) ([9], © IEEE 2010)

Parameter	Value
Rated output	3.0 kW
Rated voltage	200 V
Rated frequency	60 Hz
Rated speed	2,500 rpm
Gear ratio	10.5:1

Table 20.2 Rating of main components (coreless synchronous generator) ([9], © IEEE 2010)

Parameter	Value
Rated output	1.5 kW
Rated voltage	200 V
Rated frequency	60 Hz
Rated speed	200 rpm

minute (rpm) of the servo-motor that rotates the coreless generator. Wind turbine generated AC power is converted to DC power with the 6-pulse diode rectifier.

The parameters of the servo-motor and the coreless synchronous generator are listed in Table 20.1 and 20.2, respectively.

20.2.3 Tidal Turbine (Flywheel)

An induction generator produces electrical power when its shaft is rotated faster than the synchronous frequency of the equivalent induction motor. Induction machines are used in tidal system installations because of their ability to produce useful power at various rotor speeds. Induction machines are mechanically and electrically simpler than other generator types.

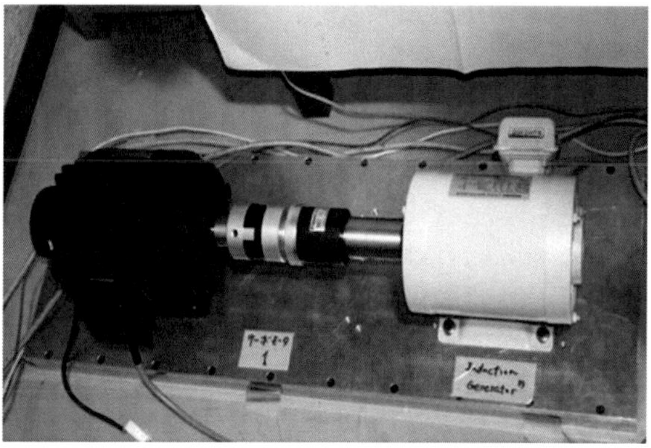

Fig. 20.3 Tidal turbine generator/motor experimental model ([9], ©IEEE 2010)

The energy scenario in the world is calling for efforts toward more efficient use of electrical energy, as well as improvements in the quality of its delivery. This issue involves the use of energy storage devices, such as a tidal turbine used as a flywheel. The demand for and use of such equipment are increasing. One type of energy storage system is a tidal turbine flywheel energy system (TTFES). Due to the advancements in machines and power electronics, the flywheel is becoming more popular. Many feasible projects employing flywheel storage systems have been implemented all over the world [7–11].

Figure 20.3 shows the experimental model of a tidal turbine induction generator/motor and a servo-motor. The main concept in this project is to apply and control a bi-directional (two way) energy flow scheme, so that energy is injected into the offshore-wind turbine or stored as kinetic energy from/to the tidal system (induction machine).

Flywheels are one of the oldest forms of energy storage, having been used for thousands of years. The potter's wheel is one of the earliest applications of a flywheel. The kinetic energy stored in the flywheel results from spinning a disk or cylinder coupled to a machine's rotor. This energy is proportional to the flywheel mass and the square of its rotational speed:

$$E = {}^1\!/_2 I\omega^2 \tag{20.1}$$

where I is the moment of inertia in Kg m^2 and ω is the rotational speed in rad/s. The induction machine (tidal system) works as a motor with almost no load, and the rotational kinetic energy is stored as a function of the square of the rotational speed. The stored energy is extracted by decelerating the induction machine.

TTFES systems, in comparison with conventional batteries, present some interesting characteristics when used as an energy source to compensate for voltage sags and momentary power interruptions. The induction machine is used for bi-directional energy conversion from/to the tidal turbine. The servo-motor is used as an input model of tidal energy to the induction generator, which converts the mechanical energy into electrical energy. The induction machine can work as a motor by using the bi-directional insulated-gate bi-polar–transistor (IGBT)

Table 20.3 Rating of main components (tidal servo-motor) ([9], ©IEEE 2010)

Parameter	Value
Rated output	1.5 kW
Rated voltage	200 V
Rated frequency	60 Hz
Rated speed	2,500 rpm

Table 20.4 Rating of main components (induction machine) ([9], ©IEEE 2010)

Parameter	Value
Rated output	7,500 W
Rated voltage	200 V
Rated frequency	60 Hz
Rated speed	1,110 rpm

converter and a one-way clutch. When the induction machine's rotational speed is larger than that of the servo-motor, the servo-motor clutch turns to the off-state.

The design parameters of the servo-motor and the induction machine are listed in Tables 20.3 and 20.4, respectively. A speed of 1,110 rpm is selected for the induction machine to store the rotational kinetic energy. In a real system, the tidal turbine rotational speed should be much lower than that of the servo-motor, and a step-up gear will be necessary.

In this application, the TTFES supports the offshore-wind turbine, supplying power to the DC load in the case of over loads or dips. This occurs when there is a low wind speed, causing an offshore-wind turbine voltage or frequency dip or overload in the hybrid side. Thus, an offshore-wind system is not capable to supply all the power needed by the DC load. The system gets help from the TTFES, which has stored kinetic energy. Therefore, the main purpose of the flywheel is to accumulate rotational kinetic energy, which can be injected into or extracted from the DC side whenever it is required.

20.2.4 Maximum Power Flow Control

The grid connection inverter is a current control type because the AC voltage is fixed by the grid. Therefore the DC link voltage in the HOTT is kept within a certain range for stable operation by controlling the AC output current of the grid inverter. The AC output current of the grid inverter is controlled so as to give the maximum output power with a certain DC voltage. This control is based on the MPPT algorithm.

In the MPPT control (Fig. 20.4), in order to search the DC link voltage which gives the maximum DC output power, small perturbation, ΔV ($\pm$ 4 V), is given to

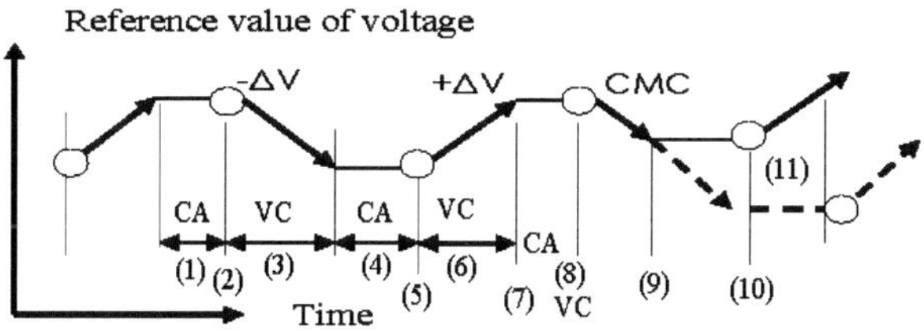

Fig. 20.4 MPPT electric power follow control algorithm

the DC reference voltage and check the DC output change. If the DC power increases, the perturbation is approved to make a new DC reference voltage. Conversely, if it decreases, the reference voltage is changed in the opposite direction. This algorithm makes it possible to find the maximum electric power point when the characteristic of the DC side changes by wind and tide speed changes, etc. [12].

MPPT Algorithm

(1)–(8) A series of processes for MPPT

- (1), (4) and (7): Period for calculation of average value of DC voltage, current and power.
- (3) and (6): Period for DC voltage change (perturbation ΔV (± 4 V)).
- (2) and (5): Check and memorize the deviation of DC output and voltage by corresponding perturbation.
- (8): New DC voltage reference is determined by the results at (2), (5) and (8).
  - When the DC output power increases with DC voltage change +ΔV, the new DC voltage reference moves to (9) and start again from (1).
  - Otherwise, it moves to (10) and restart from (1).

20.2.5 Inverter Circuit Configuration

Figure 20.5 shows the HOTT inverter circuit for this method. The output of the inverter is a single-phase three-wire system. A general domestic power supply is often a single-phase three-wire system. The inverter has a circuit configuration that combines two half-bridges. The input can share one line with the output by using the half-bridge type inverter [13]. In order to meet the grid voltage (200 V), the boost up chopper circuit is adopted in Fig. 20.5 to increase the voltage of the DC side.

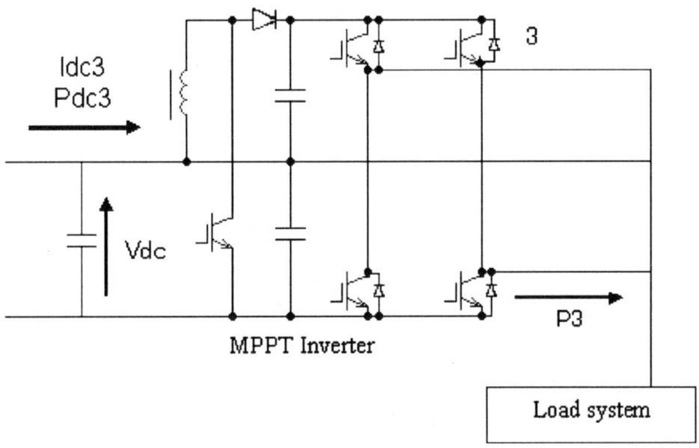

Fig. 20.5 Boost up chopper with a half-bridge inverter circuit

20.3 Hybrid System (Circuit Configuration)

This section describes the system and circuit configuration of the proposed HOTT with TTFES, which was designed and constructed based on the reviews of the alternatives and the components in the successive sections [6–9, 14].

The following is a block diagram of our TTFES with an offshore-wind system (also shown in Fig. 20.6). The offshore-wind coreless synchronous generator output is simply rectified by a 6-pulse diode bridge to charge a DC capacitor. The tidal turbine induction generator/motor output is connected to the DC capacitor through a 6-pulse IGBT dual converter. The DC link capacitor is connected to the commercial grid through a grid-connected, single-phase, 3-wire inverter. The grid-connected inverter is of a transformer-less half-bridge type with a boost up chopper circuit. The voltage-source inverter output current is controlled by a PWM controller under MPPT control. The MPPT control monitors and maintains the DC link capacitor voltage, providing the maximum output power by controlling the output AC current. It monitors the DC voltage perturbations of 4 V up and down (2 V/s) every 4 s, calculates how to change the output power caused by them, and then determines the DC voltage reference at the next stage to give more power. Several small controllers are implemented at both ends to provide the required performance for the system.

20.4 Changing Voltage Frequency 50–46–50 Hz

In order to simplify the description, an autonomous power supply with a limited capacity is substituted for the HOTT system (Fig. 20.7). A flywheel induction machine is connected to the offshore-wind system.

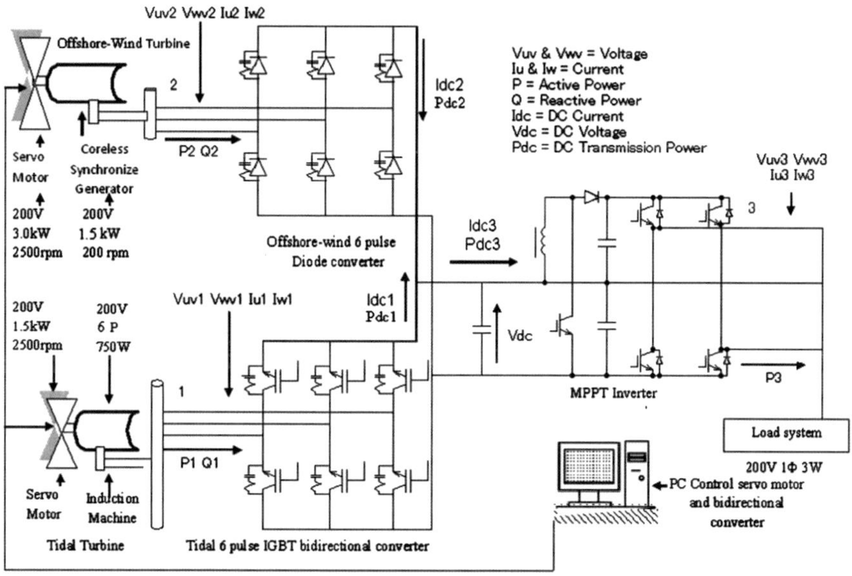

Fig. 20.6 HPB system configuration ([9], ©IEEE 2010)

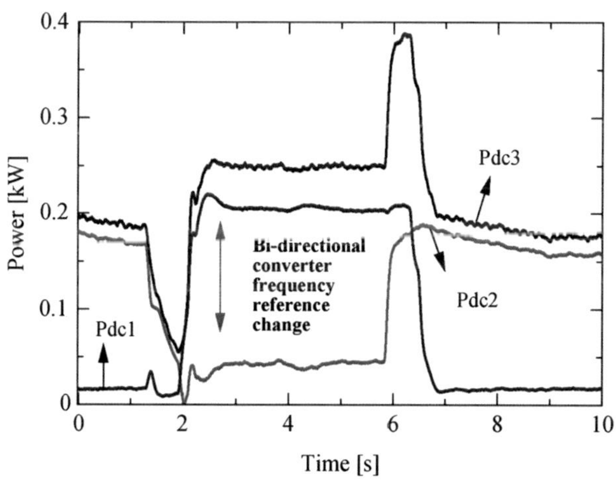

Fig. 20.7 DC power output

In order to compensate for wind power fluctuation, the output of the induction generator was controlled so that the total output was steady. The reference frequency for the PWM inverter signal of the bi-directional inverter was manually changed from 50 to 46 to 50 Hz; we assumed that the servo-motor (tidal) was controlled to maintain a stable rotating speed (1,000 rpm) because tidal flow is more stable than wind flow. This experimental model system was able to keep the

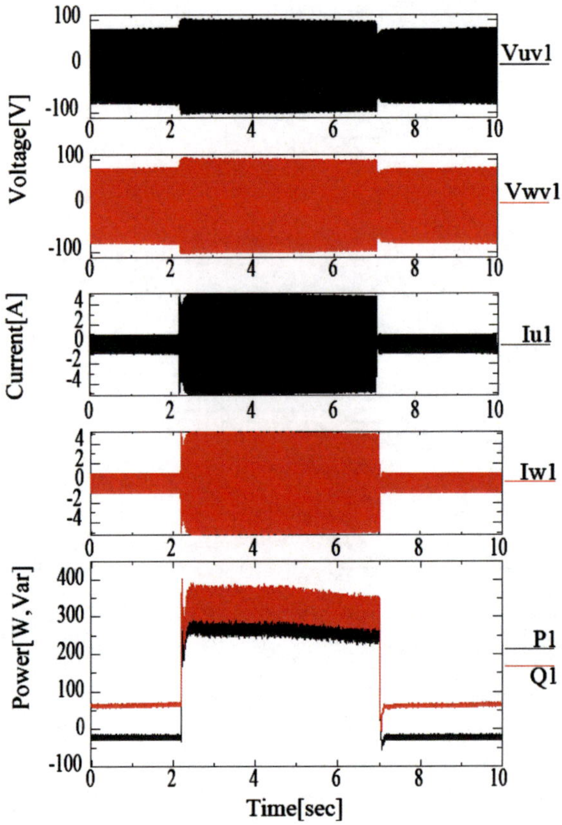

Fig. 20.8 Experimental results for induction machine (tidal turbine) voltages (Vuv1 & Vwv1), currents (Iu1 & Iw1), active power (P1), and reactive power (Q1) when the reference frequency for the PWM inverter signal of the bi-directional inverter was manually changed from 60 to 50 to 60 Hz

total output steady by controlling the induction generator by changing the AC voltage frequency input from the bi-directional converter.

20.4 Experimental Results

Figures 20.8, 20.9, 20.10, 20.11, 20.12 show the experimental results obtained from the system using a coreless generator as an offshore-wind system and an induction machine as a tidal system, the parameters of which are shown in Tables 20.3 and 20.4. A commercially available IGBT-based inverter, the vector modulation technique, was employed as the bi-directional voltage-fed-PWM inverter.

Figures 20.8, 20.9, 20.10, 20.11, 20.12 show the HPB total system experimental output for the tidal turbine (induction machine) and offshore-wind turbine (coreless generator). These figures show the tidal turbine, offshore-wind turbine, and power system (voltage, current, active power, and reactive power), as well as the DC side (voltage, current, power) and rotating speed.

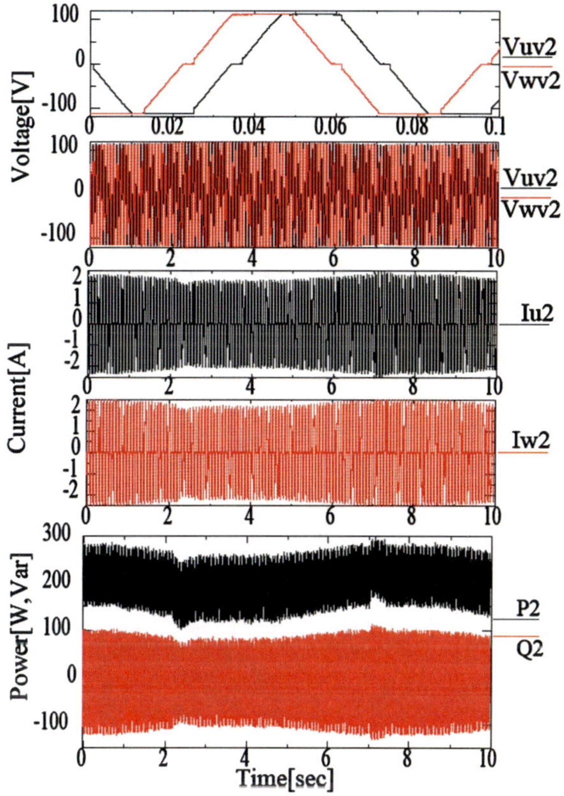

Fig. 20.9 Experimental results for coreless generator (offshore-wind turbine) voltages (Vuv2 & Vwv2), currents (Iu2 & Iw2), active power (P2), and reactive power (Q2)

The HPB bi-directional converter has been designed to be able to vary the AC voltage frequency between 20 and 100 Hz. This ability to change the stator frequency of the induction generator makes it possible to optimize the power output from the wind turbine by adjusting the frequency in relation to the tidal turbine. With HPB, the frequency was varied between 60 and 50 Hz during commissioning.

As shown in the figure for the tidal turbine (induction machine), Pdc1 works in motor mode to extract energy by changing the voltage frequency. When the voltage frequency is 60 Hz, if the tidal turbine rotating speed is less than 1,200 rpm, it works as a motor, and if its rotating speed is more than 1,200 rpm, it works as a generator. When the voltage frequency is 50 Hz, if the tidal turbine rotating speed is less than 1,000 rpm, it works as a motor, and if its rotating speed is higher than 1,000 rpm, it works as a generator.

The reference frequency for the PWM signal of the bi-directional converter was manually changed from 60 to 50 Hz at 2.2 s, and then back to 60 Hz again at 7 s. In this test, we controlled the tidal servo-motor to maintain a stable rotating speed because the tidal flow is more stable than the wind flow. The tidal turbine will help to balance the distribution of power in the power system. The offshore-wind and tidal turbines were DC side connected, and the total output load power was 0.4 kW.

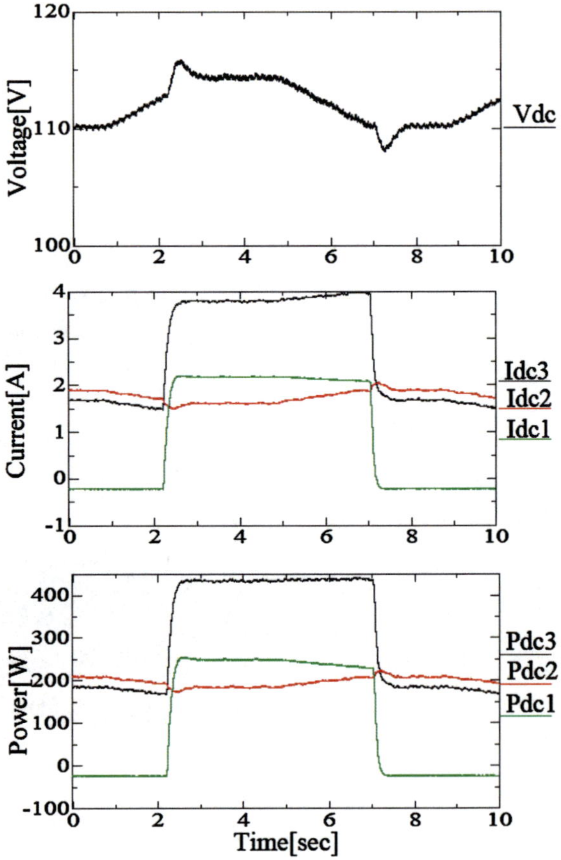

Fig. 20.10 Experimental results for DC side hybrid voltage (Vdc), currents (Idc1, Idc2, & Idc3), and powers (Pdc1, Pdc2, & Pdc3)

As shown in Fig. 20.10 for the DC side powers, the offshore-wind generated power, Pdc2, was almost a constant 190 W, while the tidal generated power, Pdc1, stepped up from −25 W (motor mode) to 250 W (at 2.2 s), and then stepped back down to −25 W (at 7 s). The total DC power (Pdc3) was 190 W, 60 Hz during the motor mode (0.0–2.2 s) of the induction machine, and stepped up to 420 W, 50 Hz (2.2–7.0 s), and 190 W (7.2–10 s).

As shown in Fig. 20.10 for the DC link voltage, Vdc, the MPPT controller gave DC voltage perturbations of 4 V up and down (2 V/s) every 4 s. The grid-connected inverter with MPPT control worked according to the generation power changes at 2.2 and 7.0 s, without any large disturbance in the DC link voltage.

Figure 20.11 shows the rotating speeds of the induction machine (tidal), servo-motor (tidal), and coreless generator (offshore-wind) of the HPB system. The rotation speed of the servo-motor (coreless) was constant, and the coreless generator output was also constant. Constant speed control was applied to the

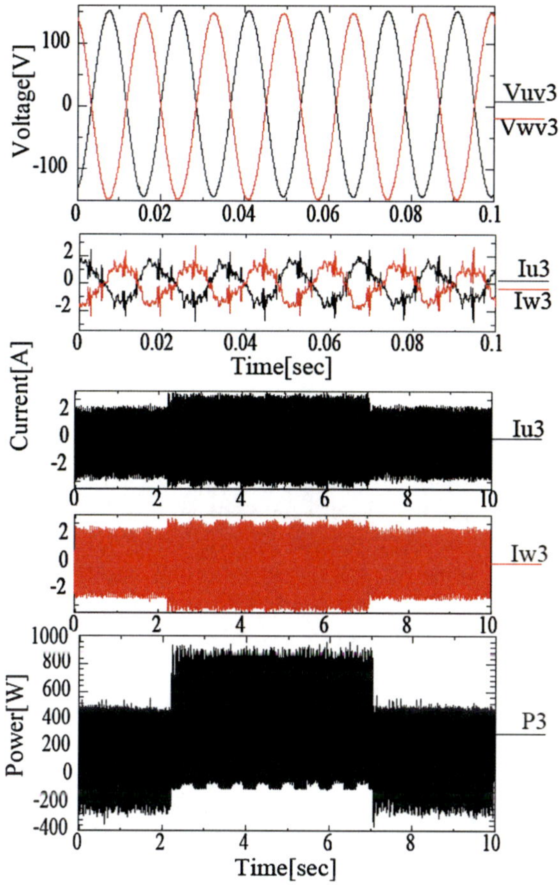

Fig. 20.11 Experimental results for AC load side voltages (Vuv3 & Vwv3), currents (Iu3 & Iw3), and power (P3)

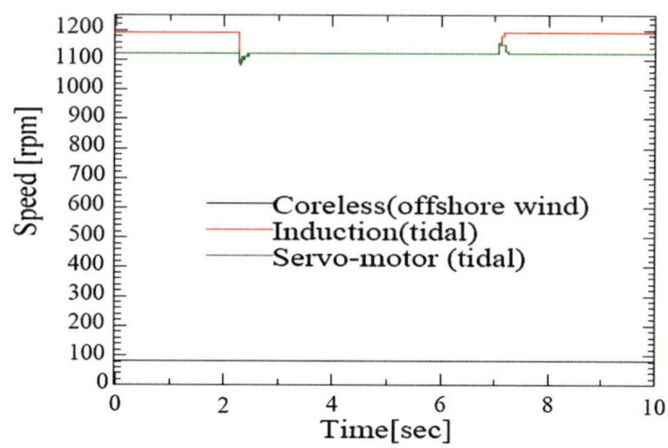

Fig. 20.12 Experimental results for rotating speeds of coreless generator (offshore-wind generator), induction machine (tidal generator), and servo-motor (for tidal generator)

servo-motor (tidal) in the HPB system. Only the transient response to the frequency change of the terminal voltage (60–50 Hz and back to 60 Hz) by the PWM control of the bi-directional inverter was observed for the rotating speed.

20.5 Discussion

The induction machine is controlled by a bi-directional converter. The converter uses IGBTs to implement the inverter bridge that provides a PWM motor/generator control waveform to the machine. The inverter's 60/50/60 Hz voltage frequency was controlled by a manual switch change.

The system's microcontroller interfaces closely with the inverter board to apply the correct waveforms to the machine during all operation modes. At the pre-specified speed of 1,190 rpm (motor mode), the system switches from the motor to generator mode. This change is coordinated by the system controller. The overall system is managed by a micro-controller. The microcontroller's main function is to control the gate drives of the IGBTs in the inverter during the motoring and generating operations. A simple user interface is provided to turn the system on and off. The micro-controller and associated circuitry are controlled via a software algorithm loaded onto the micro-controller using a PC interface.

The proto-type HPB system successfully demonstrated. The hybrid output to the grid was able to be controlled by just changing the frequency of the bi-directional inverter voltage while the rotating speeds of both turbines were constant. The tidal system was smoothly switched from/to generator and motor modes by changing inverter voltage frequency.

20.6 Conclusion

The experimental prototype model of the proposed HOTT was designed and fabricated to demonstrate its operation using a real generator/motor and power semiconductor converters. Laboratory scale basic experiments were carried out to show the feasibility of the system, and to propose new control strategies using the bi-directional converter and MPPT grid connection converter.

The method was a HPB. The tidal power generation system consists of an induction generator/motor connected to the DC capacitor link through an IGBT bi-directional converter. Using PWM control of the bi-directional converter, the speed of the rotating magnetic field (control by the AC voltage frequency) fed to the induction machine can be controlled. The generated power can be controlled quickly by just changing the frequency reference for the PWM converter. A fundamental test was carried out to clarify the ability of the proposed system. The bi-directional converter frequency was changed manually (60–50–60 Hz), while the rotational speed of the tidal turbine (servo-motor) was kept at a constant 1,100 rpm (55 Hz). The tidal generation power was changed quickly and smoothly even when switching from motor mode to generator mode. This control had little influence on the operation of the wind power generator. It was confirmed that the voltage frequency control of the tidal induction generator had a good performance to compensate for the power fluctuation of the total system. The tidal induction generator/motor system with the bi-directional converter and one-way clutch had the ability to compensate for the power fluctuation caused by wind conditions. The induction machine could switch smoothly between the generator and motor modes by controlling the rotation speed of the tidal rotor and the inverter voltage frequency. Additionally, in the motor mode of the induction machine, its rotating kinetic energy could be used for power conditioning as a flywheel system. The design of the bi-directional converter control for power control, taking the one-way clutch function into consideration, is an important issue for the next research step.

References

1. Muetze A, Vining JG (2006) Ocean wave energy conversion—a survey. Industry applications conference, 2006. 41st IAS annual meeting. Conference record of the 2006 IEEE, vol 3. pp 1410–1417
2. Thorpe TW, Picken MJ (1993) Wave energy devices and the marine environment. Sci Meas Technol IEE Proc A 140(1):63–70
3. Scottish supply firms, marine renewable (wave and tidal) opportunity review. http://www.oreg.ca/docs/marine_renewable_opportunity_review.pdf [online]. Accessed 2007
4. Akagi H, Sato H (2002) Control and performance of a doubly-fed induction machine intended for a flywheel energy storage system. IEEE Trans Power Electron 17(1):109–116
5. Wave energy conversion, University of Michigan College of Engineering. http://www.engin.umich.edu/dept/name/research/projects/wave_device/wave_device.html [online]. Accessed 2009
6. Rahman ML, Shirai Y (2008) Hybrid offshore-wind and tidal turbine (HOTT) energy Conversion I (6-pulse GTO rectifier and inverter). In: IEEE international conference on sustainable energy technologies, ICSET 2008. pp 650–655
7. Rahman ML, Shirai Y (2010) DC connected hybrid offshore-wind and tidal turbine (HOTT) generation system. Green energy and technology, Springer, Academic Journals, pp 141–150
8. Rahman ML, Shirai Y (2009) Hybrid power system using offshore-wind turbine and tidal turbine with flywheel (OTTF). Europe's offshore wind 2009, VIND2009 (eow2009, Stockholm)
9. Rahman ML, Oka S, Shirai Y (2010) Hybrid power generation system using offshore-wind turbine and tidal turbine for power fluctuation compensation (HOT-PC). IEEE Trans Sustain Energy 1(2):92–98
10. Silva Neto J, Rolim G (2003) Control of a power circuit interface of a flywheel-based energy storage system. UFRJ, Cidade Universitiria, Rio de Janeiro
11. Kim WH, Kim JS, Baek JW, Ryoo HJ, Rim GH (1998) Improving efficiency of flywheel energy storage system with a new system configuration. In: IEEE power electronics specialists conference, Fukuoka
12. Omron Industrial Automation, Japan, model number KP40H
13. Chapmann S (2005) Electric machinery fundamentals, 4th edn. McGraw Hill, New York
14. Rahman ML, Oka S, Shirai Y (2011) Hybrid offshore-wind and tidal turbine power system for complement the fluctuation (HOTCF). Green energy and technology, Springer, Academic Journals, pp 177–186

Chapter 21
Transmission of Bulk Power from DC-Based Offshore Wind Farm to Grid Through HVDC System

S. M. Muyeen, Ahmed Al-Durra and J. Tamura

Abstract Trends in growth of the wind energy is getting additional pace by offshore technology. This chapter investigates a suitable control strategy for a DC-based offshore wind farm to transmit bulk power to an onshore grid through a high voltage DC (HVDC) transmission line. The offshore wind farm is composed of variable-speed wind turbines driving permanent magnet synchronous generators (PMSG). Each PMSG is connected to the DC bus through a generator-side converter unit to ensure maximum power point tracking control. The DC voltage of the DC-bus is stepped up using a full-bridge DC–DC converter at the offshore HVDC station, and the wind farm output power is transmitted through the HVDC cable. The onshore HVDC station converts the DC voltage to a suitable AC grid voltage. Detailed modeling and control strategies of the overall system are presented. Real wind speed data is used in the simulation study to obtain a realistic response. The effectiveness of the coordinated control strategy developed for the proposed system is verified by simulation analyses using PSCAD/EMTDC, which is the standard power system software package.

S. M. Muyeen (✉) · A. Al-Durra
Department of Electrical Engineering, The Petroleum Institute,
P.O. Box 2533, Abu Dhabi, UAE
e-mail: smmuyeen@pi.ac.ae

A. Al-Durra
e-mail: aaldurra@pi.ac.ae

J. Tamura
Department of Electrical and Electronic Engineering,
Kitami Institute of Technology, Kitami 090-8500, Japan
e-mail: tamuraj@mail.kitami-it.ac.jp

S. M. Muyeen (ed.), *Wind Energy Conversion Systems*,
Green Energy and Technology, DOI: 10.1007/978-1-4471-2201-2_21,

21.1 Introduction

Recently, the generation of electricity using wind power has received a great deal of interest. As per global wind energy council (GWEC) statics the global wind capacity will reach close to 450 GW by the end of 2015 [1]. It means that more than 50% of the present installation until 2010 is going to be added to the power grid. Development of offshore technology has significant contribution in this rapid expansion of wind power. A number of leading countries in the wind energy arena are focusing increasingly on offshore technology. Offshore turbines can generate electricity for 70–90% of their total operating time. By 2007, the industry had developed 25 projects with a total capacity of approximately 1,100 MW in five countries in Europe [2], many of which are large-scale, fully commercial projects. The United Kingdom and Denmark are leading offshore trend and already successfully commissioned hundred megawatt class wind farms in the sea [3]. The EWEA forecasts that between 1,000 and 1,500 MW of new offshore wind capacity will be fully grid connected in Europe during 2011; 10 wind farms, totaling 3,000 MW, are currently under construction. When completed, Europe's installed offshore capacity will increase to 6,200 and 19,000 MW are currently fully consented. [4]. Many other countries are also focusing on installing hundred of megawatt class offshore wind farms.

Offshore wind farms are generally located a few tens kilometers or further from the onshore grid connection point. Transmitting hundreds of MW of offshore wind power over such a long distance is a great challenge for both the wind farm developer and the transmission system operator (TSO). This chapter focuses on a high-voltage DC (HVDC) system for offshore wind farm interconnection with the onshore grid. Two types of HVDC transmission topologies, i.e., HVDC with a voltage source converter using IGBTs (VSC-HVDC) and line-commutated converter HVDC (LCC-HVDC), have been proposed for offshore wind farm connectivity [5–14]. In the present study, the VSC-based HVDC transmission system is considered for offshore wind farm connection with the grid.

In previous studies, both fixed-speed and variable-speed wind generators have been considered in the offshore wind farm. The induction generator is, in general, used as a fixed-speed wind generator. On the other hand, doubly-fed induction generators (DFIGs), conventional multi-pole synchronous generator (SGs), and permanent magnet synchronous generators (PMSGs) are commercially available as variable-speed wind generators. Recently, the PMSG has become very popular as a wind generator. In the PMSG, the excitation is provided by permanent magnets instead of field windings. Permanent magnet machines are characterized as having large air gaps, which reduce flux linkage even in machines with multi-magnetic poles [15,16]. As a result, low-rotational-speed generators can be manufactured with relatively small sizes with respect to their power rating. Moreover, the gearbox can be omitted due to the low rotational speed in the PMSG wind generation system, resulting in low cost. In a recent survey, the gearbox was found to be the most critical component, because its downtime per failure is high in comparison to other components in a wind turbine generator system (WTGS) [17]. Therefore, in the present study, the PMSG is considered as the wind generator used in the offshore wind farm.

VSC-based HVDC technology has reached a mature stage [7–14, 18–23]. Recent data indicates that, along with this technology, offshore wind farms composed of variable-speed wind generators will secure a major share of power generation in the wind power industry [2–4]. In general, the variable-speed wind generator is connected to the AC-bus using a partial or full rating frequency converter followed by an AC transformer, depending on the type of generator used. For HVDC interconnection, an AC-bus-based offshore wind farm requires the use of an AC transformer in order to step up the voltage at the offshore HVDC station to a desired level, as reported in earlier studies [7–14]. On the other hand, the DC-based wind farm is another interesting, cost-effective, and relatively new technology in wind power industry [24–26]. The DC–DC converter is one of the key components in the DC-based wind farm. The fundamental concepts and efficiency analysis of a DC-based wind farm along with the DC–DC converter have been reported [25, 26]. However, a coordinated control scheme, e.g., maximum power point tracking control of an individual WTGS unit of a DC-based wind farm, or the DC voltage control strategy of the DC–DC converter, was not considered. The present study attempts to demonstrate another alluring opportunity for the DC–DC converter to step up the voltage to the level of the HVDC system that interconnects a DC-based wind farm. A coordinated control scheme is developed for the DC-based offshore wind farm connected with the HVDC system, emphasizing the control strategy for the key component, namely, the full-bridge (FB) DC–DC converter. The proposed system is simulated using PSCAD/EMTDC, the laboratory standard power system simulator package, to demonstrate the operation and control of the HVDC system connected to the DC-based offshore wind farm. The simulation analysis of this intricate system is necessary for the present and future development of offshore wind technology, because the goal is to install a capacity of more than 100 GW by 2030, as mentioned earlier.

In the proposed scheme, the generator side converter ensures the maximum power transfer from the individual VSWT-PMSG to the DC-bus used in the offshore DC-based wind farm. The control for the full-bridge DC–DC converter must adapt to the nonlinear DC-bus voltage dynamics. The fuzzy logic controller (FLC) is considered to control the duty cycle of the FB DC–DC converter, where the phase shift technique is also adopted to obtain zero voltage switching (ZVS) of the IGBT switches. The FB DC–DC converter connects the offshore wind farm with the HVDC system and boosts the DC-bus voltage to the desired level of the HVDC system. Real wind speed data measured in Hokkaido Island, Japan, are used in the dynamic analysis to obtain realistic responses. Moreover, transient analysis is performed considering a severe three-line-to-ground fault in order to observe the effectiveness of the proposed control strategies during a network disturbance. The overvoltage protection scheme at the offshore HVDC station considered herein is an effective means by which to control the DC voltage of the HVDC system without any dependency on the individual WTGS used in the offshore wind farm. Finally, the proposed FLC-based full-bridge DC–DC converter is shown to be an effective means by which to constitute the HVDC system that connects a DC-based wind farm.

21.2 System Overview

A line diagram of the proposed system composed of an offshore wind farm, two HVDC stations, and a DC cross-linked polyethylene (XLPE) cable is shown in Fig. 21.1. The system is connected to the infinite bus through an AC transmission line. In the DC-based offshore wind farm, each PMSG is connected to the DC-bus through an AC–DC converter. The offshore wind farm power capacity is considered to be 150 MW. In order to speed up the simulation, an aggregate model of the wind farm, in which multiple wind generators can be represented with a single large wind generator, is considered. Therefore, in the simulation, two PMSGs each rated at 75 MW are considered. The parameters of the PMSG are shown in Table 21.1. The cable length is considered to be 100 km. The HVDC cable parameters are available in [14]. The system has a 150 MVA base.

The offshore HVDC station is a DC–DC converter and one of the key components of the proposed system. The choice of the converter type used in the offshore HVDC station is important from a practical viewpoint. The DC–DC converter with a transformer can effectively handle the issue of higher voltage ratio, which is well suited for HVDC interconnection of the DC-based offshore wind farm. For low ratios (i.e., 1:1–1:3) between the input and output voltages, a boost converter has certain advantages, such as high efficiency and low component quantity. However, for higher voltage ratios transformer-less converters become less applicable.

In the AC-based offshore HVDC station, an AC transformer is used immediately before the VSC station. On the other hand, in the DC-based wind farm a DC–DC converter using a high-frequency transformer connects the offshore wind farm to the HVDC cable. The power electronic components used are more or less the same in both systems. The main difference in investment costs is the cost of the transformer. The transformer in the DC–DC converter operates at a higher switching frequency. Since the transformer size and weight vary inversely with the frequency, the higher switching frequency leads to a dramatic reduction in the transformer size considering the core material, windings, and insulation. The energy production cost for the DC-based wind farm is also found to be lower than that for the AC based wind farm, as reported in [24]. Based on these considerations, a full-bridge (FB) DC–DC converter, as shown in Fig. 21.2, is suitable for use in the offshore HVDC station. In the present study, a standard DC–AC voltage source converter (VSC) system is considered for use in the onshore HVDC station.

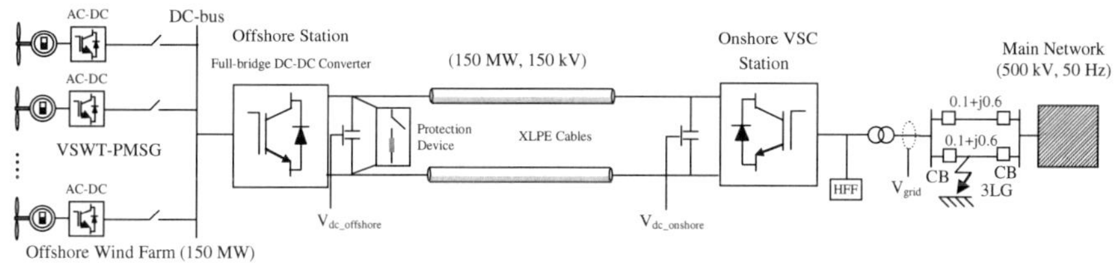

Fig. 21.1 Single line diagram of the proposed system

Table 21.1 Parameters of the PMSG

Rated power	75 (MW)	Stator resistance	0.01 (pu)
Rated voltage	6.0 (kV)	*d*-axis reactance	1.0 (pu)
Frequency	20 (Hz)	*q*-axis reactance	0.7 (pu)
Number of poles	150	Field flux	1.4 (pu)
H	3.0 (s)		

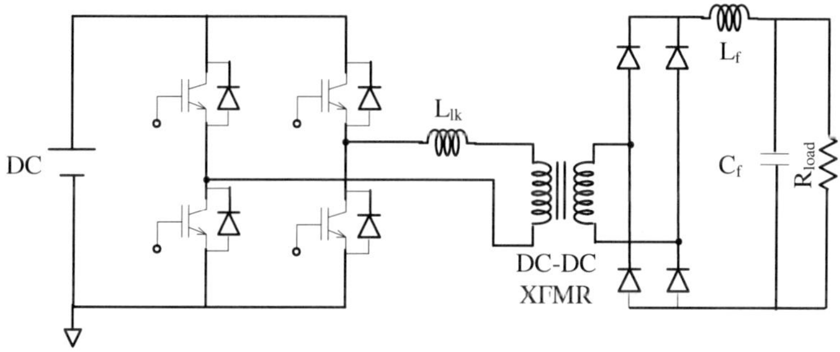

Fig. 21.2 Conventional full-bridge (FB) PWM DC–DC converter

21.3 Modeling and Control of the Individual Components of the Proposed System

21.3.1 Wind Turbine

The mathematical relation for mechanical power extraction from the wind can be expressed as follows [27]:

$$P_M = 0.5\rho C_p(\lambda, \beta)\pi R^2 V_W^3 [W] \tag{21.1}$$

where P_M is the power extracted from the wind, ρ is the air density (kg/m^3), R is the blade radius (m), V_w is the wind speed (m/s), and C_p is the power coefficient, which is a function of both tip speed ratio, λ, and blade pitch angle, β (deg). In addition, C_p is obtained as follows [28].

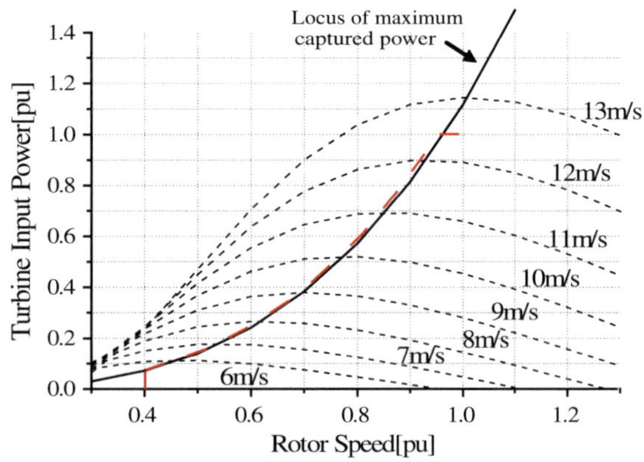

Fig. 21.3 Wind turbine characteristics for variable-speed operation ([29], © [2010] IEEE)

$$C_p(\lambda, \beta) = 0.5\left(\Gamma - 0.02\beta^2 - 5.6\right)e^{-0.17\Gamma} \tag{21.2}$$

$$\lambda = \frac{\omega_m R}{V_W}, \qquad \Gamma = \frac{R}{\lambda} \cdot \frac{3600}{1609} \tag{21.3}$$

The wind turbine parameters are given in the Appendix.

The C_p characteristic changes depending on the wind speed. Equation 21.4 shows the relationship between the optimal rotational speed, $\omega_{\mathrm{m\_opt}}$, corresponding to the optimum power coefficient $C_{\mathrm{p\_opt}}$, and the wind speed, V_W, where ω_m is the rotational speed (rad/s). This expression is obtained by differentiating C_p with respect to ω_m, assuming that β is equal to zero. Figure 21.3 shows the relationship between the wind turbine power P_M and the rotor speed ω_m for various wind speeds, in which the locus of the maximum output power $P_{\max}$ is indicated by a dashed line. Here, $P_{\max}$ is chosen as the reference power, P_{ref}, for the AC–DC converter used in each VSWT-PMSG, so that the maximum power can be transferred to the common DC-bus. Figure 21.4 shows the maximum power point tracking (MPPT) controller to obtain P_{ref} [29]. This controller is designed based on linear functions expressed in Eq. 21.5A. At the same time, the wind turbine should be operated at $\omega_{\mathrm{m\_opt}}$ as expressed in Eq. 21.4 in order to achieve the power capture corresponding to P_{ref}. In addition, the rotor speed ω_m does not fall below 0.4 pu, which is considered to be the minimum rotor speed in the present study, when the wind speed is low.

$$\omega_{\mathrm{m\_opt}} = 0.0775 V_w \text{ (pu)} \tag{21.4}$$

$$P_{\mathrm{ref\_1}} = 0.197 V_w - 1.451 \text{ (pu)} \tag{21.5A}$$

$$P_{\mathrm{ref\_2}} = 0.127 V_w - 0.749 \text{ (pu)} \tag{21.5B}$$

$$P_{\mathrm{ref\_3}} = 0.068 V_w - 0.282 \text{ (pu)} \tag{21.5C}$$

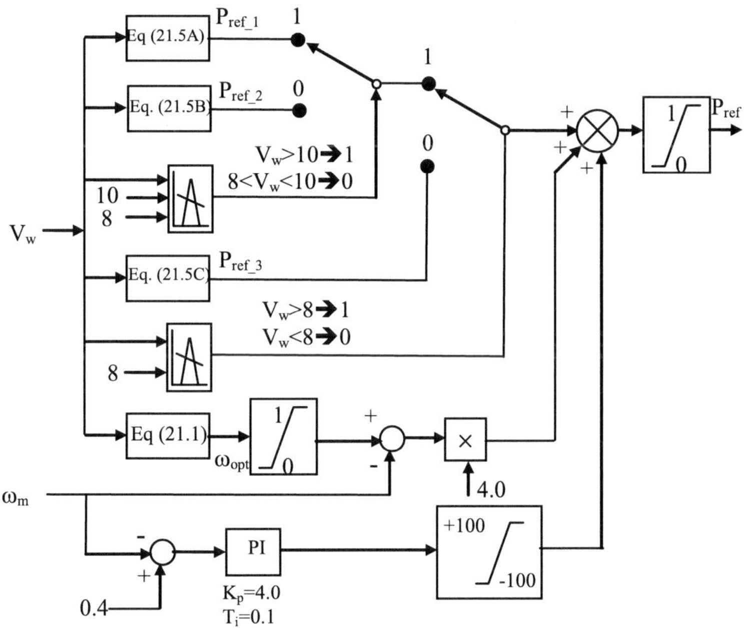

Fig. 21.4 Block diagram of the MPPT controller ([29], © [2010] IEEE)

When the rotor speed ω_r exceeds the rated speed of the PMSG, the pitch controller presented in [30] is considered to control the rotational speed. Therefore, the output power will not exceed the rated power of the PMSG.

21.3.2 DC-Based Wind Farm

In the present study, it is considered that the wind farm is composed of VSWT-driven PMSGs. Each direct drive PMSG is connected to the DC-bus through a fully controlled AC–DC converter. In the simulation analysis, the PMSG model available in the package software PSCAD/EMTDC [31] is used. The electrical scheme of the VSWT-PMSG adopted in the present study is shown in Fig. 21.5. Each converter is a standard three-phase two-level unit that is composed of six IGBTs and anti-parallel diodes. In the present study, the cascaded control technique shown in Fig. 21.6 is adopted for VSWT-PMSG operation. Since the converter is directly connected to the PMSG, its q-axis current is proportional to the active power. The active power reference, P_{ref}, is determined so as to provide the maximum power to the DC-bus, as explained in Sect. 21.3.1. On the other hand, the d-axis stator current is proportional to the reactive power. The reactive power reference is set to zero in order to perform unity power factor operation.

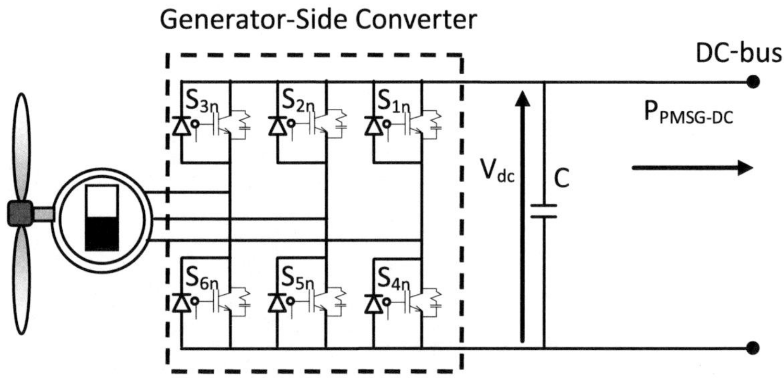

Fig. 21.5 Electrical scheme of the VSWT-PMSG

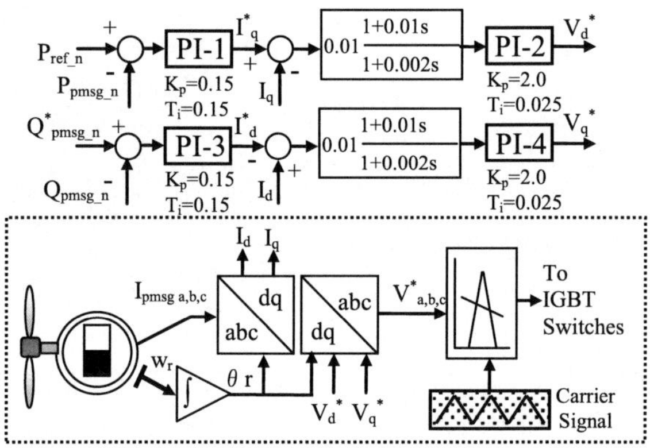

Fig. 21.6 Control block for the generator-side AC–DC converter

The angle, θ_r, for the transformation between the abc and dq variables is calculated from the rotor speed of the PMSG. In Fig. 21.6, subscript n represents the individual number of the wind generator. The nominal speed is considered to be the maximum rotor speed, ω_{r_max}. The pitch controller is activated when the rotor speed exceeds the maximum rotor speed of the PMSG.

21.3.3 Full-Bridge DC–DC Converter of the Offshore HVDC Station

In the present study, as stated earlier, a full-bridge DC–DC converter is used in the offshore HVDC station. In a full-bridge converter, the input bridge creates a high-frequency square wave at the transformer, which transforms the voltage to a higher level. The high voltage square wave is rectified using the diode bridge circuit at the secondary side of the transformer. The ripple is minimized using an output filter

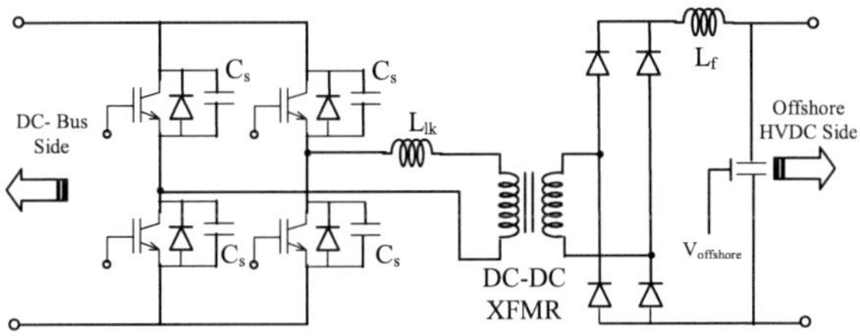

Fig. 21.7 Full-bridge (FB) PWM DC–DC converter with phase shift control

[25]. Since the output filter is current-stiff, the current in the switches and the transformer is approximately constant during the on period.

The output voltage of the FB DC–DC converter can be controlled in different ways. In all switching converters, the output voltage is a function of the input line voltage, the duty cycle, and the load current, as well as the converter circuit element values. In DC–DC converter application, it is desirable to obtain a constant output voltage despite the disturbances in the input voltage and load current and variations in the converter circuit element values. A simple pulse width modulation (PWM) based duty cycle control can be used for the control of the FB DC–DC converter. The duty cycle control provides a mean voltage across the transformer that is equal to the output voltage divided by the transformer ratio. This is achieved by the on-state, where the switches in two legs are on, applying the input voltage across the transformer. In the off-state, all switches are off and the load current free-wheels through the diodes in the output bridge, which yields zero voltage across the transformer [25, 32, 33]. However, the switching loss is higher because no snubber circuits are considered across the switches and switches are turned off at full load voltage. This issue can be handled using phase shift PWM control.

21.3.3.1 Phase Shift PWM Control of the FB DC–DC Converter

Few types of soft-switching have been considered for the reduction of higher switching loss for FB topology. A common practice is to use the phase-shifted pulse-width modulation (PWM) technique [32, 33], which requires snubber capacitors connected across the switches, as shown in Fig. 21.7. Figure 21.7 also represents a topology that is suitable for use in the offshore HVDC station, as described in Fig. 21.3. The circuit parameters shown in Fig. 21.7 are given in the Appendix. The switching scheme for the FB converter with phase shift control is as described above. The on-state with the input voltage across the transformer is achieved by turning on two switches. The difference lies in the switching and the off-state, which is achieved by turning one switch off and letting the load current flow through one switch and one diode. With this control method, the switching losses can be lowered by the capacitors across the switches, which are charged after turn-off of a device, enabling zero voltage turn-on switching (ZVS) [25, 32, 33].

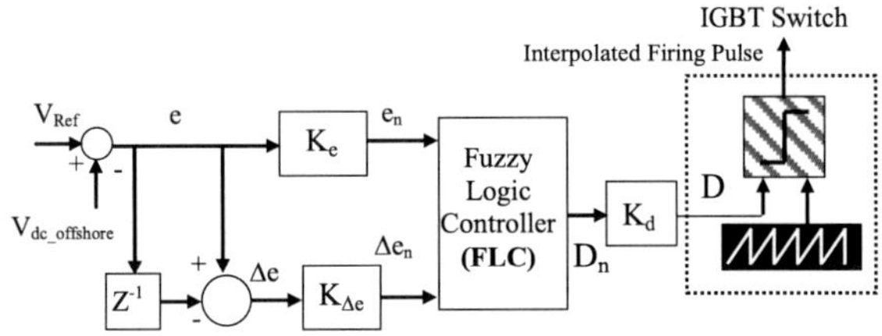

Fig. 21.8 Control block of the DC–DC buck/boost converter

In the present study, the simple voltage model control is adopted to control the output voltage of the FB phase-shifted DC–DC converter, as shown in Fig. 21.8. In the proposed control scheme, the output voltage is compared to a reference voltage to generate an error signal. The error signal is then processed through the FLC to obtain the duty cycle, which is further explained in the following section. The duty cycle is then compared with the carrier wave and its phase-shifted signal based on the PWM technique in order to generate interpolated firing pulses for IGBT switches. The interpolated firing pulse circuit is a simulation technique concerned with generating firing pulses through an interpolation procedure. This allows for exact switching between time steps based on a comparison between the sinusoidal reference and the high-frequency carrier signal, which is considered in the simulation.

21.3.3.2 Fuzzy Logic Controller

Classical control theory suffers from some limitations due to the assumptions made in designing the control systems, such as linearity and time-invariance. Essentially, the conventional proportional–integral (PI) and proportional–integral–derivative (PID) controllers have been used in a number of control applications due to the robustness of these controllers and offer a wide stability margin. However, conventional PI and PID controllers are very sensitive to parameter variations and nonlinearity of dynamic systems. The setting of the parameters of the PI controller used in a large system is cumbersome, especially in power system application, which is difficult to express by a mathematical model or transfer function. These problems can be overcome by using artificial-intelligence-based control techniques. Such control systems can also be less sensitive to parameter variation than classical control systems [34].

Fuzzy logic, in general, is an innovative technology that enhances conventional system design with engineering expertise. The use of fuzzy logic can help to circumvent the need for rigorous mathematical modeling. Unlike the reasoning based on classical logic, fuzzy reasoning aims at the modeling of reasoning schemes based on uncertain, tolerant, or imprecise information. In addition, since fuzzy reasoning has inherent abilities to deal with imprecise or noisy data, fuzzy reasoning is able to extend its control capability even to operating conditions under which linear control techniques fail (i.e., large parameter variations). Therefore, in

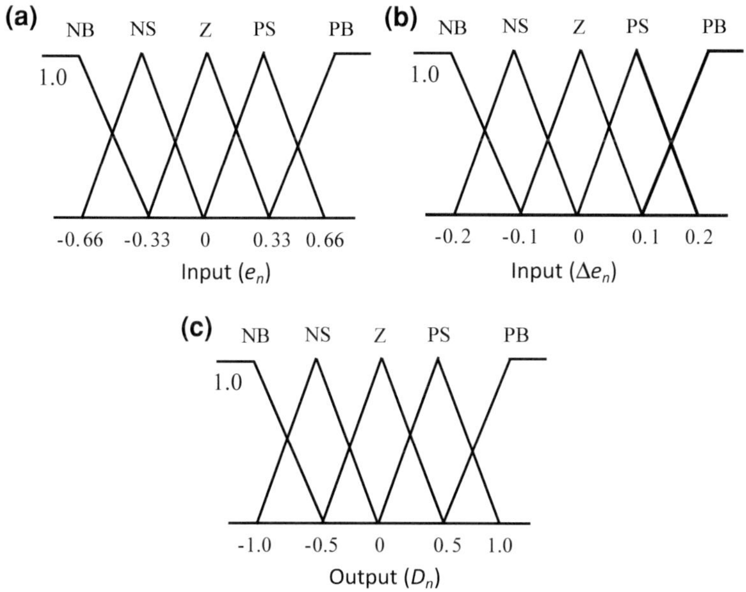

Fig. 21.9 Fuzzy sets and their corresponding membership functions

the present study, the FLC is considered to control the output voltage of the FB DC–DC converter to maintain the DC voltage of the offshore HVDC station at the desired level. The FLC is described in detail in the following section.

In order to design the proposed FLCs, the error signal, $e(k)$, and the change of the error signal, $\Delta e(k)$, are considered as the controller inputs. The duty cycle (D) is chosen as the controller output, which is compared with sawtooth carrier wave signals to generate the switching pulses for IGBT devices.

For convenience, the inputs and outputs of the FLC are scaled with coefficients K_e, $K_{\Delta e}$, and K_d, as shown in Fig. 21.8. In Fig. 21.8, Z^{-1} represents one sampling time delay. The triangular membership functions with overlap used for the input and output fuzzy sets are shown in Fig. 21.9, in which the linguistic variables are represented as NB (Negative Big), NS (Negative Small), Z (Zero), PS (Positive Small), and PB (Positive Big). The fuzzy rule table is shown in the Appendix. In the present study, for the inference mechanism, Mamdani's max–min (or sum–product) [35] method is used. The center of gravity method [35] is used for defuzzification to obtain D_n. The actual duty cycle signal, D, can be determined by multiplying D_n by the scaling factor K_d.

21.3.3.3 Overvoltage Protection Scheme

When a network disturbance occurs at the onshore grid, real power cannot be transmitted from the offshore wind farm to the onshore grid through the HVDC

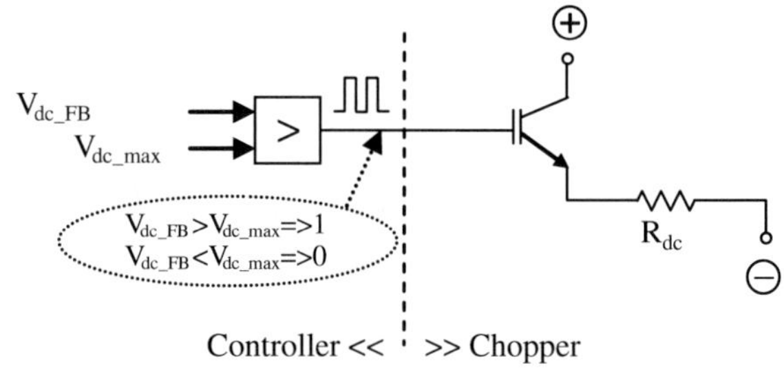

Fig. 21.10 Configuration of protective device used at offshore HVDC station

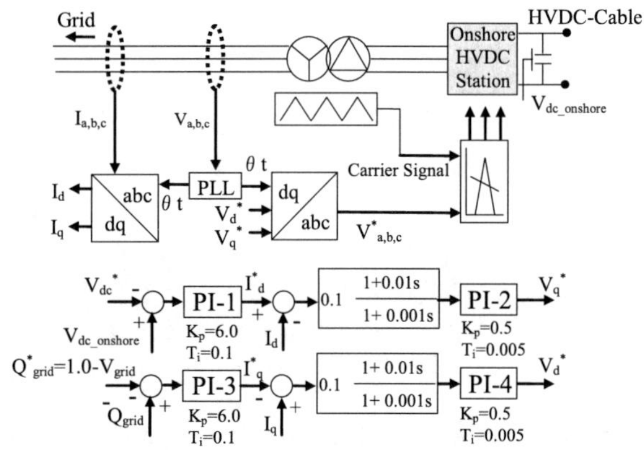

Fig. 21.11 Control block diagram of the onshore VSC station

cables. As a result, the DC voltage, which must be regulated, increases rapidly. One way to do this is to decrease the power generation from individual WTGS of the offshore wind farm by updating the power command to all MPPT control units, which requires strong coordination between the onshore and offshore stations. In the present study, as a simple protection scheme, the braking chopper, as shown in Fig. 21.10, is installed at the offshore HVDC station, which can work independently without any remote signal transmission between MPPT units and the onshore grid. The chopper is activated when a DC overvoltage is detected at the local end and dissipates the excess active power into the resistance during the voltage dip at the onshore grid.

21.3.4 Onshore HVDC Station

In the present study, the cascade control scheme with independent control of active and reactive currents, as shown in Fig. 21.11, is applied to the control of onshore VSC station. The dq quantities and three-phase electrical quantities are related to each other by reference frame transformation. The angle of the transformation is detected from the three phase voltages (V_a,V_b,V_c) at the high-voltage side of the

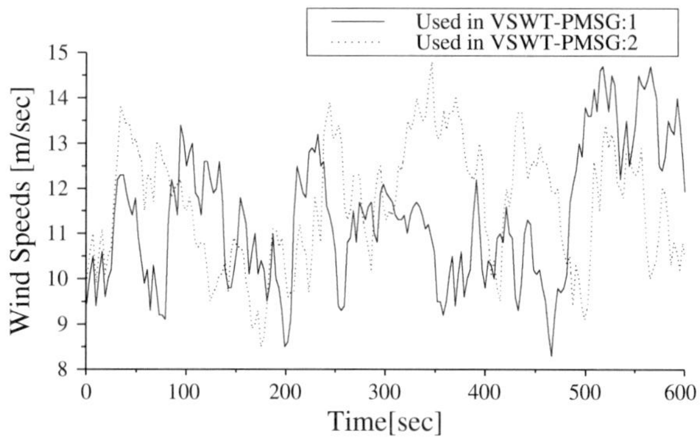

Fig. 21.12 Wind speed data

grid side transformer. The d-axis current can control the dc-link voltage. The dc voltage of the DC-link capacitor is controlled to be constant by two PI controllers. On the other hand, the q-axis current can control the reactive power of the grid-side inverter, and hence the onshore grid voltage can be maintained to be constant. The rated DC-link voltage is 150 kV.

21.4 Simulation Analysis

In the present study, both the dynamic and transient characteristics of the proposed system shown in Fig. 21.1 are analyzed. A detailed switching model is considered instead of the time average model for the sake of analysis precision. The time step is chosen to be 0.00002 s. The simulation time for the dynamic and transient characteristics analysis are chosen as 600 s and 5 s, respectively. Simulations were carried out using PSCAD/EMTDC [31].

21.4.1 Dynamic Characteristics Analysis

In order to obtain a realistic response, real wind speed data measured at Hokkaido Island, Japan, as shown in Fig. 21.12, is used in the simulation.

In the offshore wind farm, the maximum power generation from each VSWT-PMSG at a particular wind speed is ensured by the controller, as explained in Sect. 21.3.1. The responses of the real power references of the generator-side converters of the PMSGs, their rotor speeds, and the power delivered from each generator to the DC-bus are shown in Figs. 21.13, 21.14, and 21.15. The pitch controllers attached to the wind turbines are activated when the rotor speeds of the generators exceed the rated speeds, as shown in Fig. 21.16. The zoom voltage at the primary side of the FB DC–DC converter

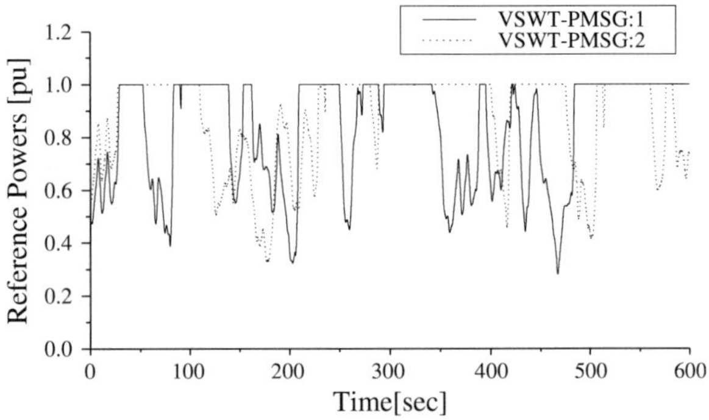

Fig. 21.13 Real power references for the generator-side converters

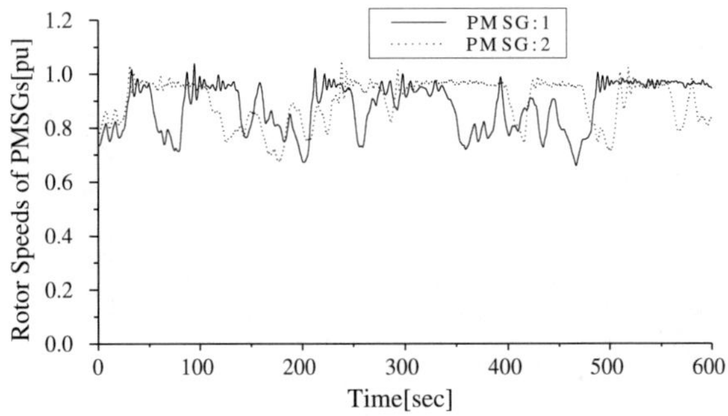

Fig. 21.14 Rotor speeds of PMSG1 and PMSG2

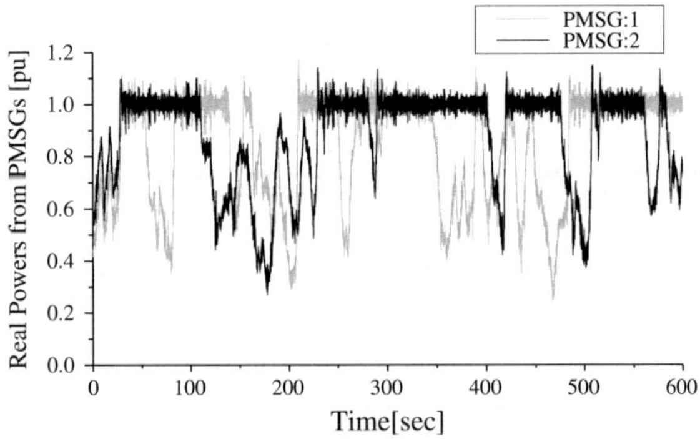

Fig. 21.15 Real power of PMSG1 and PMSG2

is shown in Fig. 21.17, which is basically a high-frequency square wave signal. The output voltage of the FB DC–DC converter, which is the offshore HVDC station voltage, is shown in Fig. 21.18. The DC voltage at the onshore HVDC station is maintained constant by the VSC, as shown in Fig. 21.19. The onshore grid voltage shown in Fig. 21.20 is also maintained constant by the onshore HVDC station. Under the proposed control strategy, the real offshore wind farm power is successfully transmitted to the onshore grid through the HVDC cable, as shown in Fig. 21.21.

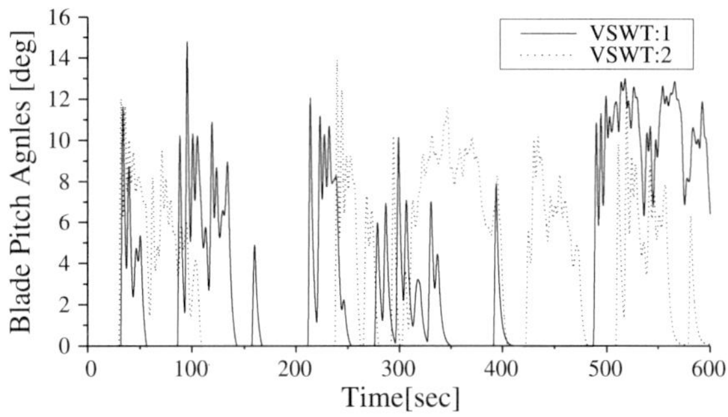

Fig. 21.16 Blade pitch angle of the wind turbines

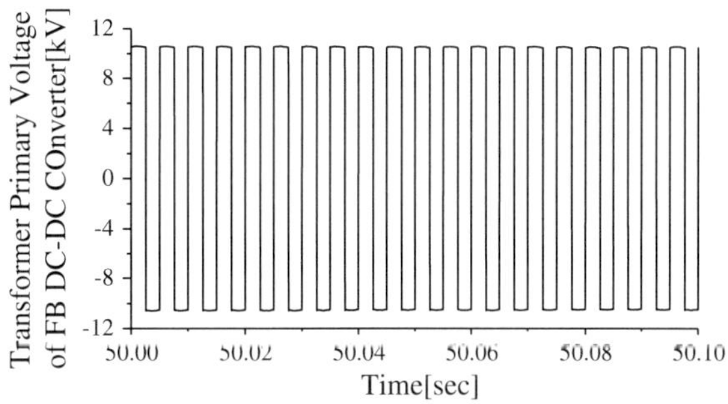

Fig. 21.17 Transformer primary voltage of the FB DC–DC converter (zoomed)

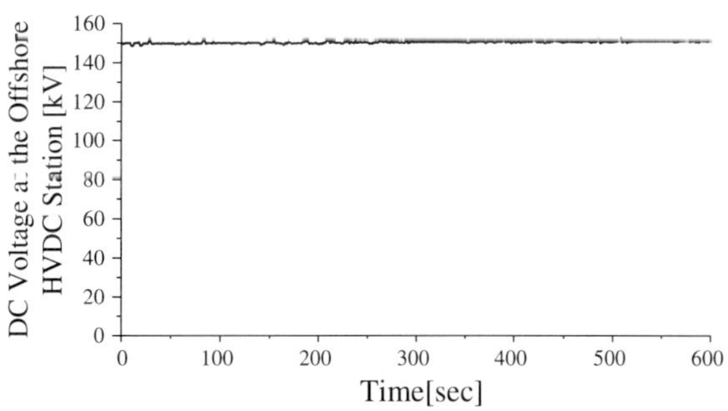

Fig. 21.18 Output voltage of the FB DC–DC converter

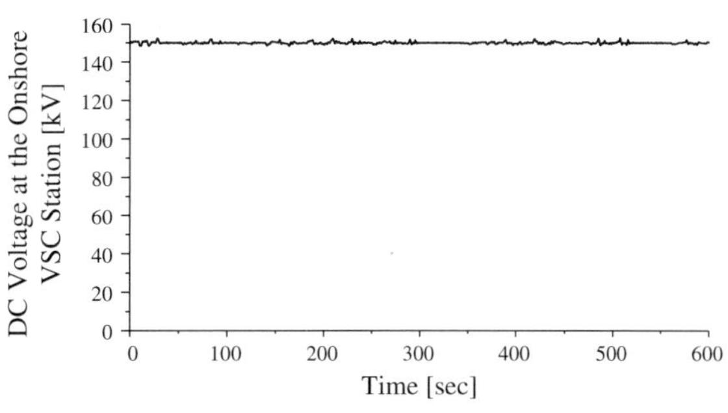

Fig. 21.19 DC voltage of the onshore HVDC station

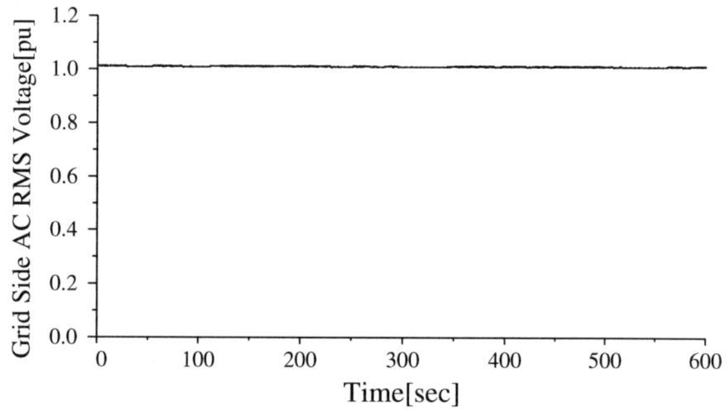

Fig. 21.20 AC voltage of the onshore grid

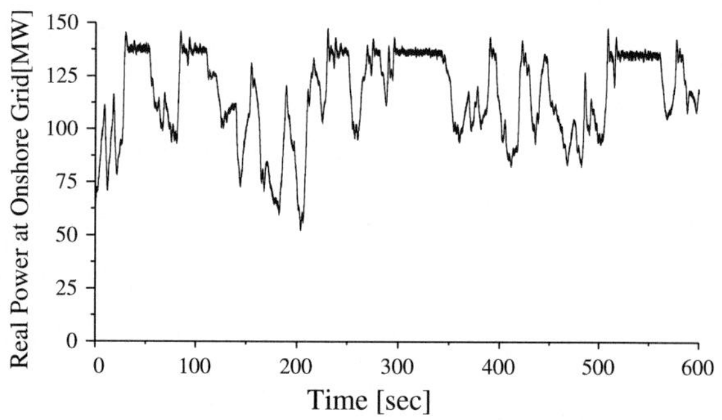

Fig. 21.21 Real power at the onshore grid

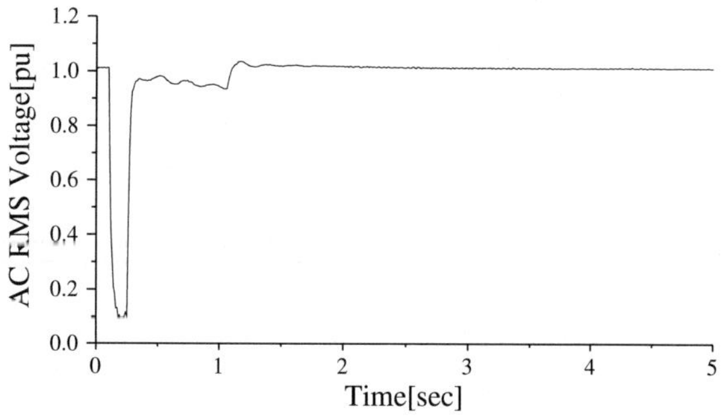

Fig. 21.22 AC-side voltages at the onshore grid (3LG)

21.4.2 Transient Characteristics Analysis

For the transient characteristics analysis, a severe three-line-to-ground (3LG) fault is considered as a network disturbance, which occurs at fault point F in Fig. 21.1. The fault occurs at 0.1 s. The circuit breakers (CB) on the faulted lines are opened at 0.25 s, and are reclosed at 1.05 s. It is assumed that the wind speed is constant and equivalent to the rated speed for the variable-speed WTGSs. This is because the wind speed may be considered not to change dramatically during the short time interval of the simulation.

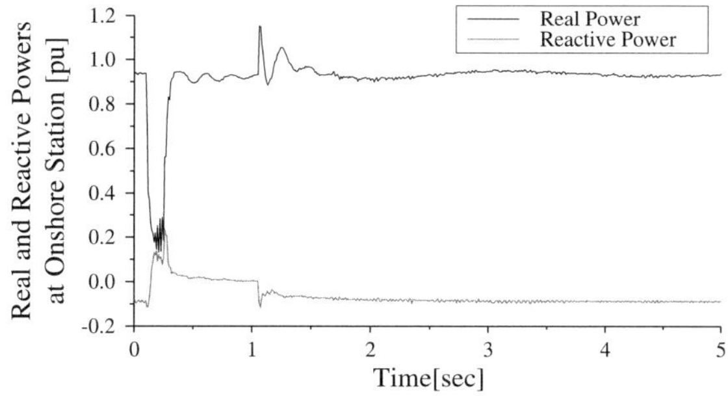

Fig. 21.23 Real and reactive powers at the onshore HVDC station (3LG)

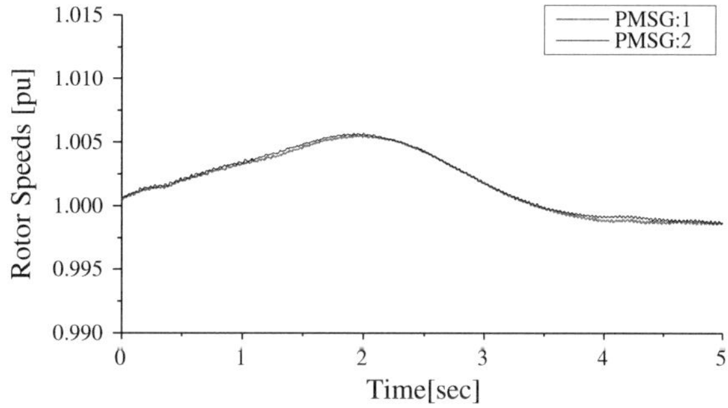

Fig. 21.24 Rotor speed of the PMSG (3LG)

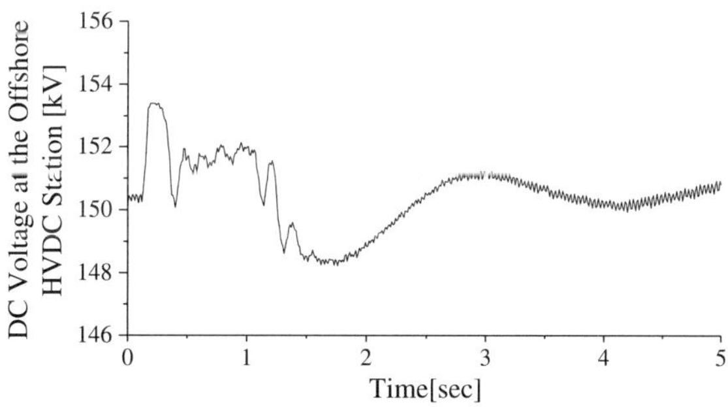

Fig. 21.25 DC voltage at the offshore HVDC station (3LG)

The responses of the AC voltage at the onshore grid, which satisfy the grid code requirement, are shown in Fig. 21.22 [36]. The real and reactive power responses at the onshore grid are shown together in Fig. 21.23. The pitch controller controls the mechanical power of the individual wind turbine to stabilize the PMSGs, as shown in Fig. 21.24. Due to the use of a protection device at the offshore HVDC station, the DC voltage cannot exceed a predefined limit, as shown in Fig. 21.25. The longer the fault duration, the greater the importance of the requirement for a protection device. The simulation results clearly show that the proposed system can overcome a severe 3LG fault under the developed control strategy.

21.5 Conclusions

This paper introduced an HVDC interconnected power transmission scheme for a DC-based offshore wind farm, in which a DC–DC converter using a transformer plays an important role in the offshore HVDC station. The wind generators of the offshore wind farm are connected to a DC-bus through fully controlled AC–DC converters, which ensures that maximum power is delivered to the DC-bus. Detailed modeling and control strategies of the FLC-based phase-shifted full-bridge DC–DC converter as well as other components of the proposed system were presented. The dynamic performance of the overall system was evaluated by simulation analysis using real wind speed data. The overvoltage protection scheme used in the coordinated control system can maintain the DC voltage of the HVDC system, even in the transient or fault condition. A full-bridge DC–DC converter can works well as an offshore HVDC station and can transmit the real power to the onshore grid through the XLPE HVDC cable.

Acknowledgments The present study was supported by a Grant-in-Aid for JSPS Fellows from the Japan Society for the Promotion of Science (JSPS). The authors would also like to thank Thomas Ackermann and Stephan Meier for providing valuable technical data.

Appendix

Wind turbine parameters

Rated power	2 MW	Air density	1.225 kg/m^3
Blade radius	36 m	Rated speed	24 rpm

Parameters of the Phase Shift FB–DC–DC converter

Transformer rating	150 MVA	C_s	10 μF
Transformer leakage reactance	0.05 pu	L_f	0.002 H
Base operational frequency	200 Hz	L_{lk}	0.001 H

Fuzzy Rules Table

D_n		Δe_n				
		NB	NS	ZO	PS	PB
e_n	NB	PB	PB	PS	PS	ZO
	NS	PB	PS	PS	ZO	NS
	ZO	PS	PS	ZO	NS	NS
	PS	PS	ZO	NS	NS	NB
	PB	ZO	NS	NS	NB	NB

References

1. The global wind energy council (2011) Global wind report-annual market update 2010. http://www.gwec.net/
2. The European wind energy association (2007) Delivering offshore wind power in Europe: policy recommendationsfor large-scale deployment of offshore wind power in Europe by 2020
3. The global wind energy council (2009) GWEC Regions-Europe-United Kingdom. Leader in Offshore Wind. http://www.gwec.net/
4. The European wind energy association (2011) The European offshore wind industry key trends and statistics 2011, EWEA Publications. http://www.ewea.org
5. Kirby NM, Xu L, Luckett M, Siepman W (2002) HVDC transmission for large offshore wind farms. IEE Power Eng J 3:135–141
6. Cartwright P, Xu L (2004) The integration of large scale wind power generation into transmission networks using power electronics. CIGRE General Session, Paris (CD-ROM)
7. Skytt AK, Holmberg P, Juhlin KE (2001) HVDC light for connection of wind farms. In: 2nd international workshop on transmission networks for offshore wind farms
8. Sobrink KH, Sorensen PL, Christensen P, Sandersen N, Eriksson K, Holmberg P (1999) DC feeder for connection of a wind farm. In: CIGRE Symposium
9. Lu W, Ooi BT (2003) Optimal acquisition and aggregation of offshore wind power by multiterminal voltage-source HVDC. IEEE Trans Power Deliv 18(1):201–206
10. Xu L, Andersen BR (2006) Grid connection of large offshore wind farms using HVDC. Wind Energy 9(4):371–382
11. Xu L, Yao L, Bazargan M, Yan A (2009) Fault ride through of large offshore wind farms using HVDC transmission. In: CD record of the IEEE PowerTech2009 conference, paper no. 308, Romania
12. Lingling F, Zhixin M, Osborn D (2009) Wind farms with HVDC delivery in load frequency control. IEEE Trans Power Syst 24(4):1894–1895
13. Vermaak R, Potgieter JHJ, Kamper JM (2009) Grid-connected VSC-HVDC wind farm system and control using permanent magnet induction generators. In: CD record of the international conference on power electronics and drive systems (PEDS09)
14. Muyeen SM, Takahashi R, Murata T, Tamura J (2009) Integration of hydrogen generator into wind farm interconnected HVDC system. In: CD record of the IEEE PowerTech 2009 conference, paper no. 271, Bucharest, Romania
15. Vas P (1992) Electrical machines and drives—a space vector theory approach. Oxford University Press, New York
16. Miller TJE (1989) Brushless permanent-magnet and reluctance motor drives. Oxford University Press, New York
17. Ribrant J, Bertling LM (2007) Survey of failures in wind power systems with focus on Swedish wind power plants during 1997–2005. IEEE Trans Energy Convers 22(1):167–173
18. Povh D, Thepparat P, Westermann D (2009) Analysis of innovative HVDC control. In: CD record of the IEEE PowerTech 2009 conference, Bucharest, Romania
19. Latorre HF, Ghandhari M, Soder L (2009) Use of local and remote information in POD control of a VSC-HVDC. In: CD record of the IEEE PowerTech 2009 conference, Bucharest, Romania
20. Hazra J, Phulpin Y, Ernst D (2009) HVDC control strategies to improve transient stability in interconnected power systems. In: CD record of the IEEE PowerTech 2009 conference, Bucharest, Romania
21. Zhang L, Nee H-P (2009) Multivariable feedback design of VSC-HVDC connected to weak AC systems. In: CD record of the IEEE PowerTech 2009 conference, Bucharest, Romania
22. Prabhu N, Padiyar KR (2009) Investigation of subsynchronous resonance with VSC-based HVDC transmission systems. IEEE Trans Power Deliv 24(1):433–440

23. Flourentzou N, Agelidis VG, Demetriades GD (2009) VSC-based HVDC power transmission systems: an overview. IEEE Trans Power Electron 24(3):592–602
24. Lundberg S (2004) Evaluation of wind farm layouts. In: Nordic workshop on power and industrial electronics (NORPIE), poster no. 2693, Norway
25. Max L, Lundberg S (2008) System efficiency of a DC/DC converter-based wind farm. Wind Energy 11(1):109–120
26. Prabhakar AJ, Bollinger JD, Hong MT, Ferdowsi M, Corzine K (2008) Efficiency analysis and comparative study of hard and soft switching DC–DC converters in a wind farm. In: IEEE 34th industrial electronics conference (IECON2008), pp 2156–2160
27. Heier S (1998) Grid integration of wind energy conversion system. Wiley, Chicester
28. Wasynczuk O et al (1981) Dynamic behavior of a class of wind turbine generators during random wind fluctuations. IEEE Trans Power Apparatus Syst PAS-100(6):2837–2854
29. Muyeen SM, Takahashi R, Tamura J (2010) Operation and control of HVDC-connected offshore wind farm. IEEE Trans Sustain Energy 1(1):30–37
30. Muyeen SM, Murata T, Tamura J (2008) Stability augmentation of a grid-connected wind farm. Springer, UK
31. PSCAD/EMTDC manual (2005) Manitoba HVDC research center, April 2005
32. Cho J-G, Sabate JA, Guichao H, Lee FC (1996) Zero-voltage and zero-currentswitching full bridge PWM converter for high-power applications. IEEE Trans Power Electron 11(4): 622–628
33. Mihalache L (2004) A modified PWM control technique for full bridge ZVS DC–DC converter with equal losses for all devices. In: Proceedings of IEEE IAC'04, vol 3. pp 1776–1781
34. Rashid MH (2007) Power electronics handbook. Reference book, 2nd edn. Amsterdam, Elsevier
35. Driankov D, Hellendoorn H, Reinfrank M (1993) An introduction to fuzzy control. Springer, Heidelberg
36. E.On Netz, Grid Code, High- and Extra-High Voltage (2006). www.eon-netz.com/

Index

S. M. Muyeen (ed.), *Wind Energy Conversion Systems*,
Green Energy and Technology, DOI: 10.1007/978-1-4471-2201-2,

I

K

L

M

N

O

P